D0876456

WITHDRAWN

# Microcomputer Programs for Chemical Engineers

## CHEMICAL ENGINEERING BOOKS

Sources and Production Economics of Chemical Products
Calculator Programs for Chemical Engineers—Volumes I and II
Calculator Programs for Multicomponent Distillation
Controlling Corrosion in Process Equipment
Effective Communication for Engineers
Fluid Movers: Pumps, Compressors, Fans and Blowers
Industrial Air Pollution Engineering
Industrial Wastewater and Solid Waste Engineering
Modern Cost Engineering: Methods & Data—Volumes I and II
Practical Process Instrumentation & Control
Process Energy Conservation
Process Heat Exchange
Process Piping Systems
Process Technology and Flowsheets: Volumes I and II
Safe and Efficient Plant Operation and Maintenance
Selecting Materials for Process Equipment
Separation Techniques 1: Liquid-Liquid Systems
Separation Techniques 2: Gas / Liquid / Solid Systems
Skills Vital to Successful Managers
Solids Handling
Supplementary Readings in Engineering Design
Synfuels Engineering
You and Your Job

## BOOKS PUBLISHED BY CHEMICAL ENGINEERING

Fluid Mixing Technology: James Y. Oldshue
Industrial Heat Exchangers: G. Walker
Physical Properties: Carl L. Yaws
Pneumatic Conveying of Bulk Materials: Milton N. Kraus

# Microcomputer Programs for Chemical Engineers

Edited by
David J. Deutsch
and the Staff of Chemical Engineering

Program Translations by
William Volk
of Princeton, New Jersey

McGraw-Hill Publications Co., New York, N.Y.

Printed in the United States of America.

**Library of Congress Cataloging in Publication Data**

Main entry under title:

Microcomputer programs for chemical engineers.

Includes index.
1. Chemical engineering—Computer programs.
2. Microcomputers—Programming. I. Chemical engineering.
TP149.M5 1984 660'.028'542 83-6386
ISBN 0-07-606067-5 (Chemical Engineering Magazine)
ISBN 0-07-010852-8 (McGraw-Hill Book Co.)

# CONTENTS

**Section III Physical Properties Correlation**



**Section IV Fluid Flow**



**Section V Heat Transfer**



**Section VI Mass Transfer**



**Section VII Engineering Economics**

# PREFACE

Microcomputers offer easier/better access to computing power than do mainframe units. They are "friendlier" than scientific calculators, too. But they are not going to replace either—that's expecting too much. In 1978, CHEMICAL ENGINEERING introduced its first book of calculator programs for chemical engineers. But, just as the scientific calculators replaced slide rules for engineering computation, they were destined to be virtually replaced in just a few years by microcomputers.

Now, CHEMICAL ENGINEERING is proud to present its first volume of microcomputer-program listings for chemical engineers. But, not only are program listing offered here. In addition this book provides an introduction to microcomputers, and articles on how to develop technical software, the use of spreadsheet software to solve engineering problems, and applications of microcomputers for process design, energy management, simulation and process-systems engineering.

The BASIC microcomputer programs presented here are interactive and easy to use. They are intended as a single-source library of programs, specifically designed to solve chemical engineering problems, and are based on material carefully selected from the pages of CHEMICAL ENGINEERING. The contents of the programs listed cover the full range of chemical engineering principles from engineering mathematics to physical-properties correlation to design-oriented programs for solving fluid flow, heat transfer, mass transfer, and engineering-economics problems.

The programs are written for the Apple II microcomputer, but are easily adaptable for use on other popular personal computers. See the next section for further details on program listings.

# PROGRAM LISTING NOTES

The BASIC programs in this book are written for the Apple II microcomputers. But owners of other microcomputers should not despair—the programs are easily adaptable. In fact, the only variations in BASIC dialect are for display commands, such as those to clear the screen ("HOME" for the Apple, but "CLS" for the IBM PC), or to move the cursor vertically (for instance, "VTAB 4" moves the cursor down four lines on the Apple II microcomputers). Accordingly, users of other microcomputers should change programs listings for these commands.

The programs are interactive and easy to use. For each program, the user is given instructions via the program menu, and at that point told which data should be entered and how to proceed from there to the solution. In many cases, a variety of engineering units can be used.

The program listings are in BASIC, and the programming logic should be understandable to most readers. For those more proficient in programming, the listings can easily be modified for customizing to the individual problem.

# Section I
# Introduction

# Microcomputing for chemical engineers

*Computer power in the U.S. is doubling every few years, largely due to the boom in microcomputers. Engineers who want to boost their personal productivity should know what computing tools are available, how they work, and what they cost. The following article and the one that begins on p. 12 tell what is at hand in microcomputing hardware and software.*

Originally published May 31, 1982

# Microcomputing hardware

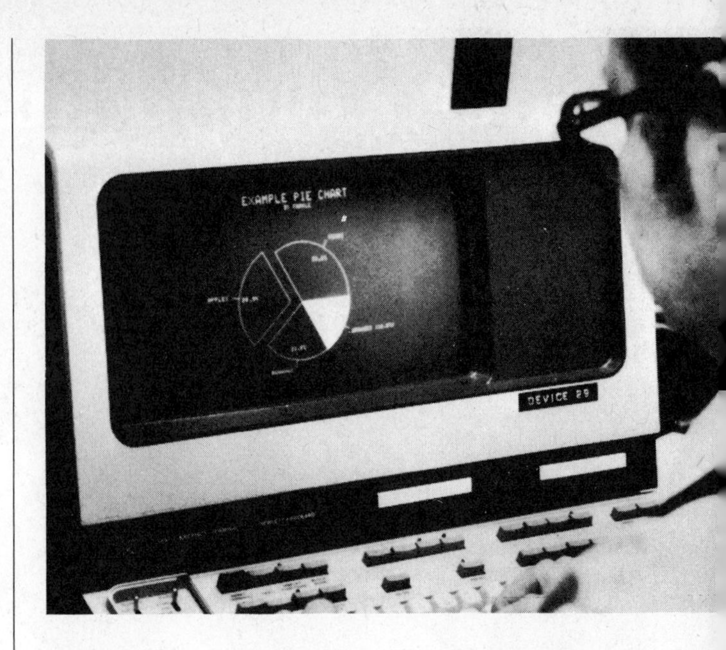


*Stewart Goldfarb* and *Steve Griffin,* *Dow Corning Corp.*

☐ Today's microcomputer is powerful enough to solve real engineering problems, yet inexpensive enough that every engineer could have one. Such technology means improved productivity for those who take advantage of it.

However, making decisions about computers is difficult because the technology is changing so fast. Perhaps the best approach is to solve today's problems with today's computing tools, and to leave tomorrow's problems for the tools that will be available then.

This article focuses on handheld and desktop microcomputers, which an engineer can own and use professionally because they are powerful and relatively inexpensive. But first a brief look at the spectrum of computers is in order. Since some of the terms may be unfamiliar, a glossary is provided on p. 11.

## Micro, mini, maxi

Engineers use a variety of computers in their work: programmable calculators, handheld and desktop microcomputers, minicomputers and mainframe maxicomputers. These machines frequently overlap in capabilities, and distinctions among them will blur even more in the future, but the capabilities of those that remain on the scene will undoubtedly expand greatly.

Fig. 1 compares the various types of computers in terms of calculation speed—number of floating-point (decimal, as opposed to integer) multiplications per second. Not shown is the variation in data storage and handling capability, which is at least as important as sheer number-crunching in many applications. On such a scale, "business" microcomputers would outshine "scientific" ones, by design. We can now describe each of the computer types briefly, focusing on differences in their capabilities and costs.

*Programmable calculators,* such as the Texas Instruments TI-59 and the Hewlett-Packard HP-41 series, are widely used by chemical engineers in design and operations because they are compact, inexpensive, and handle engineering math well. Though limited in the past by their modest memory and few non-math functions, programmable calculators are now becoming more versatile. For example: The HP-41C calculator can be augmented by: a link for communicating with compatible instruments; a memory-expanding module or cassette recorder; a card or bar-code reader; and a printer/plotter. Retail prices for programmable calculators range from $50 to about $500.

*Handheld microcomputers* are about the same size as calculators, yet offer alphanumeric display (24 characters), a miniature typewriter keyboard, and the high-level BASIC language. With BASIC, programs are easier to develop, debug and transfer to other computers than are calculator programs. Table I shows a simple heat-transfer calculation in BASIC and in a calculator language. Early models (such as the Radio Shack TRS-80 PC-1 shown on the cover) are roughly comparable with calculators in speed and storage capability, and cost about $200–250. But recent improvements offer greater speed, plus useful accessories such as color printing.

*Desktop computers* are called by several names. Here we will separate them into three types: personal computers for general math and data manipulation; business microcomputers, with greater storage, manipulation and display capability; and scientific computers, with powerful calculation, graphics and data-acquisition. Desktop computers range from about $300 to over $3,000, depending on the type of microprocessor, storage capacity, and built-in accessories. A complete system, however, can run from $1,000 to $20,000 or more, depending on the types of peripheral devices (for storage, printing, etc.) attached to the computer.

*Minicomputers* are bigger and faster than desktop computers, which allows them to do any calculation a chemical engineer might need except for simulations of complex processes. The difference between minicomputers and microcomputers is somewhat hazy, in that today's microcomputers are as powerful as some of yesterday's minis. In fact, Hewlett-Packard and Digital Equipment now sell their bottom-of-the-line minicomputers as desktop computers. Minis are still the workhorses of industrial applications, however, because they

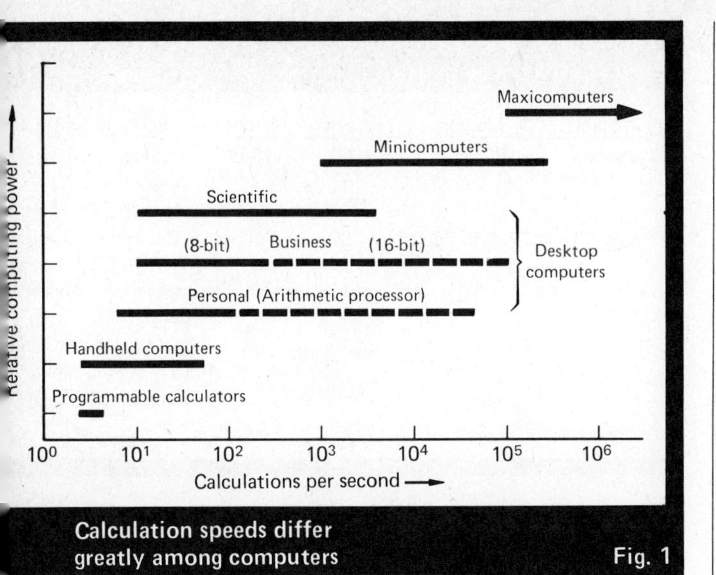


**Calculation speeds differ greatly among computers** Fig. 1

can handle several users and they calculate faster. For example: A minicomputer might do a tray-by-tray distillation calculation in several minutes, while with a microcomputer such a nested iterative calculation could take hours. Costs for minicomputer systems range from about $20,000 to $250,000.

*Maxicomputers* are the most powerful computers built today: the fastest number-crunchers, with the greatest memory and peripheral-device capability, and able to handle the most users. Compared with the business microcomputer, the maxi has up to 256 times as much memory, and performs calculations 1,000 times as fast. But it is also 100 to 1,000 times as expensive, with costs ranging between $250,000 and $10 million. Note that a maxi is not necessarily any less "personal" than a desktop computer, since current designs allow each user to think he or she has the whole computer alone. But in practice, the user will usually be aware of other users slowing down the computer's response.

Though chemical engineers employ all these computing machines, for the rest of this article we will look only at handheld and desktop microcomputers. These can handle many engineering problems that do not require lengthy iterations, and are so priced that an engineer can expect to obtain one for personal use.

**High-level language vs. programmable-calculator language** Table I

**Equations for heat-exchange calculation:**

$\Delta T = (T_1 - T_2 - T_3 + T_4)/\ln((T_1 - T_2)/(T_3 - T_4))$

$Q = UA\Delta T$

**Programmed in BASIC:**

```
60  T(5) = (T(1) – T(2) – T(3)+ T(4))/LOG (T(1) – T(2))
    /(T(3) – T(4))
70  Q = U * A * T(5)
```

**Programmed in calculator language**

| | | |
|---|---|---|
| RCL | 06 | ) |
| 01 | RCL | ) |
| – | 05 | = |
| RCL | – | STO |
| 02 | RCL | 05 |
| = | 06 | x |
| STO | ÷ | RCL |
| 05 | ( | 07 |
| RCL | LN | x |
| 03 | ( | RCL |
| – | RCL | 08 |
| RCL | 05 | = |
| 04 | ÷ | STO |
| = | RCL | 00 |
| STO | 06 | |

## Handheld microcomputers

At the heart of a handheld computer is a microprocessor. Earlier models, such as the Radio Shack TRS-80 PC-1, and the similar Casio and Sharp computers, use microprocessors that handle relatively small amounts of information (4 bits) at a time. They therefore run programs about as fast as programmable calculators do. Recent models, such as the TRS-80 PC-2 or Sharp PC-1500, use an 8-bit microprocessor. This means that they perform calculations nearly as fast as desktop computers, and much faster than calculators.

The ability to run programs in the BASIC language is an important advantage of the handheld computer, because BASIC is relatively easy to write and can be transferred to other computers. There are differences, however, in the versions of BASIC used by the various computer manufacturers. Therefore, some modification of programs may be needed. Transferring to other microcomputers is generally not a problem, since the handheld-computer version of BASIC is derived from Microsoft BASIC, a *de facto* standard for micros. But extra effort is needed when transferring to computers that use a different version of the language.

In a recent article,* an engineer compares the TRS-80 PC-1 computer with the HP-67 programmable calculator. For his circuit calculation, the program for the handheld computer took one-fourth as long to write, had better prompting features, was easier to follow and more easily modified. The author was pleased that the display on the handheld computer is not used for computation, because this allows him to interrupt easily. But he preferred the calculator's magnetic-strip storage to the computer's cassette recorder.

The early handheld computers now cost about $200, and the newer models are about $250. Peripherals for printing and program storage are generally comparable to those available for programmable calculators, with a few exceptions. For example, Panasonic provides add-on modules to expand its HHC computer into a desktop computer (the whole unit fits into a briefcase), and Radio Shack's new TRS-80 PC-2 offers a color printer/plotter. Table II summarizes features and prices for a few of the available handheld computers.

## Desktop computers

Desktop computers are becoming a familiar sight in the chemical process industries, finding uses as word processors, terminals for communicating with other

*Lewart, Cass R., Pocket Computer Solves for LC Resonance Using BASIC, *Electronics,* June 16, 1981.

computers, data-acquisition systems, and computational tools. Manufacturers tend to use the term "personal computer" for many desktop computers, resulting in some confusion about capabilities and applications. Here we will be more specific, defining three categories of desktop computers:

- Personal microcomputers, inexpensive enough for home and hobby applications.
- Business microcomputers, dedicated to professionals for their day-to-day office work.
- Scientific desktop computers, for high-powered calculation and laboratory work.

All desktop computer systems include a typewriter keyboard, some type of multiline display such as a cathode-ray tube (CRT) or television set, and a means of saving and loading data and programs. We will now look at each of the three types of systems in more detail.

## Personal microcomputers

Personal microcomputers most familiar to engineers are models such as the Radio Shack TRS-80 Model-III, Commodore PET, and Atari 400. Such a computer can be used for job-skill improvement (learning about programming), recreational uses such as games, educating children, and a variety of hobby applications such as keeping track of personal records.

Such computers are "personal" because they are so inexpensive: A system with a few needed peripherals often costs $1,000 or less. Table III lists several computers that might be used in such systems.

Those willing to spend more can get a more powerful processor and greater memory in a computer such as an Apple II, HP-85 or IBM Personal Computer. With a few peripherals, a system based on such a computer might cost $1,500–2,500, and would offer the hobbyist enough power for most home or recreational applications. However, such computers can also form the heart of a business microcomputer system—what differs is the amount of storage and the type and quality of peripheral devices. Such computers are also listed in Table III.

**Handheld computers offer calculator-like portability** — **Table II**

| Maker | Model | List price | Memory | Remarks |
|---|---|---|---|---|
| Panasonic | HHC | $500 | 2K-4K | Includes printer. Wide range of peripherals available. |
| Radio Shack | TRS-80 PC-2 | $280 | 2.6K | New model. Original TRS-80 PC-1 is the same as the Sharp below. |
| Sharp | PC-1211 | $230 | 1.9K | Comparable in speed to programmable calculators. |
| Casio | FX-702P | $200 | 1.7K | |

## Business microcomputers

The distinction between personal and business microcomputers is not perfectly clear, because most of the differences are those of degree. The business microcomputer system has more memory, more mass storage, a larger display, and greater communication capability than the personal microcomputer system, because the computer itself may be somewhat different and because the system includes more peripheral devices of greater quality and capability.

To highlight the differences between the computers themselves and between the types of systems, let us look at the six basic hardware areas: computers, terminals, mass-storage devices, printers, plotters and communication devices. The computer itself typically accounts for only 20% of the cost in a business microcomputer system, so the peripherals deserve as much attention.

*Computers:* The heart of the computer is the microprocessor, an integrated silicon chip that performs the arithmetic and logic operations needed for all computer tasks. Table IV lists some of the microprocessors used in today's microcomputers.

**Personal microcomputers vary in capability and cost** — **Table III**

| Maker | Model | List price | Memory | Remarks |
|---|---|---|---|---|
| Apple Computer | Apple II | $1,300 | 16K | Expandable, includes BASIC and graphics, most supported in hardware and software |
| Atari | 400 | $330 | 16K | Includes BASIC and graphics, cassette games |
| | 800 | $1,100 | 16K | Expandable, includes BASIC, sound and graphics |
| Commodore Business Machines | VIC-20 | $300 | 5K | Expandable, includes graphics |
| | PET 4000 | $1,000 | 16K | Expandable, includes BASIC, graphics and display |
| | CBM | $1,500 | 32K | Includes BASIC, graphics and display |
| International Business Machines | IBM Personal Computer | $1,750 | 48K | Latest technology, very little software yet |
| Radio Shack | TRS-80 Model-III | $1,000 | 16K | Expandable, includes BASIC, graphics and display |
| | TRS-80 Color Computer | $400 | 4K | Principally for graphics, expandable |
| Texas Instruments | TI-99/4A | $525 | 16K | Expandable, includes graphics and sound |

The relative computing power of a microprocessor is proportional to the number of bits it can handle in a single step. Today, the 8-bit microprocessor dominates the desktop-computer field, but by the end of 1982 the 16-bit processor, comparable to a minicomputer, will take over. In the future, the 32-bit processor will bring another leap in performance.

An important consideration in choosing among computers is the potential for expansion. This depends on the design of the particular computer: In some, the microprocessor and other components are on a single circuit board, and cannot be altered or expanded. Other computers (such as the Apple II or the TRS-80 Model-II) provide a means for attaching other printed-circuit boards with added memory, language capability, and communication features. These systems can be upgraded as technology improves, which protects one's investment.

*Terminals:* The terminal links the user with the computer and other elements of the computer system. It consists of a keyboard and a CRT display, which may be included with the computer in a single unit in some cases (like the TRS-80 Model-III). Desirable features for the keyboard are non-glare sculptured keys, positive feedback when a character is entered, user-defined function keys, and ability to detach the keyboard from the computer for convenient use. The CRT display should have a non-glare screen, sharp and clear characters, and no less than 24 lines of 80 characters each. Other desirable features are reverse video, highlighting, cursor addressing and editing commands. A typical stand-alone terminal costs between $700 and $2,400.

*Mass storage* is used instead of random-access memory (RAM) to hold programs and data permanently. While personal applications may use cassette tapes, business applications usually demand the much faster disk drives. These can be either floppy disks (flexible plastic disks covered with magnetic media inside a protective cardboard cover) or rigid Winchester-type disks. Floppy disks are inexpensive ($3–8), can store from 70K to 1.25M bytes or more, and require a simpler drive than do Winchesters. But Winchesters can access data faster, meaning that the computer spends less time looking for information. Winchester-type disks can hold from 5M to 26M or more bytes of data.

Table V lists more information about disk drives. Note that performance specifications for the drives do not indicate system performance, because the actual speed of access depends on the computer and software and how they work together.

*Printers* come in two basic types: dot-matrix and impact. Dot-matrix printers use a column of wires in a print head to produce a matrix of dots on the paper as the print head moves across the carriage. The matrix density determines the sharpness of the characters and the cost of the printer: A 5x7 matrix is the least expensive, and least sharp, while a 9x24 matrix approaches letter quality. Dot-matrix printers can be purchased with various options, such as graphics, business-forms control, programmable character sets, and variable print density. Recent models that provide multicolor graphics and printing can be substituted for plotters, but to date there is little software to take advantage of such features.

Impact printers provide letter-quality printing, using either a flat wheel (e.g., Daisy Wheel) or a thimble (e.g., Spinwriter) as a printing head. Such printers are usually used with word-processing software to produce reports and memos. Dot-matrix printers are generally faster, and are therefore used for applications where print quality is less critical and speed is important. Table VI summarizes the various types of printers.

*Plotters* provide hard copies of video-graphics displays, generated by special software. Software is a key concern here, because a computer's ability to generate graphic displays and use a plotter depends on how it is programmed. A video-graphics software package should at least emulate a Tektronix 4010 graphics terminal, since it can then be used with a wide range of large-computer graphics packages such as SAS, Tell-A-Graph, and DISSPLA. One widely used emulation is the TEKSIM program for the Apple II, which costs about $475 and works with the Tell-A-Graph or DISSPLA software package. Hewlett-Packard and Tektronix plotters are supported by all three of the mentioned packages.

The plotter itself typically produces either an 8½ x 11-in. or 11 x 17-in. plot, and costs about $1,000–2,000 for the smaller and $2,000–6,000 for the larger format. Features available include intelligence, automatic pen selection, communications buffers and paper feed.

*Communication* links provide access to other computers and information systems. Many computers include a cassette and printer interface, but serious users generally add on another type of link. The most widely used communications interface is the RS232C serial interface, which enables the computer to communicate over telephone lines (via a modem), and links the computer with compatible peripheral devices such as printers, plotters and analog-to-digital data-acquisition systems for instrument applications.

**Microprocessors used in desktop microcomputers — Table IV**

| Type | Model | Original maker | Remarks |
|---|---|---|---|
| 8-bit | 8080 | Intel | CP/M compatible |
| | 6800 | Motorola | |
| | 6502 | Mostek | Apple II, Atari, PET |
| | Z80 | Zilog | CP/M |
| | 8085 | Intel | CP/M |
| | 6809 | Motorola | TRS-80 Color Computer |
| | 8088 | Intel | IBM Personal Computer |
| 16-bit | 9900 | Texas Instruments | TI-99/4A |
| | 8086 | Intel | IBM Display Writer |
| | Z8000 | Zilog | |
| | 68000 | Motorola | HP-9826 |
| | 99000 | Texas Instruments | |
| 32-bit | iAPX 432 | Intel | Special, new |

Modems are telecommunication links, accessible through an RS232C or other interface. Their cost depends on their speed: A 300-baud (30 character per second) modem costs less than $250, while a 1,200-baud modem costs about $1,000. Of course, the faster modems cut down the time wasted transferring data or programs. Besides speed, other desirable modem features are: direct connection, to avoid the noise problems caused by acoustic couplings, and originate-and-answer capability, to allow communication between microcomputer systems.

Communication can get very complex, since greater speed and access to certain devices require more than a single modem and RS232C. The IBM Personal Computer offers software (BISYNC, for about $900) that allows it to communicate with IBM 360/370 and Series 30XX processors directly through a synchronous modem. With this capability, the Personal Computer can be added to existing IBM networks. Likewise, Radio Shack has expanded its TRS-80 Model II to be compatible with BISYNC, and has announced plans for compatibility with the ARC network from Datapoint.

Another desirable feature in business microcomputers for engineering applications is a floating-point arithmetic processor, which plugs into a compatible computer. This executes floating-point calculations 10–100 times faster than floating-point software packages. There are several such processors available, but it is difficult to use them because software for standard microcomputers is not designed to take advantage of them. New computers such as the IBM Personal Computer do offer such processors, and programming languages that use them are expected soon. With this innovation, iterative calculations that are too lengthy for today's microcomputers will become practical.

## Using a business microcomputer system

Fig. 2 illustrates a typical business microcomputer system for an engineering office. Besides the microcomputer and key peripherals already discussed, the system shown includes a graphics plotter, and software to perform many basic tasks. Some of the programs that have been written for such a system (which includes a TRS-80 Model II microcomputer) at Dow Corning are:

- Preliminary design and cost-estimating for a distillation column and auxiliary equipment.
- Manpower planning, based on capital-spending data.
- Physical properties.
- Plotting and curve-fitting, used with a Houston Instruments intelligent plotter.
- Thermosiphon reboiler evaluation.
- Flash calculation.

The distillation-column program uses the Underwood-Fenske method, and will estimate 40 different cases in 15 minutes to find the minimum-cost design. Written in FORTRAN, the program is compiled so as to run faster.

A computer system that will perform all the functions needed in an engineering office will cost anywhere from $10,000–15,000; the one in Fig. 2 adds up to $11,000. A typical system may include a computer with a Z-80 microprocessor, currently the most popular for business applications, which can handle a memory block of up to 64K bytes. Two double-density 8-in. floppy disks will store at least 1M bytes of information. The letter-quality printer is needed for word processing, especially in offices that produce many reports and proposals. If communication with another computer is important, a modem will be required. A plotter is handy for preparing charts and graphs.

Since software constitutes a major cost, let us look briefly at the program packages included in the Fig. 2 system. There is a tremendous variety of software available for microcomputers, so these are, of course, just a few of the many options:

*Electronic spreadsheets* ease the analysis and presentation of numerical data and text, for applications such as financial statements, statistics and cash-flow projections. The T/Maker II package listed in Fig. 2 runs on the CP/M operating system, a *de facto* standard for microcomputers with the Intel 8080 or 8085 and Zilog Z-80 microprocessors. Other packages are VisiCalc and CalcStar. Note that CP/M capability itself is not always included in a microcomputer, but must be

**Disk drives add mass-storage capability — Table V**

| Type | Storage, bytes | Access time, milliseconds | Transfer speed, bits/s (baud) | Cost |
|---|---|---|---|---|
| 5¼-in. floppy-disk | 70K | 250 | 125K | $400 |
| | 140K | 250 | 250K | $600 |
| | 300K | 200 | 250K | $750 |
| | 930K | 180 | 250K | $1,000 |
| 8-in. floppy-disk | 250K | 70-200 | 250K | $750 |
| | 500K | 70-200 | 500K | $950 |
| | 1,000K | 70-200 | 500K | $1,250 |
| Winchester hard-disk | 5M-26M | 25-150 | ≈5M | $3,000-8 |

**Dot-matrix and impact printers for microcomputers — Table VI**

| | Dot-matrix printers | | |
|---|---|---|---|
| | 8½-in. carriage | 14-in. carriage | Impact print |
| Characters per line | 80-132 | 132-220 | 132-158 |
| Character type | 5 x 7 matrix | 5 x 7 to 24 x 9 matrix | Impact |
| Speed, characters per second | 30-120 | 80-220 | 26-75 |
| Cost range | $400-1,000 | $1,000-2,500 | $1,700-5,0 |
| Remarks | For light to medium duty in hobby or office applications. | For heavy-duty business printing. May include color graphics and/or word-processing features. | For word processin May have graphics. |

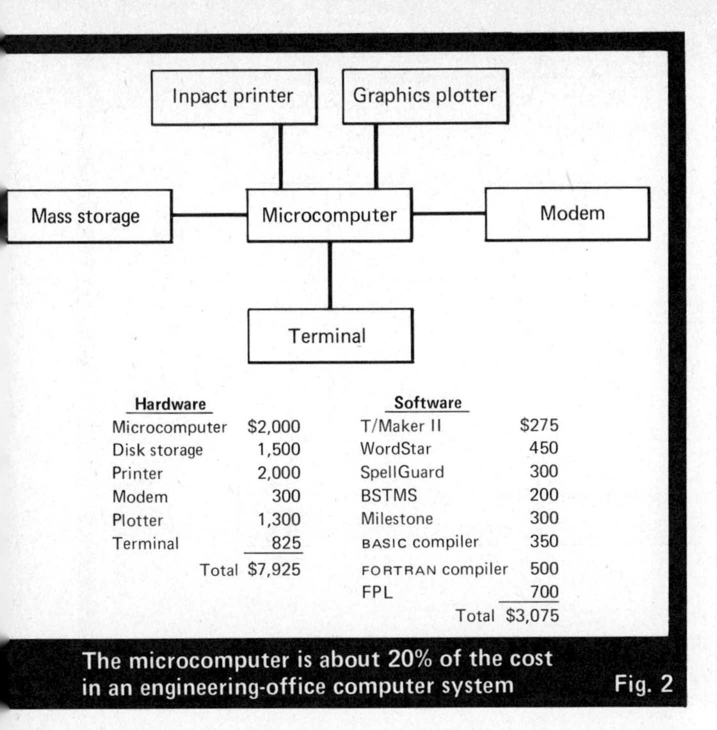


| Hardware | | Software | |
|---|---|---|---|
| Microcomputer | $2,000 | T/Maker II | $275 |
| Disk storage | 1,500 | WordStar | 450 |
| Printer | 2,000 | SpellGuard | 300 |
| Modem | 300 | BSTMS | 200 |
| Plotter | 1,300 | Milestone | 300 |
| Terminal | 825 | BASIC compiler | 350 |
| Total | $7,925 | FORTRAN compiler | 500 |
| | | FPL | 700 |
| | | Total | $3,075 |

**The microcomputer is about 20% of the cost in an engineering-office computer system** **Fig. 2**

bought. Among current microcomputers, the Xerox 820 is supplied with CP/M, and the IBM Personal Computer emulates CP/M.

*High-level languages* needed for scientific programming (BASIC, FORTRAN, APL, etc.) are implemented either as interpreters or compilers. A BASIC interpreter is included with many microcomputers. On an 8-bit microcomputer, this runs programs about 10 times as fast as a calculator does—rather slow because the interpreter translates and executes the program line by line. Compiling a program (translating the entire high-level-language program into rapidly-executed machine code) before execution will speed up the calculation considerably. For example: A compiled FORTRAN program runs on a microcomputer about 100 times as fast as a calculator program. But using a compiler has its costs: investment in a compiler software package; delay in running a program while it is being compiled; less-direct contact between the user and the program. Overall, compiled high-level language pays off when a lengthy program is to be repeated a number of times.

*Word-processing* systems for business microcomputers compete favorably with stand-alone word processors. In Fig. 2, a word-processing program (WordStar 3.0) is combined with a dictionary program (SpellGuard) to produce error-free reports.

*Special-purpose software* is available for a variety of business and scientific applications, but engineering software must generally be developed by the user. The system in Fig. 2 includes: BSTMS, a package that enables the microcomputer to communicate with other computers; Milestone, for project scheduling and critical-path analysis; and FPL, for financial planning and analysis.

Two goals in putting together a microcomputer system are speed and integration. For engineering purposes, desktop microcomputers are acceptable for simple iterative calculations; but, even with a compiied

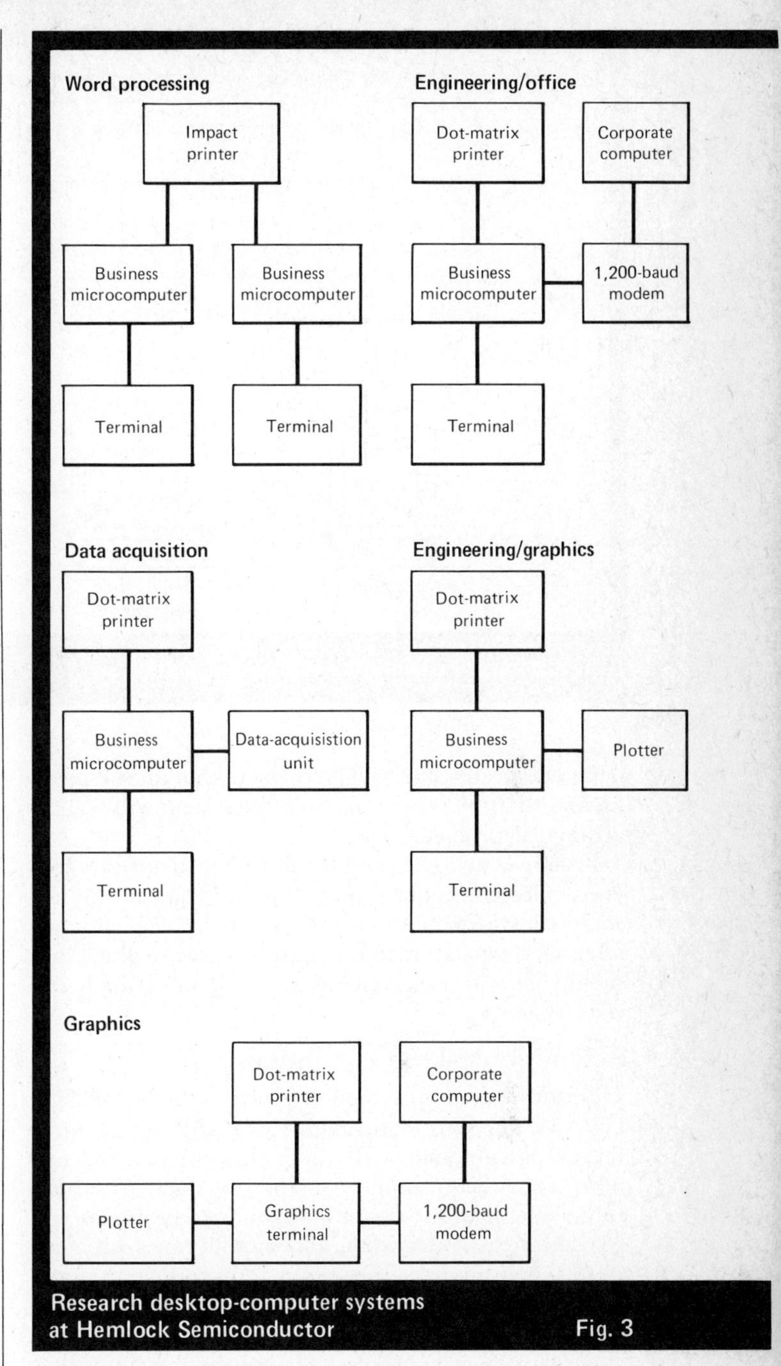


**Research desktop-computer systems at Hemlock Semiconductor** **Fig. 3**

high-level language, such a small computer has only one-thousandth the speed of a maxicomputer. Therefore, complex nested calculations, such as steady-state nonideal distillation-column designs, are simply not to be done on microcomputers. One will wait hours for answers to such problems.

Even more important is making sure that the computer system will work as expected when all of the cables are plugged in and the power is turned on. The term "integrated system" describes the desired situation where hardware and software are compatible in design and capacity, and will work together as an efficient package. Since an engineer is primarily concerned with solving engineering problems, and not with checking specifications for computers, it is recommended that a system be purchased from a local vendor who knows

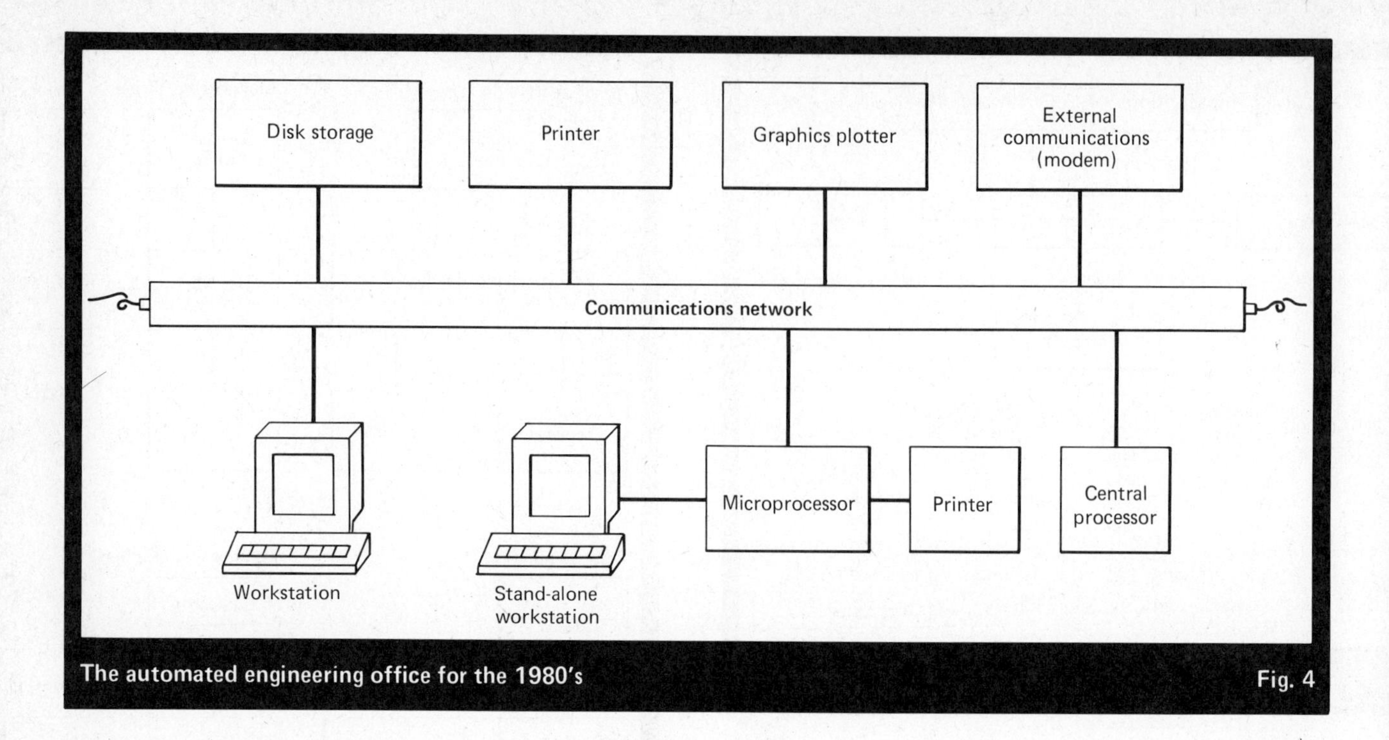


The automated engineering office for the 1980's — Fig. 4

what one needs. This will help assure that all the pieces fit, and that the person who sold the system will be near at hand if problems arise.

If there is no local vendor competent to aid in selecting a microcomputer system, then it is best to buy the complete system from one manufacturer who provides integrated packages. In purchasing software alone, one should likewise seek vendors who sell integrated software systems.

## Scientific desktop computers

Scientific desktop computers, such as the HP-9826 or DEC-MINC, were designed for powerful mathematics and graphical display—the functions most useful to engineers—plus acquisition of laboratory data. Manufacturers offer a wide array of hardware and software, designed for easy integration, that simplify measurement, control and analysis in research and production applications. Such computers may also offer computer-aided design capability.

Hewlett-Packard, Digital Equipment and Tektronix, among others, manufacture such computers. These computers are generally desktop packages with high-quality graphics capability. Some models, like the HP-9826, include built-in graphics display, disk drive and instrument interface. Such a package costs about $9,000, and saves the trouble of finding and attaching peripherals from other manufacturers. Generally, the types of peripherals available for business microcomputers (disk drives, large-format printers and plotters, etc.) are also available for scientific computers.

The 16-bit scientific microcomputers can handle most complicated calculations, including matrix manipulation and polynomial evaluation. All are supplied with a BASIC interpreter, and interpreters for other high-level languages are also available. Manufacturers provide extensions to high-level languages, which take advantage of their systems' advanced graphic and data-acquisition capabilities. There is no standard for these extensions, however, so one manufacturer's language is not likely to run on another's computer.

Language interpreters are useful in that they make programming as interactive and simple as possible, but compilers are more useful in performing extensive calculations quickly. While the more expensive scientific computers can be many times as fast as a business microcomputer with a compiler, they are not as fast as they could be, because no compilers are available for them. Also, the number-crunching capability of scientific desktop computers may be more devoted to computer-aided design functions—rotation, scaling, zooming, etc.—than to fast execution of programs. That is why these computers are used in computer-aided design systems, but this is no advantage in getting calculations performed.

Manufacturers of scientific desktop computers provide extensive software support for their systems. Especially helpful are the user groups, which over the years have created large libraries of nominal-cost programs useful in engineering work.

Overall, scientific desktop computers are well-suited to engineering problems because of their number-crunching, graphics and data-acquisition capabilities. They are also rugged, reliable and easy to integrate with other system components. Largely because of their special functions, scientific computer systems tend to be much more expensive than other desktop systems—ranging from about $8,000–45,000, depending on the type of computer and peripherals.

## The automated engineering office

Fig. 3 illustrates the multiple-computer system in use at Hemlock Semiconductor Corp., a Dow Corning subsidiary. Here the company began in 1978 with one North Star computer used for data-acquisition, and has since added:

- A graphics workstation based on the HP-2647 system, with a software package that prepares multicolor graphs, charts and slides for reports, memos and presentations. One application program written by the users produces process flow-diagrams.
- A word-processing workstation, used to prepare letter-quality text for the same reports and memos.
- Two engineering workstations, one of which is linked with the corporation's mainframe computer.

In the near future, we can expect to see more desktop computers linked together by data-communications networks—the automated office of the future. Most products being introduced for such applications are aimed at business offices, but fortunately business has become interested in graphics. Once graphics are integrated into the workstation, the engineer will have access to an extremely useful tool for day-to-day work. Fig. 4 shows what such a system involves.

Introducing the automated office system will prove to be a sensitive issue, as engineers will have to learn new ways of doing some of their work—typing by themselves, for example—to keep up with the technology. We expect that management will have to lead the way in using such systems if it expects to see cooperation and productivity increases.

## Future trends

It is impractical to speculate past 1990, but in the next five years or so, certain trends are likely to have an impact on the computer world: The automated office system will probably be widely used, as software developers invent the tools that make automation practical and efficient. The hardware for the computer of 1986 is already available or on the drawing boards: Intel's 32-bit microprocessor can be compared with an IBM 370/145 maxicomputer, and other manufacturers have developed prototypes that are just as powerful.

This means that the desktop computer five years from now will compare to today's maxicomputer, except that it will cost only $20,000 or so, and will be able to handle any calculations an engineer does today. Likewise, today's desktop microcomputers should be available in handheld versions by the mid to late 1980s.

In conclusion, let us restate that the engineer should aim to solve today's problems with today's microcomputer systems, and leave tomorrow's to the systems then available. Technology is changing too rapidly to plan otherwise.

## Glossary of terms

**APL.** A language used for interactive high-level programming, including operations on sets.

**BASIC.** Easy-to-learn language similar to FORTRAN.

**Baud.** Binary units of information (bits) per second.

**Bit.** Contraction of "binary digit." A bit is the universal unit of information, equal to either a "0" or a "1".

**Bug.** A mistake. Getting out the bugs is known as debugging.

**Byte.** Set of 8 bits, used universally to represent a character. 1K bytes means 1,024 bytes.

**Computer.** System that incorporates a CPU, memory, input/output facilities, power supply and enclosure.

**CPU.** Central processing unit. This part of a computer fetches, decodes and executes instructions.

**Compiler.** Translation program that converts high-level instructions into a set of binary-coded instructions for execution by the CPU.

**Floppy disk.** A mass-storage device that uses a soft (floppy) disk to record information.

**FORTRAN.** Early high-level language devised for numerical computations.

**High-level language.** Programming language resembling human language in its complexity. Examples are FORTRAN, BASIC, APL, PASCAL.

**Interpreter.** Translation program used to execute high-level language statements one at a time.

**Memory.** CPU storage area for binary data and programs. Memory is accessed directly, therefore quickly, while mass storage is slower.

**Microprocessor.** CPU on a single chip.

**Modem.** Modulator-demodulator, used to connect a digital device with an analog telephone line.

**Operating system.** Software that manages the hardware resources of the computer system, including peripherals, and also the logic resources such as storage space.

**PASCAL.** A high-level programming language.

**Plotter.** Mechanical device for drawing lines under computer control.

**RAM.** Random-access memory. Denotes the read-and-write memory on a large-scale-integrated chip.

**ROM.** Read-only memory. This may include built-in software such as a BASIC interpreter.

**RS232C.** Popular-standard interface for connecting computers with terminals, displays, modems and other components.

## The authors

**Stewart M. Goldfarb** works in the process engineering and research group of Hemlock Semiconductor Corp., a wholly-owned subsidiary of Dow Corning Corp., at 12334 Geddes Rd., Hemlock, MI 48626. He is currently working on ASPEN simulations of new technology, and on manufacture of polycrystalline silicon. Mr. Goldfarb holds a B.S. degree in chemical engineering from Wayne State University, and was employed by Dow Corning Corp. in facilities engineering and capital projects before taking his present position. He belongs to AIChE.

**Steve J. Griffin** is a project engineer working on control systems technology for the Facilities Engineering Dept. of Dow Corning Corp., Midland, MI 48640. He holds B.S. and M.S. degrees in chemical engineering, both from the University of Michigan. Before joining Dow Corning, Mr. Griffin worked in controls technology and simulation for the Chemicals and Plastics Div. of Union Carbide Corp. He is a member of AIChE.

# Microcomputing software

*David P. Bloomfield,* *Physical Sciences, Inc.*

☐ Strictly speaking, a microcomputer is nothing without software. Developed by the user or bought from a vendor, the programs that tell the computer what to do determine its capabilities and performance.

Though this article follows one on microcomputer hardware, hardware does not necessarily precede software when putting a microcomputer system together. For example, computer retailers say that people who want the capabilities of VisiCalc or other electronic spreadsheets will make sure to buy a computer system that can implement that software.

This article will cover what is available in software for chemical engineering—there isn't very much yet, unfortunately—plus general business and engineering. And we will try to cover how to get the most out of microcomputers by choosing, writing and using software effectively.

## Software capabilities

The figure shows the flow of information in a microcomputer system. Software controls all of it, telling the computer what logic and arithmetic functions to perform, the plotter how and when to plot, the disk where to find needed data and how to retrieve it.

To engineers, the most important capabilities in a computer system are engineering analysis, word processing and business analysis. In resolving engineering problems, microcomputer solutions are limited only by the creativity of the programmer and the speed of the computer. In my own case, the application has been systems research in advanced fuel-cell powerplants. Previously, this type of analysis had been performed only on large computers such as the IBM 370.

With the present revolution in printers and word-processing software, one can use the same system for writing reports and for performing engineering analyses. One of the many advantages of this approach is that proofreading is done only once, and corrections are performed electronically; this means that you can develop an outline and then expand it into a report without the drudgery of cutting and pasting. In fact, you can even obtain a program that will check spelling.

For the engineer/manager, a wealth of software is ready to assist in manpower planning, proposal costing, and accounting. Perhaps the most versatile software for such applications is VisiCalc, and software like it, which are like electronic scratchpads with many easy-to-use formatting features. The table lists this and other general-purpose software.

While the capabilities of microcomputer software are astounding, and growing extremely fast, microcomputers are still limited. Programs run considerably slower than on larger machines, making some applications impractical, and memory is also considerably smaller. Nevertheless, it is both possible and cost-effective to develop engineering programs on a microcomputer, running them later on a larger computer if necessary to speed up the operation. In my own case, I developed a modular systems-analysis program on a microcomputer, using BASIC, and translated it to FORTRAN for use on a faster minicomputer. And recently, I acquired a software package that translates such BASIC programs to FORTRAN automatically.

## Chemical engineering software

At the present time, few programs are directly relevant to chemical engineering problems. But the chemical engineer is not totally without software resources.

Several excellent programs are available through the National Energy Software Center (at Argonne National Laboratory in Illinois); COSMIC in Athens, Georgia maintains a library of NASA programs; and CHEMICAL ENGINEERING can provide a reprint of "Computer programs for chemical engineers."* If you have access to a large machine through an RS232C interface, you can load some of these programs from the large ma-

*By Jeffery N. Peterson and others, 1978, plus "More computer programs for chemical engineers," by Chau-Chuyn Chen and Lawrence Evans, May 21, 1979.

**General-purpose software of use to engineers**

| Type of application | Examples of available software |
|---|---|
| Database management | Data Base Manager, dBase II, DataStar, FMS 80, VisiDex |
| Electronic spreadsheet | CalcStar, SuperCalc, T/Maker, VisiCalc |
| Graphics | Data Plot, Logo, VisiPlot |
| Language compiler | APL, BASIC, C, FORTH, FORTRAN, PASCAL |
| Project management | MicroFinesse, Milestone, Project Manager |
| Spelling | SpellGuard, SpellStar |
| Word processing | Apple Writer, Easy Writer, Scripsit, LazyWriter, Magic Wand, Text Writer, WordStar |
| Other | BSTMS (communications), MuMath (algebra and calculus, FPL (financial planning) |

Originally published May 31, 1982

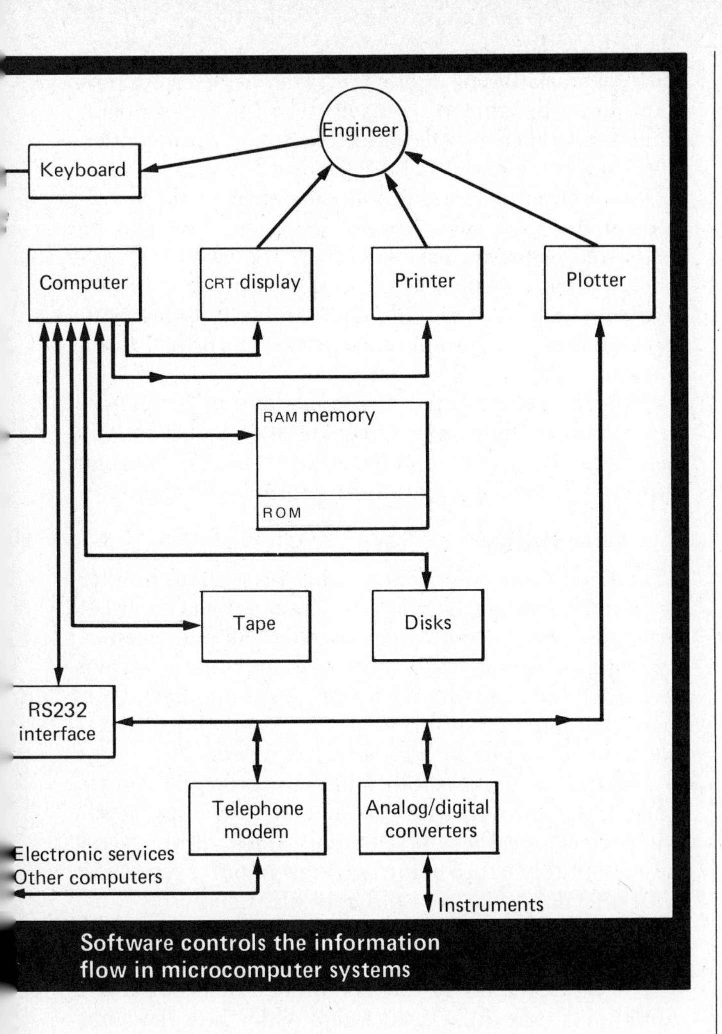

Software controls the information flow in microcomputer systems

chine to your microcomputer. Most of the time, though, you will need a FORTRAN capability, or you will have to translate programs to BASIC.

Among the general-purpose programs, those most useful to engineers are:

*Mathematical* programs for regression analysis, Fourier transforms and statistics (generally costing $10–150). Of particular note is a package that does algebra and calculus: Mu-Math by Microsoft.

*Structural or electrical-network analysis* programs. Many chemical engineering problems are analogous to these, so such programs can be valuable guides. Places to find them include manufacturers' books (such as the "TRS-80 Applications Sourcebook"), magazines such as *Byte,* and various local (especially college) bookstores. However, do not expect too much; go for additional help to large software houses.

The current lack of chemical engineering software is probably not going to last, and is not critical because engineers familiar with computers can write their own software without much difficulty. For the moment, the key thing to know is how to get the most out of the general-purpose software that is available at low cost.

## Interactive programming

Today's microcomputers are superior to programmable calculators and mainframe computers in their ability to interact with the user; that is, to provide immediate feedback during programming. This speeds up development, eliminates the need for a professional programmer, and gives engineers a better feel for what a program is really doing.

BASIC and other high-level languages can run either line-by-line, through an interpreter, or all at once, through a compiler that translates the high-level program to the computer's language. The interpreter mode increases the run time of a program—compiled programs run considerably faster—but speeds program development and debugging. Using interpreter BASIC is truly interactive.

Debugging is especially important because both syntax (computer-language) and analytical errors occur almost inevitably. An interpreter BASIC program can be interrupted at any point in its execution, so that variables can be checked and syntax problems fixed. In addition, the user can change the program completely at any time. With the interactive approach, changes and corrections are simple and fast, while compiled programs need to be recompiled completely and rerun to find their errors.

Overall, programs to be used once or only a few times are more efficiently developed and run using the interpreter mode, while programs to be run repeatedly are better compiled to take advantage of the higher speed of execution. Another distinction between the two modes is the amount of storage space needed for a program, important because storage is short in microcomputers. In general, a compiled program (now in the computer's simple language) takes about three times as much space as a high-level-language program, depending on the power of the compiler.

## High-level languages

The most popular microcomputer language today is undoubtedly BASIC, especially the interpreter BASIC that interacts with the user. Easy to learn, the versions of BASIC provided on microcomputers are generally superior to those found on larger machines.

More-powerful languages such as FORTRAN, APL and FORTH are generally available through compiler software. These offer some advantage over interpreter BASIC in execution speed, at some penalty in required storage space. For example, a recent article by J. Gilbreath in *Byte* magazine (September 1981) compared several high-level languages in terms of their execution times: C, PASCAL, FORTRAN, FORTH, BASIC, and others. The article showed that execution times for the same program ranged from 14–930 s for compiled programs, and pointed up some interesting aspects of BASIC: Compiled integer BASIC took about 18 s, floating-point BASIC about 715 s, and uncompiled floating-point BASIC about 1,920 s (at home, it took 90 min on my microcomputer).

This comparison should be of interest to engineers who regularly use floating-point (decimal) rather than integer numbers, and explains why some manufacturers are offering floating-point processors to replace such cumbersome software. For example: Intel's new 8087 floating-point processor (about $400)—which can plug into an IBM Personal Computer—speeds up floating-point calculations dramatically.

One should also note that the "benchmark" program

used to check execution time involved only addition. If division were included, times for the same number of calculations would have been much slower.

This is important, but the execution time is only part of the total time needed to run a program; it also takes time to compile and load the program. If you are crunching numbers for optimization or sensitivity analysis, then perhaps compile-and-load time can be ignored. But for many applications, the extra time needed to get the program ready to run (50–90 s for the benchmark program in the article) is more important than the execution time, and in any case the difference between compiled and uncompiled overall times is less than execution times indicate.

In my opinion, the advantages of interpreter BASIC outweigh its execution-time drawbacks for most users. Remember that the interpreter mode requires no compiler time, less memory, and is also likely to produce better programs that run faster inherently. One can program badly in any language, but interactive programming is likely to produce a better result.

## Programming techniques

To exploit the microcomputer fully, and get the most mileage out of the programs you write, it is important to be sophisticated about how you put programs together. The key concepts here are structured programming and modular design.

Structured programming means writing programs that you and other people will be able to understand, so that you can use them with ease and change them readily when needed. Many of the newer high-level languages were specifically designed to force the programmer to develop structure—PASCAL, for example. You should use their capabilities to write programs that are clear and efficient. Otherwise:

- Someone else who tries to use your poorly written program will not understand what the program does, and will be annoyed.
- You will feel the same way when you read your own program later.
- Sloppy programs force the computer to thrash around and waste time trying to compile or run them.

Good structure means: separating data statements from calculations; using variable names that relate to the physical variables being described; using comment statements to explain what a program is doing. Developing good structure is a skill that can be learned, just like writing good English, and it is worth the effort.

Modular programming is especially important in writing long and complicated programs. Basically, a program module is a subroutine, complete with needed comments and definitions, that performs some operation on an input. By carefully specifying inputs and outputs, and being consistent over time in designing modules and naming their variables, the programmer can quickly develop a library of subroutines. These can then be linked to do complex tasks. For example, subroutines that calculate physical properties or solve for friction factors can be used repeatedly in chemical engineering applications.

Once a library of modules is available, complex programs can be written quickly. All that is required is an overseeing program that: relates subroutine variables with the variables of immediate interest; describes how data are to be entered; controls the sequence of operations; and describes the objectives—e.g., optimization of a particular variable by varying another.

Many potential modules already exist in the literature in the form of calculator programs that can be translated to BASIC. Likewise, there is a wealth of software written for the mini- and maxicomputers of yesterday, which were roughly comparable to today's microcomputers. Government agencies often provide such software at very low cost.

Problem-solving capability grows rapidly as one develops new modules, since they are always on hand for new jobs. By using modular programming, one can expect to handle programming problems efficiently.

## Transferability

The microcomputer industry has been clamoring for years about standardization of languages, but as of yet programs developed on one computer will not necessarily work on another. Most manufacturers use a version of BASIC developed by Microsoft, but some have their own. Compiled programs tend to be even less transferable than programs in high-level languages.

The lack of transferability has two principal causes: First, the set of instructions that each microprocessor will perform is somewhat different. Among 8-bit microprocessors, for example, the two most popular types (the 6502 and the Z-80) have different instruction sets. This means that any given function may have to be described uniquely for each type of microprocessor.

The other reason is that operating systems, and the communication rules associated with those systems, vary widely. The CP/M system enjoys wide popularity today; in the future, new systems will be developed to take advantage of improved microprocessors. This means that transferability will continue to be a problem in developing software.

## Conclusions

The definitions and descriptions in this article are transitory at best, since software capabilities are changing so rapidly. Today, we see that microcomputers can be valuable tools in the analytical end of engineering, and that their value in both analysis and control is likely to grow. It is certainly an interesting time to be an engineer.


### The author

**David P. Bloomfield** is vice-president of the PSI/Systems Div. of Physical Sciences, Inc., 30 Commerce Way, Woburn, MA 01801. In charge of microcomputer software, hardware and control systems, he recently developed and commercialized a systems-analysis program for chemical and power plants. Previously, he did research in fuel cells at Physical Sciences, Inc. and at Giner Inc. and United Technologies Corp.; he holds fuel-cell patents. Mr. Bloomfield earned a bachelor's degree from the Polytechnic Institute of New York, and a master's from Rensselaer Polytechnic Institute, both in mechanical engineering.

# Computer vistas

**This report examines three aspects of computer technology that are having an ever-increasing impact on engineering in the chemical process industries:**

# Integrated computing

The latest move is to integrate the computers, programs and data that engineers use. Today, the field is still in flux. But the trend is clearly toward systems that talk to each other better, and give engineers greater and more convenient access.

***Mark A. Lipowicz**, Associate Editor*

☐ Experts say that the computerized tools engineers use will not be integrated for a few years yet, but companies planning ahead have to think about integration now. Says one manager about system vendors, "We don't want to know what they're doing about today; we want to know what they're doing about 1988."

What vendors are doing is developing systems that allow engineers to share data, and have better access to it, through database and networking technology. And engineers are integrating their own systems, and putting the new technology to work, step by step. The need is there because most of today's computer aids for engineers serve only one or a few purposes—which has resulted in a multiplicity of hardware, software and data that is difficult to manage and put to work.

With integration, engineers can expect easy access to a variety of programs and data through "intelligent workstations." One of the things to come is previewed in Fig. 1, a display from a workstation being evaluated at Exxon Research and Engineering Co. (Florham Park, N.J.). With a design program that runs under a Unix operating system, engineers can work on several problems at once.

How and how much to integrate is not universally agreed upon. And more and more "personal" computer applications are springing up (see the article following, and *Chem. Eng.*, Sept. 19, 1983, p.14). But companies are putting integration into practice now, in systems for process engineering, plant design, and process control.

## The process engineering environment

Today, process engineers can call on all kinds of programs that estimate physical properties, draw flowsheets, simulate unit operations and flows, size and specify equipment, calculate heat and mass balances, and estimate costs. Fig. 2 outlines the scope of process design activities.

While the activities are all connected, the programs and data may be distinctly apart. Since the object is an

Originally published January 9, 1984

**☐ *Integrated computing* — There is a trend toward tying together hardware, software and systems so that engineers can share computer resources and data more easily. Yet "personal" computing applications are also growing.**

**☐ *Desktop software (p. 23)* — Engineering software for microcomputers is finally becoming widely available. Over 70 sources and references are listed here.**

**☐ *Online information (p. 26)* — Taking full advantage of advanced technology requires access to more and better information. Bibliographic and numerical databases put literature and data at the engineer's fingertips.**

integrated design, there is a clear need for tying together more of the systems and data that process engineers use.

Complicating this, companies may be maintaining a hundred or more programs. The corporate engineering department at Monsanto Co. (St. Louis) is responsible for some 100–125 programs, according to Ronald Morris, an engineering superintendent. "On top of that," says Morris, "there are probably 50 or so other program packages—such as statistics, word processing, and database management."

And the total computing environment is still more complex. At Kao Corp. (Tokyo), the Kao Soft Bank includes some 200 technical programs, plus 265 others. The company's 4,800 workers have at their disposal some 30 general-purpose computers, 100 office computers, 385 terminals, and 70 word processors—plus 270 desktop computers and 1,200 handheld ones.

Over the years, the company programs have been documented, to ensure their reliability, and made accessible through terminals connected to mainframe computers. There has also been an effort to combine process-design capabilities into single, powerful programs, notably commercial flowsheeting programs such as: Aspen Plus, based on the U.S. Dept. of Energy's Aspen (*CE*, May 31, 1982, p. 59), and extended and marketed by Aspen Technology, Inc. (Cambridge, Mass.); Concept (CADCentre, Cambridge U.K.); Design/2000 (ChemShare Corp., Houston); Process (Simulation Sciences, Inc., Fullerton, Calif.); and Flowtran (Monsanto Co.).

These are large programs, made up of hundreds of subroutines—over 1,500 in the case of Aspen—that carry out calculations on data files set up by the user: physical properties; steady-state mass and energy balances; simulation of unit operations such as distillation, flash, reaction and heat transfer; recycles; equipment sizing and costing. Newer capabilities include modeling of processes involving solids and electrolytes.

Still, engineers turn to other tools in screening processes, designing heat-exchanger networks, specifying or costing equipment to inhouse standards, or drawing piping and instrumentation diagrams (P&IDs). At each step, work may have to be reviewed, and data transferred, manually. Thus, the engineering computing environment is still characterized as involving many programs, many databanks and files, and many computers, with little sharing of resources.

At the recent AIChE Diamond Jubilee Meeting, Prof. R. L. Motard of Washington University (St. Louis) pointed out the shortcomings of this fragmented approach to computing: difficulty in tracking large quantities of programs and data; limited capability to define data outside of the program that processes it; duplication of data, with inconsistencies in multiple copies; limited computer facilities for transferring data to and from central storage; data not being available in formats required by programs that need it. Others have pointed out the high cost—including engineering time—of maintaining all these resources.

## Integrating process data

One large step in tying process systems together is the integration of process-engineering data; i.e., setting up files or database-management systems (DBMS, see box on p. 18) that allow engineers and programs to share data more easily.

For example, a taskforce at Monsanto is taking existing stand-alone programs, integrating them and their data, and making them interactive. According to Ken Kettler, manager of computer-aided engineering, a current task is to link economic and technical evaluation.

At the AIChE meeting, Leonard A. Barnstone, a senior engineering associate of Exxon Research & Engineering Co., explained how a facility called the Engineering Toolkit allows interactive process-development

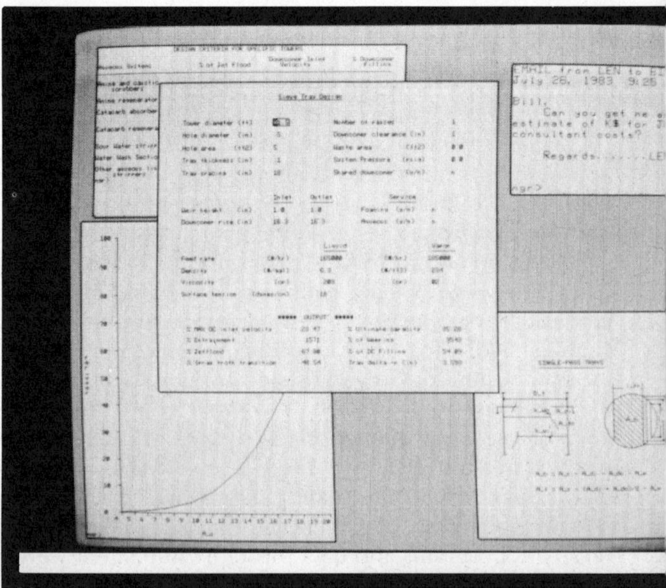

Multiple windows on process data are supported by Unix operating system Fig. 1

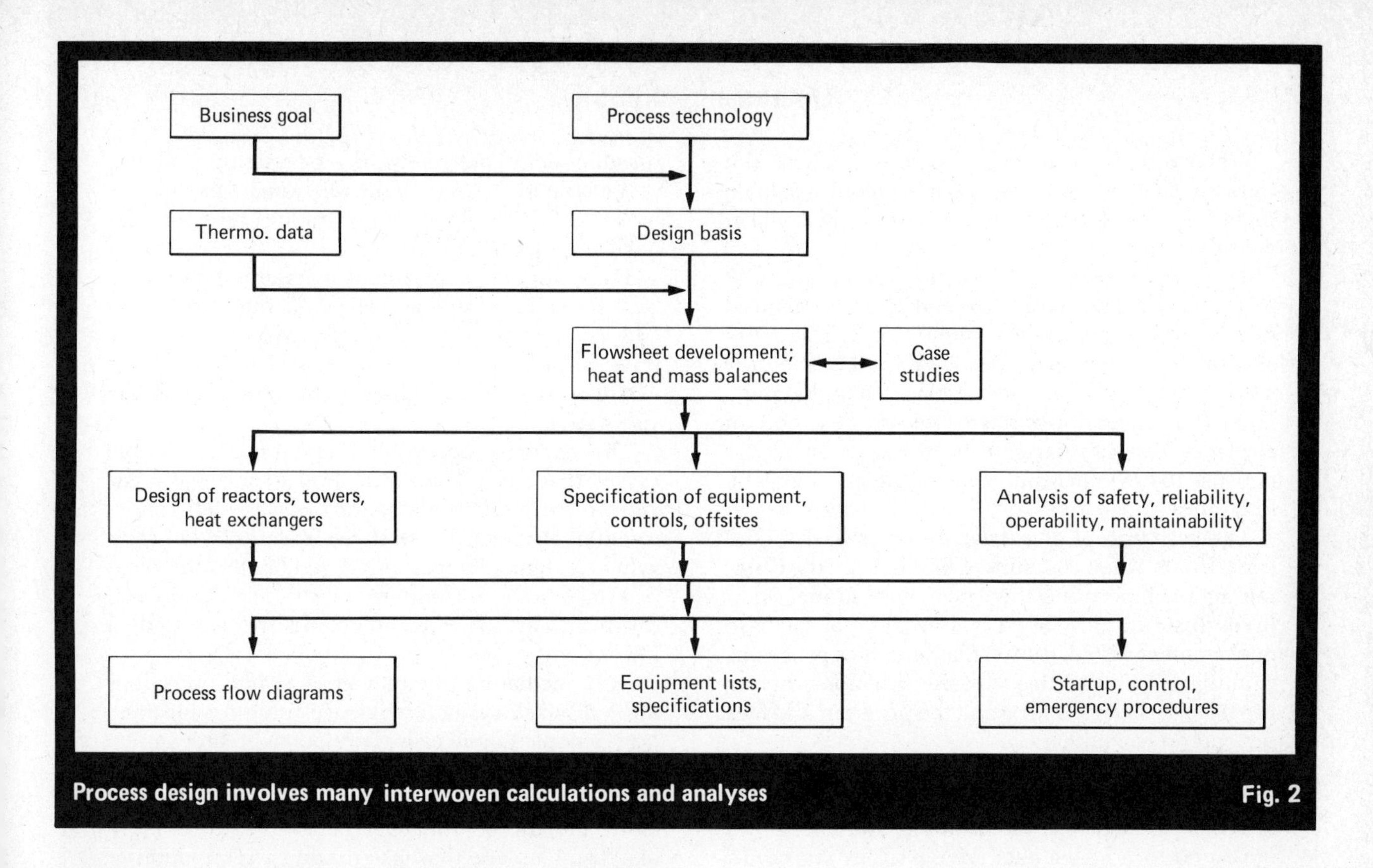


**Process design involves many interwoven calculations and analyses** **Fig. 2**

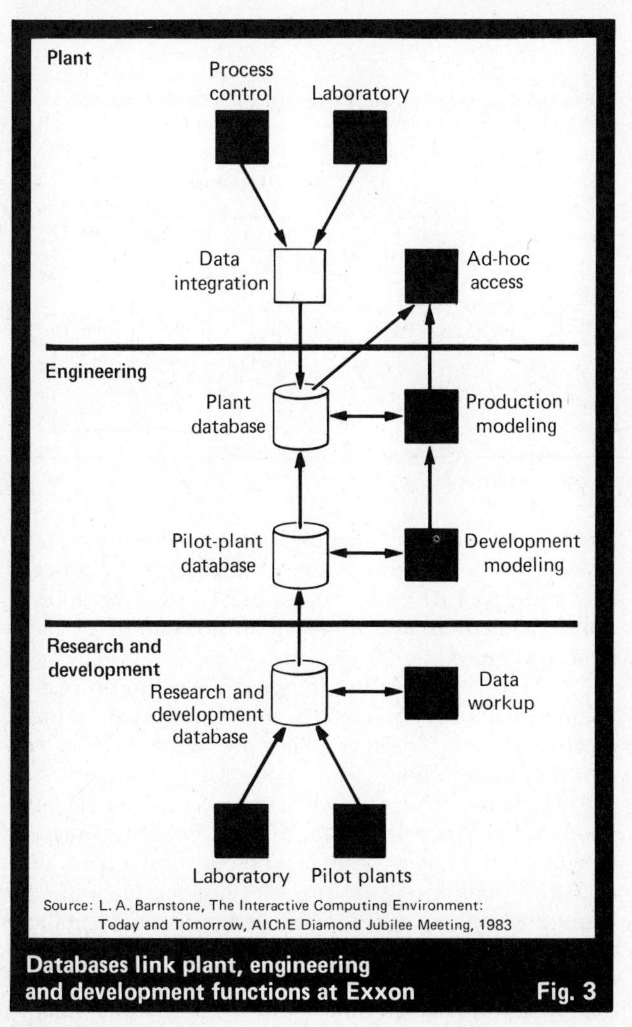


Source: L. A. Barnstone, The Interactive Computing Environment: Today and Tomorrow, AIChE Diamond Jubilee Meeting, 1983

**Databases link plant, engineering and development functions at Exxon** **Fig. 3**

programs to share data, program modules and utility programs. The Toolkit is a database manager implemented under Nomad, a DBMS used in business.

According to Barnstone, the Toolkit now supports four interactive programs—vapor-liquid equilibrium, tower internals, distillation, and heat exchangers—and three or four more will be added this year. "We're not yet at a point where we have the core design programs on the system, but we're heading that way."

Exxon engineers also have access to plant, lab and pilot-plant data from remote locations. These data are collected locally, and transmitted periodically to engineering locations, where engineers can use them for analysis, troubleshooting, or design work. Fig. 3 illustrates the flow and storage of data in this system.

Even as individual applications are integrated, companies are aiming at bigger targets. Since database technology started catching on in the early 1970s, companies such as Chiyoda Engineering and Construction Co. (Tokyo) and ICI PLC (Runcorn, Cheshire, U.K.) have been aiming at systems that tie together all process-engineering data and programs.

## Process database ties-in flowsheeting

One system built for that purpose is DesignMaster, a joint venture of ICI and ChemShare, due to be released in February according to Larry Lesser, ChemShare's vice-president of sales. Fig. 4 outlines the product, an outgrowth of ICI's inhouse process engineering that encompasses a just-announced ChemShare flowsheeting system—Design/2, which combines Design/2000 and Refine.

According to the developers, DesignMaster takes the process engineer through a stepwise buildup of the

## Database technology

When you phone an airline to reserve a seat, the clerk can tell you in a few seconds about available flights, routes, seats, meals and fares, and confirm your plans just as fast.

The airline reservation systems were among the first interactive databases. Here, data are structured as networks of passengers, flights, airplanes, crews and other entities, plus details about each entity. That is, a passenger is associated with a flight; and a flight is associated with passengers, a crew, and an airplane. You can navigate from any point in the network to any other, but some of the paths may be long ones.

Today's concept of a database dates back to the early 1970s, when committees of the Assn. for Computing Machinery and the American National Standards Institute developed a concept of a database management system (DBMS) that separates programs from data. Contrast this with the situation where a program sets up data files, and the program and files depend on each other.

The generic DBMS moves data three levels away from the user and his or her programs. The user sees only some part of the database, and a structure unrelated to how the data are physically stored. In fact, the user may see data that are not stored at all; e.g., specific volume may actually be calculated from a density value.

From the user's point of view, the data are structured in hierarchies, networks, or relations (tables), as illustrated below.* The relational DBMS is the most flexible, as its structure implies no particular navigation paths.

The computer department maintaining the database sees all the data, as well as the built-in security and validation checks. The lowest-level view is the physical one; i.e., the computer's view of how and where it stored everything. The DBMS keeps the views straight.

The database approach can eliminate multiple copies that waste space and lead to inconsistency; allow many to share data; and allow administrators to enforce standards. All these features are of clear value to engineers working together. But the database takes a lot of computer power, and imposes an overhead of restrictions on engineers trying to do a job.

Also, the nature of engineering applications does not lend itself easily to databases built for business transactions: Engineering has data of greater variety and complexity, and there is often a time factor. In process control, almost everything is time-dependent, and the data-management situation is more like flying a plane than like making a reservation to fly in one.

*Based on C. J. Date's "An Introduction to Database Systems," 3rd. ed., Addison-Wesley, Reading, Mass., 1977.

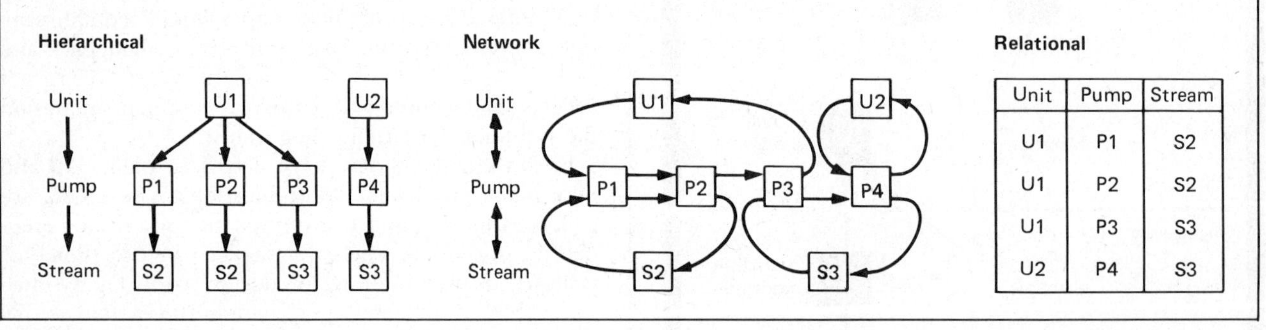


| Unit | Pump | Stream |
|---|---|---|
| U1 | P1 | S2 |
| U1 | P2 | S2 |
| U1 | P3 | S3 |
| U2 | P4 | S3 |

design, from selecting data to producing equipment specifications and drawing process flow diagrams. But it goes beyond flowsheeting in providing a single window to all data and programs for several projects. Fig. 5 shows the data structure.

ICI is also looking ahead to a design database having a more flexible structure—i.e, that does away with fixed paths for data storage and retrieval. Such a concept is being researched by Motard and coworkers under sponsorship of ICI and other companies.

Looking ahead at DesignMaster, the developers plan to open its process window to plant-design databases—i.e., to tie in computer aids used by other engineering disciplines. But for 1984 the scope is process data.

"We wanted to help the process engineer get his bit of the action right, before we considered anything else," says John Liles, of ICI's Design Systems Group. Part of the action is an ability to apply fullblooded simulation to process synthesis. Liles points out that an actual process isn't going to work quite the way that hand and graphical synthesis methods tell you—e.g., if nonstandard equipment is required.

Liles adds, "One of the things you have to do is give engineers time to be creative and conceptual." More-creative process schemes that work are seen as a key payoff of integration.

At last summer's Conference on Foundations of Computer-Aided Process Design, Stanley Proctor, Monsanto's director of engineering technology, pointed out that in a $100-million capital project, engineering will run about $20 million, and process design about $4 million of that. So there is more leverage in improving the design, where a 1% capital reduction is worth $1 million, than in cutting design time.

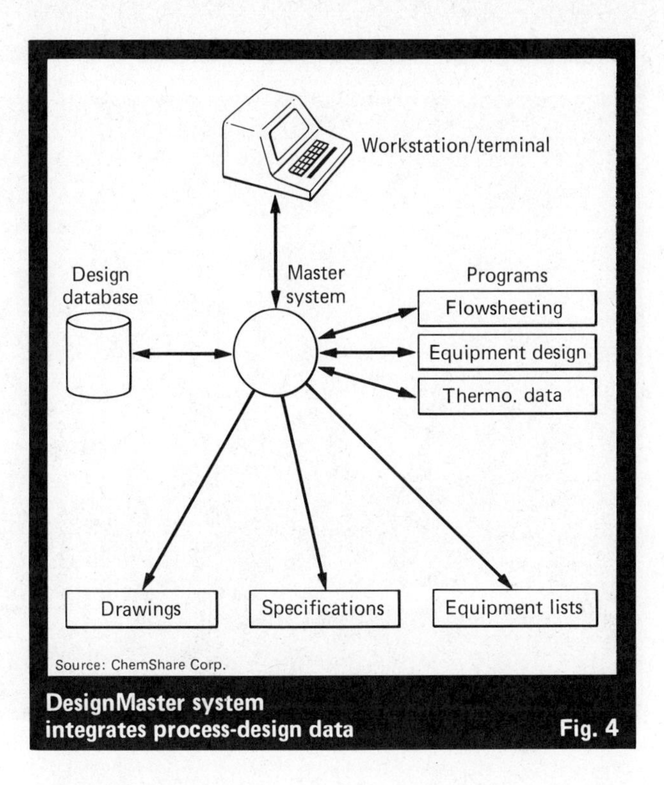


**DesignMaster system integrates process-design data** **Fig. 4**

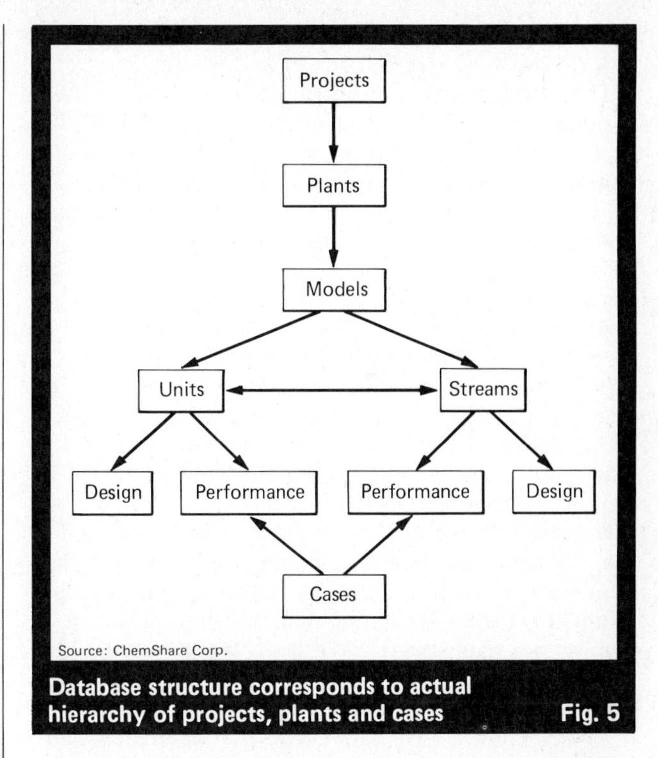


**Database structure corresponds to actual hierarchy of projects, plants and cases** **Fig. 5**

The leverage on the total engineering costs is not being ignored, however. Firms that have been using computer-aided drafting tools are integrating more design functions—from P&IDs through pipe routing to electronic models of all the plant systems—on their computer-aided design (CAD) systems.

## Computered-aided modeling and design

At Union Carbide's Linde Div. (Tonawanda, N.Y.), a CAD system is already doing "complete process plant design," says Robert Knight, manager of computer-aided design engineering. "We'll be phasing out plastic models in 1984." Two air-separation plants were designed on the system since its installation in 1983—Fig. 6 is a CAD view of one—and Linde is expanding and upgrading.

The original setup, from GE Calma Co. (Santa Clara, Calif.), was six terminals sharing a VAX 11/780 computer (Digital Equipment Co., Marlboro, Mass.). All these terminals are being replaced by Calma's new distributed-processing workstations, which will free the VAX computer to do analysis and other functions. This year, Knight expects to at least triple the number of stations, and to add capability for analysis of structural, flow, and process-control systems.

Today's CAD systems, from companies such as (Huntsville, Ala.), Computervision (Bedford, Mass.) and Prime Computer, Inc. (Natick, Mass.), are capable of building complete three-dimensional data models of plants (*Chem. Eng.*, Aug. 8, 1983, p. 19). Each user can have a different view of the data—say, as a piping isometric—but the views can be overlaid to give the whole picture. And the model is smart, including specifications along with graphic data. When modeling is done, any view can be drawn automatically—complete with dimensions. Specifications, too, can be viewed differently, and used to generate lists of equipment, piping, valves, materials, instruments, etc.

Such systems are growing out of the 2-D drafting systems that have been used by engineering firms since the mid-1970s. Those paid their way in saving drafting errors and speeding up schedules. "Now," says Joe Barba, director of management systems at C-E Lummus (Bloomfield, N.J.), "since we need less time to draw, we can do more process-design planning." And the new systems, by tying data together, are cutting out costly design errors. "We find that the accuracy of design is phenomenal," says Barba of Lummus's PDMS system.

### What's happening in Japan?

"We are using the 3-D CAD technology after a fashion," notes a source at the chemical plant and engineering headquarters of Mitsubishi Heavy Industries Ltd. (Tokyo), "but we are using lots of personal computers." While CAD is employed to the fullest by Japan's automobile makers, there is little evidence that engineering firms use the latest technology extensively.

Two top firms, Chiyoda and JGC Corp. (Tokyo), decline to divulge details of computer usage. But they privately concede that perhaps none of the big three engineering companies stands out in computer technology. Meanwhile, at Toyo Engineering Corp. (Tokyo), a 3-D CAD system is used for at least piping as part of detailed design.

In the U.S., engineering firms see Japan (and Korea) as improving their competitive posture through computer technology. And technology vendors see fertile markets in Asia. For instance, Japan has been the biggest buyer of IBM's Advanced Control System.

"Where historically we found 8% errors in field piping, now we see only 1–2%."

The starting point of plant design is the process designers' process flow diagram, as illustrated in Fig. 7. At Lummus, line lists from diagrams generated on a graphics system are sent electronically to a PDMS CAD system. The company is now looking at a flowsheeting system that will transfer its design data that way.

Then, the design work of the various engineering disciplines is integrated to varying degrees. At M. W. Kellogg Co. in Houston, "the thrust for 1984 is a 3-D modeling system integrated on a multidiscipline basis," says Leonard A. Neidinger, Manager of the Computer Graphics Group. To support 3-D modeling, Kellogg is taking delivery of a VAX 11/780 at presstime, and "for 1984 you're safe in multiplying the number of workstations we have now by two."

Kellogg's Intergraph CAD system, which replaced a drafting-oriented system, generates process flow diagrams, P&IDs, wiring diagrams, instrumentation loops, plot plans, exchanger setting plan drawings, mechanical details, and data sheets. "We use a mainframe for the real tasks—heat exchanger rating is an example—and transfer the data to the graphics-system computer."

Says Neidinger, "We have clients that sit down with us and lay out a plot plan in a matter of hours. This used to take weeks; recycling the comments alone took 5–10 days. It's almost unbelievable how we can shorten schedules." Meanwhile, at Kellogg's London office, a PDMS CAD system already in place has additional capabilities such as dynamic rotation of a 3-D model.

Ultimately, process- and plant-design systems are going to be linked with systems operating a plant. A systems vendor called this "cradle-to-grave CAD," and it is being put into practice at a Tennessee Valley Authority coal-to-methanol plant to be constructed starting this fall. According to a TVA spokesman, all the planning and modeling of the plant, and all operation and revamp for its expected life, will be supported by integrated CAD and control systems—in this case, both running on mainframe computers.

## Integrating process data

In process control, integration is a different class of problem: There many sources of data, even tens of thousands, and process data change by the second.

Yet there are strong incentives for integration. Says Cecil L. Smith, a consultant in Baton Rouge, La.,"In terms of information, one of the needs I see is an ability to integrate systems from the regulatory level right through management information. The ones to do that the most now are the oil refineries, but the need is there more and more in chemical plants—to tie in inventory control, order entry, raw-material management, business forecasting, and all the various factors that impact production."

This kind of integration implies a hierarchy of control: from regulatory controllers in contact with the process, to supervisory computers, to systems supervising plant-wide functions, to systems providing management information. To achieve it means keeping track of data, and making it available to the various computers and users, via database and networking technology.

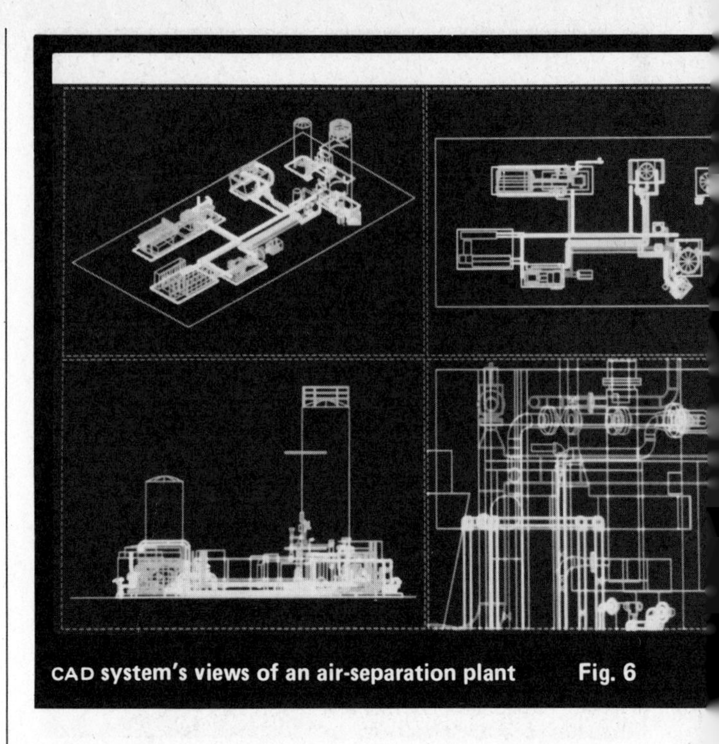

CAD system's views of an air-separation plant Fig. 6

## Control networks

In 1983, several vendors announced control systems that link controllers and computers throughout a plant, and provide a single window through a generic workstation. Among them are Vision 2002 from Measurex Corp. (Cupertino, Calif.), and TDC-3000 from Honeywell Inc. (Minneapolis), which is illustrated in Fig. 8. Meanwhile, Bailey Controls Co. (Wickliffe, Ohio) is tying its Network 90 system to office systems via an AT&T office communications network.

According to Honeywell, its integrated network links the process and its controllers with: workstations that serve operators, managers or engineers; computers supervising lower-level systems; higher-level computers performing optimization functions; programmable logic controllers (PLCs) handling discrete operations; and other computers or networks supporting engineers and administrators.

This kind of integration has grown out of the distributed control systems introduced in the last eight years by vendors such as Honeywell, Bailey, Foxboro Co. (Foxboro, Mass.), Fischer & Porter Co. (Warminster, Pa.), Fisher Controls (St. Louis), Toshiba Corp. (Tulsa, Okla.), and Westinghouse Electric Corp. (Pittsburgh). Here, the microprocessor-based controllers each handle several loops, and the supervisory computers handle several controllers—plus the operator interface.

But now the data highways are faster, up to millions of bits per second (bps); they extend farther; and computers, controllers and PLCs of various vendors can sometimes be linked through off-the-shelf interfaces. For instance, Westinghouse and Bailey provide interfaces for IBM Personal Computers. This makes it more convenient to implement integrated applications. For example: putting data from the lab into process-control schemes; or using data on inplant energy transfers as input to cost-allocation and energy-audit systems.

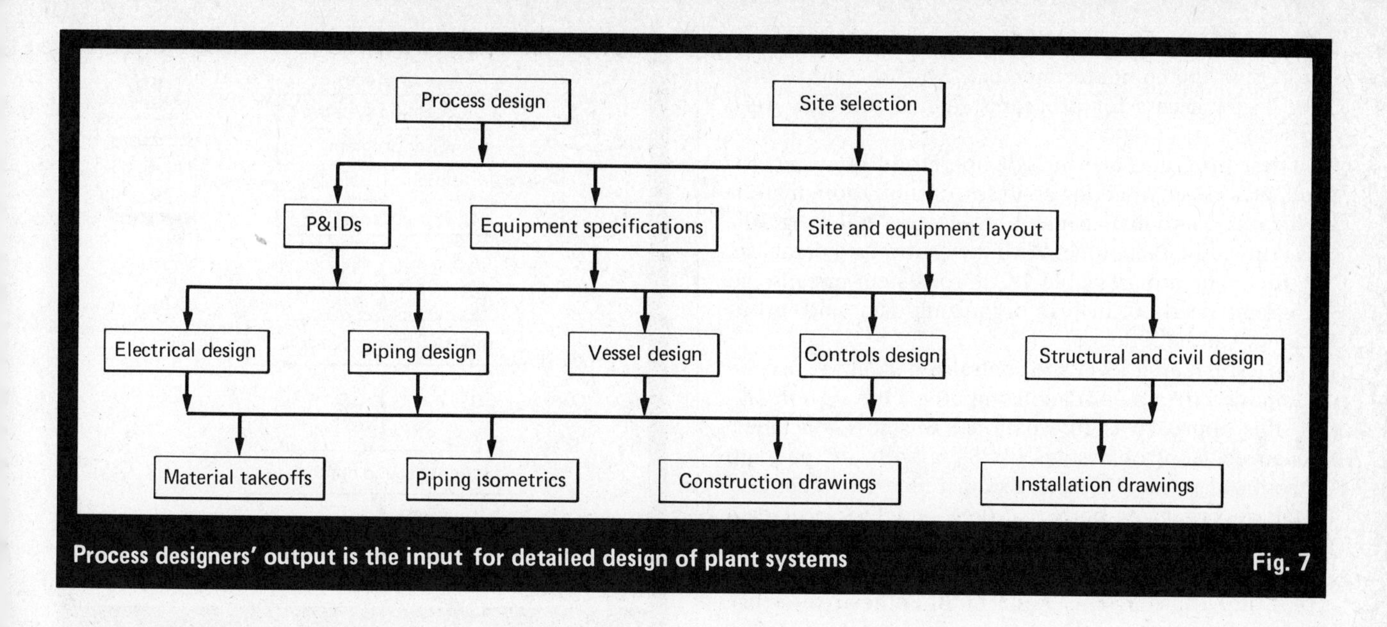


**Process designers' output is the input for detailed design of plant systems** **Fig. 7**

Standing in the way of integration, systems of different vendors still do not speak the same language. There are standards for local area networks based on token-passing communication around a two-way ring—from the IEEE 802.4 committee—and vendors have provided standard interfaces to distributed-control or PLC systems for their supervisory-level computers. These interfaces involve both hardware and software; one of the complexities is to provide security.

## Databases in process control

Distributed-control systems traditionally store data at the lowest practical level. If a higher-level computer is using average values for management information or advanced control, the averaging is typically done at the controller, or at a computer devoted to historical data from which data can be summoned via the data highway

One problem with this approach is that there may be inconsistencies in data—say, alarm limits—that are used by several machines. As companies take more interest in plantwide control, this may be a limiting factor.

Now, high-powered supervisory computers take on some of the work of organizing data and making it available. One system, the Optrol 7000 from Applied Automation Inc. (Bartlesville, Okla.), has a process database described as the collection and distribution center

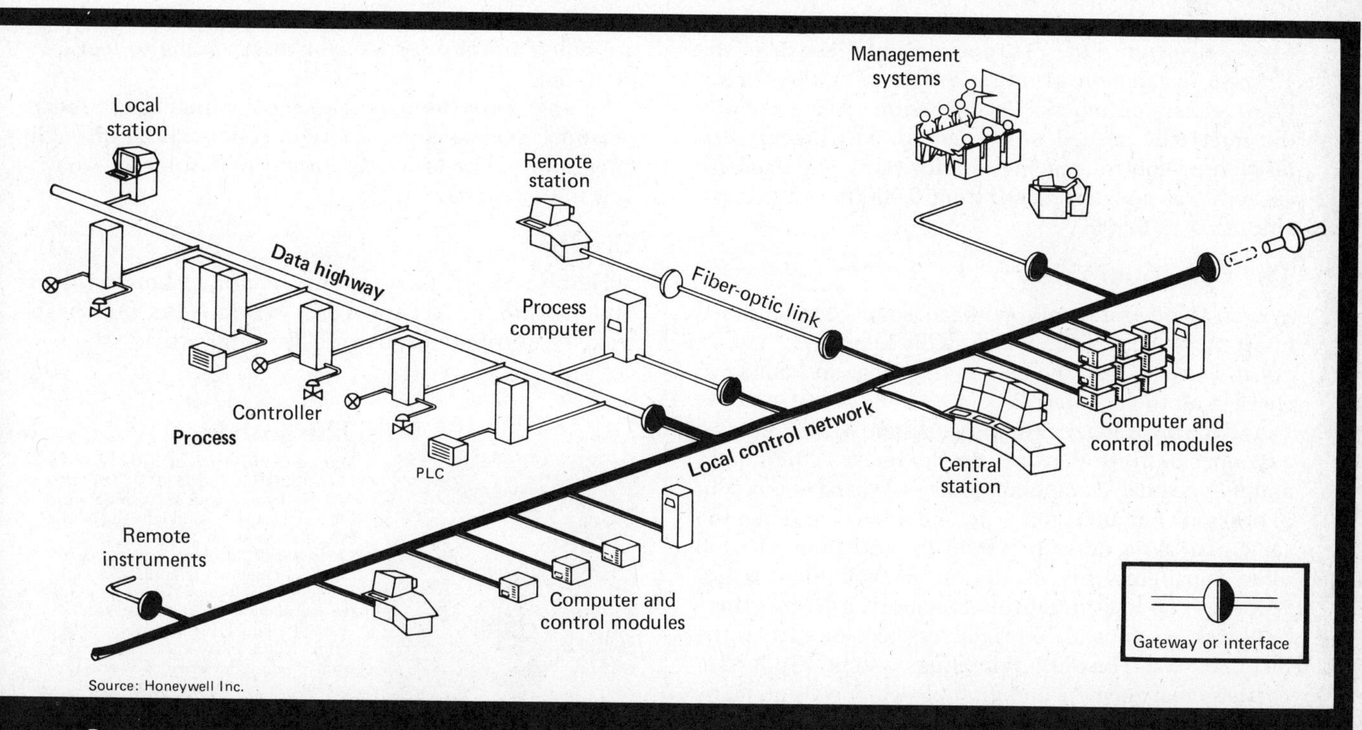


**Process-connected devices communicate with control network via data highway** **Fig. 8**

for all the data in the process. The database resides entirely in the computer memory, which is said to provide faster access for programs than if the data were distributed.

These programs may include operator display generation, process engineering analyses, optimization, material balances, and management reports; several programs may run at the same time. With the centralized database, all the data are available by a consistent means, so programs need not involve organizing data, and inconsistency can be reduced.

On a still higher level, the centralized database may be maintained by a mainframe computer. One system taking this approach is International Business Machines' Advanced Control System (ACS), a software package introduced in the 1970s at Exxon refineries. The ACS database organizes both real-time and historical data from all over a plant, providing a consistent view of everything going on in process units, material-storage areas, and utility systems. Fig. 9 outlines the architecture of the ACS system.

Lynn Holmes, senior market support administrator for the ACS, explains that the data are structured as variables of seven basic types—full-value process variables, status variables, time variables, material-balance variables, triggers, switches, and calculated variables. Each variable is made up of several attributes, which may be stored in different places—i.e., measured values and setpoints in main memory, other attributes on disk.

But this is of no concern to the user, who can write control or analysis programs and generate reports based on all the attributes of any variable. And the database approach means that the user may simply expand the database, rather than create a new one, to accommodate added equipment or instruments.

Such a centralized system may sit atop another, decentralized one. At last October's Instrument Soc. of America International Conference and Exhibit, Glenn Bodie of Esso Petroleum Ltd. (Toronto, Ont.) described the 1982-83 installation of an ACS system at the Sarnia, Ont., refinery of Imperial Oil, an Exxon affiliate. Among the interfaces needed were ones for instruments; for distributed-control systems; and for PLCs. The database capacity was also expanded, from 6,000 full-value variables to over 32,000.

## The "personal" view

Even as integration moves ahead, some companies see limits on how far it should go. ICI's Liles says "You've got to leave a lot of the decision-making, and ability to check, with the engineer."

And having to obey a computer system's rules may put a damper on fresh thinking. So, companies continue to adopt "personal" computing tools. ICI plans to buy a lot of desktop computers this year, and others are doing the same. Today, a desktop system for less than $10,000 offers capabilities found only on minicomputers a few years ago: up to 1 megabyte of memory, a 10-megabyte hard disk, a powerful 16-bit microprocessor such as the MC-68000, a Unix-like operating system supporting database management and multiple windows, and many support features for graphics and communications.

And the software has gotten better. Today, the desktop computer user can call on integrated packages (e.g., Context MBA, Lotus 1-2-3) that allow spreadsheet-analysis and graphic-presentation software to share data. Sophisticated database management is also available on the desktop, through systems such as dBase II.

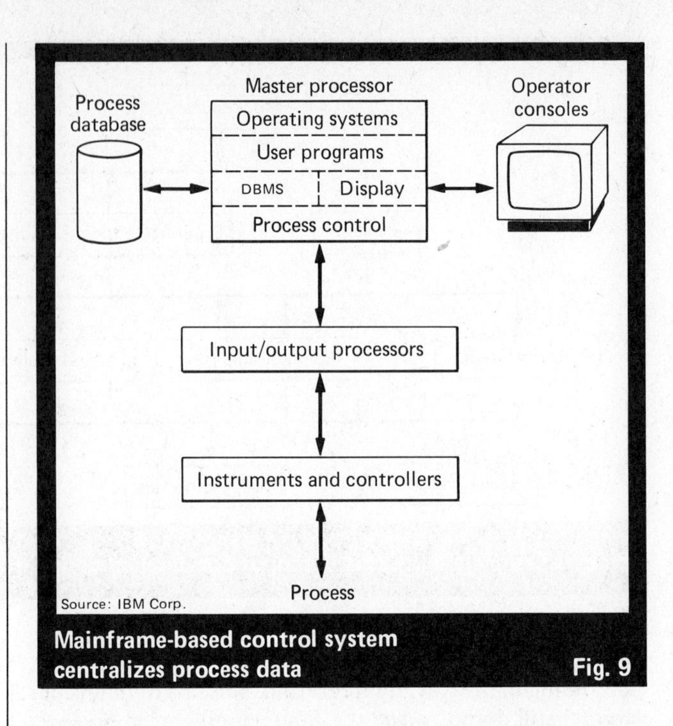


Mainframe-based control system centralizes process data — Fig. 9

Ralph Love, information services manager at Bechtel Petroleum Inc. (San Francisco), is now implementing a dBase II system for tracking engineering records on a network of three IBM Personal Computers—among some forty in the office. Says Love, "We've found an incredible number of things personal computers can handle around here."

John Hale, manager of online systems in Du Pont's engineering department, says "If personal computers are doing a useful service, you aren't going to legislate them out."

So engineers can expect to see computing get more personal, even as it gets more institutionalized through integration. The article following tells some of what's new in that arena.

## Credits

Herbert Short, *CE*'s regional editor in London, and Shota Ushio, of McGraw-Hill's World News Bureau in Tokyo, contributed substantially to this article.

## The author

**Mark A. Lipowicz** is an associate editor of CHEMICAL ENGINEERING (Tel. 212-512-6749), where he seeks out authors for feature articles and edits their work for publication. He also handles the *Chemputers* department, which covers chemical-engineering applications of computer technology. When he joined the magazine in 1980, he had three years of engineering experience. He has a B.S. degree in chemical engineering from Washington University, and is working on his M.B.A. at New York University. He belongs to AIChE.

# Desktop software

Where can you find microcomputer software for engineering applications? This article lists sixty companies that sell such software, and tells where to go for more information.

**Mark A. Lipowicz,** *Associate Editor*

☐ If you want to do some real engineering work on that desktop computer, the first question is where do you get the software—i.e., the programs.

One choice is to write it yourself. This is what many engineers have done—41% in an April 1983 survey of *CE* readers. Another is to use some that another person in the company has written—53% have done this. But there might exist a software package outside the company that does just what you want, or close to it, and can be cost-justified as a timesaver. Only 21% of the respondents to the survey had bought software for use on the job, but as more and better engineering software becomes available this is likely to change.

This article lists a number of vendors of desktop-computer software for chemical engineering and related applications, and provides references to publications and exchanges that list still other sources.

## Sources of chemical engineering software

The table lists commercial vendors of software, the applications they say they cover, and brands of computers (or operating systems) that their software is claimed to run on. This list covers software of interest primarily to chemical engineers—programs for distillation or flash calculations, flowsheeting, flow analysis, and such. It also covers software of interest to engineers in general—e.g., project management, simulation, drafting and statistics.

Please note that software offered for one computer may run on other, compatible ones and may be available in a variety of formats. And note that some packages require a particular setup—say, a minimum amount of memory. Finally, there are of course many other sources, such as engineers who want to sell or exchange their own software. Some of these may be found in the references at the end of the article.

## Also of interest

There are also software packages that come "bundled" with hardware; instrument interfacing is an example: The Pro-BASIC of Hewlett-Packard Series 200 computers includes 31 input/output and control statements. Vendors such as Cyborg Corp. (Newton, Mass.) and Inconix Corp. (Natick, Mass.) package data-acquisition software along with interface circuitry.

One special bundle is a PPDS physical-property data module, to be introduced in late 1984 by the U.K.'s National Engineering Laboratory (East Kilbride).

And, of course, the general-purpose software packages have been widely applied in engineering work: A spreadsheet program is handy for comparing design cases; database managers can keep track of laboratory or project information; presentation packages can create charts and graphs that help sell an engineering idea; word processors are handy in writing proposals and reports; integrated software packages can do all these things together.

What is expected for 1984? More software for chemical engineers, and a greater variety, as programs are written for the newer computers and operating systems. Some of the things to look for are programs that can make use of plug-in mathematics chips, and operating systems that permit several programs to run at once.

## Acknowledgements

Richard J. Bigda of Technobyte (Tulsa, Okla.), and Herbert Short, *CE*'s regional editor in London, contributed substantially to this article.

## Sources of desktop software for engineering applications

| Company | Applications | Computers/ operating systems |
|---|---|---|
| **ACCI Business Systems** 4625 N. Freeway Houston, TX 77022 | Project management; specifications | CP/M; MS–DOS; PC–DOS |
| **Applied Business Technology Corp.** 76 Laight St. New York, NY 10013 | Project management | IBM |
| **Applied I** 200 California Ave., Palo Alto, CA 94306 | Dynamic simulation | Apple; IBM; CP/M |
| **AutoDesk, Inc.** 150 Shoreline Hwy., Bldg. B Mill Valley, CA 94941 | 2–D design and drafting | IBM; NEC; Victor; Zenith; CP/M–80 |
| **Bluebird's Inc.** 2267 23rd St. Wyandotte, MI 48192 | Linear programming; statistics | Radio Shack |
| **Gordon S. Buck** 5801 Parkhaven Dr. Baton Rouge, LA 70816 | Centrifugal pumps; mechanical seals | HP |
| **CAETech** Maple House, Maple Rd. Bramhall, Cheshire SK7 2DL, U.K. | Steam and water properties | Epson |
| **CET Research Group, Ltd.** P.O. Box 2029 Norman, OK 73070 | Equation solver; statistics; curve fitting | Apple |
| **COADE** 8552 Katy Freeway Houston, TX 77024 | Flowsheet simulation; distillation; flash; thermodynamics; flow networks; dynamic simulation; reactors; absorption; extraction; heat exchange; flow simulation; process control; economics; project management; maintenance | IBM |
| **Compco Computer Specialists** 7110 W. Fond du Lac Ave. Milwaukee, WI 53218 | Data display | CP/M |
| **Computer Aided Design** 764 24th Ave. San Francisco, CA 94121 | 3–D design and drafting | IBM |
| **Computer Analysts International** 4544 Post Oak Place, Houston, TX 77027 | Maintenance management | IBM; CP/M |
| **Constech Inc.** P.O. Box 610663 DFW Airport, TX 75261 | Construction cost management | Altos |
| **Delftse Uitgevers Maatschappij** Postbus 2851, 2601 CW Delft, The Netherlands | Book of BASIC separation-process programs | Sharp |
| **Digital Marketing Corporation** 2363 Boulevard Circle Walnut Creek, CA 94595 | Project management | IBM; Victor; CP/M; CP/M–86; UCSD Pascal |
| **Dynacomp, Inc.** 1427 Monroe Ave. Rochester, NY14618 | Heat exchangers; fuel cells; reactors; flow, pressure-drop analysis; thermodynamics; statistics; graphics | Apple; IBM; Radio Shack; CP/M |
| **Engineered Software** 2420 Wedgewood Dr. Olympia, WA 98501 | Design of piping systems, orifices, pumps, valves, flow networks | CP/M; CP/M–86 |
| **Entek Scientific Corp.** 4480 Lake Forest Dr. Cincinnati, OH 45242 | Rotating-machinery monitoring, analysis | HP |
| **Harvard Software, Inc.** Software Park Harvard, MA 01451 | Project management | IBM |
| **Hayden Book Co.** 10 Mulholland Dr. Hasbrouck Heights, NJ 07604 | Book of BASIC engineering programs | |
| **Hewlett-Packard Co.** 1501 Page Mill Rd. Palo Alto, CA 94309 | Statistics; engineering mathematics; graphics | HP |
| **Institute for Scientific Analysis, Inc.** 36 E. Baltimore Pike, Media, PA 19063 | Project management; plant utilities | IBM; Radio Shack; CP/M |
| **Institute of Industrial Engineers** 25 Technology Park/ Atlanta Norcross, GA 30092 | Project management; statistics; quality control; operations research; mathematics | Apple; Radio Shack |
| **Interactive Microware, Inc.** P.O. Box 771 State College, PA 16801 | Data display; curve fitting | Apple |
| **Kelix Software Systems** 11814 Coursey Blvd. Suite 220 Baton Rouge, LA 70816 | Chemical engineering; distillation; heat exchangers; flow networks; ventilation systems | Apple; HP; IBM |
| **Kern International** P.O. Box 1029 Duxbury, MA 02331 | 3–D drawing; data plotting; engineering | Apple; IBM; Zenith |
| **L.N. Engineering, Inc.** P.O. Box 2462 Kalamazoo, MI 49003 | Heat exchangers; batch distillation; boilers; vessel jackets; drying | Apple; IBM; Radio Shack |
| **Magellan Systems, Inc.** P.O. Box 4161 Bryan, TX 77805 | Distillation; flash; absorption; heat exchange; evaporation, compression; fluids handling; phase equilibrium; curve-fitting; mathematics | Apple; IBM; Radio Shack |
| **MCS, Inc.** 17942 Cowan Irvine, CA 92714 | 2–D design and drafting | Data General |
| **McGraw-Hill Book Co.** 1221 Ave. of the Americas New York, NY 10020 | Books of BASIC and Pascal programs, and BASIC subroutines, for engineering, science | Apple; IBM; NorthStar; Radio Shack |
| **The Mesa Co.** Drawer M Kemblesville, PA 19347 | Energy systems analysis | Apple; DEC; HP; IBM; Radio Shack; CP/M–80 |
| **Microcomp** Ashurst Lodge, Ashurst, Southampton SO4 2AA, U.K. | Pipe friction, networks, plotting; open-channel flow; project management | Apple; CP/M |
| **Micro Mode, Inc.** 322 Graycliff Dr. San Antonio, TX 78223 | Project management | CP/M |
| **North America Mica, Inc.** 11722 Sorrento Valley Rd. San Diego, CA 92121 | Project, resource, materials management; bill of materials | CP/M; CP/M–86 |

(continued on next page)

**Sources of desktop software for engineering applications (continued)**

| Company | Applications | Computers/ operating systems |
|---|---|---|
| **Omicron Software** 57 Executive Park S. Atlanta, GA 30329 | Project management; statistics; data display | IBM |
| **Osborne/McGraw-Hill** 2600 Tenth St. Berkeley, CA 94710 | Books of programs for engineering, science, mathematics | Apple; Radio Shack |
| **Process Systems International** RD#3 Foxchase-Friendship Rd. Vincentown, NJ 08088 | Flash; thermodynamics | IBM |
| **PSI Systems Div.** P.O. Box 3100 Andover, MA 01810 | Flowsheet analysis; heat exchangers; fuel cells; fans, cyclones, turbines; thermodynamics; reformers; burners | Apple; IBM; Radio Shack |
| **Radio Shack** One Tandy Center Fort Worth, TX 76102 | Statistics; job costing | Radio Shack |
| **Scitor Corp.** 256 Gibraltar Rd., Bldg. 7 Sunnyvale, CA 94089 | Project management | DEC; Grid; IBM; TI |
| **Serendipity Systems Inc.** 225 Elmira Rd. Ithaca, NY 14850 | Statistics; mathematics | Apple |
| **Sheppard Software Co.** 4750 Clough Creek Rd. Redding, CA 96002 | Project management | IBM; Tektronix |
| **Ski Soft Computer Service** Bryony, Westwood Ave. Woodham, Weybridge, Surrey, U.K. | Pipe sizing; relief valves | Apple; ITT |
| **Software Arts, Inc.** 27 Mica Lane Wellesley, MA 02181 | Equation solver; mechanical engineering; basic science | DEC; IBM; Wang; others |
| **Software Systems Corp.** 5766 Balcones Dr. Austin, TX 78731 | Energy conservation; combustion; fluid flow; steam properties; cogeneration; statistics | IBM; TI |
| **Spectrum Software** 690 Fremont Ave. Sunnyvale, CA 94087 | Statistics; mathematics; matrices | Apple; IBM |
| **Statistical Graphics Corp.** P.O. Box 1558 Princeton, NJ 08542 | Data analysis, display | IBM |
| **Sybex Computer Books** 2344 Sixth St. Berkeley, CA 94710 | Books of VisiCalc, BASIC, FORTRAN engineering programs | |
| **Systems Design Lab** 2612 Artesia Blvd. Redondo Beach, CA 90278 | Regression; forecasting | Apple |
| **T&W Systems, Inc.** 7372 Prince Dr., Huntington Beach, CA 92647 | 2–D design and drafting | Apple; HP; IBM |
| **Technobyte** 6216 S. Lewis Ave. Tulsa, OK 74136 | Distillation; flash; equipment sizing, costing; heat exchangers; physical properties; process design; mathematics | Apple; HP; IBM; Osborne; Radio Shack; Sanyo; Xerox |
| **Turnaround Planning Services** 823 Bradwell Houston, TX 77062 | Turnaround management | HP |
| **United Networking Systems, Inc.** 7007 Gulf Freeway, Houston, TX 77087 | 2–D design and drafting | IBM |
| **VisiCorp.** 2895 Zanker Rd. San Jose, CA 95111 | Project management | Apple; IBM |
| **Wadsworth Electronic Publishing Co.** 20 Park Plaza Boston, MA 02116 | Statistics | Apple; IBM |
| **Western Software** P.O. Box 953 Woodland Park, CO 80863 | Engineering mathematics; statistics | IBM; Sage |
| **Westico** 25 Van Zant St. Norwalk, CT 06855 | Project management; statistics; graphics | CP/M–80; MP/M–80; CP/M–86; MP/M–86; MS–DOS; PC–DOS |
| **Westminster Software Inc.** 3000 Sand Hill Rd., Menlo Park, CA 94025 | Project management | CP/M; MP/M; PC–DOS; MS–DOS |
| **York Research Consultants** 938 Quail St. Denver, CO 80213 | Environmental, regulatory-compliance database | IBM |
| **Zimpro, Inc.** Military Rd. Rothschild, WI 55475 | Preventive maintenance | Apple; IBM |

Notes: "CP/M" and "MP/M" refer to operating systems of Digital Research, Inc. "PC–DOS" and "MS–DOS" refer to operating systems of Microsoft Corp.

## For more information

**Directories**

"Chemical Engineering Applications Software for Personal Computers," AIChE, 345 E. 47th St., New York, NY 10017. To be published in early 1984.

"Datapro Directory of Microcomputer Software," Datapro Research Corp., Delran, NJ 08075. Other directories for Apple, CP/M, IBM; still others for larger computers and systems.

"Data World Scientific and Engineering Software Directory," Auerbach, 6560 N. Park Dr., Pennsauken, NJ 08109.

"HP Third Party Technical Systems Software Solutions," Hewlett-Packard Co., Cupertino, Calif.

"HP Series 80 Software Catalog," Hewlett-Packard Co., Corvallis, Ore.

"PC Clearinghouse Software Directory," PC Clearinghouse, Inc., 11781 Lee Jackson Highway, Fairfax, VA 22033.

**Periodicals (besides *CE*) and exchanges**

*Byte*, 70 Main St., Peterborough, NH 03458.

*Computers & Chemical Engineering*, Pergamon Press, Inc., Maxwell House, Elmsford, NY 10523.

*Computers for Design and Construction*, Meta Data, 441 Lexington Ave., New York, NY 10017.

*Computers in Mechanical Engineering*, American Soc. of Mechanical Engineers, 345 E. 47th St., New York, NY 10017.

*The Engineering Software Exchange*, 41 Travers Ave., Yonkers, NY 10705.

*Journal of Chemical Information & Computer Sciences*, American Chemical Soc., 1155 16th St. N.W., Washington, DC 20036.

Research Corp./Research Software, 6840 E. Broadway Blvd., Tucson, AZ 85710-2815.

# Online information

Online information systems, accessible through a terminal and a telephone connection, put the engineer in touch with extensive files of numerical data, text, and bibliographic references.

***Monica E. Baltatu,*** *Fluor Engineers, Inc.*

☐ The information explosion in science and technology has resulted in a mass of published materials, more than we are able to store or assimilate. But advances in computer and telecommunication technology are bringing about great improvements in our ability to locate and retrieve data pertinent to an engineering problem.

This article is about online databases that provide access to technical literature and numerical data. As engineering work gets more demanding, such systems are being used more and are becoming more important.

## Online databases

Online information retrieval has become familiar to many engineers through their use of bibliographic systems available via Dialog, Orbit and other services. Such systems allow rapid, interactive searching of many files, and minimize the need for complex instructions.

A bibliography, however, provides only potential sources of information. Other online systems provide full text (e.g, of articles, patents, regulations) and numerical data such as physical properties. We will cover all these kinds of online services, and explain what they do, but first a note on terminology is in order.

In the published literature [*1*–*3*], the term "database" is frequently used to describe bibliography and text as well as numerical data. Here, a database is considered an organized collection of information on a well-defined topic [*4*]. By the nature of the information, a database is classified as numerical or non-numerical.

The "machine readable" databases are the ones stored on disk or tape that may be searched and drawn from by a computer, via specialized software. The information may be distributed on tape, or it may be available "online"—i.e., from a central computer, via a terminal that communicates with the computer through a telephone line or other channel.

## Bibliographic databases

Online bibliographic databases provide references to technical articles, conference papers, patent disclosures and statistical data. These references are available to the engineer through commercial services such as Dialog, Bibliographic Retrieval Service (BRS), System Development Corp. (SDC), Medline, Toxline and others.

Table I lists twelve such databases of interest to chemical engineers. Each of these, and of the dozens of others available, offers some unique characteristic or research capability. An efficient and thorough search of some topic requires a well-trained user, as a search may involve several databases and search techniques [*1*]. As an example, the patent literature—a top source for technical developments—can be searched thoroughly only by a combination of files from several searching services.

A newer feature in bibliographic databases is the capability of printing the full text—though not the drawings or photos—of technical articles. One such database is Nexis, a service of Mead Data Central. Information enters the Nexis database shortly after the printed version becomes available. This database includes: newspapers (e.g., *The New York Times*); business magazines (e.g., *Business Week*); technical magazines (e.g., *Chemical Engineering, Oil & Gas Journal*); newsletters. Among the other sources available is the *Code of Federal Regulations*, a compilation of U.S. regulations in force.

Because the right information at the right time is valuable, new and more-creative online systems are becoming available. Here, it is new capabilities, rather than easier access or more-compressed information, that is

Originally published January 9, 1984

Table I

**Online bibliographic databases for chemical engineers**

| Database/coverage | Supplier | Field/description | Charges | Vendors |
|---|---|---|---|---|
| APILIT (1964–) | American Petroleum Institute | Petroleum refining. Worldwide journals, papers, reports. | $70/h | SDC |
| APIPAT (1964–) | American Petroleum Institute | Petroleum patents from U.S. and 8 other countries. | $70/h | SDC |
| Biotechnology (1982–) | Derwent Publications, Ltd. | Biotechnology. Over 1,000 journals, and patents. | $100/h | SDC |
| Compendex (1970–) | Engineering Information, Inc. | Engineering worldwide. Over 3,500 sources. | $98/h | BRS, Dialog, SDC |
| Chemdex (1972–) | Chemical Abstracts Service | Chemical dictionary. All compounds cited in literature since 1972; formulas. | $132/h | SDC |
| Claims/Uniterm (1950–) | IFI/Plenum Data Co. | U.S. chemical patents. Patent index based on full text. | $300/h | Dialog |
| Energyline (1971–) | Environmental Information Center | Energy information, including research and production. | $90/h | BRS, Dialog, SDC |
| Enviroline (1971–) | Environmental Information Center | Air, land, water environment. Resource management. Full text available. | $90/h | BRS, Dialog, SDC |
| Environmental Bibliography (1973–) | Environmental Studies Institute | Environmental and related fields. Over 300 publications. | $60/h | Dialog |
| Federal Index (1977–) | Predicasts, Inc. | Federal regulations, plus rules, hearings, contract awards. | $90/h | Dialog |
| Paperchem (1968–) | Institute of Paper Chemistry | Paper industry raw materials, processes, chemistry, equipment. | $65/h | Dialog |
| Pollution Abstracts (1970–) | Cambridge Scientific Abstracts | Leading source on environmental literature. | $73/h | BRS, Dialog |

Notes: Listed charges are subscriber rates per hour connected. There are additional charges for offline printing. The vendors' addresses are: Bibliographical Retrieval Services, 1200 Route 7, Latham, NY 12110; Dialog Information Services, Inc., 3460 Hillview Ave., Palo Alto, CA 94304; System Development Corp., 2500 Colorado Ave., Santa Monica, CA 90406.

driving development. One of the exciting features is the ability to combine bibliographic databases with numerical ones and with other related information.

A pioneer in this field is the Chemical Information System (CIS), a service initiated by the National Institutes of Health (NIH) and the U.S. Environmental Protection Agency (EPA). CIS is a set of online databases containing bibliographic and descriptive information, plus numerical data, on over 192,000 chemical substances.

Table II

| Purpose of database | Machine-readable databases Total | Online |
|---|---|---|
| Identification of unknown substance | 19 | 9 |
| Properties of pure substances and mixtures | 20 | 10 |
| Metallurgical calculations | 5 | 5 |
| Thermodynamic and thermochemical properties of individual substances | 6 | 3 |
| Properties of plastics | 2 | 0 |
| Chemical process simulation and design | 11 | 5 |
| Engineering data on materials | 58 | 3 |
| Total | 121 | 35 |

The CIS system is built around a central electronic file known as the Structure and Nomenclature Search System (SANSS). This allows the user to search more than 60 different files; the search may be for a known or unknown substance, by structure, index name, trade name, or Chemical Abstracts Service Registry Number. Then, the user can search for environmental records, physical properties, and other data in other files. For example:

- The Registry of Toxic Effects of Chemical Substances, provided by the National Institute of Occupational Safety and Health (NIOSH), contains over 40,000 toxicological measurements.
- The Partition Coefficients database, provided by Pomona College (Calif.), has coefficients for about 5,000 chemicals, and allows searching by solvent or by the compound of interest.

## Numerical databases

Bibliographic databases can provide references to data, but then it is up to the engineer to get the document or text, extract the data, and rearrange them in the

**Online numerical databases for chemical engineers** — **Table III**

| Database | Materials/data | Supplier |
|---|---|---|
| **Identification of unknown substances** | | |
| C–13 NMR | 33,000 nuclear-magnetic-resonance spectra. Part of NIH/EPA CIS. | Interactive Sciences Corp.<br>918 16th St. N.W.<br>Washington, DC 20006 |
| FIRST–1; Infrared Retrieval Program | 143,000. Matches up infrared spectra. Includes API, NRC/NBS, DMS/Butterworths, and other files. | DNA Systems, Inc.<br>1258 S. Washington St.<br>Saginaw, MI 48605 |
| IRIS; Infrared Information System | 110,00. Matches up infrared spectra. Files as for FIRST–1. | Sadtler Research Laboratories<br>3316 Spring Garden St.<br>Philadelphia, PA 19104 |
| MSSS; Mass Spectral Search System | 38,000. Matches up infrared spectra. Joint effort of NIH, EPA, and MSDC (U.K.) | Interactive Sciences Corp.<br>918 16th St., N.W.<br>Washington, DC 20006 |
| PBM/STIRS | 74,000 spectra for 58,000 materials. Compares and interprets spectra. | Dept. of Chemistry<br>Cornell University<br>Ithaca, NY 14853 |
| **Physical properties of pure substances and mixtures** | | |
| AESOPP; An Estimator for Physical Properties | Retrieval of stored data, and estimation of properties for pure components and mixtures. | Institute of the Union of Japanese Scientists and Engineers<br>5-10-11 Sendgaya<br>Shibuya-Ku, Japan |
| CPPDMS; Celanese Physical Properties Data Management System | Data correlation and retrieval. Data belong to Celanese Chemical Co. | University Computing Co.<br>1930 Hi-Line Dr.<br>Dallas, TX 75207 |
| DETHERM; Dechema Thermophysical Property Data Bank | 3,000 for retrieval, 550 for estimation of mixture properties. Joint effort of industry and universities. | Deutsche Gesellschaft für Chemisches Apparatewesen e.V.<br>Postfach 97<br>Frankfurt 97, West Germany |
| Keydata | 500, including PPDS (see below). Data retrieval and calculation. | CADCentre Ltd.<br>Madingley Rd.<br>Cambridge CB3 0HB, U.K. |
| Databank; System for Binary and Ternary Phase Diagrams | Phase diagrams for 45 binary systems and 120 ternaries. Includes Al, C, Co, Cr, Fe, Mo, Nb, Ni, Ti, W. | Manlabs, Inc.<br>21 Erie St.<br>Cambridge, MA 02139 |
| NEL-APPES; A Physical Properties Data-Estimation System | 200. Data retrieval and calculation. Based on AIChE APPES system. | National Engineering Laboratory<br>East Kilbride<br>Glasgow G75 0QU, U.K. |
| PPDS; Physical Properties Data System | 430 compounds, largely hydrocarbons. Calculates pure-component and mixture properties. | The Institution of Chemical Engineers<br>165/171 Railway Terr.<br>Rugby CV21 3HQ, U.K. |
| Thermodata Data Bank | 2,500 inorganics. Metallurgy, properties, reactions. A cooperative effort by European metallurgists. | Bibliothèque Universitaire des Sciences<br>B.P. 22<br>38402 Saint Martin D'Heres<br>France |
| TRL Correlation Package | 350 compounds. Pure-component data. Available to sponsors. | Thermodynamics Research Laboratory<br>Washington University<br>St. Louis, MO 63130 |
| **Properties of individual substances** | | |
| Fluids Pack | Properties of cryogenic fluids. Also ammonia properties. | U.S. National Bureau of Standards<br>Washington, DC 20234 |
| **Process simulation and design** | | |
| Aspen Plus | Generates pure-component and mixture properties. Includes data on 100 ionic solutions and 100 solid materials. | Aspen Technology, Inc.<br>251 Vassar St.<br>Cambridge, MA 02139 |
| Chemtran | 850 materials. Generates pure-component and mixture properties. Interfaces with process-design programs. | ChemShare Corp.<br>P.O. Box 1885<br>Houston, TX 77001 |

**Online numerical databases for chemical engineers (continued)** **Table III**

| Database | Materials/data | Supplier |
|---|---|---|
| **Process simulation and design (continued)** | | |
| GPA*SIM | 63 compounds, hydrocarbons plus water and hydrogen. Calculates thermodynamic properties, and multicomponent flash. | Gas Processors Assn.<br>1812 First Place<br>Tulsa, OK 74103 |
| Process | 650 including PPDS database. Interfaces with process-design and data-regression programs. | Simulation Sciences Inc.<br>1400 W. Harbor Blvd.<br>Fullerton, CA 92635 |
| **Engineering data on materials** | | |
| Alloy | 2,000 metals, alloys. Physical, mechanical, electronic properties. | U.S. National Bureau of Standards<br>Materials Building B-150<br>Washington, DC 20234 |
| EPIC; Electronic Properties Information Center | 40,000 materials. Electronic and thermophysical properties. | TEPIAC<br>Purdue Industrial Research Park<br>2595 Yeager Rd.<br>West Lafayette, IN 47906 |
| Polyprobe | 8,500 commercial plastics. Mechanical, electrical, other properties. | The International Plastics Selector, Inc.<br>9889 Willow Creek Rd.<br>San Diego, CA 92131 |

Source: Ref. 3

format required—e.g., for a process-design program.

Numerical databases, on the other hand, provide the actual data, and in some cases automatically make data available for use in interpolation, regression, or design programs. Having automatic access to high-quality data is clearly of value in finding better solutions to industrial problems, and is a major change in the way engineers work. However, to develop, maintain and extend such a database requires a high level of scientific knowledge, plus considerable workhours and funding.

Companies that do process simulation on their own computer systems typically have their own computerized files of experimental data and correlation methods—for convenience and consistency. Some also have systems that calculate or retrieve physical properties. But numerical databases are available outside, too.

The first directory to such databases was compiled by the Lawrence Livermore National Laboratory, under sponsorship of the U.S. Dept. of Energy [*4*]. Now, there are over 120 machine-readable databases containing numerical data of interest to chemical engineers, but only some 35 are online, as indicated in Table II.

Table III describes selected databases in some detail. Here, we follow a breakdown used by Hampel et al. [*3*], and previously by Hilsenrath [*5*]. A different breakdown, included in the latest Cuadra Directory of On-Line Data Bases, is by types of properties [*6*]—e.g., thermochemical, toxicological. Table III does not list the online services that carry these databases, nor the data networks that link services within and among countries—e.g., Telenet or Tymnet in the U.S., Euronet in Europe. Hampel et al. provide an excellent summary of these [*3*].

Some of the new capabilities of machine-readable databases include [*7,8*] graphical display and comparison, statistical analysis, and easy expansion and updating. However, there is still some lack of confidence in the quality of stored data, though some databases contain only critically-reviewed data. And there are still gaps in the coverage of materials and properties.

## References


1. Stanley, G. W., Unique Information Resources for the Chemical Engineer, *Chem. Eng. Prog.*, June 1981, pp. 80-82.
2. Buck, Evan, Data Banks in the United States, Proceedings of the 2nd International Conference on Phase Equilibria and Fluid Properties in the Chemical Industry, Berlin, March 1980, pp. 771-779.
3. Hampel, Viktor E., et al., A Directory of Databases for Material Properties, Lawrence Livermore Laboratory UCAR 10099, August 1983.
4. Lide, David R., Jr., Critical Data for Critical Needs, *Science, 212* (1981), pp. 1343-1349.
5. Hilsenrath, Joseph, Summary of On-line or Interactive Physico-Chemical Numerical Data Systems, U.S. Dept. of Commerce, October 1980.
6. Westbrook, J. H., Cooperation in Developing Computerized Material Databases, "Data for Science and Technology," Proceedings of the 8th International CODATA Conference, Jachranka, Poland, October 1982, North Holland Publishing Co., 1983, pp. 91-98.
7. Westbrook, J. H., Review of Existing Material Properties Compilations, AIChE Summer National Meeting, Denver, August 1983.
8. Brown, W. F., et al., Materials Property Data Management—Approaches to a Critical National Need, NMAB Report 405, 1982.


## The author

**Monica E. Baltatu** is a senior process engineer in the Advanced Technology Div. of Fluor Engineers, Inc., 3333 Michelson Dr., Irvine, CA 92730. She is responsible for data acquisition, analysis, and development of physical-properties correlations, and has published papers in these areas. Ms. Baltatu has a B.S. in chemical engineering, and an M.S. in physical chemistry, from the University of Bucharest. She is a member of AIChE, and is active in the Gas Processors Assn. enthalpy steering committee.

# Developing a computer program

**Engineers who have their own computers must take on tasks once left to computer specialists—software development and evaluation, and even programming.**

*Carolyn Hardee, Engineered Software*

☐ It pays to write a computer program—or have one written for you—only when you know how to solve the engineering problem, know that a computer can solve the problem, and cannot find an already-written program. Writing a good program requires planning, documentation and testing. This will assure that the program works, that you can change it later, and that other people can use it.

## Where to begin

Define the problem first, then decide whether a computer program is the best way to solve it. While computers are very capable, they are not suited for every application. For example: A calculation involving many numbers but simple mathematics—e.g., an average—might be solved best by a calculator. If you need an analysis of variance as well, then a simple computer program should be more efficient, especially if you expect to use it often.

If the problem calls for a computer, the next question is where to get the software.* The first place to look is the "public domain"—programs listed in magazines, books and government publications. For example: Mathematical and scientific subroutines are listed and documented in "Basic Scientific Subroutines," by F. R. Ruckdeshell, Byte/McGraw-Hill, 1981. Numerical-analysis and economic-analysis subroutines are listed in "Scientific and Engineering Programs," ed. by John Heilborn, Osborne/McGraw-Hill, 1981.

If you can find such a program, it would be wonderful to simply put it into your machine and run it. But this may not be easy:

*Languages* such as FORTRAN or BASIC have several variations, each with its own syntax—e.g., valid variable names, punctuation, spacing, mathematical functions, array-indexing rules. If your version of the language is different, you will need to modify the listed program.

*Documentation* is another problem. If the program does not work as expected the first time, you may have little information with which to launch a second attack. Even if the program works, lack of knowledge about its method of solution may cause trouble later. For example: A value falling outside the range of the algorithm may "crash" the program or, worse yet, produce a result that seems normal but is actually erroneous.

Users' groups and networks are other sources of software—and of assistance if a program fails. A users' group is a group of people having similar computers who gather to share information. Many groups have software libraries available to members. There are also networks of computer users who have access to libraries and maintain bulletin boards of programs.

## Purchasing software

If the software you need is not in the public domain, perhaps you can buy it. The popular electronic-spreadsheet packages may be the best way of solving repetitive calculations that involve only numbers. Word processing and statistics packages are also widely available.

If you need extensive calculations, or a detailed printed report, you may have to search out specialized software. If you do find a package, you must determine whether it solves the problem with the required degree of accuracy, provides results in a convenient form, and is easy to use. These things can be answered only by close inspection of the documentation, which should be available separately. If documentation is not offered, consider looking at another package.

If the software passes inspection, look next at its cost. You might avoid this cost by having software developed in-house, but this takes time, has its own costs, and requires that you know more than the mechanics of a programming language.

## Plan your approach

If you decide to develop a program, first determine which tasks the computer will do, and which will be done manually. The computer should obviously do the calculations, but you can also apply its speed and logic in other ways. For example, looking up values from a table is repetitive, and you might save time by storing the table on a disk file and letting the program do the lookups. Similarly, some graphs can be reduced to mathematical functions and incorporated as such. And of course the program can create printed output in whatever format is needed.

Next, consider how often the program is going to be used, and who will be using it. If it is to be used only a few times by the person who wrote it, a "quick and dirty" program will suffice. But you should write a spec-

*It may be advisable to consider software before getting a computer system, but here we will assume that you already have the system.

Originally published May 2, 1983

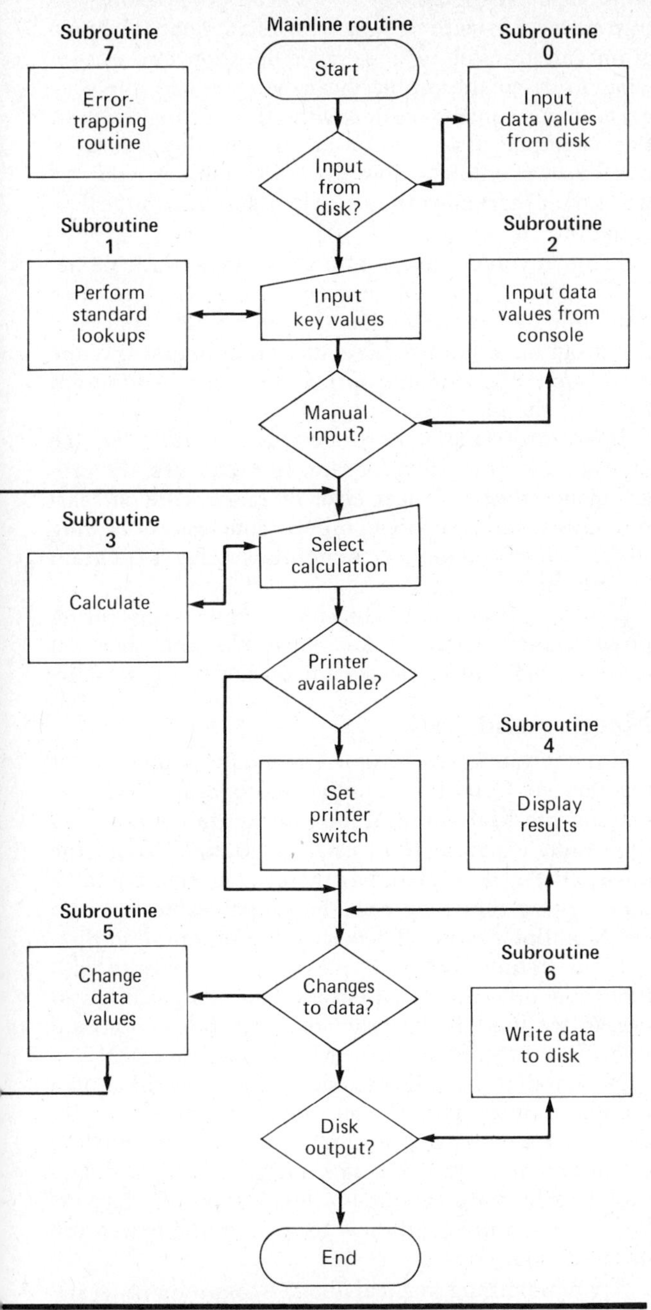


**A structured program has a main routine that calls several subroutines**

ification if the program is to be used often by a number of people. The specification should have a detailed outline of exactly what the program is to do—this pins down the size of the job and prevents surprises.

What language should you use? BASIC (Beginners' All-purpose Symbolic Instruction Code) is the easiest language to learn. Other languages offer more-advanced data structure, precision, and efficiency, but BASIC can handle most engineering problems. And a compiled version of BASIC can increase speed and save program storage space.

What about output? Displaying final results on the monitor is the fastest way, but it may be better to create a printed report showing intermediate values as well as final results. This provides an audit trail—a way to trace what the program has done.

Output to a disk file is a less obvious option that has two advantages: You can use the output of one program as the input for another without doing any error-prone transcription; and you can feed data and results stored on disk back to the program for repeat runs. For example: In figuring head loss, you can repeat the calculation quickly for a variety of flowrates and pipe sizes if the rest of the inputs are maintained on the disk. Disk files are not difficult to use; BASIC requires only five statements to handle one.

## Designing the program

You start with an outline of the engineering method—including assumptions and equations—that solves the problem. This should be written down and agreed upon by all parties involved, so as to avoid misunderstandings later on.

Next, decide how to provide the required data. The program should take data from standard tables whenever possible, to avoid excessive keying by the user. This may require another program that creates a data file on disk. Data that must be entered manually should be in the units most convenient for the user. If a variety of units are required, the program should do the conversions. Remember to convert the results back into the input units before displaying them.

At this point, break the program up into a series of logical subroutines to be called in turn by a controlling program, as in the figure. This is known as structured programming. The tactic is simple—divide and conquer. No matter how complicated the problem, it can be tackled if it can be divided into manageable parts.

You must clearly define the interfaces between the controlling program (Mainline in the figure) and the subroutines. The definition includes inputs and outputs: values that must be established before entering a subroutine, and values passed from the subroutine. This is important in developing the flow pattern, and in testing the program later on.

The overall flow pattern is: input, calculation, output. But you must order the subroutines so that data to be passed from one to another are available when needed. Also, the program should access disk files before asking for manual input. This minimizes the re-keying required in case the user has to exit and check a file name. The program should request manual inputs in a logical order, and should print reports in an understandable form.

## Documentation

Now is the time to begin formal documentation. This may seem like putting the cart before the horse, but documenting early can prevent problems later on.

The first step is a flowchart, as in the figure, that shows the flow pattern. Subroutines need little detail at this stage; individual blocks will suffice. The Mainline routine should be more detailed, showing the sequence of subroutines and the conditions for skipping a routine. A handy way to index the subroutines is by assigning a set of BASIC line numbers to each one—e.g., 2100–2199 for Subroutine 1.

You should write an index of all the variables to be used, and their units and range of values. If you do this before writing any code, you can establish descriptive prefixes and suffixes. For example: Most BASICs allow you to use periods in variable names, so you can distinguish pipe diameter from valve diameter by using "P.DIA" and "V.DIA". This may seem like window dressing, but only until you have searched through seventeen pages of code to find what the symbol "UZQ" stands for.

Most languages let you define variables (sometimes just those beginning with particular letters) as being of a particular type—e.g., integer, floating-point, or array. The type determines how a number is stored in the computer, which affects storage required and execution time. In general, you will save both space and execution time by defining integer variables as integers. And it is easier to define a variable once than to append a special character (usually a "%") every time you key it in.

Advance documentation should also include:

*User functions*—frequently-used calculations that can be invoked throughout the program by name. This is handy for unit conversions.

*Layouts* for disk files, input and output screens, and printed reports. On screen layouts, allow space for signs, error messages and the largest possible numbers. And remember to provide a means of entering report-heading information.

Like any technical job, a program requires advance planning, and the more complex the program the more planning is needed. It is also important that the intended users be involved in the development process. This fosters a feeling of team effort, and makes acceptance of the finished program much more likely.

## Programming

Only now do you start coding the program. Begin with the Mainline routine. (This is known as top-down programming.) Code it, then test the flow pattern by using dummy print statements instead of actual subroutines. When the routine works, save it on disk. Then code and test each subroutine individually, saving it on disk, and merge the tested routines to form the complete program. Structured programming simplifies not only testing but also future changes: Changes in one subroutine can be made and tested independently of the rest of the program.

Catching erroneous data as they are entered is known as input editing. At a minimum, every program should check for inadvertent "return" keystrokes. In BASIC, this produces a zero value, which causes fatal problems in many equations. Further, the program should check the entered value against the valid range (e.g., nonnegative numbers), if you can specify such a range.

In case of error, the program should ask the user to reenter the data immediately, while the source is close at hand. If the program uses input data to look up values in a table, the lookup should be performed as soon as the key value is entered. Then the user will immediately know if the table value is unavailable.

Even the best input editing cannot guarantee successful results: Problems such as division by zero or running out of memory should still be expected. BASIC helps you trap such errors with its "On error" statement: The program can transfer to a specified location and decide what to do. If the problem can be corrected, the program may resume execution without troubling the user. For example: If a friction-factor equation causes a numerical overflow at low Reynolds number, you can write the program so that it switches to a laminar-flow equation.

Errors involving disk files include: invalid file name, file not found, or file in the wrong format. In all of these cases, the user probably has the wrong file name, and the program should request the name again. But the user should also have the option of exiting the program to check the file directory.

If an error cannot be corrected automatically, the user has to restart the program. In such cases, the program should issue a clear error message, write out partial results, and come to an orderly conclusion. Nothing shakes a user's confidence more than seeing a program "bomb out."

You should of course continue documenting during programming. Changes are inevitable, and they are easier to manage if the documentation is up to date.

## Merging and testing

If the routines have unique line numbers, they will fit together to form the completed program. Begin by loading the Mainline routine and assigning values to the results expected from each subroutine. Verify the flow pattern, then merge-in the subroutines, one at a time, testing that they pass the proper values back to the Mainline routine. Then, save the program on disk.

Final testing compares the program's results with independently calculated ones. Include impossible values, to see that the program does not bomb out, and remember to simulate all the error conditions.

Now, pull some of the documentation together into a manual that the typical user can follow: operating instructions, step-by-step example, and detailed method of solution, plus error messages, units and standard data tables. If in doubt about the user's level of expertise, begin the example with how to turn on the power and proceed from there.

There is nothing magical about a computer program. If you plan properly, write the program in a logical and orderly manner, and provide documentation that explains it to the user, your program is sure to succeed.

### The author

**Carolyn Hardee** is currently developing an integrated set of programs to handle piping-design calculations on desktop computers, at Engineered Software, 2420 Wedgewood Dr., Olympia, WA 98501 (Tel. 206-786-8545). Her background includes analysis and programming work on a variety of computer systems, and two years as an IBM systems engineer. Ms. Hardee holds a B.S. degree in mathematics, cum laude, from St. Bonaventure University.

# Developing technical software

**What do you do when you are responsible for developing technical software? Start by defining the problem, and stay involved throughout the entire process.**

*David H. Chittenden and Walter B. Dulak*

☐ Whether you develop technical software or have it done by someone else, its success or failure is your responsibility. So you need to approach the project with as much background and savvy as possible.

The amount of effort you devote to such a project should of course depend on the money involved—both costs and benefits. But you will always need to specify your needs, define your requirements, consider alternatives, and justify your course of action.

## Do you need a computer program?

Not all technical problems are best solved by a computer program. Those that are tend to be:

*Solvable by some known method.* If you do not know how to solve a problem, or do not have the necessary data, a computer is obviously no help.

*Highly repetitious.* For example: Multicomponent vapor-liquid equilibrium calculations, which require trial-and-error solutions. Sometimes a problem becomes repetitious when adapted for computer solution. An example is a numerical, trial-and-error solution to a problem that could be solved graphically by hand.

*Often encountered.* If a problem is encountered only rarely, developing a computerized solution may not be worthwhile. The exception is the problem that cannot be solved at all without a computer. For example: Analyzing rocket-launch data by hand would take too long to be of use during the flight.

## Specifying your requirements

The primary requirement for any software is that it solve your problem. But what exactly is the problem? If the general problem, for example, is figuring the volume of fluid in a tank, will the operator input the fluid level? Or will the program interpret a signal from a level detector? You need to define carefully, and consider the whole computer system (as in Fig. 1): the user (you), the computer, the software, the data and the output.

At this point, you should think about the kind of operation you need. Batch operation involves submitting a job and then waiting minutes or even hours for the results—this may be suitable for regular weekly or monthly reports. Timesharing operation means talking to your company's computer or an outside computer through a terminal. This speeds up results, and so is more convenient for an engineer who needs frequent access. Desktop computers are inexpensive and can handle moderate-size problems, but they are slower than larger computers—especially when programmed in interpretive BASIC.

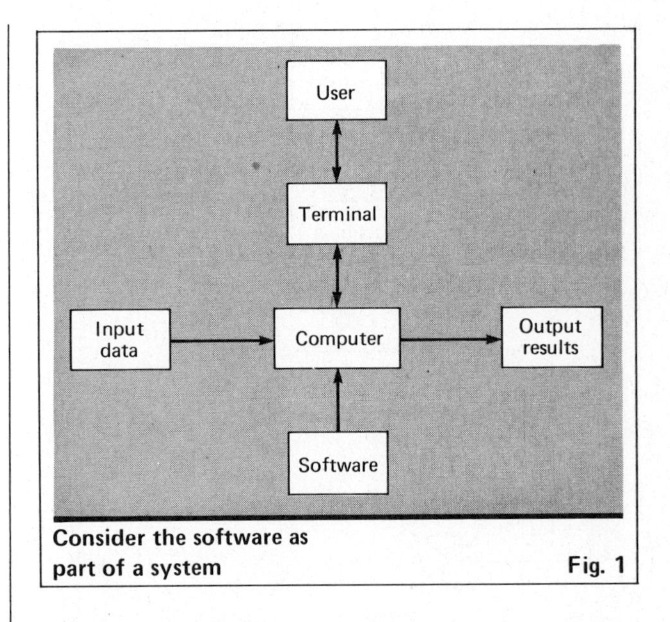


**Consider the software as part of a system** **Fig. 1**

When thinking about other requirements (e.g., input format), always consider how much they contribute to solving the problem. For example: Programmers may say that something has to be done a particular way for the computer's sake. This may be so, but it may also be a coverup for using an unsuitable computer.

You will have to justify the software cost, so you should begin thinking about payoffs early. The clearest payoff is a reduction in the engineering time needed to produce some result. Secondary payoffs might be timelier or more-accurate results.

Once you have specified the problem and mode of operation, it is time to consider how you will have it developed. Your first concern is finding the expertise you need.

## Where to find qualified people

Do you have people qualified to develop technical software for you? The first place to look is in your own department. Engineers there have probably studied programming, but they have probably forgotten it, too. To be practical, only a few engineers who have not programmed recently will remember enough to be efficient programmers. It may be possible to send someone to a short course in using the company's mainframe comput-

Originally published May 30, 1983

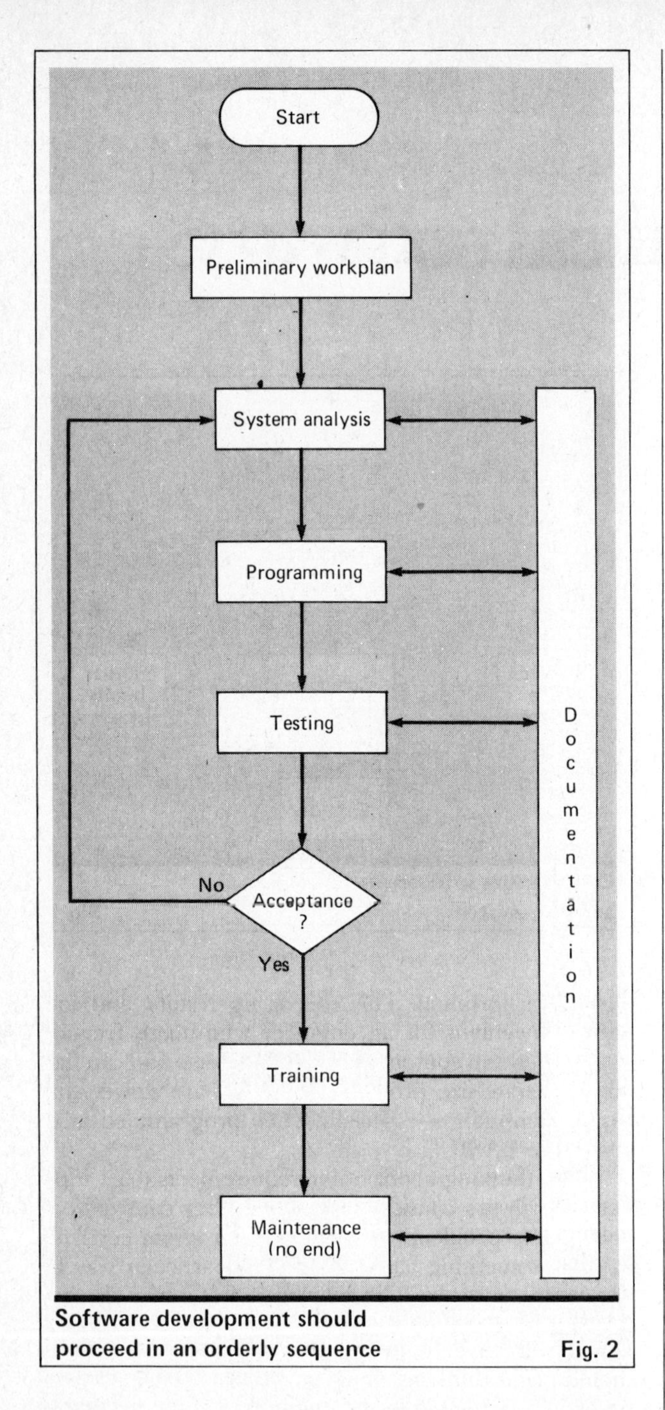


**Software development should proceed in an orderly sequence** **Fig. 2**

er; this would be adequate qualification for small problems. But major development projects are best handled by full-time professional programmers, since the complications that arise in extensive programming lead to bad results in inexperienced hands.

Your company's computer department may be able to help. However, such a group is often devoted mainly to business applications, and the techniques and computer languages useful in chemical engineering may be outside its expertise. If the computer department has people who specialize in technical applications, discuss your problem with them to see if they are able to handle it. Of course, you will still have to define the problem and solution method very carefully.

There might be engineers and scientists in other parts of your company who would be happier doing programming or systems analysis. Looking for them is worthwhile, but you have to make sure that they have or can acquire the expertise needed to do your project.

Outside your company, there are a host of people available. Local university professors, perhaps assisted by their more-able students, are a relatively inexpensive source of expertise. Timesharing companies will do development under contract, as will private consultants. Since your contact with people outside will be limited, you will have to make an extra effort to specify your requirements precisely and to see that they are met.

To get a schedule and cost estimate for the development project, you need to specify requirements in detail. This is a substantial task in itself.

## Detailed requirements

To develop your software, your experts need a detailed specification of input (e.g., data, forms, hardware), calculation techniques, and output (e.g., displays, reports), plus a clear idea of how the software is to be used. Some of the key specifications:

*Input data* can be: (1) written into the software, if they are not going to change (e.g., conversion factors); (2) read by the software from a disk or tape, if they are to be changed or extended occasionally (e.g., correlation constants); (3) or entered for each run, if they are pertinent only to the run (e.g., measured flowrates in a material-balance program). Data specific to a run might be sent to the computer automatically (e.g., instrument readings), or entered manually. In the latter case, the software could be set up to ask for the data—i.e., provide a menu.

Remember that data must be available. If you do not know where your data are coming from, you should reconsider your problem before proceeding.

*Calculation methods* need to be written down, and units, decision rules and terms carefully defined. Knowing the engineering method is the first and easiest part; the hardest is explaining it to the programmers. The more carefully you specify, and the better you explain, the more likely you are to get what you need.

*Output* can be of several types. Immediate output on the user's terminal should include only the essential or summary results, especially those that are needed to determine whether the results are usable (e.g., checksums on a material balance). Detailed results that are lengthy and require study can be printed out, perhaps at a central location. Results that are needed for future runs (e.g., for summary reports or comparisons) can be sent to disk or tape, from which the computer can obtain them quickly.

At this point, you can get an estimate of time and cost, from the people who are going to do the work, and proceed with development. If developing software will take too long or cost too much, you should consider purchasing available software. If you can find software that meets your needs (i.e., you have tested it and it works), and the cost is reasonable, you can proceed to the justification stage. Otherwise, you will be involved in a development project.

## The development project

You have to consider what *qualities* you want the software to have, and what *activities* will get you there. Fig. 2

maps the activities involved in software development:

*A preliminary workplan* divides the project into phases, and estimates the duration and cost of each phase.

*System analysis* works out the input, calculation and output techniques and verifies that they are suitable to both the user and the programmer. User dissatisfaction with input and output is a common cause of failure, so this is a key step.

*Programming and testing* will be done by the computer people, but you will want to be involved throughout. Testing might include: hand calculation, using the program's techniques; independent verification of the techniques; checking limiting cases; sensitivity analysis. Chances are that the software will not do exactly what you want at first, but you can get what you want easily if you request changes early on. When all is done, the ultimate test is comparing the software's results with experimental ones—to see whether the model is adequate.

*Documentation and training* are not left until last in successful projects. Documentation should go on continually, in case a programmer suddenly leaves. A way to get documentation is to request informal memos describing progress in detail.

The above activities may have different names and may be combined or modified. The important thing is that you and the programmers cooperate throughout the process, to make sure your objectives are clearly understood and ultimately met.

While your basic goal is getting correct results, there are also matters of quality to consider:

*Convenient operation* means that the software is easy to use and provides timely results. Otherwise, users may shun the system, or may find that results are never on hand when needed. For example: A plant inventory manager who needs a daily update will not be satisfied with a system that delivers a fat book of tables three days after the data are collected.

Convenience to the user is basically a matter of input and output—e.g., clear data requests and reports. Of course, not all users are the same, and you will want to allow for this. For example: Provide novice users with menus and informational messages, but give experienced ones the option of bypassing such aids.

*Quality programming* leads to easy maintenance. A good program will be structured—i.e., constructed of independent subroutines—to allow a programmer to change part of it without changing the whole. Likewise, complete documentation helps someone other than the author of the software to understand and maintain it.

*Self-checking* helps assure correct results. This may involve: input checking, to catch erroneous data; closure checks on mass and energy balances; checks on convergence in trial-and-error calculations.

*Written documentation,* which is essential for using and maintaining the software, should be available in several forms: a user manual containing sample input, calculation techniques, and sample output; a system manual containing details of the programming; and perhaps a manual for hardware belonging to the system.

## Justifying the project

Whether you develop software or buy it, you have to justify the cost. The effort you put into justification should be in proportion to the cost—a $50 program deserves little effort, while a development project that will cost thousands of dollars rates a thorough analysis.

Estimates of development cost and duration unfortunately tend to be inaccurate: system analysis and programming are not very predictable activities, and individual capabilities vary widely; changes in the middle of a programming project add cost and delay; computer people may feel, perhaps subconsciously, that a given project would not be approved if they were to quote a conservative estimate.

Technical software can be considered an investment, so the well-known return-on-investment tools are the ones you use to evaluate it. Measurable benefits include reduced person-hours to do a given job, and more-efficient operation of a manufacturing or inventory system. Like costs, the benefits can be hard to estimate accurately.

## Where to turn for help

When something goes wrong with the software, the developers are the logical ones to turn to for help. If they are outside the company, a maintenance fee or retainer may be paid to assure their support. Before having software developed, you should ask about the quality of maintenance provided by the developers on prior projects for other customers.

High-quality documentation becomes essential when errors occur, since the programmers will probably have forgotten the original intention of their code, and may have moved on to new jobs. Some system-development groups give responsibility for software in use to two people—the original programmer, and a backup person who is expected to know the software. If the programmer leaves the company, the backup person can take over and train a new backup person.

## The authors

**David H. Chittenden** is Senior Process Engineer at Irvine Sensors Corp., 3001 Redhill Ave., Bldg. 3, Irvine, CA 92626; Tel. (714) 549-8211. When this article was written, he was employed by Occidental Systems Corp. Now, he is responsible for developing process models, and designing and analyzing experiments. Dr. Chittenden has taught chemical engineering at the University of New Hampshire and the University of Santa Barbara. He holds a B.S. degree from the Illinois Institute of Technology, and a Ph.D. from the University of Wisconsin, both in chemical engineering.

**Walter B. Dulak** is Supervisor of the Computer Sciences Department at Occidental Research Corp., 2100 S.E. Main, Irvine, CA 92714, where he is responsible for providing computer support to all business and engineering functions. He holds a B.S. in chemical engineering, and an M.S.B.A. in management science, both from the University of Massachusetts. He is a member of AIChE.

# Spreadsheet software solves engineering problems

**Engineers equipped with computers can save time and tedium by using a spreadsheet to do their calculations.**

***Steve Selk,*** *Gulf Canada Products Co.*

☐ Electronic spreadsheets can do whole tables of calculations over and over, right before your eyes, once you set up the table headings and formulas and put in some data. This is very convenient for business applications such as sales forecasting and accounting, where you have to try things several times, and it is just as convenient for repetitive engineering calculations such as figuring the pressure drop through a piping system.

Spreadsheet software is friendly, in that you need little programming skill and can see what the computer is doing. For this reason, and for its unusual capabilities, spreadsheet software is credited as a major reason why tens of thousands of businesspeople have bought desktop computers that can run it.

This article explains what a spreadsheet is, how it works, and how to apply it to engineering problems.

## What is an electronic spreadsheet?

A spreadsheet is a computer program (usually on disk) that you buy, load into your computer's memory, and then use as an analytical tool. Not all computers can use one, but most desktop computers having 48K or more of random access memory (RAM) can use one or more different brands. *(Editor's note: Spreadsheet software on the market in the U.S. includes VisiCalc, CalcStar, Multiplan, SuperCalc, T-Maker, and many others.)*

From the user's point of view, an electronic spreadsheet is a *grid,* or table, displayed on a video screen. This grid is made up of *cells,* arranged in rows and columns (typically 254 rows by 63 columns), into which you can enter *text* (e.g., headings), *numbers,* or mathematical *formulas.* The information in the grid is stored in the computer's memory, and is easy to change or update. Such a filled-in grid is called a *template.*

Rather than catalog a spreadsheet's capabilities, let us look at a typical engineering problem and see what the spreadsheet can do with it.

## Frictional pressure drop

Table I shows a VisiCalc template that calculates the frictional pressure drop for fluid flow through a pipeline. Rows 4–9 of Columns A–C contain labels for the data: pipe diameter, mass flowrate, specific gravity, viscosity, equivalent length, absolute roughness. Below, in Rows 11–16, are labels for results that the computer evaluates: fluid velocity, Reynolds number, relative roughness, friction factor, pressure drop.

The formulas for the results do not appear on the screen. They are entered in Column D, Rows 11–16, as expressions in which cell locations are the variables. For example, the formula for velocity (ft/s) is:

$$(M)/(\rho(62.4)(3{,}600)\pi(D/24)^2)$$

where $M$ is mass flowrate (lb/h), $\rho$ is specific gravity, and $D$ is pipe diameter (in.). This formula is entered into Cell D11 (Column D, Row 11) as follows:

D5/D6/62.4/3600/3.14/((D4/24)^2)

**Table heading take three columns** — **Table I**

|  | A | B | C | D | E |
|---|---|---|---|---|---|
| 1 |  |  |  |  |  |
| 2 | LIQUID-P | HASE PRES | SURE DROP |  |  |
| 3 |  |  |  |  |  |
| 4 | PIPE DIAM | ETER | (IN.) |  |  |
| 5 | FLOWRATE |  | (LB/H) |  |  |
| 6 | SPECIFIC | GRAVITY |  |  |  |
| 7 | VISCOSITY |  | (CP) |  |  |
| 8 | EQUIVALEN | T LENGTH | (FT) |  |  |
| 9 | ROUGHNESS |  | (IN.) |  |  |
| 10 |  |  |  |  |  |
| 11 | VELOCITY |  | (FT/S) |  |  |
| 12 | REYNOLDS | NUMBER |  |  |  |
| 13 | RELATIVE | ROUGHNESS |  |  |  |
| 14 | FRICTION | FACTOR |  |  |  |
| 15 |  |  |  |  |  |
| 16 | PRESSURE | DROP | (PSI) |  |  |

Originally published June 27, 1983

**The user enters data, and the spreadsheet does the necessary calculations** **Table II**

| | A | B | C | D | E |
|---|---|---|---|---|---|
| 1 | | | | | |
| 2 | LIQUID-PHASE PRESSURE DROP | | | | |
| 3 | | | | | |
| 4 | PIPE DIAMETER | | (IN.) | 3.068 | |
| 5 | FLOWRATE | | (LB/H) | 100000 | |
| 6 | SPECIFIC GRAVITY | | | 1 | |
| 7 | VISCOSITY | | (CP) | 1 | |
| 8 | EQUIVALENT LENGTH | | (FT) | 1000 | |
| 9 | ROUGHNESS | | (IN.) | .0018 | |
| 10 | | | | | |
| 11 | VELOCITY | | (FT/S) | 8.675510 | |
| 12 | REYNOLDS NUMBER | | | 205960.7 | |
| 13 | RELATIVE ROUGHNESS | | | 5.867E-4 | |
| 14 | FRICTION FACTOR | | | .0192176 | |
| 15 | | | | | |
| 16 | PRESSURE DROP | | (PSI) | 38.06721 | |

**You can replicate a column or row with a simple command** **Table III**

| | A | B | C | D | E |
|---|---|---|---|---|---|
| 1 | | | | | |
| 2 | LIQUID-PHASE PRESSURE DROP | | | | |
| 3 | | | | | |
| 4 | PIPE DIAMETER | | (IN.) | 3.068 | 4.026 |
| 5 | FLOWRATE | | (LB/H) | 100000 | 100000 |
| 6 | SPECIFIC GRAVITY | | | 1 | 1 |
| 7 | VISCOSITY | | (CP) | 1 | 1 |
| 8 | EQUIVALENT LENGTH | | (FT) | 1000 | 1000 |
| 9 | ROUGHNESS | | (IN.) | .0018 | .0018 |
| 10 | | | | | |
| 11 | VELOCITY | | (FT/S) | 8.675510 | 5.038000 |
| 12 | REYNOLDS NUMBER | | | 205960.7 | 156951.7 |
| 13 | RELATIVE ROUGHNESS | | | 5.867E-4 | 4.471E-4 |
| 14 | FRICTION FACTOR | | | .0192176 | .0190464 |
| 15 | | | | | |
| 16 | PRESSURE DROP | | (PSI) | 38.06721 | 9.695586 |

where D5 is mass flowrate, D6 is specific gravity, and D4 is pipe diameter.

In the same manner, other formulas for calculated values are written into the appropriate cells. A typical spreadsheet allows you to use mathematical functions such as log, sine and cosine, and even more-complex ones such as net present value. But iterative formulas and branching commands are typically not allowed. For example: The single formula for friction factor is*:

$$\left\{-2 \log\left[\frac{\epsilon/D}{3.7} - \frac{5.02}{Re} \log\left[\frac{\epsilon/D}{3.7} + \frac{14.5}{Re}\right]\right]\right\}^{-2}$$

where $\epsilon$ is absolute roughness. If you wanted to use a different formula for laminar flow, you would have to enter it in another cell.

Once the original template is entered, you can store it on a floppy disk with only a few keystrokes, and use it anytime by loading it into memory from the disk.

## How to use the template

To figure the pressure drop, you fill in each of the blank data cells—by moving the cursor to the cell and typing in the number. Data for a sample problem are shown in Cells D4–D9 of Table II. When you tell the computer to calculate, it fills in the other cells, in this case D11–D16. If you change the data, and recalculate, the values in Cells D11–D16 change accordingly.

Because recalculating is so easy, the template lets you answer "What if?" questions very quickly. For example: What if you had 4-in. pipe instead of 3-in.? Move the cursor to D4, enter the new diameter, recalculate, and within seconds read the new pressure drop in Cell D16.

Even better, you can replicate column D—i.e., copy the data and formulas—into whatever column (or columns) you want. Then you can change the diameter in the other column, recalculate, and look at both columns simultaneously—as in Table III, which shows the results for 3-in. and 4-in. pipe side by side. You can also replicate rows. This is something that you cannot do with pencil and paper, nor with a programmable calculator.

Likewise, trial-and-error calculations are very easy to do. Say that you want to figure out what flow you can expect in a 4-in. pipe with 15 psi of available pressure drop. Simply enter the basic data (as in Table III, Column E), then put in guesses for flow and recalculate until pressure drop is sufficiently close to 15 psi.

## Useful features

To further improve this simple template, you can add additional features. Table IV, for example, shows an additional grid (on the same template) that calculates the equivalent length of fittings in a run of pipe. You fill in the number of each kind of fitting, and the length of straight-run pipe. The computer multiplies the number of each fitting by a length factor to get the "L/D" values, then sums these and multiplies by pipe dia. to get equivalent length. If you wanted, you could set up the pressure-drop section so that the computer would enter the total equivalent length where needed (e.g., D8).

**You could also set up a reference grid (see Table V) showing inside diameters for various pipe sizes—eg., 3.068 in. for 3-in. Sch. 40 pipe. If you use the LOOK-UP**

*From Shacham, M., *Ind. Eng. Chem. Fund.*, May 1980, pp. 228–229.

**A separate area of the spreadsheet can do a different calculation** — **Table IV**

| Row | | Quantity | L/D |
|---|---|---|---|
| 18 | | | |
| 19 | EQUIVALENT-LENGTH DETERMINATION | | |
| 20 | | | |
| 21 | FITTING TYPE | QUANTITY | L/D |
| 22 | | | |
| 23 | 90-DEG STD ELBOW | 10 | 300 |
| 24 | 45-DEG STD ELBOW | 0 | |
| 25 | 90-DEG LONG RADIUS ELBOW | 0 | |
| 26 | 90-DEG STREET ELBOW | 0 | |
| 27 | 45-DEG STREET ELBOW | 0 | 0 |
| 28 | STD TEE FLOW THRU RUN | 1 | 20 |
| 29 | STD TEE FLOW THRU BRANCH | 0 | 0 |
| 30 | GLOBE VALVES | | |
| 31 | CONVENTIONAL FULLY OPEN | 1 | 340 |
| 32 | Y PATTERN 45-DEG FULLY OPEN | 0 | 0 |
| 33 | GATE VALVES | | |
| 34 | FULLY OPEN | 5 | 65 |
| 35 | 3/4 OPEN | 0 | 0 |
| 36 | 1/2 OPEN | 0 | 0 |
| 41 | BUTTERFLY VALVES | | |
| 42 | CONVENTIONAL >6-IN. | 0 | 0 |
| 43 | | | |
| 44 | LENGTH DUE TO FITTINGS | 185.3583 | |
| 45 | | | |
| 46 | STRAIGHT-RUN LENGTH | 500 | |
| 47 | | | |
| 48 | TOTAL EQUIVALENT LENGTH | 685.3583 | |

**Template can also hold reference tables** — **Table V**

| Row | | |
|---|---|---|
| 51 | | |
| 52 | DIAMETER OF STEEL PIPE, SCHEDULE 40 | |
| 53 | | |
| 54 | | |
| 55 | NOMINAL SIZE, IN. | INSIDE DIA., IN. |
| 56 | | |
| 57 | 1/8 | 0.269 |
| 58 | 1/4 | 0.364 |
| 59 | 3/8 | 0.493 |
| 60 | 1/2 | 0.622 |
| 61 | 3/4 | 0.824 |
| 62 | 1 | 1.049 |
| 63 | 1 1/4 | 1.38 |
| 67 | 3 | 3.068 |
| 68 | 3 1/2 | 3.548 |
| 69 | 4 | 4.026 |
| 70 | 5 | 5.047 |
| 71 | 6 | 6.065 |

function, you can even retrieve these diameters automatically. But Table V is set up for reference only.

Together, the three grids take care of calculations and eliminate the need for reference tables. This example template takes less than 4% of the storage space on one side of a 5¼-in. single-density floppy disk, so about four dozen such templates could be stored on a single disk, using both sides.

If your computer has a printer, you can get a hard copy of your filled-in grid with a simple command from the computer keyboard. The spreadsheet software takes care of formatting the printout.

## Conclusions

You can reduce the time you spend on often-encountered problems by creating a number of templates to do the calculations and store the basic reference data. If you are fortunate enough to have your own computer on your desk, you can use these templates as needed and free more time for thinking.

If you are considering buying a computer, remember that the software is what makes a computer so powerful. So, first determine what software you need—e.g., spreadsheet, language compiler, graphics—and only then look for the hardware required to run it.

## The author

**Steve Selk** is a senior engineer in the process development dept. of Gulf Canada Products Co., 800 Bay St., Toronto, Ont. M5S 1Y8, Canada, where he is responsible for studies relating to crude units, alkylation, thermal cracking, and sulfur plants. Prior to joining Gulf, he worked in the thermoplastics dept. of Dow Chemical Canada. Mr. Selk holds a B.A.Sc. degree in chemical engineering from the University of Toronto, and is licensed.

# Microcomputer data-acquisition

You can save labor and get faster access to lab or pilot-plant data by installing a computer-based data-acquisition system.

*Stewart M. Goldfarb*

☐ A computer improves people's productivity when it takes over repetitive work such as equilibrium-stage calculations or data acquisition. It may also improve the quality of the work, or enable people to do things that would otherwise be impossible.

This article shows how an automated data-acquisition system can be built around a microcomputer; a system for a polycrystalline-silicon (polysilicon) pilot plant serves as an example. The payoff here is a labor saving equal to one person's full-time effort.

## Collecting data manually

Chemical engineers need data from sources such as laboratory instruments, pilot plants and production units. The simplest way to collect and record such data is manually: An operator, technician or engineer carries a piece of paper (e.g., a log sheet) to the instrument or process, makes an observation, and writes it down. After the data are collected, they may be manipulated so that their meaning can be extracted.

In a research department, engineers and technicians have responsibilities that are often more important than collecting data. Thus, readings may be missed when a person is busy elsewhere. Also, there may be no one on hand overnight, so some data on 24-hour-a-day processes may be lost. In such cases, there is a clear need for automation.

## Automated data-acquisition

The first level of automation is the chart recorder, which takes a signal from an instrument and traces its relative value on a moving piece of paper. The user reads values from the chart, figures when they were recorded, writes them down, and perhaps performs calculations to learn what they mean. Such recorders often include hardware that converts the input signal to engineering units (e.g., milliamp current to °F).

At the next level are electronic data-loggers built around microprocessors. They may also be called process monitors or input modules. All such devices can read a variety of input signals and convert them to other units according to the user's instructions. But output capabilities and cost vary widely.

Printed, tabular output is typical for low-cost units. Costlier ones may have other output options: multi-

color printing; graphics; color-CRT display; mass storage of data on a floppy disk. They may also be able to do trend analysis, alarming, or simple control.

The data-logger can communicate with computers and other devices if it has a compatible communication port—e.g., an RS-232C serial interface. Without such a link, the user might have to enter data manually.

In the final stage of automation, a computer is an integral part of the collection and storage process—signals go to a computer either directly or through a connected data-logger. Analysis is all that is left to the user. However, the user must be confident that the data are valid, and this is a greater concern in an automated system because there is less human involvement.

Unlike a data-logger, a general-purpose computer need not be devoted solely to data-acquisition. It may be able to do other tasks at the same time, and can be put to use elsewhere once a project is finished. Such flexibility is a computer's key advantage.

How does a computerized data-acquisition system work, and why would you want one? Rather than look at general principles, let us focus on a particular system and see how it worked before automation, why it was automated, how equipment was selected, how it worked after automation, what it cost, and what it was worth.

## Example: a polysilicon pilot plant

Polysilicon is the starting raw material for the electronics industry, where it is melted and then recrystallized into long, single-crystal rods. These are cut into wafers and used to make semiconductor devices such as transistors and integrated circuits.

The pilot-plant process is chemical vapor deposition of silicon from trichlorosilane ($HSiCl_3$) in the presence of hydrogen. This takes place at 1,000°C in a quartz bell-jar reactor like that shown in Fig. 1. Polysilicon is deposited on rods in the reactor; unreacted $HSiCl_3$ and $H_2$ exit through the vent along with other reaction

Originally published April 4, 1983

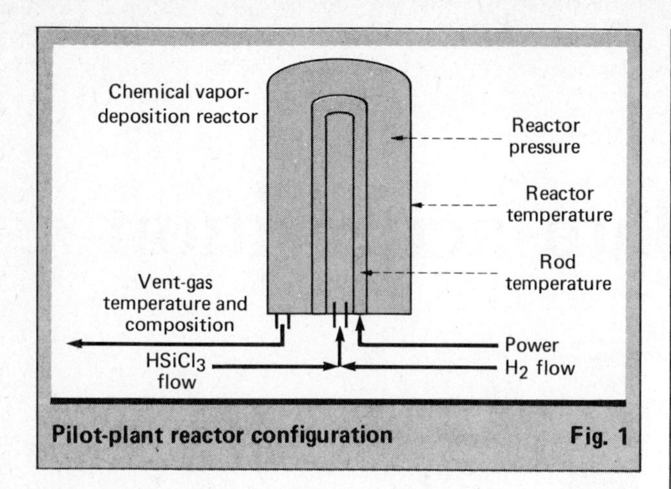


Pilot-plant reactor configuration **Fig. 1**

products ($SiCl_4$ and HCl). Material and energy balances require all of the data points indicated in Fig. 1.

## The semiautomated pilot plant

Before computerization, data collection was partly manual and partly automatic. Two computer-controlled gas chromatographs took samples continuously and automatically, and stored composition data on cassette tapes. After a reactor run, these tapes were read by a microcomputer, which stored the data. Flowrate, temperature, pressure and power data were recorded manually on log sheets or automatically on a strip chart, then entered manually into the computer. Once all the data were in, a computer program calculated the material and energy balances and summarized the results of the run.

This system worked, but there were problems. One was a difference in the timing of readings. For example, a power reading would be taken at 10:00 a.m., flowrates and temperature at 10:10, and gas chromatograph samples at 10:20. Also, data were lost when people responsible for readings were interrupted. Finally, expansion of the pilot-plant project made it impossible to collect data on all reactors simultaneously.

Options for improving the system were to:

1. *Install a data-logger* that would collect and print out the data. The user would still have to enter data into the computer manually, and coordination with the gas chromatographs would be difficult.

2. *Install a general-purpose microcomputer.* The department already owned several, so new-equipment cost would be low. A microcomputer could be programmed to collect the chromatograph data, but process data would have to be recorded and entered manually.

3. *Install both.* A microcomputer connected to a data-logger could collect and store both chromatograph and process data. There would be no manual recording or data-entry, and timing could be coordinated. This option was chosen because it solved the problems and appeared to be cost-effective.

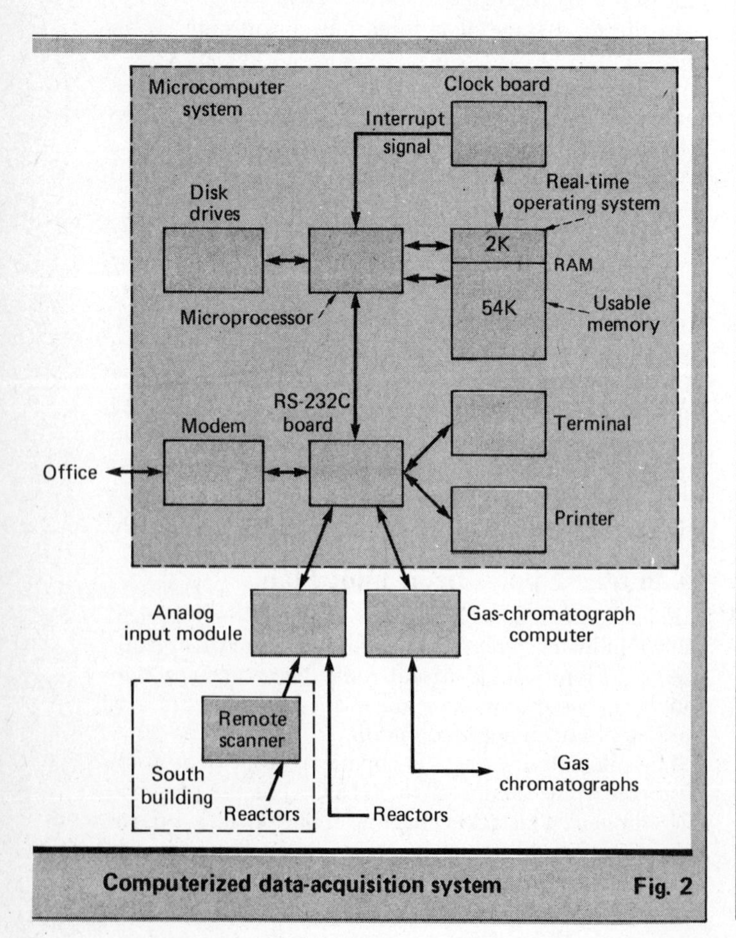


**Computerized data-acquisition system** **Fig. 2**

## The new system

Fig. 2 shows the computerized data-acquisition system. The microcomputer, shown in detail, is connected to the gas-chromatograph computer and the analog-input module via an RS-232C serial-communication link. It collects data, stores it on floppy disks, analyzes it, and generates reports.

The analog-input module (AIM) collects analog signals from the process instruments and converts them to digital ones. The gas-chromatograph (GC) computer controls sampling and calculates compositions. A modem connected to a dedicated phone line allows communication with a microcomputer located in the office.

The system makes use of an existing microcomputer and GC computer to keep costs down. However, the same or similar equipment would have been selected if the system had been designed from scratch. Let us now look at the system in detail:

*The computer* is configured with a Z-80 microprocessor, a 56K RAM, three floppy-disk drives, a terminal, a printer and a modem. The operating system is a modified version of CP/M that incorporates a time-of-day clock board and a real-time operating system.

The user can do computations or other tasks on the computer until the clock board tells the operating system to interrupt with a warning message. The user must then let the system run the data-acquisition program.

This system was put together in early 1982. Microcomputers introduced since then include ones that support multiple users and multiple tasks and ones designed specifically for data-acquisition. Such computers might not have to interrupt a user.

*The analog-input module (AIM)* is built around a microprocessor and designed for independent operation. It collects signals from over 90 process instruments (e.g.,

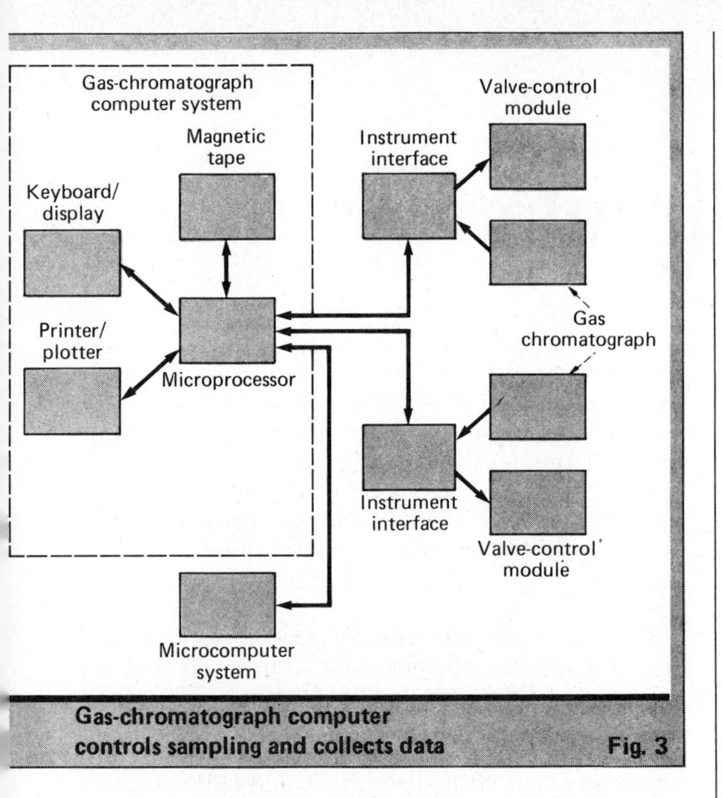


**Gas-chromatograph computer controls sampling and collects data** **Fig. 3**

thermocouples); bringing 90 wires into the microcomputer would have been impractical. The AIM converts analog signals (i.e., voltage or current) into numbers and converts some of the numbers to the desired units—complex conversions are performed by the computer, later. When the computer requests data, the AIM transmits them via the RS-232C link. The AIM can be programmed from the computer terminal via the link.

Some of the pilot-plant reactors are located far from the AIM. Analog signals from these reactors go to a remote scanner, which transmits them to the AIM. This saves on cable and conduit—two long cables rather than sixty—and reduces the chance that noise will interfere with the analog signals.

*The GC computer* can control, collect and report data for up to four chromatographs. Its console includes a keyboard and printer/plotter, so that programs can be entered and listed; this can also be done from the computer terminal. The GC computer controls sampling and analyzes data via BASIC programs, several of which can run simultaneously. Instrument-interface modules select streams, and draw and inject GC samples at prescribed time intervals. The valve sequence is controlled by the software. Fig. 3 shows the block diagram for the GC computer.

## How the system works

The programs that run the system are designed to be flexible. The main program refers to definition files that tell how each reactor is configured, so the user can specify a new configuration without changing the program. For example: A new pressure-cell calibration requires only one new parameter in the definition file. The definition program is menu-driven—i.e., provides a menu of choices. The system is ready to start collecting data once the definition files are generated.

Data collection is as follows: First, a program tells the clock board to generate an interrupt signal every hour. Two minutes before the hour, the terminal displays a message telling the user to give up the computer. At the hour, the clock board interrupts.

Then the computer loads a program from disk storage and executes it. The program first determines which reactors are running. If a reactor is running, the program determines whether it is time to read data, and if so it looks up the definition file for that reactor. The program then collects the data from the AIM and the GC computer and stores it in a reactor-run file. Once all the data are collected, the program ends execution and returns control to the operating system.

Another program, also menu-driven, allows the user to specify which reactors are to be sampled and how often. This program also lets the user scan the collected data; this is important because a reactor run might last hundreds of hours.

At the end of a run, all the data are stored on a floppy disk. The user then checks the data for continuity and consistency by means of a checking program. Once the bad data are set aside, the computer can perform material- and energy-balance calculations. The material balance is complicated, involving numerical integration, because material accumulates within the reactor. Finally, a program generates a report that summarizes the results of the run.

## Costs and benefits

Approximate costs of the system's components are: $6,000 for the microcomputer and peripherals; $9,500 for the GC computer; $15,500 for the analog-input module and the remote scanner. In our case, the microcomputer and GC computer were on hand already, as were the two gas chromatographs ($15,000 each) and their sampling system ($4,000). Installation labor cost $8,000; programming cost $9,000.

The key payoff is a labor saving comparable to one person's full-time effort in data-collection and analysis. Also, researchers can plan new reactor runs faster because the run report is available within minutes, vs. two weeks by the semiautomated method.

## Acknowledgements

The data-acquisition system was the work of many people. I want to thank Bill Albe, Mark Richardson, Dave Sawyer and John VanSickle for their help and support.

## The author

**Stewart M. Goldfarb** is currently working toward a Ph.D. degree in chemical engineering at the University of Michigan. His address is 810 W. Huron, Apt. 1, Ann Arbor, MI 48103. When this article was written, he was at Dow Corning Corp., where he worked first in process design and later in pilot-plant development and process simulation. Mr. Goldfarb has a B.S. degree in chemical engineering from Wayne State University, and belongs to AIChE.

# Microcomputer simulation

*Douglas L. Denholm**

☐ While developing a pollution-control process, my colleagues and I needed a material-balance program that would allow us to size and cost equipment, determine stream compositions, and calculate system dynamics in case of process upsets. After struggling with "canned" software, which never did run satisfactorily, we obtained a small business computer that costs less than what we had already spent on software royalties.

The Basic program we developed—written by an engineer and not a professional programmer—was tailored to our specific needs, and served our purposes well. Also, it was easy to modify, and faster to use than the commercial software. While canned software is justified for many situations, it is probably more economical—and just as fast—to use your own software on your own microcomputer when you need material balances to compare several processes or process options, and expect to make frequent modifications.

## The software situation

The process, which removes hydrogen sulfide from a variety of gas streams, involved five reactors, forty streams and seven components as we originally modeled it. During the evaluation of process alternatives, we found it desirable to use computerized material-balance and flowsheet-development programs for sizing, compositions and flowrates. In our first attempt to do this, we used a process-simulation package available on a commercial timesharing system. The arrangement involved calling the timesharing system, connecting our terminal, accessing the simulation package, and interacting with it to program and run our material balance. Our costs to use the service were fees for computer connect time and disk storage of our program, and a royalty fee each time we ran a program, successfully or not.

Modeling a process in the initial stages of development is an interactive, iterative procedure that requires continual modifications and reruns of the simulation program to reflect changes in the process concept. With the commercial package, we found this stage to be difficult because the package was not well documented and we had no access to the source program. It was also expensive. Over a two-month period, our attempt to use the package cost: three man-months of time, $3,000 in royalties, and about the same amount in connect-time, storage and printing fees.

At this point, management looked for an alternative. We obtained a Radio Shack Model II small business computer (which includes 64 kilobytes of read-and-write memory and built-in disk drive, video screen and keyboard) plus a printer for program listings and hard copies of the program output. Within two or three weeks, a young engineer developed a 2,000-statement Basic program that simulated the process successfully. We used this program during the development-and-design process, later adapting it to provide a dynamic simulation of the process, which was needed to calculate surge-tank capacity.

## Tradeoffs

Developing our own program took about one man-month of effort, while our work with the commercial package took three man-months. Of course, some of this time was spent learning the rules of the commercial system, and presumably we could have used the package for construction of new flowsheets and balances more efficiently after that. But our experience with the microcomputer indicated that we would also benefit from a learning curve there, and could write new programs for completely different processes in much less than a month's time.

While it is undoubtedly true that a mainframe computer would execute a complete modeling program faster than a microcomputer, the expense and inconvenience associated with developing the model outweighs the speed of execution—as long as the microcomputer program runs in a reasonable length of time. In our case, execution of a complete balance required 10 minutes, with another 5 minutes for printout. A total time of half an hour to get a complete material balance printed out, or an hour to get a dynamic simulation, was quite acceptable to us. In fact, we got printed copy faster, because our output had been printed at a computer center and mailed to us when we were using the timesharing system.

In some respects, our experience may not be typical. Our system is aqueous and involves three phases, while commercial simulation software is generally aimed at two-phase hydrocarbon processes. Also, development of a process model during the transition from pilot-plant to commercial scale requires frequent program changes, while this might not be the case for established processes with set flowsheets.

Based on our experience, we believe that commercial software packages can be justified when: the system thermodynamic properties are fully described in the package's database; a large number of similar processes are to be evaluated; the use of ancillary programs for equipment design, costing, or optimization is being considered. But for new process developments, where budgets are tight and commercial simulation software is inappropriate or overly costly, using your own microcomputer can be a cost-effective alternative.

*EIC Laboratories, Inc., 55 Chapel St., Newton, MA 02158.

Originally published April 5, 1982

# BASIC, FORTRAN and Pascal

**These general-purpose programming languages are suitable for engineering computations and available on many computers. Which one should you use for your programs?**

***Namir C. Shammas**, Infilco Degremont, Inc.*

☐ How well can your desktop computer handle your chemical-engineering problems? This depends partly on your choice of a programming language, since that affects how you write and modify programs, and how well they work.

Three of the most widely available general-purpose languages are BASIC, FORTRAN and Pascal. A version of BASIC is included with almost every desktop and portable computer sold in the U.S.; FORTRAN and Pascal are add-ons, available for many but not all such computers.

This article explains important features of these three languages, and shows what these features mean to programmers and users. Short example programs serve to illustrate the similarities and differences, and references for more information are listed.

## Languages and your computer

BASIC is either built into the computer (e.g., Applesoft BASIC on an Apple II), in which case the language is ready to use as soon as you turn the computer on, or loaded from a floppy disk (e.g., Microsoft BASIC). Both FORTRAN and Pascal are always loaded from disk.

If a language is not built in, you must get a version compatible with your computer's operating system. Also, FORTRAN and Pascal may require more memory than is included with small desktop computers—a minimum of 64K RAM is typical.

What does an add-on language cost? It is impossible to quote a single number, since prices vary widely, but interpreter versions of BASIC typically cost about $100–$150, and compilers for BASIC, FORTRAN or Pascal cost about $200–$500.

## Interpreter vs. compiler

BASIC is essentially an interpreter language, which means that the computer translates and executes programs one line at a time. Once the interpreter is done with a line, the line is forgotten. This means that program lines within a loop (e.g., FOR...NEXT in BASIC) are interpreted over and over, and that you find an error in a program only when the line containing the error is executed.

Thus, interpreter BASIC is slow, but you can always see what is going on, and can make changes if something goes wrong. In this sense, interpreter BASIC is interactive. It is also interactive in the sense that it monitors what you type, and knows whether a string of characters is meant as a program line or an immediate command (e.g., RUN).

FORTRAN and Pascal are always compiled (and there are compilers for BASIC). This means that you write a program on a word processor or text editor, save it on disk, then load a compiler program that translates the program into an equivalent string of machine language (i.e., the 1s and 0s that the computer understands). This is done all at one gulp, before execution. The program's syntax and branching are then checked, and if they are correct you can run it.

A compiled program runs several times as fast as an interpreted one, but may require more random-access memory (RAM). For example, a compiled version of an Applesoft BASIC program takes about 25% more RAM. This can be a problem in trying to compile large BASIC or FORTRAN programs on a computer having little memory.

Pascal, however, is designed to "bite more than it can chew." At your request, it can break down big programs into segments and swap compiled portions between RAM and disk. This means you can run bigger programs, but such swapping slows down the execution because going back and forth from disk is slow.

## Programming

Pascal requires structure and a systematic approach, while BASIC and FORTRAN will tolerate casual programming. Overall, Pascal is probably a better language for multi-purpose and interrelated programs, while the less-structured languages may be easier to use for single-purpose and stand-alone programs. Let us go over some of the important features:

*Program listings*: BASIC lines begin with line numbers that tell the order of execution and organize the branching. If space allows, you can insert new lines between existing ones. A disadvantage is that when writing a branching statement you may not know what line number to go to.

FORTRAN listings are not necessarily numbered, but the programmer can label them with numbers for branching purposes. If the labels are chosen unsystematically, however, reading such a program can be very confusing—i.e., you might have to look through most of the program to find a given label.

In Pascal, a set of lines performing a given task is arranged in physical sequence. Such a set, known as a

Originally published July 25, 1983

block, is attached to others to handle a bigger task. Thus, Pascal *demands* that you examine the task, break it down into subtasks, and build the program systematically. It is not a language for sloppy programmers.

*Branching*: BASIC uses line numbers as destinations for branching commands (e.g., GOTO 550); FORTRAN uses numeric labels as destinations in a branch (e.g., GO TO 550) or loop (e.g., DO 550). A program having many such commands is usually hard to read.

Pascal was designed to discourage the use of branching statements, though it does allow branching to a numbered line via a GOTO command. Statements to be repeated a number of times are sandwiched between two statements that declare the beginning and end (i.e., BEGIN and END), with the overall loop controlled by a logical expression—FOR...DO...,REPEAT...UNTIL..., or WHILE...DO.... Statements to be executed conditionally are made part of a logical structure (IF...THEN... ELSE...), as in FORTRAN but without jumping to numbered lines.

*Subroutines, user functions and recursion*: Before comparing the languages, let us go over the difference between global and local variables. A global variable is one known and understood throughout a program, while a local one is known only within a particular subroutine or function. Thus, you can use the same local-variable name in several places without causing any confusion.

A subroutine can lie anywhere in a BASIC program; it is called by a GOSUB command followed by a line number telling where the subroutine begins. Thus, it is simple to write a subroutine but it may be difficult to find it later. In BASIC, subroutine variables are global, so the subroutine and main program have to share variable names.This is a problem if you want to add a routine to a program, since the variable names must be the same.

User-defined functions are limited to one-line definitions in most desktop-computer versions of BASIC. Here, variables are defined locally.

FORTRAN and Pascal handle subroutines and user functions about the same way: separate from the main program, using local variables. Thus, you do not have to rewrite them to use them in new programs.

BASIC and FORTRAN allow for no recursion—i.e., a subroutine or function cannot call itself. In Pascal, a function can call itself a number of times in order to yield a result.

*Variable types*: BASIC variables may represent integers (e.g., 88), floating-point numbers (e.g., 88.088) of single or double precision, or alphanumeric strings (e.g., UNIT PROCESS). FORTRAN variables may be one of those types, or logical (i.e., true or false). In either language, variables may also represent arrays of numbers.

Both BASIC and FORTRAN have conventions for naming variables. For example, FORTRAN variables beginning with the letters I–N are integers, unless you declare them as being of another type.

Pascal has all the variable types that BASIC and FORTRAN do, but you must declare every variable. This takes some time, but it adds order. Pascal also lets you define new *types* of variables.

For example: You can declare a variable—call it CHEM-DATA—composed of critical temperature, critical pressure, compressibility factor, boiling point, freezing point and critical density. The set of these data for one chemical is known as a record; all the records together are a file. In Pascal, you can access and use an element of a record (e.g., critical pressure for one chemical), a record (i.e., data for one chemical), or a file (i.e., data for all chemicals). This type of structure simplifies data-handling in programs doing complex tasks.

*Translating programs*: You can convert programs from Pascal to FORTRAN or BASIC, or between FORTRAN and BASIC, with relative ease. However, converting to Pascal may require that you redraw the flowchart, and structure the tasks and data, before writing any Pascal code.

## Linking programs together

Chemical-engineering calculations involve physical and thermodynamic properties and unit operations. In any language, you can write subroutines and functions, and set up reference tables, so that writing a program for a given problem is simply a task of linking already-written modules. Along the same lines, you could even set up a library of process designs.

In Pascal, you can store subroutines and functions separately. Then, instead of recoding them in a new program, you tell the Pascal compiler to include them, as if you did code them, in the compilation. Some FORTRAN compilers also have this capability. BASIC programs can be chained, so that one program passes its variables to another program that replaces it in RAM. This can be useful, but it does not allow for putting new programs together without recoding.

## Example programs

Consider the problem of calculating the settling velocity of particles, following Stokes' law, for a range of temperatures. Let us see how the program for this problem looks in each language.

The basic equation for terminal settling velocity ($V$, m/s) is:

$$V = D^2\, g\, (d_p - d_a)/(18\, \mu)$$

where $D$ is particle dia., m; $d_p$ is particle density, kg/m$^3$; $d_a$ is air density, kg/m$^3$, a function of temperature; $g$ is gravity, m/s$^2$; and $\mu$ is air viscosity, cP, a function of temperature.

Let the particle be of 0.0001 m dia., with a density of 2,570 kg/m$^3$. The temperature range is 10–50°C, and we are interested in seeing $V$ at 10° increments. The air is at 1 atm, and is assumed to behave as an ideal gas.

Tables I–III list BASIC, FORTRAN and Pascal programs that compute $V$ and print it out for five temperatures. The specific versions of the languages used are Applesoft BASIC, Apple FORTRAN, and UCSD Pascal. The variable names have been kept short, which would be necessary in some versions of these languages. A few features should be pointed out:

- The BASIC program (Table I) is the shortest, and has the simplest formatting.
- The FORTRAN program (Table II) has only three line numbers: two for formats, and one for the DO loop. FORTRAN has powerful formatting features, but the FORMAT statements can be complicated. Note that the variable name $Y$ appears in each function. Function variables are local, so names can be duplicated.

**BASIC program is short and simple** — **Table I**

```
1    REM FNV is the air-viscosity function
2    REM FND is the air-density function
100  DEF FNV (X)=( ( ( -4.50966E-9*X)+4.41E-5)*X+5.181E-3)/1000
110  DEF FND (X)=353.4/X
120  PRINT "TEMP." , "VELOCITY"
130  G=9.81  : REM G is gravity constant
140  D=0.0001: REM D is particle diameter
150  FOR T=10 TO 50 STEP 10
160       U=G*D*D*(2570-FND(T+273.15) )/18/FNV(T+273.15)
170       PRINT T, U
180  NEXT T
190  END
```

**FORTRAN program has user functions at end** — **Table II**

```
      WRITE (3,1)
1     FORMAT ("TEMP.", 10X, "VELOCITY")
C     G is gravity constant
      G=9.81
C     D is particle diameter
      D=0.0001
      DO 2 T=10, 50, 10
         U=G*D*D* (2570-DENSITY (T)/18/VISCOS (T)
         WRITE (3, 3) T, U
2     CONTINUE
3     FORMAT (E3.4, 10X, E13.4)
      STOP
      END
C     VISCOS is air-viscosity function
      FUNCTION VISCOS (X)
      Y=X+273.15
      VISCOS=( ( (-4.50966E-9*Y)+4.41E-5)*Y+5.181E-3)/1000
      RETURN
      END
C     DENSITY is air-density function
      FUNCTION DENSITY (Y)
      DENSITY=353.4/(Y+273.15)
      RETURN
      END
```

**Pascal program declares all variables** — **Table III**

```
FUNCTION DENSITY (X: REAL) : REAL;
BEGIN
     DENSITY := 353.4/(X+273.15)
END;

FUNCTION VISCOSITY (X:REAL) : REAL;
VAR Y: REAL;
BEGIN
     Y := X+273.15
     VISCOSITY :=( ( (-4.50966E-9*Y)+4.41E-5)*Y+5.181E-3)/1000
END;

PROGRAM VELOCITY; VAR T,U,G,D: REAL;
(*$I PROP.CHE)
BEGIN
     G := 9.81 ; D :=0.0001; T :=10;
     WRITELN ('TEMP.    VELOCITY');
     REPEAT
               U := G*D*D*(2570-DENSITY(T) )/18/VISCOSITY(T)
               WRITELN (T,'      ',U); T := T+10
     UNTIL (T > 50)
END.
```

■ In the Pascal program (Table III), the functions for viscosity and density are assumed to be stored separately, in a file named PROP.CHE, and called from this file in the main program.

## Choosing a language

Overall, BASIC is adequate for many engineering problems, but is probably not the best choice for complex problems that require fast calculation and data-handling. FORTRAN is somewhat more powerful, and has the advantage that many existing engineering programs are written in this language. Pascal seems the best choice for interrelated design and calculation programs, since subroutines and user functions can be stored and used readily, and data are handled easily once the structure is established. The references offer more information on these three languages.

There are of course other languages to consider, though we cannot cover them in detail. Two of them are:

■ FORTH is an interactive, structured language similar to the RPN-type calculator languages used on Hewlett-Packard machines. The language lets you define procedures (called words), test them on the spot, and then use them immediately to define new words.

■ C is similar to Pascal. Its key plus is that it can be a replacement for machine language in writing systems software. Bell System people developed it, and used it to write their Unix mainframe-computer operating system.

## References


1. Albrecht, R., Finkel, L., and Brown, J., "BASIC: A Self-Teaching Guide," 2nd ed., John Wiley & Sons, New York, 1978.
2. Peckham, H. D., "BASIC: A Hands-On Method," 2nd ed., McGraw-Hill, New York, 1981.
3. Presley, B., "A Guide to Programming in Applesoft," Van Nostrand Reinhold, New York, 1982.
4. Coan, J., "Basic FORTRAN," Hayden, Rochelle Park, N.J., 1980.
5. Friedman, J., Greenberg, P., and Hoffberg, A., "FORTRAN: A Self-Teaching Guide," 2nd ed., John Wiley & Sons, New York, 1981.
6. Meissner, L., and Organick, E., "FORTRAN 77 Featuring Structured Programming," Addison-Wesley, Reading, Mass., 1980.
7. Wagener, G., "FORTRAN 77 Principles of Programming," John Wiley & Sons, New York, 1980.
8. Grant, C., and Butah, J., "Introduction to the UCSD p-System," Sybex, Berkeley, Calif., 1982.
9. Koffman, E., "Pascal: A Problem-Solving Approach," Addison-Wesley, Reading, Mass., 1982.
10. Luehrman, A., and Peckham, H., "Apple Pascal: A Hands-On Approach," McGraw-Hill, New York, 1981.
11. Schneider, G. M., Weingart, S., and Perlman, D., "An Introduction to Programming and Problem Solving with Pascal," John Wiley & Sons, New York, 1978.
12. Tiberghien, J., "The Pascal Handbook," Sybex, Berkeley, Calif., 1981.


## The author

**Namir C. Shammas** is a chemical engineer at Infilco Degremont, Inc., 2828 Emerywood Parkway, Richmond, VA 23229, where he performs pilot-plant research and development on wastewater treatment. He uses a desktop computer for statistical data analysis and design calculations, and describes himself as a language hobbyist. Mr. Shammas has a B.Sc. from Baghdad University, and an M.Sc. from the University of Michigan, both in chemical engineering. He belongs to the PPC calculator and computer club, which is based in California.

# Microcomputer systems for process design

**Microcomputers can solve substantial problems such as distillation design, but do not have the speed needed for more complex calculations and flowsheets.**

***Ian Sucksmith,*** *Arco Chemical Co.*

☐ For about $3,000–5,000, a chemical engineer can set up a microcomputer workstation that will solve many design problems. Besides being inexpensive, such a system will be more convenient and easier to use than a larger system. But it will not handle the most complex problems, and may be too slow for extensive calculations, such as those involving recycle streams and multiple unit operations.

Design programs discussed in this magazine have included flash, combustion and shortcut distillation calculations [*1,2*]. In general, current microcomputer programs have a single use, rather than being integrated into larger programs having many uses, and do not involve extensive calculations. It is possible, however, to run integrated software on a microcomputer, and this is where advantages and limitations become evident.

This article specifies a microcomputer system for process-design work, and explains its pros and cons. A steady-state-simulation of a distillation column by a software package called MICROCHEM (written by the author) serves as an example application and a basis for evaluating performance.

## A process-design workstation

Microcomputers and peripherals vary in capability, so it is important to specify requirements carefully. For design work, a chemical engineer should have a system with at least the following features:

*64K of RAM (random-access memory).* RAM is the computer's working memory, which determines the problems one can solve. Most microcomputers have at least 16K, and 64K is usually available. One should know that machine utilities and advanced versions of BASIC typically take about 25K of RAM, so memory available to user programs is reduced.

*A 16-bit microprocessor.* This gives greater speed and memory capability than an 8-bit microprocessor. Many microcomputers (e.g., IBM Personal Computer, Apple III and TRS-80 Model 16) are built around 16-bit microprocessor chips.

*A monochrome high-resolution screen of 25 80-character lines.* Many systems have 40 characters as a standard, but this restricts output severely. Color screens have inferior resolution, and color-graphic display is not very useful in design unless one has a color plotter. A home television set is not adequate.

*An 80-character-width printer.* It should be capable of printing upper- and lower-case letters at 80 characters per second or faster. A dot-matrix printer should provide "correspondence" quality.

*Two disk drives.* Total storage space should be at least 300K; this can be achieved with either 5.25-in. or 8-in. diskettes (floppy disks).

*A 1,200-baud modem.* This allows the system to communicate with other computers over a telephone line.

How capable will such a system be? DeVoney and Summe [*3*] provide a vivid illustration by comparing the IBM Personal Computer with the full-size IBM 360, which was widely used in the 1960s and early 1970s: The 360 cost up to $500,000, while an equivalent Personal Computer costs about $5,000; the Personal Computer's processor is four times as fast as the 360's; both machines have about 262K maximum RAM; the Personal Computer can operate almost anywhere, while the 360 needed a controlled environment and was definitely not portable; diskettes and displays are faster and more convenient than the 360's card readers and punches.

What about programming languages? Almost all microcomputers use an interpretive (i.e., translated line-by-line rather than compiled) version of BASIC as their standard. This is handy for the programmer, who can see how a program is running and fix problems quickly. But interpretive BASIC is not so good for the user, because it is slow. To make BASIC programs more usable, one can convert them to compiled languages such as FORTRAN, Pascal or a compiled version of BASIC. A compiled program will require more RAM, but will probably increase speed by a factor of three or more—this is detailed in [*4*].

## Microcomputer pros and cons

The key quantitative advantage of a microcomputer-based system over a larger one is simply low cost. The equipment is inexpensive compared with owning or sharing a large computer, and every engineer could have a personal system.

Originally published January 10, 1983

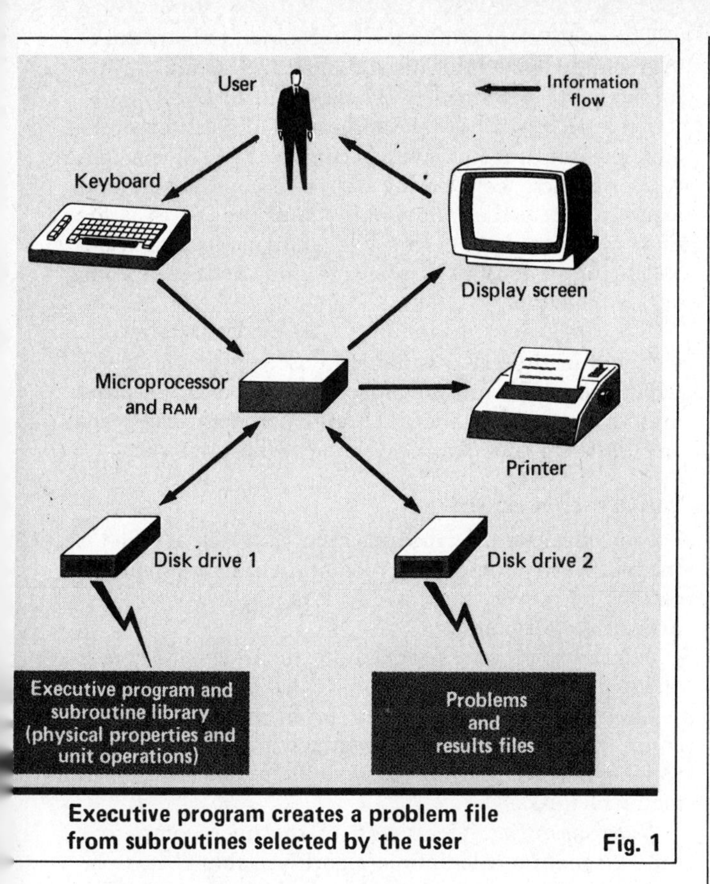


**Executive program creates a problem file from subroutines selected by the user** **Fig. 1**

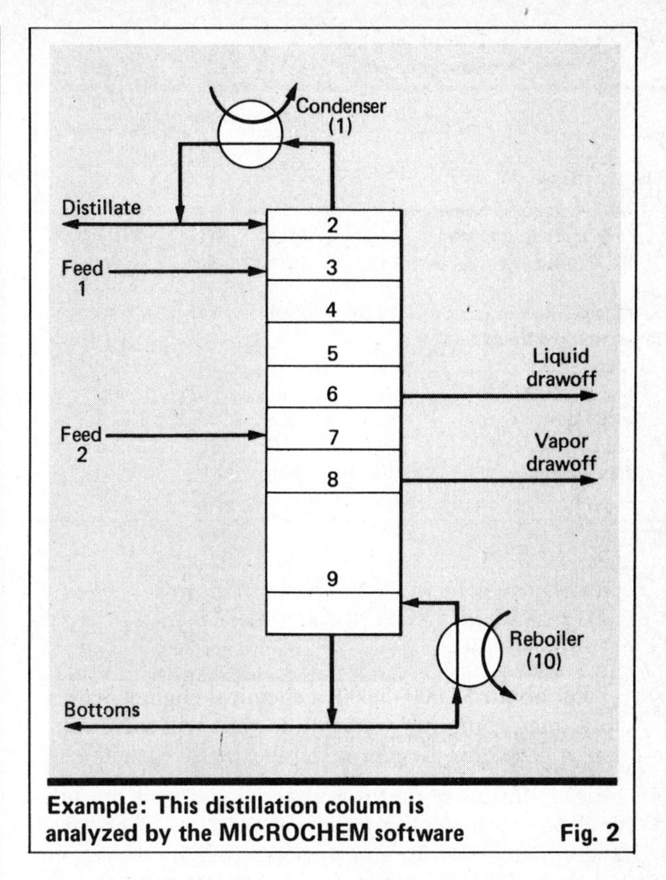


**Example: This distillation column is analyzed by the MICROCHEM software** **Fig. 2**

For the working engineer, the microcomputer system also has qualitative advantages:

1. The engineer controls it completely, and does not have to depend on a central system that may be down or may delay a critical piece of work.
2. Most microcomputers are portable, so it is practical to work at another company location, at a jobsite, or at home. If equipped with a modem, an engineer can telecommunicate with the base location. Some microcomputers (e.g., the Osborne 1) are quite portable, being about as large and heavy as a suitcase.
3. The microcomputer can be used for office tasks such as filing and word-processing.

**Data required for example problem** **Table I**

| Components (1-5) | Ethane, propane, *n*-butane, *n*-pentane, *n*-hexane | | | |
|---|---|---|---|---|
| **Feedstreams** | **Flow, lbmol/h** | **Temp., °F** | **Pressure, psia** | **Component 1-5 mol fractions** |
| 1 (Stage 3) | 100 | 150 | 300 | 0.05, 0.10, 0.30, 0.20, 0.35 |
| 2 (Stage 7) | 200 | 100 | 300 | 0.05, 0.15, 0.50, 0.25, 0.05 |
| Distillate | Vapor 10 lbmol/h; liquid 100 lbmol/h | | | |
| Drawoffs | Vapor 50 lbmol/h; liquid 10 lbmol/h | | | |
| Column | 10 stages; 1 million Btu/h heat removal from Stage 2; Reflux ratio 3; Condenser and reboiler pressure 300 psia | | | |

Key disadvantages are low power and lack of speed. A microcomputer system will probably not solve full flowsheets or make sophisticated unit-operation analyses, and will take longer to execute a program. Execution is especially slow if the program is in interpretive BASIC: In an example given later, a moderately complex distillation calculation takes more than five hours.

But note that run time is only part of the total time needed for a job. On the basis of total time, the microcomputer may not compare too badly with a larger system: Though a large computer may execute a program in seconds, the engineer doing the work may wait for a terminal, the program may wait in a job queue, and the printout may have to be retrieved from some remote location. For many small programs, run time becomes trivial when compared with the delays inherent in a large system.

## A steady-state simulator

To get a clearer picture of pros and cons, let us look at how a microcomputer-based system handles a fairly sophisticated simulation program. This program, called MICROCHEM, is designed for the workstation described, and is built around the following constraints:

*One unit operation at a time.* The program includes several unit-operation and thermodynamic subroutines (e.g., bubble-point distillation and ideal-liquid activity), but it does not do flowsheeting. Such capability could be built in, but problems involving recycle and several operations would probably take too long.

*Maximum of five components.* Most real-life problems involving hydrocarbons reduce to light, light-key, intermediate-key, heavy-key and heavy components.

**Example: Most-complex problem took over five hours to solve** **Table II**

| | Run number | | | |
|---|---|---|---|---|
| | 1 | 2 | 3 | 4 |
| **Specifications** | | | | |
| Number of components | 2 | 5 | 5 | 5 |
| Liquid-activity method | Ideal | Ideal | NRTL | NRTL |
| Vapor-fugacity method | Ideal | Ideal | Ideal | Redlich-Kwong |
| **RAM used, K** | 19.0 | 20.1 | 21.0 | 21.8 |
| **Time, min (rounded)** | | | | |
| Input | 1.5 | 1.5 | 2 | 2 |
| Preprocessing | 3 | 3 | 3.5 | 3.5 |
| Calculation | 5 | 22.5 | 295 | 328.5 |
| Output | 2.5 | 2.5 | 2.5 | 2.5 |
| Total time | 12 | 29.5 | 303 | 336.5 |

*Distillation is limited to 20 theoretical stages.*

*Program is user-friendly.* MICROCHEM is menu-driven, meaning that it prompts the user to answer questions, and it provides default values for missing data, where possible.

*Results are stored on disk and easy to retrieve.*

*Program is written in interpretive BASIC.*

Fig. 1 illustrates that disk space is used to store subroutines, component data, results, and an "executive" program that sets up a particular problem. To see the sequence of events in running a MICROCHEM simulation, let us look at the four phases of simulation as defined by Motard *et al.* [*5*]:

*Input.* The user loads the executive program from Disk 1 to the RAM, then runs it. The executive asks for information about the problem such as project number, components, and which unit-operation and thermodynamic subroutines are needed.

*Preprocessing.* The executive program builds a problem file (program) by copying the needed subroutines from Disk 1 to Disk 2 in the correct order. Then the executive loads the problem file into the computer's RAM. Since the RAM can store only one program at a time, this erases the executive program.

*Calculation.* The user runs the problem file, which asks for more information such as number of stages for a distillation. Once all the questions are answered, calculation begins.

*Output.* When calculation is completed, the problem file creates an output file on Disk 2 and stores all results there. Then the problem file loads the output file into the computer's RAM, erasing itself. Results can then be printed or displayed.

Although the executive and problem files are erased during a simulation, they are still stored on disk and may be recalled at any time. In fact, Disk 2 can be used to store many problem and output files indefinitely.

To show the performance of MICHROCHEM, let us consider analyzing the distillation column shown in Fig. 2. The column, not based on an actual case, is quite complex, having ten theoretical stages, two feedstreams, a partial condenser, one vapor and one liquid side-draw stream, and heat removal at one stage. Table I shows the data inputs.

The analysis method is the bubble-point algorithm of Wang and Henke [*6*]. This and other tridiagonal matrix methods are described by Henley and Seader [*7*].

Table II shows how the program handles this problem, for four different levels of complexity. The first run is the simplest, considering only two components and assuming ideal thermodynamics, and the fourth is the most complex, considering five components and using NRTL liquid-activity coefficients and Redlich-Kwong vapor fugacities.

The conclusions? Maximum RAM used was only 22K, well within the 41K available to user programs in this 64K system. The most complex run took over 5 h total, and run time was about 11 min per iteration. Non-calculational time was only about 8 min per run.

## Future directions

From the example, it seems clear that lack of speed is the biggest weakness of microcomputers in design applications. However, there are at least three ways to address this shortcoming:

*New microprocessors.* For example, the Intel 8087 chip is claimed to speed floating-point calculations tenfold.

*Compiled programs.* Once the problem file is created, compiling it into machine language is almost certainly worthwhile because this can cut execution time by a factor of three.

*Better algorithms.* Since most of the microcomputer's time is spent on calculation, algorithms that reduce the number of iterations can speed execution greatly. For example, the distillation problem discussed could have been solved faster by using a simultaneous-correction method.

Note that this MICROCHEM program does not tax the RAM or disk capacity of our workstation, as long as the one-operation constraint is met. But more-complex flowsheeting would probably overload the memory of this and most microcomputer systems.

## References

1. Goldfarb, S., and Griffin, S., Microcomputing hardware, *Chem. Eng.,* May 31, 1982.
2. Schmidt, P. S., Microcomputers in energy management, *Chem. Eng.,* July 12, 1982.
3. DeVoney, C., and Summe, R., IBM's Personal Computer, Que Corp., 1982.
4. Bloomfield, D. P., Microcomputing software, *Chem. Eng.,* May 31, 1982.
5. Motard, R. L., Sacham, M., and Rosen, E. M., Steady State Chemical Process Simulation, *AIChE J.,* Vol. 21, No. 3, 1975.
6. Wang, J. C., and Henke, G. E., *Hydrocarbon Proc.,* Vol. 45, No. 8, 1975.
7. Henley, E. J., and Seader, J. D., "Equilibrium Stage Operations in Chemical Engineering," John Wiley & Sons, New York, 1981.

## The author

**Ian Sucksmith** is a senior process engineer with Arco Chemical Co., Engineering and Environment S.W. Office, P.O. Box 777, Channelview, TX 77530, where he is involved in plant technical service as well as process design. He earned his B.Sc. degree in chemical engineering at the University of Nottingham, and is presently working on his master's at the University of Houston. He belongs to AIChE, and is registered in the state of Texas.

# Computer horizons in process-systems engineering

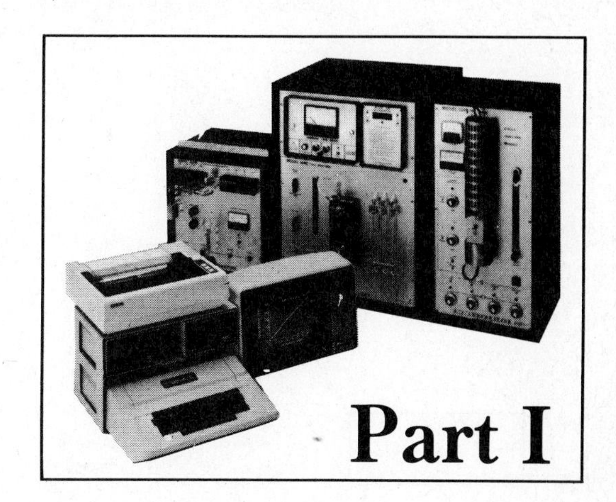


**This two-part article scans the near future of computer-aided process-systems engineering, telling where computer technology is headed and how advances will affect engineering practice.**

*R. L. Motard, Washington University*

☐ The expanding role of computers has brought about a minor revolution in process analysis, design and simulation over the past fifteen years. And the future holds tremendous promise as computer technology continues to evolve and chemical engineers apply it in process-systems engineering.

What computer advances can we expect to see in the next several years? This article looks at what is ahead in: computer and memory chips; software development; computer design; data management and data-bases; and computer languages. Part II, appearing on p. 53, will tell how these developments may be applied in process analysis, design and simulation.

## Silicon chips still improving

Many of us have read projections for increases in packing density of silicon chips that will reduce hardware cost. As of now, we hear that silicon memory chips may ultimately reach a density of 1–10 megabytes per chip, vs. 256K today, and computer chips may have 2–10 million transistors by 1990, vs. half a million today. The semiconductor industry projects a 15%/yr decrease in cost/performance (dollars per million instructions per second) for large general-purpose computers and 25%/yr for small computers [1].

However, silicon technology will probably reach its limits in a few years. Where it peaks will depend on the ability to produce ultrapure silicon economically, which is a chemical engineering problem in its own right. Impurities affect the yield of acceptable chips in mass production, and so critically limit the ability to shrink electronic elements and connections onto silicon.

The cost of chip design and the problems of building working computers from chips also affect computer cost. Chip design is analogous to process design: a system architect produces a block diagram meeting the functional requirements; a logic designer translates the block diagram into a logic-and-gate specification; a circuit designer then produces a circuit diagram; this is converted into a mask (template) for manufacture by a layout designer and drafter [2]. Design of very-large-scale integrated (VLSI) elements is so complex that progress depends on development of computer-aided design tools.

Today, the design of a new chip costs from $1 million for a simple device to about $20 million for a complex microprocessor chip. Therefore, chip makers have to sell very large volumes to recover their investment.

New technologies such as optical-electronic digital systems promise further cuts in hardware size, and biocomputers may eventually shrink computer elements to molecular scale, but such alternatives to silicon chips will have to fight the market power of an entrenched, mature technology.

## The software problem

The value of rapid progress in computer hardware is limited by the slower progress in software development. Cost of computer hardware will have dropped a millionfold between 1955 and 1985, but programmer productivity will have increased only fourfold [3]. Another view of the problem is that by 1985 the cost of leasing one million instructions per second of computer capability will be less than the salary of a professional programmer for the first time [1]. Today, programming costs approach 85% of users' total computing costs.

Growth in computer use will depend on increased programmer productivity, which may happen in several ways. For one, low-cost personal computers have spawned a cottage programming industry, whose new software has sometimes captured surprising market share. Two examples are VisiCalc, a spreadsheet planning tool, and CP/M, an operating system for the Z80 microprocessor.

This article is based on a paper presented at the International Symposium on Process Systems Engineering, Kyoto, Japan (August 1982).

Originally published February 7, 1983

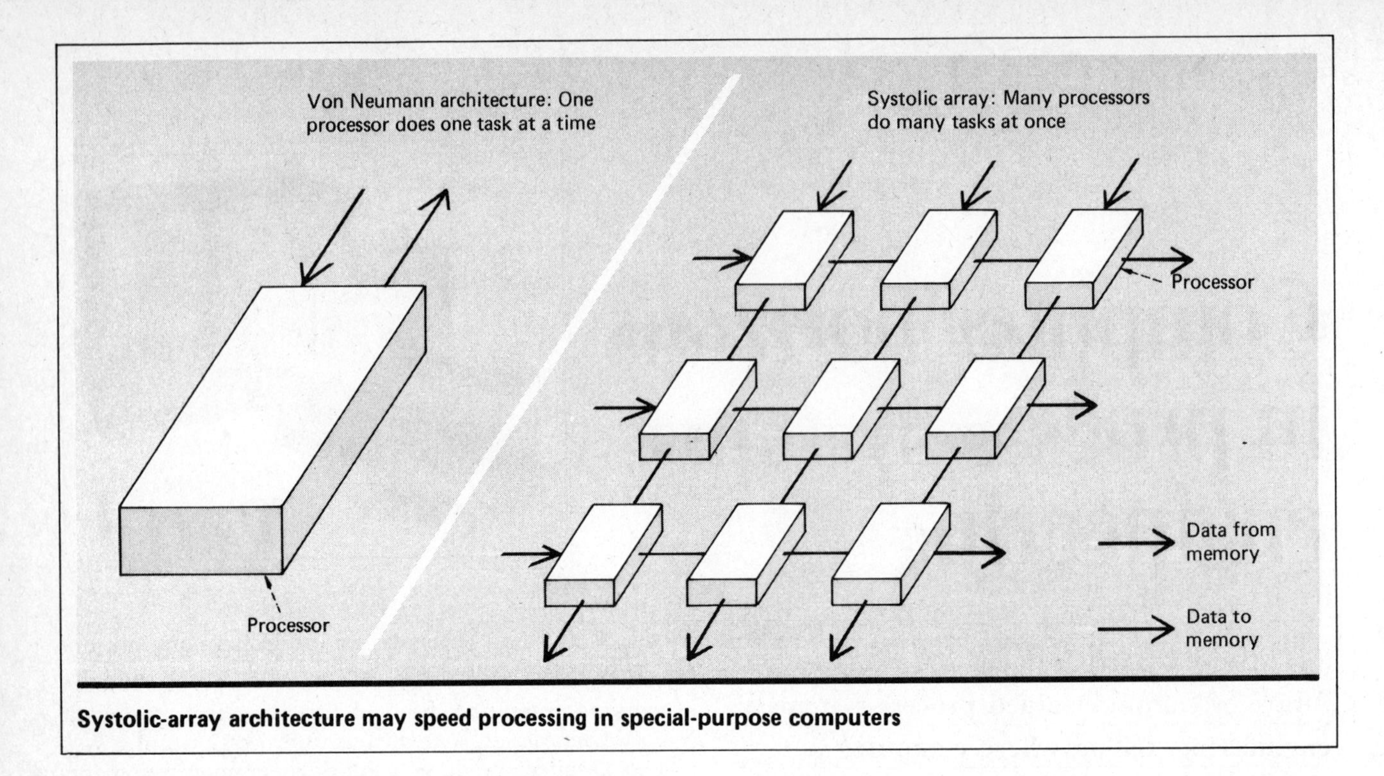


**Systolic-array architecture may speed processing in special-purpose computers**

Use of standard software blocks in building system software will also improve programming. Because blocks can be verified and tested rigorously, high-quality programs can be assembled quickly at low cost. This approach has been successful in application programming (e.g., process-design programs) but not yet in systems programming (e.g., operating systems for computers) because of problems with determining the right set of blocks and providing proper interconnections [*4*].

However, relatively standard operating systems for microprocessors are being provided as ROM (read-only memory) chip sets [*5*]. Some of these are even processor-independent, requiring only a few external parameters to define their interface with users' programs. Such an operating system typically adds twenty or more high-level instructions to a processor's basic set, and may be able to implement high-level languages such as Pascal. Expect to see more such software-in-silicon, in applications such as telecommunications, networking, high-level-language compilers, and graphics.

One tool that improves productivity is layering of languages: A high-level program written in a low-level language can be used on diverse computers. For example: The Unix operating system [*6*] is portable because it is written in the lower-level C language [*7*]. For any given computer, Unix is the same; all one needs is a C compiler for the particular computer (the compiler program is itself written in C). Other development tools include automated documentation, testing, debugging, and software-project management [*8*].

A general software problem is the inability to write programs to fit the application rather than the computer. Today's programs handle data storage and retrieval, which means that data must be structured to fit a particular program. But relational database-management systems that decouple programs from data may remove this restriction. On a higher level, programs that build new programs are limited by the architecture of today's computers [*4*].

## Chemical engineers designing computers?

Special-purpose computers for process-systems engineering might prove economically feasible with new computer architecture and user involvement in the computer-design process. Today, most computers still have the architecture developed decades ago by von Neumann—one memory, one input channel, one processor—and processing proceeds in a single sequence. This architecture limits the potential of VLSI chips because cross-chip communication becomes a bottleneck.

One solution is to break up processing into several parallel sequences that can be performed concurrently. Another is to use a systolic array, as shown in the figure, in which processing may proceed in multiple directions at multiple speeds.

The key to either approach is to find the appropriate regularity in the application problems. Then each type of problem could have its own special-purpose machine, with VLSI chips designed to do exactly what the users want. For example: A systolic array having thousands of processors could process images very quickly.

To develop problem-oriented computers, the computer architect needs automated aids and input from the users. What better opportunity for chemical engineers to get involved in designing computers for their work? The idea of embodying process analysis, design and simulation in silicon is not so farfetched. Caltech has a graduate course that teaches computer-science students to design VLSI chips—these are usually designed by electronics engineers. If the design process is automated, engineers without an electronics background can quickly produce chip designs nearly as good as those the experts do.

Already, prototype automated systems design chips

**Relational database allows selection, projection and joining operations on data**

**Relation: Streams**

| Stream | Temperature | A% | C% |
|---|---|---|---|
| 1 | 25 | 60 | 15 |
| 2 | 85 | 60 | 15 |
| 3 | 108 | 60 | 15 |
| 4 | 122 | 5 | 90 |
| 5 | 100 | 70 | 1 |
| (etc.) | | | |

**Relation: Pumps**

| Pump | Stream | Flowrate | Pressure |
|---|---|---|---|
| 82 | 2 | 90 | 40 |
| 83 | 3 | 90 | 60 |
| 87 | 5 | 70 | 25 |
| (etc.) | | | |

**Database operations:**

**Joining:** Join (Streams, Pumps) on (Stream)

| Stream | Temperature | A% | C% | Pump | Flowrate | Pressure |
|---|---|---|---|---|---|---|
| 2 | 85 | 60 | 15 | 82 | 90 | 40 |
| 3 | 108 | 60 | 15 | 83 | 90 | 60 |
| 5 | 100 | 70 | 1 | 87 | 70 | 25 |
| (etc.) | | | | | | |

**Projection:** Project (Stream, Temperature) over (Streams)

| Stream | Temperature |
|---|---|
| 1 | 25 |
| 2 | 85 |
| 3 | 108 |
| 4 | 122 |
| 5 | 100 |
| (etc.) | |

**Selection:** Select (Temperature, A%) from (Streams) where (Stream = 5)

100, 70

only 15–20% less densely packed than expert-designed chips. With greater automation, the cost of chip design will drop drastically, and short-run production of special-purpose VLSI chips may be economically feasible.

## Data are more important than programs

Computers will be more useful when today's program-centered approach to large-scale applications is replaced by a data-centered approach. In the past, the program and its developer dominated the application, and data had a secondary role. More attention was paid to the elegance of the program than to the value of the data; indeed, most data ended up as a pile of printouts.

Process-systems engineering has a proliferation of stand-alone computer programs, able to solve particular problems (e.g., process synthesis, physical-property estimation, automated drafting) but not to interact with other programs. To the typical project manager, concerned with deadlines and budgets, the entire mix of programs and data may seem almost unmanageable.

The solution is a data-centered approach to program development. After all, decisionmaking depends on the information content of the data, and not on the program. The business community discovered this long ago, and moved to database-management techniques. Now it is the engineering community's turn.

## Database management systems

When a program assigns and manipulates data storage in detail, it can operate only on data having a particular structure. Therefore: programs take excessive effort to write; any change in data structure requires expensive reprogramming; data are duplicated, and so may be inconsistent. The key goal of a database-management system (DBMS) is to decouple application programs from data [9]. Then programs can be free of detailed storage commands, several programs can operate on one consistent set of data, and program maintenance is reduced. Today's database-management systems generally do not provide complete data independence.

With a DBMS, the application program is less concerned with how or where data are stored. This is because data structure is expressed by associations among data rather than by arbitrary addresses. In other words, the DBMS considers the meaning of data rather than where it is located and how it is arranged.

For example: Consider a set of data (e.g., temperature, pressure, composition) about a number of process streams. Each stream may be considered an entity; each type of data (e.g., temperature) an attribute; and each description (e.g., 100°C) a value. A table of all such (entity, attribute, value) sets is considered a relation (a set of such relations is a relational database). The DBMS description of one stream temperature is:

$$100°C \leftarrow \text{Temperature (Stream 5)}$$

More generally:

$$\text{Value} \leftarrow \text{Attribute (Entity)}$$

Temperature data for all streams may be retrieved by

a simple command such as "Select Temperature From Streams," with no iteration required. In contrast, retrieving such data from conventional files is iterative, and requires a knowledge of where files are located (e.g., on a disk or tape) and which bits represent stream number and temperature. The table shows a relational representation of process data, and illustrates three set operations (selection, projection and joining) that a DBMS can perform.

Database management systems that provide greater data independence will allow construction of more powerful and versatile process-systems application programs. But there are still problems. One is that very large collections of data lead to slow storage and retrieval. A typical chemical-process capital project uses one to two gigabytes (billion bytes) of data per billion dollars of investment [*10*]. Such a volume of data has led to speculation that a complete process-engineering database would be composed of separate databases for each discipline [*11*].

## Whither computer languages?

The DBMS language must be usable both interactively and embedded within other programs. Then the programmer can write and debug data-related commands before incorporating them into application programs written in a host language such as FORTRAN or Pascal.

What about the future of host languages for process-systems engineering? It seems probable that a language like Pascal or ADA will be as widely used as FORTRAN has been for the past two decades. Pascal is well-structured, with a simplicity that reduces errors and makes them easier to find when they do occur. It also encourages creation of portable and modular programs, and has features that allow a programmer to verify that a program matches its designer's intent. ADA, which has its roots in Pascal, was the result of a five-year effort by the U.S. Dept. of Defense to create a universal system-development language. While there are concerns about some of its features, it will probably remain viable because of the support behind it.

Of course, new languages do not mean that old programs must be discarded. In a DBMS environment that provides data independence, statements for data allocation and manipulation in old FORTRAN programs can be replaced with new DBMS access statements. But new programs will be written in the new languages.

## Operating systems

It is difficult to disentangle operating and database-management systems, because they both handle data storage and retrieval. However, we should examine trends toward standardization because this affects prospects for software building blocks as well as software-in-silicon. If there is one candidate emerging as a standard, it is the Unix operating system developed at Bell Labs in the 1970s. Perhaps future operating systems will all be similar to Unix, especially those for the computer workstations that chemical engineers will be using.

Unix [*6*] is basically an electronic filing system. Each file is treated simply as a stream of characters (or bytes) without the reference to hardware characteristics that most operating systems require. Associated with files is a hierarchy of directories; these contain information that helps to organize large collections of files (including other directories). Input and output devices are handled the same way as files, so that an application program is not concerned with the source or destination of data. This decoupling of programs and data makes Unix valuable for building high-level software systems.

Unix permits multiprocessing, which means that a user may have several programs running at one time. It also supports modular programming, which in turn promotes flexibility and evolutionary changes in programs. Another virtue of Unix is its small size and clean structure, which makes it very popular in university computer-science departments.

This operating system has found broad application in text processing, software development, laboratory automation, computer-science education and small databases. Already, Unix has some 8,000 installations [*12*]. A relational database called INGRES is available with a Unix interface [*13*]; an interface for Pascal is also available.

What will these advances in computer technology mean to process-systems engineering? Part II, appearing March 7, will tell how new hardware and software will enhance productivity in process synthesis, design, simulation, and engineering-office functions.

## References


1. Branscomb, L. M., Electronics and Computers: An Overview, *Science,* Vol. 215 (1982), pp. 755–760.
2. LaBrecque, M., Faster Switches, Smaller Wires, Larger Chips, *Mosaic,* National Science Foundation, January/February 1982, pp. 26–30.
3. Birnbaum, J. S., Computers: A Survey of Trends and Limitations, *Science,* Vol. 215 (1982), pp. 760–765.
4. Bacon, G., Software, *Science,* Vol. 215 (1982), pp. 775–779.
5. Lettieri, L., Software-in-Silicon Boosts System Performance, Cuts Programming Time, *Mini-Micro Systems,* March 1982, pp. 93–95.
6. Kernighan, B. W., and Morgan, S. P., The UNIX Operating System: A Model for Software Design, *Science,* Vol. 215 (1982), pp. 779–783.
7. Kernighan, B. W., and Ritchie, D. M., "The C Programming Language," Prentice-Hall, Englewood Cliffs, N.J., 1978.
8. Howden, W. E., Contemporary Software Development Environment, *Comm. Assn. for Computing Machinery,* Vol. 25 (1982), pp. 318–329.
9. Codd, E. F., Relational Database: A Practical Foundation for Productivity, *Comm. Assn. for Computing Machinery,* Vol. 25 (1982), pp. 109–117.
10. Perris, F. A., Imperial Chemical Industries Ltd., personal communication, 1981.
11. Cherry, D. H., Grogan, J. C., Knapp, G. L., and Perris, F. A., Use of Data Bases in Engineering Design, *Chem. Eng. Prog.,* Vol. 78 (1982), pp. 59–67.
12. Rosenblatt, A., 1982 Award for Achievement, *Electronics,* Oct. 20, 1982, pp. 109–111.
13. Weiss, H. M., INGRES: A Data-Management System for Minis, *Mini-Micro Systems,* January 1982, pp. 231–237.


## The author

**R. L. Motard** is Professor and Chairman of the Dept. of Chemical Engineering at Washington University, St. Louis, MO 63130, where his research interests include process synthesis and design-data management. He taught at the University of Houston for 21 years, and served as a research engineer at Shell Oil Co. before that. He received his D.Sc. degree from Carnegie-Mellon University. Dr. Motard developed the CHESS flowsheeting package, which is used to train students in engineering design. He is a member of AIChE, and a trustee of CACHE Corp.

# Computer horizons in process-systems engineering

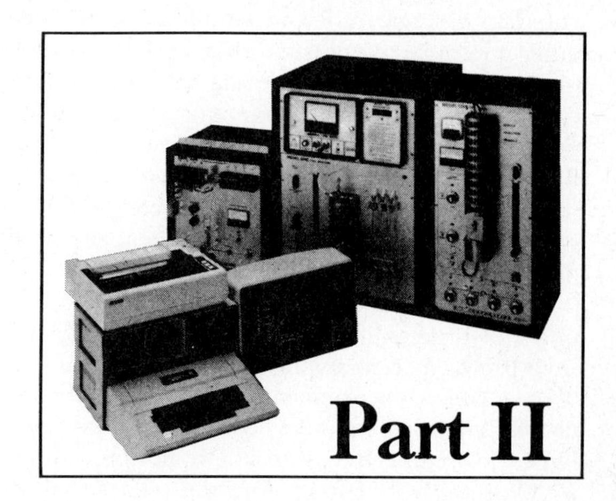


**Advances in computer technology will result in hardware and software better suited to the problems of process synthesis, analysis and design.**

*R. L. Motard, Washington University**

☐ Process-systems engineering typically involves a number of people, large amounts of data, and extensive calculations. Engineers already enjoy the speed and power of computerized calculation, but advances in communication and data-handling are not as well developed or widely applied.

In the next few years, however, advanced communication and data-handling capability—as well as greater computing power—should be available at the engineer's desk. Farther out, "expert" systems that can solve problems without being given a step-by-step method may become available.

## Hardware linked in networks

Most of today's computers have the traditional von Neumann architecture, in which processing proceeds in a single sequence, and are designed for general applications. Future hardware, in contrast, is likely to have a distributed architecture—either parallel or systolic-array. Also, special-purpose computers for process-systems engineering will be feasible when computer architects fully apply computer design aids to special-purpose VLSI (very-large-scale-integrated) chips. Such special-purpose hardware will be integrated with general-purpose computers, graphics systems, storage, input/output devices and other elements in the computer system.

The workstation linked to others in a network will be the means of access to computing power and data. A local network can support a great array of computers, storage devices and peripherals, so a single workstation having access could call on considerable computing power and other resources. The workstation may be in the office, but access from home or other locations will also be possible.

Communication via cable or satellite will combine voice, message, data and video signals; local networks will be connected to national and international ones by such channels. Not only will computing power be distributed, but the notion of an engineering enterprise being located in one place will become irrelevant when people can work together through telecommunication.

Large-scale data storage is likely to become less expensive with the advent of new media such as laser-optical videodisks. Anticipated recording density for such a disk is about 3 billion bytes (about 750,000 text pages) per side [*1*]. This means that whole libraries of reports, design data and instructional materials could be stored compactly and retrieved quickly.

The VLSI age will also provide affordable graphics processors; color displays with 1,024 × 1,024-pixel (dot) or better resolution will grace every engineer's workstation; and engineers will be able to communicate with process-systems software via electronic sketchpads. Distributed service centers will provide access to computers capable of 100 million instructions per second, but most computing will be done within the local network.

## Software in silicon

Process-systems software will be highly modular, data-independent, and structured for ease of maintenance or modification. Through networking, it will be supported on personal workstations as well as large host computers. Such software will make liberal use of building-block architecture, and some subsystems will be embedded in special-purpose VLSI chips. Here, the distinction between hardware and software is blurred. Operating systems, graphics processors, language compilers, and database-management systems will probably be available as chips.

Intermediate-level process-design operations may also be cast in silicon. Candidates include physical-property determination and two- and three-phase equilibrium calculation. Another chip, having the systolic-array architecture, might handle all matrix operations.

Originally published March 7, 1983

## Database management systems

Today, process-systems programs are frequently devoted to a single task and cannot be combined with other programs to operate on a single set of data. A key improvement will be the decoupling of programs from data via database-management-system (DBMS) software. This leads to long-term survival of programs, data-independent programming, lower software-maintenance costs, and higher programmer productivity.

How would this enhance the chemical engineer's productivity? In the first place, DBMS software would eliminate the need for coding and recoding data from one phase of process engineering to the next, which would save time, cut errors, and aid project management.

In a DBMS environment, any mix of old and new process-software modules could be assembled into larger programs by means of a nonprocedural software-assembly language. Or, the user could interact with modules piecemeal, as DBMS allows that option.

In a design project, it would be possible to work on one part of the project at a time, solving process problems bit by bit, without losing the coherence required for the overall effort. As parts of the design are completed, they are added to the database and made available to other designers. Discipline specialists could be alerted to changes as they occur, without the delay inherent in bureaucratic organizations. Reports could be generated quickly, and based on up-to-date information. There would be no need to stockpile printouts, since reports could be tailored and produced as needed.

Such a system would really be a computer aid to the process-systems engineer—supporting rather than inhibiting experimentation and case studies; continuously recording data traffic; filing multiple examples; and providing a way to compile and transmit information.

With regard to hardware systems, the DBMS environment offers great flexibility in distributing data and computing power in a computer network, and great ability to recombine decentralized information. With such support, the distributed approach to large-scale design projects becomes the natural mode of execution.

## Process-flowsheeting software

The architecture of process-flowsheeting systems will need redefinition to fit the computer technology of the future. In one respect at least, flowsheeting packages will be easier to develop: They will no longer have the internal data manipulation that today's stand-alone flowsheeting systems do. MIT-ASPEN, which is likely to become a U.S. industrial standard in the near future, probably represents the last generation of stand-alone flowsheeting systems. The next generation will be databased; it may also partition flowsheeting problems to harness the power of non von Neumann computers.

We have long been burdened with the dichotomy between equation-oriented and simultaneous-modular flowsheeting, with a spectrum of alternative models in between the two. The distinction will begin to blur in the future, when a hierarchy of approaches might be executed on the same process-flowsheeting software.

At one level, the engineer will be able to use the simplest kind of sequential model for process synthesis and for screening alternative flowsheets. Another level might be a hybrid equation-solving approach, containing both equations and discrete process modules. At a third level, modular architectures and equations might be used interchangeably to optimize the process via the "infeasible path" algorithms now being developed [2].

## Expert systems

Expert procedure in process systems involves developing a structure heuristically, analyzing it, optimizing it, and finding its weaker points. From this knowledge, the structure can then be improved, using evolutionary rules. One school of thought prefers to keep the rules simple, for easy application via pencil and paper, but computerized expert systems provide ample scope for improving the simple logic of expert procedure.

The most important successes in computerizing rules of process synthesis, as Newell [3] has emphasized, have been in problem areas that lend themselves to uniform representation—e.g., heat-exchanger networks and unintegrated distillation sequences for sharp separations. There has also been substantial progress in applying computer-search techniques to organic-synthesis problems. Here, a uniform representation is available in the structural model of organic molecules.

Other areas of chemical-process knowledge may yield to uniform representation and thereby uniform procedure, but problem-specific knowledge is for the most part nonuniform. One alternative, as recently proposed for energy-integrated sharp distillations, is to decompose the problem into uniform parts—here, heat exchange and distillation [4].

The challenge for chemical engineers is to encode chemical-process knowledge in a form suitable for computerized decisionmaking. This is difficult because any decision about a chemical process usually has a global as well as local impact. So, as expert programming matures we will see a growing emphasis on adaptive-learning procedures from the field of artificial intelligence.

Chemical engineering, however, does not lend itself to the symbol-manipulation systems often used in artificial intelligence. Our knowledge base requires extensive computation to extract meaningful information: Every sensitive variable must be evaluated in the context of mixture-dependent properties; every reaction phenomenon depends on the environment.

Thus, the expert systems of chemical engineering will have to be supported by substantial computation, and pure list-processing languages will not suffice. Nevertheless, the capabilities of relational DBMS, of VLSI chips tuned to the math of chemical-processing systems, and of multimode flowsheeting systems, raise exciting prospects for computer-aided process design.

## References

1. Goldstein, Charles M., Optical Disk Technology and Information, *Science*, Vol. 215, 1982, pp. 862–868.
2. Biegler, L., and Hughes, R. R., Infeasible Path Optimization with Sequential Modular Simulators, Paper 50a at AIChE Annual Meeting, Nov. 1981.
3. Newell, Allan, How to View the Computer, in Mah, R. S. H., and Sieder, W. D., eds., "Foundations of Computer-Aided Process Design," Vol. I, AIChE, New York, 1981, pp. 1–25.
4. Stephanopoulos, G., Linnhoff, B., and Sophos, A., Synthesis of Heat Integrated Distillation Sequences, The Institution of Chemical Engineers Symposium Series No. 74, London, 1982, pp. 111–130.

# Microcomputers in energy management—I

**Small general-purpose microcomputers can do most of the calculations and produce many of the reports that take up so much of the energy manager's time.**

*Philip S. Schmidt* , *The University of Texas at Austin*

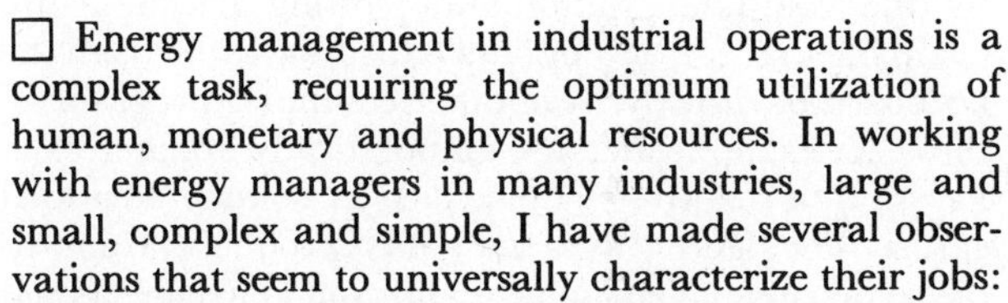

☐ Energy management in industrial operations is a complex task, requiring the optimum utilization of human, monetary and physical resources. In working with energy managers in many industries, large and small, complex and simple, I have made several observations that seem to universally characterize their jobs:

*Observation 1*—Limits on their time and that of their technical staffs are often a major impediment to rapid implementation of energy conservation measures.

*Observation 2*—Much energy management and engineering time is spent in repetitive computational tasks, such as the compilation of data for reports to management and government, financial analysis of project alternatives, and estimation of energy losses in existing and proposed systems (in order to justify expenditures to reduce the losses).

*Observation 3*—Access to large central computers is often difficult, and sometimes impossible. In small or relatively simple plants, there may be no computer at all. In large, sophisticated operations, the bureaucracy of computer management may be such that it is too inconvenient and time-consuming to use the central computer when needed.

The extraordinary rise of the microcomputer industry over the last two years has placed computer capability of significant power in a price range comparable to other individual office equipment. To date, however, the application of this hardware to energy management has been virtually nonexistent. This article will discuss some of the areas in which microcomputers can be used to make the energy manager's job easier. Part II will describe hardware and software available now or in the near future for energy conservation applications.

## Capabilities of modern microcomputers

Before going into the specifics of applying microcomputers to energy management, it will be well to outline

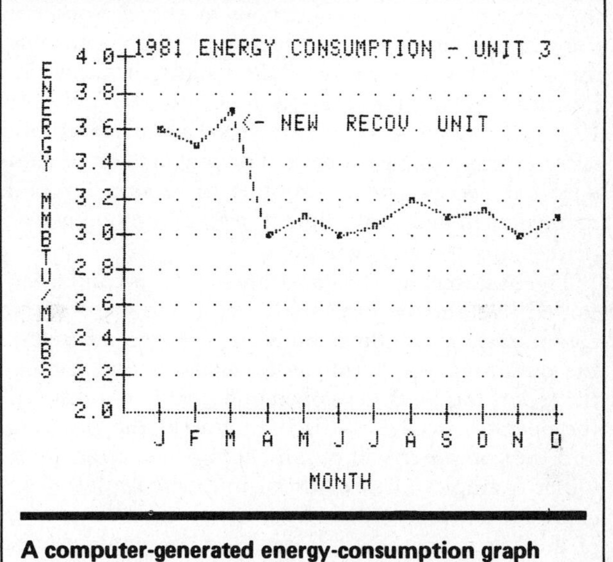


**A computer-generated energy-consumption graph**

what microcomputers can and cannot do. First, it is essential to dispel the commonly held notion that microcomputers are nothing more than overgrown desk calculators.

A programmable calculator may provide storage for about 10 numerical variables and handle simple programs of perhaps 100 or so computational steps in length. By comparison, a modern microcomputer can access hundreds of variables, using programs 1,000 or more lines long. For example, 100,000 to 200,000 data points or lengthy calculation procedures, or both, can be permanently stored on small magnetic disks and retrieved quickly, even in the midst of a program. Today, $2,000 will put on the engineer's desk a computer comparable to, or better than, machines that filled an office and cost $50–60,000 five years ago.

Another important feature of microcomputers is their ability to communicate with the user in plain English. BASIC (for Beginner's All-purpose Symbolic Instruction Code), the language common to virtually all microcomputers on the market, is "conversational" in nature. That is, the computer can request information from the user in easy-to-understand terms, process the information, ask for a decision or more data, and produce an easily readable result.

BASIC can process words as well as numbers. For example, the computer may ask, "WHAT IS THE OUTSIDE AIR TEMPERATURE?" when it needs a piece of numerical data. Having processed the data and produced an answer, it may query, "DO YOU WISH A FULL REPORT OR A SUMMARY?, and then print the

Originally published June 14, 1982

result accordingly, based on whether the operator types in "full report" or "summary."

Most of the microcomputers now on the market can provide output in graphic as well as numerical and textual form. Thus, it is possible to produce finished tables and graphs suitable for direct inclusion in reports. An example will be given later.

## Microcomputers in energy management

The powerful capabilities of microcomputers open up a broad range of applications in energy conservation-engineering and management. Perhaps the most useful is the performance of the lengthy, routine and repetitive calculations that often absorb so much time. Reporting seems to be the universal albatross around every energy manager's neck. The preparation of standardized energy and environmental reports for both management and government regulatory bodies is a perfect task for microcomputers.

The basic technical data for these reports can be submitted in standard form by a technician or clerk. No computer-programming knowledge is necessary once the program to perform the calculations and compile the report has been developed and saved on a disk. All the operator need know is the name of the program, and the computer will prompt him or her from there. Other examples of this type of routine calculation include fuel, steam, and air-flow metering, and estimates of fluid-flow and heat-transfer losses in plant systems.

The ability to permanently store data in both numerical and textual form can also be a valuable asset. For example, energy-consumption data for a number of plant units can be stored, and new data can be continually added to the data file. Daily, weekly or monthly energy-consumption plots can be generated and automatically checked against target values or normal operating limits.

This storage capability can also be used to eliminate duplication of effort. For example, basic unit operating data can be input to a microcomputer at the operating department level to produce the detailed energy reports needed by unit engineers and operators. The departmental results or unit results can be stored on another disk, and sent to the plant or corporate energy-conservation coordinator, where they can be read directly into a corporate summary program without requiring any additional manual input.

Data-base management programs commercially available for microcomputers will rapidly access and sort large amounts of directory-type information on disks. These systems have a virtually unlimited range of application in energy management. The data-base management system permits the user to find all data meeting a particular specification, e.g., all units within 25% of their annual energy conservation targets, or all days on which the boiler efficiency was under 70%. Most of the available programs for this purpose are flexible enough to allow the data to be set up in any desired format and to be sorted or retrieved in many different ways.

Because much of an energy manager's time is spent in written and graphic communication, this discussion of microcomputers would be incomplete without mention of their capabilities for word-processing and graphic output. Excellent programs are available that permit the computer keyboard and the video display to be used as an electronic typewriter, comparable to dedicated word-processing systems costing thousands of dollars. These systems allow for easy editing of text on the video display to correct errors, change wording, and even move sentences and paragraphs around at will. About 50 to 100 pages of type can be saved on one disk and retrieved later for editing.

While few energy managers are interested in taking over their secretaries' jobs, this added capability can be a tremendous time-saver in producing draft letters, memoranda, and reports; error-free typing is not really necessary since corrections are so easily made.

I find that I can produce drafts of written material about four to five times faster this way than by writing in longhand. Editing is easy and fast. Final production of the resulting report is greatly facilitated if a printer is available to produce it directly from the computer. (It should be added that lack of typing skill is no excuse; a program to teach touch-typing on the computer can be bought for about $25!)

The figure illustrates a graph of monthly energy consumption for a typical plant production unit, printed by the computer. Automatic plotting packages are available that can accept data either manually from the keyboard or directly from other computer programs—those, for example, that are automatically updated each month in a reporting system such as the one described earlier. The figure was actually first produced in multiple colors on a video monitor and could have been used to produce color slides for presentation.

Microcomputers do, of course, have their limitations. They are not well suited for storing and rapidly accessing very large data bases (those having more than a few thousand pieces of data), such as detailed long-term corporate energy and production records for a large organization. Most of the small microcomputers currently on the market use 8-bit microprocessors (eight binary digits per data unit). They have inherent limitations on speed (about 1 million operations per second) and accuracy (about 8 significant figures), which make them unsuitable for complex numerical operations such as the large finite-difference and finite-element codes that are used in some fluid-mechanics and structural calculations.

These limitations generally present no problem in the more routine kinds of calculations involved in energy management, where an extremely high degree of numerical precision is usually not critical. It should be added that interfaces are available for most microcomputers that allow them to be coupled (via telephone line) to large time-sharing computers. If, however, access to large data bases or complex numerical codes is a primary requirement, the user will probably find that a conventional computer terminal, which does not have the stand-alone computational capability of a microcomputer, is more cost effective.

# Microcomputers in energy management—II

**Here is what you will need in the way of hardware and software to employ microcomputers for energy management.**

*Philip S. Schmidt, The University of Texas at Austin*

☐ The previous article explained the advantages of microcomputers to the energy manager. Now we will consider the equipment and programs that will be needed.

The primary differences between microcomputers are in the amount of programmable memory internal to the unit (called random access memory, or RAM), the conveniences built into the resident language, and the degree of compatibility with external devices. For energy management applications, 32 kilobytes (32K) of RAM should be considered a minimum, and 48K is highly desirable. (A byte is 8 bits, the amount of memory required to store a single letter or number.)

The incremental cost of random access memory is so low that it is wise to purchase the maximum RAM available for the particular machine. This is typically 32K, 48K, or 64K, but machines are now coming on the market with 128K or more. In actuality, 48K is adequate for most of the tasks of an energy manager, and problems requiring much more than that will probably justify solution on a large mainframe computer.

Another important feature of the computer is which programming language it uses. All of the microcomputers now available use a version of the computer language called BASIC. One of the most significant differences between the various versions of BASIC is in the handling of graphic output. High-resolution graphics can be quite convenient for the production of data displays; it is unnecessary for the generation of standard numerical and textual tables. It is important that the version of BASIC used be able to handle decimal data (so-called "floating point" BASIC). While this is available for almost all machines, it is an item that should be explicitly specified, since some models incorporate integer BASIC as the standard language.

## Input and output devices

A computer, of course, is useless unless it can communicate information easily to the user. All general-purpose microcomputers use a computer-terminal keyboard, almost identical to a typewriter, as the standard input device. These keyboards differ only slightly from machine to machine, and the primary requirement is that they be rugged and reliable. Typewriter-style button keyboards are greatly preferable to the flat capacitance-type boards—found on some of the cheaper machines—for inputting commands to the computer.

Two forms of output will generally be required in energy management applications: a good video display, and a printer to generate permanent copy.

The potential buyer should look at several displays in actual operation, since there is a great deal of variation in the resolution available from different monitors and TV receivers. This can make a significant difference in terms of readability and eye fatigue.

A good printer should be considered a must for energy management purposes. While it is technically possible to transcribe data by hand from the screen, the cost of a printer will quickly be recovered in time savings and convenience.

Two basic types of printer are available, the "dot matrix"-type, and the typewriter-type, which usually incorporates a daisy-wheel printing element. The first uses a matrix of dots to form the characters, while the

Originally published July 12, 1982

second forms each character directly by striking the page with a metal or plastic element. Dot-matrix printers, using either thermal or mechanical printheads, are generally much less expensive than typewriter-quality printers but the characters are obviously printed by computer. For preparation of reports in finished form, the typewriter-quality printer is more satisfactory.

Regardless of the printer used, a suitable interface will be required to allow it to exchange data properly with the computer. The price of the interface circuit board is often not included in the printer price itself, and the buyer should be certain that a suitable interface is available for the particular computer being considered and is included in the purchase specifications.

## Mass storage

The random access memory (RAM) residing within the computer is "volatile," i.e., it will retain data and program instructions only so long as the power is on; when the computer is turned off, all information in memory is lost. Some means must be provided for permanent storage of programs and data. This "mass storage" takes the form of magnetic media, such as tape cassettes or disks. For energy-management purposes, the "floppy disk" will generally be found most suitable. A program that might typically take a minute or more to load or unload from a tape cassette will require about 3 to 4 s from a small $5\frac{1}{4}$-in. disk. Furthermore, data or programs on disk can be retrieved randomly, i.e., from any part of the disk, even in the course of executing a program. Thus, data may be retrieved from a file too large to store in RAM, and used in the computation. This flexibility is not available with most tape-cassette storage devices.

For most purposes, one single-disk drive (the device that reads and writes data on the disk), with a suitable interface for communication with the computer, is adequate. A second disk drive, which can usually be connected to the same interface board, is convenient for such things as reproducing backup copies of critical data or programs but is not an absolute necessity. Copies can be made with one drive, but at the expense of time and convenience.

## Other useful accessories

A variety of other features and equipment are available to facilitate data handling, expand the range of programmability, and permit communication with other computers. Several of the most popular microcomputer systems can be adapted to be programmed in FORTRAN, a standard scientific-and-technical language, and COBOL, a business-oriented one.

Several graphic input devices allow data to be directly input from graphs or drawings. For some computers, graphic output devices are also available. Additional disk drives can be added to minimize manual handling of disks, and "hard" disks have recently come on the market that permit storage and accessing of very large quantities of data (typically 5 to 10 megabytes).

In companies that already have large time-sharing computers online, a microcomputer can be used as a "smart" terminal to communicate with the main computer for access to databases or for performing complex calculations. Plug-in boards are available that can be connected to the mainframe computer either through an acoustic coupler or a direct wire, to send and receive data through telephone lines.

Most of these accessories are not really necessary in fulfilling the basic requirements of energy-management computing. It is recommended that the new user start with a basic system—computer, video display, printer, and one or two disk drives, and gain some experience to determine what additional features could be worthwhile. As will be emphasized in the next section, good software is a more valuable investment than many glamorous but less-useful hardware items.

## Software for energy management

The most elaborate computer system in the world is worthless without programs to tailor it to the job at hand. The author recently made a study of the market to determine what was available in the way of programs (in computer jargon, "software") oriented to the specific requirements of industrial energy management. The answer, in a word, was nothing.

A survey was conducted of members of the Gulf Coast Energy Conservation Soc., a group of industrial energy managers and engineers in the Houston area, to determine what programs were considered most useful in their work. Following up on this survey, a project was undertaken at the Center for Energy Studies at the University of Texas at Austin to develop a package of programs especially aimed at energy-management applications. Partial support for the project was provided by the Texas Energy and Natural Resources Advisory Council. The programs developed in this project are described briefly below:

*Combustion calculations*—Computation of boiler and heater efficiency and stack emissions for a variety of fuels and firing conditions.

*Economic analysis of energy-conservation projects*—A discounted-cash-flow comparison calculation for alternative projects, taking into account energy costs and escalation rates, capital and labor costs, various financing and depreciation options, and differing profiles of equipment utilization (capacity versus hours used).

*Heat-loss calculations*—Estimates of heat loss by radiation and convection from pipes and flat surfaces, accounting for temperature drops through layers of insulating material as well as bare surfaces.

*Evaluation of alternative mechanical drives*—Sizing, cost estimation, and comparative economic analysis of steam turbines and electric motors to meet given horsepower requirements.

*Fluid losses in pipes*—Determination of pressure drop and pumping horsepower for the flow of fluids in piping runs, including losses in standard fittings.

*Fluid network analysis*—Determination of flow, heat loss, and pumping power in piping runs, including losses in standard fittings.

*Curve-fitting of data*—Determination of formulas to approximate a given set of data, using any of several mathematical models (e.g., logarithmic, linear, inverse).

*Steam properties*—Determination of steam enthalpy, entropy and specific volume under any given set of thermodynamic conditions.

**Typical microcomputer-system costs (early 1982)**

| | |
|---|---|
| **Computers (including video monitor)** | |
| Brand A: 48K RAM, 1 disk drive | $2,100 |
| Brand B: 32K RAM, 1 disk drive | 2,150 |
| Brand C: 32K RAM, 2 disk drives | 2,600 |
| Brand D: 32K RAM, 1 tape drive | 2,750 |
| Brand E: 64K RAM, 2 disk drives | 1,800 |
| **Printers (including interface boards)** | |
| Dot-matrix type (thermal or mechanical) | 400-900 |
| Daisy-wheel type (letter quality) | 1,500-3,000 |
| **Other accessories** | |
| Additional floppy-disk drive | 300-500 |
| Graphic input board | 700-1,500 |
| Hard-disk drive | 3,000-6,000 |
| **Software (per program)** | 25-600 |

*Heat-exchanger design*—Estimation of size and cost of a shell-and-tube heat exchanger to meet a given set of steam heat-transfer requirements.

*Backpressure turbine/generator feasibility analysis*—Determination of steam flow requirements, turbine/generator cost, and overall make-or-buy economics of in-plant cogeneration between a given set of inlet and exhaust steam conditions.

The software described above is available for distribution through the Center for Energy Studies at the University of Texas at Austin.

All of the programs are written in BASIC for the Apple II microcomputer. Detailed documentation is provided that should permit easy adaptation to any other microcomputer using a version of BASIC. The programs are "conversational" in character; they prompt the user for input information, and require no computer programming knowledge. Data can be provided to a technician or clerk on standard input forms and run with no intervention by an engineer or computer programmer.

Most of the required technical data, such as thermodynamic or material properties, are built into the programs; and in those cases where external technical-data input is required, the necessary tables are provided in the program documentation—no time should be required to hunt down input information in handbooks or technical journals. The programs also help the operator avoid mistakes, by checking for internal consistency of the data. For example, stack-gas components in a boiler test must add up to 100%. If not, the computer prompts the user and asks that the data be reentered.

Some general-purpose programs developed by commercial software houses are also available for use in energy management. The graph shown in the previous article, for example, was generated by a general plotting program. Some very powerful tabular calculation procedures have also been developed that permit the user to set up any repetitive calculation in a matrix format. While originally conceived for financial analysis, these can often be used for such purposes as energy-reporting systems, by setting up the reporting scheme in a format compatible with the program.

## Microcomputer system costs

The microcomputer market is characterized by tremendous diversity in costs and capabilities. Technical development in this area can only be described as spectacular and, contrary to the energy situation, costs seem to be dropping almost daily, rather than rising. The result is that the energy manager's computing needs can now be met with table-top systems well within the price range of even the smallest organization.

The table shows some representative cost figures for several of the most popular systems on the market as of early 1982. Because each of these systems has "standard" features that are optional on the others, a precise point-by-point cost comparison is difficult. In general, however, the table shows that the hardware for the basic system discussed earlier—including the computer, a video display, and one or two disk drives—will typically run in the $2,000–$2,500 price range. Depending on the type of printer selected, the complete system hardware will range from $3,000–$6,000.

Software is also quite varied in price, but from an industrial user's viewpoint, it is remarkably cheap. A good program, even one costing several hundred dollars, can easily pay for itself in a single use. It is not difficult to estimate the payback potential of a microcomputer system. One costing $4,500 will pay for itself in 18 weeks if it saves a $50/h energy manager five hours a week—a handsome rate of return.

A question often asked is: "Why shouldn't I wait a year and get the same product at half the cost?" As with any energy conservation investment, a year can represent literally millions of dollars in lost energy-saving opportunities. When computer systems were priced in the tens or hundreds of thousands, the question was a legitimate one. In today's microcomputer market, however, putting off a low-cost investment in the hopes of securing an even lower-cost one in the future is false economy. When the cost of lost time is considered in the same way as the cost of lost production, the answer is obvious.

## Acknowledgement

The author wishes to express his appreciation to the Texas Energy and Natural Resources Advisory Council for support of the software development project described above, and to Ms. Maggie Keeshen of the University of Texas at Austin for her expert assistance in preparing this article for publication.

## The author

**Philip S. Schmidt** is professor of mechanical engineering, University of Texas at Austin, Austin, TX 78712. He is presently on sabbatical leave at Electric Power Research Institute, Palo Alto, Calif. (until Aug. 1982), carrying out a planning study on industrial electrification. He has been a consultant to several companies on energy conservation and on developing computer software for this purpose. He holds a B.S. from M.I.T. and an M.S. and Ph.D. from Stanford University, all in mechanical engineering.

# Section II
# Engineering Mathematics

# Program correlates data

A program written for use with the Apple II computer, fits $(x,y)$ pairs to linear, exponential, natural-log and power-curve models, and computes correlation coefficients for each.

☐ Often, we want to find some relationship between two sets of numbers, so that interpolation and extrapolation can be done. If there is enough information to draw a graph, interpolation is possible. However, extrapolation cannot be accomplished unless the relationship between $x$ and $y$ is known.

This relationship is, in fact, often found by plotting data on various graphs—i.e., linear, log-log—to see which one gives the best fit. Here we will present a program that uses a similar approach, but does so numerically on the Apple II.

## Program approach

The method uses linear regression analysis to find the best fit of the data to a curve. Four curves are used, and the data can be fit to each. The program determines $r$, the correlation coefficient, for each model. The curve that fits best has the highest absolute value of $r$. The equations to which data are fit are:

| | *Equation* | *Code* | |
|---|---|---|---|
| Straight line | $y = a + bx$ | 1 | (1) |
| Exponential curve | $y = ae^{bx}$ | 2 | (2) |
| Log curve | $y = a + b \log x$ | 3 | (3) |
| Power curve | $y = ax^b$ | 4 | (4) |

Explanation of the "code" is given later on.

The data do not have to be entered separately for each curve. Any required transformation is done by the program, thus making errors unlikely.

Eq. (2) and (4) are linearized as follows:

$$y = ae^{bx} \qquad \ln y = \ln a + bx \tag{5}$$

$$y = ax^b \qquad \ln y = \ln a + b \ln x \tag{6}$$

The program accumulates the required sums of $x$ and $y$ or functions of them, and then uses the internal linear-regression-functions to handle the computations.

Once the data have been entered, the values of $r$ can be found for Eq. (1) through (4). The equation of the best fit is then selected.

Note that three of the equations use logs of one or more variables, and that this can place limits on $x$ and $y$. There are no restrictions in Eq. (1). In Eq. (2), $y$ must be positive; in Eq. (3), $x$ must be positive; and in Eq. (4), $x$ and $y$ must both be positive.

Should negative values be used for the forementioned variables, the user is prompted to either "USE SAME DATA FOR ANOTHER EQUATION," "ENTER DIFFERENT DATA," or "END THE PROGRAM".

The codes listed in the above equations (which are the same as the equation number) are used to identify which model is currently being used. The program displays the model's code and specific equation in the first display. This will be demonstrated in the example. The program is given in Table I.

## Example

Given the following data, select the best-fitting curve. A printout of the results is given in Table II. The data are:

$$x = 10, y = 0.1007$$
$$x = 20, y = 0.1790$$
$$x = 40, y = 0.2960$$

The program is interactive. As can be seen in Table II, the first display indicates what the program can do, as well as asking the user to select (by number) a correlation equation. For each of the four correlation models, the program will supply the values of $a$, $b$, and $r$ (the correlation coefficient). Once the correlation equation is calculated, new values of $x$ or $y$ can be determined for that specific curve fit.

The data are saved in the computer's memory, allowing more than one correlation to be run with the same set of data. Further, by using the correlation equations each in turn, the user can determine the best-fit curve—the correlation have the highest value of $r$.

*Entering data*—Once a correlation curve is selected (by inputing that equation's code number), the second display requests that data be entered in $x,y$ pairs (separated by a comma). When all the data are inputed, the user is asked to enter "END,0".

*Changing data*—At this point, there is an opportunity to check the accuracy of the entered data, and make any necessary changes. Any change is made by noting the row and column of the wrong data and then the new value. In the example, this is shown by entering "3,X,40" to change the $x$ value in the third row from "30" to the correct value of "40". The user is then once again asked if there are any further changes necessary. If the answer is "no", input "N". But, if there are further changes to be made, input "Y", and make the variable changes by row, $x$- or $y$- column, and new value of $x$ or $y$. Once all the correct data are entered, the user inputs no change in data by putting in "0,0,0".

**Adapted from an article by C. S. Payne, Thames Valley (England) Police, originally published July 28, 1980.**

*Correlation values*—After the data are entered, the next display supplies the correlation constant ($r$ = 0.9971 in the given example for the straight line), and values for the equation coefficients (i.e., $a = 0.0422$ and $b = 0.0064$ or "6.4E-03", as shown).

*Program options*—At this point, the user may (1) estimate a new value for $y$ for a given $x$, using this correlation equation; (2) estimate a new value for $x$ for a given $y$; (3) use the entered data for a new correlation equation; (4) enter new data; or, (5) end the program.

*Best fit*—By selecting the third option above (using the entered data for a new correlation equation), the user can determine the best-fit curve—the equation having the highest value for the correlation coefficient. As can be seen in Table II, the correlation coefficient for the power curve ($r = 0.9993$) has the highest value of the four equations in this program. Accordingly, this equation (code number 4, $y = ax^b$ or "Y = A*X B") is the curve that best fits the data supplied.

*New values*—Using the power-factor equation as shown in the example, new values for $y$ for a given $x$ can be quickly estimated, by entering "1" as the program option when asked to "MAKE YOUR CHOICE BY NUMBER". In this instance, upon entering a value of $x$ of "15", the next display will be "YOUR ESTIMATE OF Y IS .1397".

## The author

**C. S. Payne** is a police constable for the Thames Valley Police, Room 8, Single Men's Quarters, Slough Police Station, Slough, Berkshire, U.K. He studied mathematics at Exeter College, Oxford University. Officer Payne graduated from Oxford in 1979, and has been in his present job since then. He is a member of the British Texas Instruments Users Club.

**Program fits (*x*, *y*) pairs to any of four models: linear, exponential, natural log or power curve** **Table I**

```
10  REM  PROGRAM CORRELATES DATA
20  REM  FROM CHEMICAL ENGINEERING, JULY 28, 1980
30  REM  BY C. S. PAYNE
40  REM  TRANSLATED BY WILLIAM VOLK
50  REM   COPYRIGHT (C) 1984
60  REM  BY CHEMICAL ENGINEERING
70  CLEAR
80  DIM X(100),Y(100),XI(100),YI(100)
90  REM  SET DISPLAY TO FOUR DECIMALS.
100  DEF  FN P(X) =  INT (1E4 * (X + .00005)) / 1E4
110  HOME : PRINT
120  PRINT "WITH INPUT OF PAIRS OF X-Y DATA, THE"
130  PRINT "PROGRAM CALCULATES ANY OF THE FOLOWING"
140  PRINT "CORRELATING EQUATIONS:"
150  PRINT
160  PRINT "1. Y = A + B * X"
170  PRINT "2. Y = A * EXP(B * X)"
180  PRINT "3. Y = A + B * LOG(X)"
190  PRINT "4. Y = A * X ^ B"
200  PRINT
210  IF F1 = 1 GOTO 320
220  PRINT "THE PROGRAM GIVES THE VALUES OF A, B,"
230  PRINT "AND THE CORRELATION COEFFICIENT, R."
240  PRINT
250  PRINT "AFTER THE EQUATION IS CALCULATED, THE"
260  PRINT "PROGRAM PERMITS CALCULATION OF Y FROM"
270  PRINT "INPUT VALUE OF X, OR X FROM Y."
280  PRINT
290  PRINT "THE DATA ARE SAVED IN COMPUTER MEMORY"
300  PRINT "SO THAT MORE THAN ONE CORRELATION MAY"
310  PRINT "BE RUN WITH ONE SET OF DATA INPUT."
320  PRINT
330  INPUT "SELECT YOUR EQUATION BY NUMBER.   ";OP
340  PRINT
350  IF OP < 1 OR OP > 4 GOTO 370
360  GOTO 400
370  PRINT "YOUR CHOICES WERE 1 TO 4, YOU ENTERED"
380  PRINT OP;".  TRY AGAIN."
390  GOTO 320
400  IF F1 = 1 GOTO 760
410  HOME : VTAB 8
420  PRINT "ENTER THE DATA IN PAIRS: X,Y"
430  PRINT "SEPARATED WITH A COMMA."
440  PRINT
450  PRINT "WHEN ALL THE DATA ARE IN, ENTER"
460  PRINT "'END,0'."
470  PRINT
480  PRINT "                          X     Y"
490  PRINT
500  PRINT "ENTER FIRST PAIR NOW   ";N + 1;: INPUT "     ";
     X$,Y
510 N = N + 1
520  GOTO 550
530  PRINT "NEXT PAIR OR: 'END,0'   ";N + 1;: INPUT "     ";
     X$,Y
540  GOTO 510
550  IF X$ = "END" GOTO 610
560  REM  INPUT DATA ARE SAVED AS XI(I) AND YI(I)
570  REM  CALCULATIONS ARE MADE WITH X(I) AND Y(I) AS
     FUNCTIONS OF INPUT DATA
580 XI(N) =  VAL (X$)
590 YI(N) = Y
600  GOTO 530
610 N = N - 1
620  PRINT
630  PRINT "CHECK THE DATA AND MAKE ANY CHANGES"
640  PRINT "BY ROW AND COLUMN:  '3,Y,45.3'  WILL"
650  PRINT "CHANGE THE THIRD Y VALUE FROM ";YI(3);" TO"
660  PRINT "45.3"
670  PRINT "IF THERE ARE NO CHANGES, ENTER '0,0,0' ."
680  INPUT "ENTER YOUR CHANGES NOW ";A1,B$,C1
690  IF A1 = 0 GOTO 760
```

**Program fits (*x,y*) pairs to any of four models: linear, exponential, natural log or power curve (continued) Table I**

```
700  IF B$ = "Y" GOTO 730
710 XI(A1) = C1
720  GOTO 740
730 YI(A1) = C1
740  INPUT "ANY OTHER CHANGES?  ANSWER Y OR N ";X$
750  IF X$ = "Y" GOTO 670
760  FOR I = 1 TO N
770  ON OP GOTO 780,810,920,1030
780 X(I) = XI(I)
790 Y(I) = YI(I)
800  GOTO 1150
810 X(I) = XI(I)
820  HOME : VTAB 4
830  IF YI(I) > 0 GOTO 900
840  PRINT
850  PRINT "Y(";I;") = ";YI(I)
860  PRINT
870  PRINT "EXPONENT FUNCTION CAN NOT BE USED."
880  PRINT
890  GOTO 1970
900 Y(I) =  LOG (YI(I))
910  GOTO 1150
920  IF XI(I) > 0 GOTO 1000
930  HOME : VTAB 4
940  PRINT
950  PRINT "X(";I;") = ";XI(I)
960  PRINT
970  PRINT "LOG FUNCTION CAN NOT BE USED."
980  PRINT
990  GOTO 1970
1000 X(I) =  LOG (XI(I))
1010 Y(I) = YI(I)
1020  GOTO 1150
1030  IF YI(I) > 0 AND XI(I) > 0 GOTO 1120
1040  HOME : VTAB 4
1050  IF YI(I) > 0 GOTO 1100
1060  PRINT "Y(";I;") = "YI(I)
1070  PRINT
1080  PRINT "POWER FUNCTION CAN NOT BE USED."
1090  GOTO 1970
1100  PRINT "X(";I;") = ";XI(I)
1110  GOTO 1070
1120 X(I) =  LOG (XI(I))
1130 Y(I) =  LOG (YI(I))
1140  REM  SUMS AND SUMS OF SQUARES
1150 SX = SX + X(I)
1160 XQ = XQ + X(I) ^ 2
1170 SY = SY + Y(I)
1180 YQ = YQ + Y(I) ^ 2
1190 XY = XY + X(I) * Y(I)
1200  NEXT I
1210  REM  SUMS OF DEVIATIONS FROM MEANS
1220 XP = XQ - SX ^ 2 / N
1230 YP = YQ - SY ^ 2 / N
1240 PP = XY - SY * SX / N
1250  REM  CALCULATION OF EQUATION CONSTANTS.
1260 B = PP / XP
1270 A = SY / N - B * SX / N
1280  IF OP = 2 OR OP = 4 THEN A =  EXP (A)
1290 R =  SQR (B * PP / YP)
1300  HOME : VTAB 4
1310  PRINT "CORRELATION COEFFICIENT, R = "; FN P(R)
1320  PRINT
1330  ON OP GOTO 1340,1360,1380,1400
1340  PRINT "FOR EQUATION Y = A + B * X"
1350  GOTO 1410
1360  PRINT "FOR EQUATION Y = A * EXP(B * X)"
1370  GOTO 1410
1380  PRINT "FOR EQUATION Y = A + B * LOG(X)"
1390  GOTO 1410
1400  PRINT "FOR EQUATION Y = A * X ^ B"
1410  PRINT
1420  PRINT "  A = "; FN P(A)
1430  PRINT "  B = "; FN P(B)
1440  PRINT
1450  PRINT " YOU MAY:
1460  PRINT
1470  PRINT " 1. ESTIMATE A VALUE OF Y FROM X."
1480  PRINT " 2. ESTIMATE A VALUE OF X FROM Y."
1490  PRINT " 3. USE SAME DATA FOR ANOTHER EQUATION."
1500  PRINT " 4. ENTER DIFFERENT DATA."
1510  PRINT " 5. END THE PROGRAM."
1520  PRINT
1530  INPUT "MAKE YOUR CHOICE BY NUMBER.  ";CH
1540  PRINT
1550  IF CH < 0 OR CH > 5 GOTO 1570
1560  GOTO 1600
1570  PRINT "YOUR CHOICES WERE 1 TO 5, YOU ENTERED"
1580  PRINT CH;".  TRY AGAIN."
1590  GOTO 1440
1600  HOME : VTAB 4
1610  ON CH GOTO 1620,1740,1860,10,1930
1620  INPUT "ENTER YOUR VALUE OF X        ";X
1630  PRINT "YOUR ESTIMATE OF Y IS "
1640  ON OP GOTO 1650,1670,1690,1710
1650 Y = A + B * X
1660  GOTO 1720
1670 Y = A *  EXP (B * X)
1680  GOTO 1720
1690 Y = A + B *  LOG (X)
1700  GOTO 1720
1710 Y = A * X ^ B
1720  PRINT  TAB( 10); FN P(Y)
1730  GOTO 1440
1740  INPUT "ENTER YOUR VALUE OF Y        ";Y
1750  ON OP GOTO 1760,1780,1800,1820
1760 X = (Y - A) / B
1770  GOTO 1830
1780 X = ( LOG (Y) -  LOG (A)) / B
1790  GOTO 1830
1800 X =  EXP ((Y - A) / B)
1810  GOTO 1830
1820 X =  EXP (( LOG (Y) -  LOG (A)) / B)
1830  PRINT
```

**Program fits (*x,y*) pairs to any of four models: linear, exponential, natural log or power curve (continued) Table I**

```
1840  PRINT "YOUR ESTIMATE OF X IS        "; FN P(X)
1850  GOTO 1440
1860  HOME : PRINT
1870  PRINT "THE ORIGINAL DATA CAN BE USED AGAIN"
1880  PRINT "ON THE SAME EQUATIONS.  THESE ARE:
1890  PRINT
1900 F1 = 1
1910 SX = 0:XQ = 0:SY = 0:YQ = 0:XY = 0
1920  GOTO 150
1930  PRINT : VTAB 7
1940  PRINT
1950  PRINT  TAB( 13);"END OF PROGRAM"
1960  END
1970  PRINT : PRINT "YOU MAY EITHER:"
1980  PRINT : PRINT "1. USE SAME DATA FOR ANOTHER EQUATION,"
1990  PRINT "2. ENTER DIFFERENT DATA,"
2000  PRINT "3. END THE PROGRAM."
2010  PRINT : INPUT "MAKE YOUR CHOICE BY NUMBER.  ";CH
2020  ON CH GOTO 1860,10,1930
2030  END
```

**Example for: Program correlates data** **Table II**

**(Start of first display)**

```
WITH INPUT OF PAIRS OF X-Y DATA, THE
PROGRAM CALCULATES ANY OF THE FOLOWING
CORRELATING EQUATIONS:

1. Y = A + B * X
2. Y = A * EXP(B * X)
3. Y = A + B * LOG(X)
4. Y = A * X ^ B

THE PROGRAM GIVES THE VALUES OF A, B,
AND THE CORRELATION COEFFICIENT, R.

AFTER THE EQUATION IS CALCULATED, THE
PROGRAM PERMITS CALCULATION OF Y FROM
INPUT VALUE OF X, OR X FROM Y.

THE DATA ARE SAVED IN COMPUTER MEMORY
SO THAT MORE THAN ONE CORRELATION MAY
BE RUN WITH ONE SET OF DATA INPUT.

SELECT YOUR EQUATION BY NUMBER.   1
```

**(Start of next display)**

```
ENTER THE DATA IN PAIRS: X,Y
SEPARATED WITH A COMMA.

WHEN ALL THE DATA ARE IN, ENTER
'END,0'.

                                 X    Y

ENTER FIRST PAIR NOW   1    10,.1007
NEXT PAIR OR: 'END,0'  2    20,.179
NEXT PAIR OR: 'END,0'  3    30,.296
NEXT PAIR OR: 'END,0'  4    END,0

CHECK THE DATA AND MAKE ANY CHANGES
BY ROW AND COLUMN:  '3,Y,45.3'  WILL
CHANGE THE THIRD Y VALUE FROM .296 TO
45.3
IF THERE ARE NO CHANGES, ENTER '0,0,0' .
ENTER YOUR CHANGES NOW 3,X,40
ANY OTHER CHANGES?  ANSWER Y OR N N
```

**(Start of next display)**

```
CORRELATION COEFFICIENT, R = .9971

FOR EQUATION Y = A + B * X

  A = .0422
  B = 6.4E 03

 YOU MAY:

 1. ESTIMATE A VALUE OF Y FROM X.
 2. ESTIMATE A VALUE OF X FROM Y.
 3. USE SAME DATA FOR ANOTHER EQUATION.
 4. ENTER DIFFERENT DATA.
 5. END THE PROGRAM.

MAKE YOUR CHOICE BY NUMBER.  3
```

**(Start of next display)**

```
THE ORIGINAL DATA CAN BE USED AGAIN
ON THE SAME EQUATIONS.  THESE ARE:


1. Y = A + B * X
2. Y = A * EXP(B * X)
3. Y = A + B * LOG(X)
4. Y = A * X ^ B


SELECT YOUR EQUATION BY NUMBER.   2
```

**(Start of next display)**

```
CORRELATION COEFFICIENT, R = .9739

FOR EQUATION Y = A * EXP(B * X)
```

**Example for: Program correlates data (continued)** **Table II**

```
  A = .0783
  B = .0344

 YOU MAY:

 1. ESTIMATE A VALUE OF Y FROM X.
 2. ESTIMATE A VALUE OF X FROM Y.
 3. USE SAME DATA FOR ANOTHER EQUATION.
 4. ENTER DIFFERENT DATA.
 5. END THE PROGRAM.

MAKE YOUR CHOICE BY NUMBER.  3
```

**(Start of next display)**

```
THE ORIGINAL DATA CAN BE USED AGAIN
ON THE SAME EQUATIONS.  THESE ARE:


1. Y = A + B * X
2. Y = A * EXP(B * X)
3. Y = A + B * LOG(X)
4. Y = A * X ^ B


SELECT YOUR EQUATION BY NUMBER.   3
```

**(Start of next display)**

```
CORRELATION COEFFICIENT, R = .9935

FOR EQUATION Y = A + B * LOG(X)

  A = -.2301
  B = .1409

 YOU MAY:

 1. ESTIMATE A VALUE OF Y FROM X.
 2. ESTIMATE A VALUE OF X FROM Y.
 3. USE SAME DATA FOR ANOTHER EQUATION.
 4. ENTER DIFFERENT DATA.
 5. END THE PROGRAM.

MAKE YOUR CHOICE BY NUMBER.  3
```

**(Start of next display)**

```
THE ORIGINAL DATA CAN BE USED AGAIN
ON THE SAME EQUATIONS.  THESE ARE:


1. Y = A + B * X
2. Y = A * EXP(B * X)
3. Y = A + B * LOG(X)
4. Y = A * X ^ B


SELECT YOUR EQUATION BY NUMBER.   4
```

**(Start of next display)**

```
CORRELATION COEFFICIENT, R = .9993

FOR EQUATION Y = A * X ^ B

  A = .017
  B = .7778

 YOU MAY:

 1. ESTIMATE A VALUE OF Y FROM X.
 2. ESTIMATE A VALUE OF X FROM Y.
 3. USE SAME DATA FOR ANOTHER EQUATION.
 4. ENTER DIFFERENT DATA.
 5. END THE PROGRAM.

MAKE YOUR CHOICE BY NUMBER.  2
```

**(Start of next display)**

```
ENTER YOUR VALUE OF Y       .131

YOUR ESTIMATE OF X IS       13.8089

 YOU MAY:

 1. ESTIMATE A VALUE OF Y FROM X.
 2. ESTIMATE A VALUE OF X FROM Y.
 3. USE SAME DATA FOR ANOTHER EQUATION.
 4. ENTER DIFFERENT DATA.
 5. END THE PROGRAM.

MAKE YOUR CHOICE BY NUMBER.  1
```

**(Start of next display)**

```
ENTER YOUR VALUE OF X       15
YOUR ESTIMATE OF Y IS
         .1397

 YOU MAY:

 1. ESTIMATE A VALUE OF Y FROM X.
 2. ESTIMATE A VALUE OF X FROM Y.
 3. USE SAME DATA FOR ANOTHER EQUATION.
 4. ENTER DIFFERENT DATA.
 5. END THE PROGRAM.

MAKE YOUR CHOICE BY NUMBER.  5
```

**(Start of next display)**

```
             END OF PROGRAM
```

# Curve fitting for two variables

This section introduces (MULTFIT), a program for finding a relationship between one dependent and one independent variable.

☐ If data fall on curves such as the ones shown in Fig. 1a and 1b, program MULTFIT can help select the curve of best fit and will also give a measure of how good that fit is. The program tests the data against equations of the form $Y = a + b \cdot F(X)$, and gives the equation constants and correlation coefficients for four relationships so that the engineer may select the best equation. It is possible to use functions of the independent variable other than those written into the program, and there is provision for correcting data entries before the program is run.

## Basis of program MULTFIT

A common way to handle data with one dependent and one independent variable is to plot the data and, if there appears to be an orderly relationship between the variables, try to fit an equation to the points. Fig. 1 may help in selecting a suitable equation.

If the data fall on a straight line, then the linear equation

$$Y = a + b \cdot X \tag{1}$$

will apply.

If the $Y$ values increase with increasing $X$ values, and the data form a curve that is concave downward, then a linear expression may apply:

$$Y = a + b \cdot F(X) \tag{2}$$

with one of the following functions of $X$: $F(X) = 1/X$ with a negative coefficient, $b_1$; $F(X) = \log(X)$ or $\ln(X)$; $F(X) = \sqrt{X}$, or $X$ to some other exponent less than 1.

If the $Y$ values increase with increasing values of $X$, and the data follow a trend that is concave upward, the function of $X$ in Eq. (2) may be: $F(X) = e^X$, $F(X) = X^2$, or $X$ to some other exponent greater than 1.

Curves of these types are shown in Fig. 1a, with coefficients chosen to make the curves fit on the same figure. The values of $a$ displace the curves in the $Y$ direction, and the values of $b$ affect the slope or the curvature of the lines.

If the $Y$ values decrease with increasing $X$ values, the figure is simply inverted, as shown in Fig. 1b. If, in this case, a plot of the data appears to be concave downward, the $e^X$ or the $X^2$ function, or a function of $X$ to some other exponent larger than 1, might fit. If the data appear to be concave upward, the $1/X$, $\log(X)$ or $\sqrt{X}$ functions might fit.

The MULTFIT program calculates the constants for the simple linear Eq. 1, and also for three equations of the type shown in Eq. 2, where the functions may be any three selected by the person running the program. The calculation gives the equation constants and the correlation coefficient for each equation, so that the best one may be selected.

The **correlation coefficient** is a measure of the amount of variation in the dependent variable (measured as the sum of the squares of the deviation from the mean) that is accounted for by the correlating equation. If the correlation accounts for all of the variation (i.e., gives a perfect fit), the correlation coefficient is 1.0. If the equation accounts for none of the variation (i.e., the deviation from the equation is no less than the deviation from the mean), the correlation coefficient is 0. The better the fit, the larger the correlation coefficient, but it cannot be greater than 1.0 or less than 0. The square of the correlation coefficient is the fraction of the variation accounted for by the correlating equation. The correlation coefficient is usually designated $r$. If $r = 0.920$,

Adapted from an article by William Volk, Consultant, originally published April 23, 1979.

$r^2 = 0.846$, indicating that 84.6% of the variation of the dependent variable is accounted for.

## The MULTFIT program

The program is given in Table I. It is written for an Apple II computer. Table II offers the computer display.

## Nitric acid example

Fig. 2 is a plot of some data showing the vapor-phase-volume to total-volume ratio for nitric acid at different pressures. The actual data are tabulated on the figure. The $Y$ values decrease with increasing $X$, and the trend is concave upward, most nearly corresponding to the functions $1/X$, log ($X$) and $\sqrt{X}$ in Fig. 1b. These are functions already in MULTFIT, but which offers the best fit of this data? MULTFIT will supply the answer.

Table II illustrates, for this example, how MULTFIT works. As seen in the first display, the program will correlate $x$-$y$ data for the following equations:

$Y = a + b \cdot X$
$Y = a + b/X$
$Y = a + b \cdot \log(X)$
$Y = a + b \cdot \sqrt{X}$ (function denoted as "SQR(X)" in program listing)
$Y = a + b \cdot e^X$ (function denoted as "EXP(X)" in program listing)
$Y = a + b \cdot X^2$

The program will calculate the regression constants ($a$ and $b$) and the correlation coefficient ($r$) for each of the above equations. The equation having the highest $r$ value is the one that offers the best fit of the data.

Data are entered (with prompting from the program) in $x,y$ pairs. When all the data are inputed, the user should enter "END,0".

After all the data are entered, the next display offers the calculated results. For each equation (as indicated by the F(X) part of the curve), the intercept (the value of $a$), the slope (the value of $b$), and the correlation coefficient ($r$) are displayed.

For this example, the curve, $Y = 1.0896 - 0.3751 \cdot \log(X)$, has the highest correlation coefficient ($r = 0.9977$). Accordingly, this equation offers the best fit, and this function very closely represents the data.

## Additional remarks

Note that zero and negative values of $X$ cannot be used with the Log function; that zero values of $X$ cannot be used with the $1/X$ function; and that negative values cannot be used with the $\sqrt{X}$ function or with other fractional exponent functions.

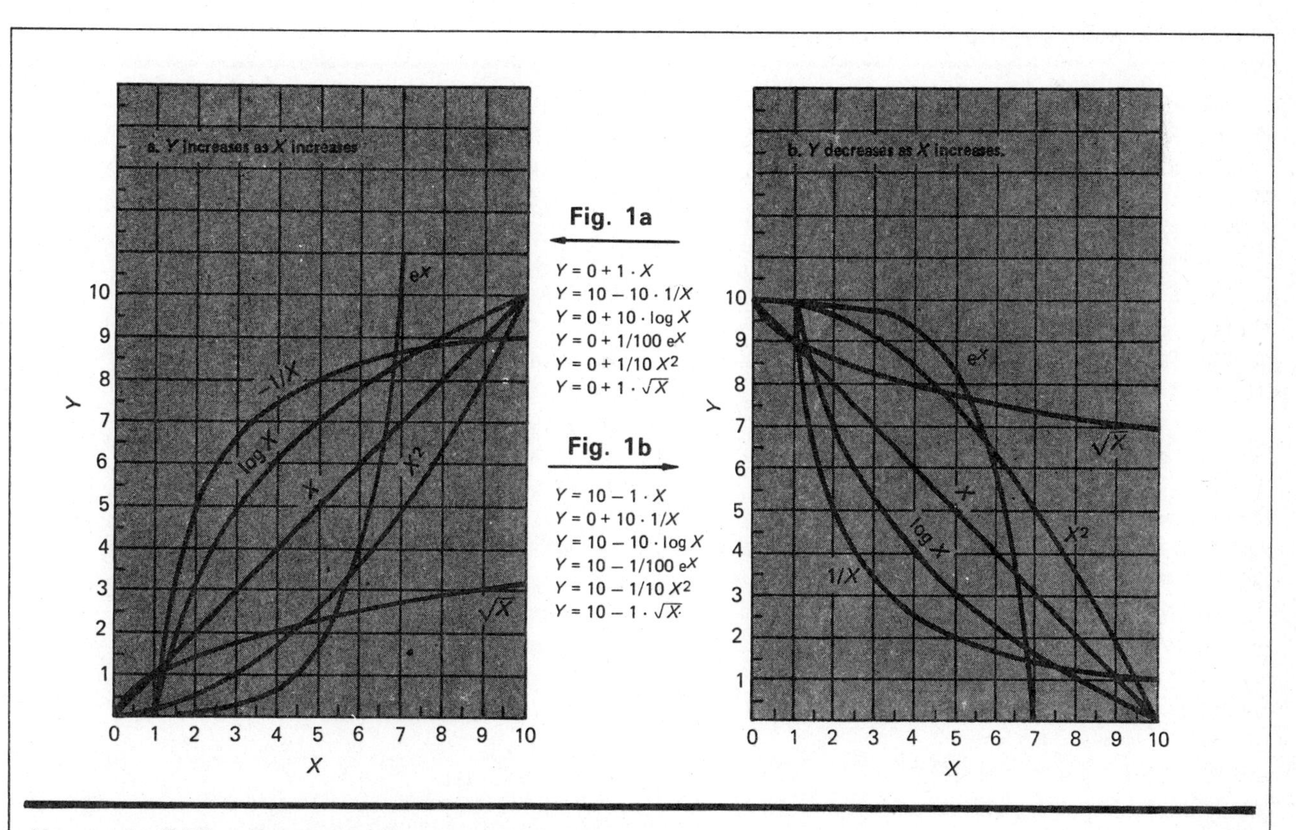


$Y = a + b \cdot F(X)$: **Representative curves of functions of $X$** Fig. 1

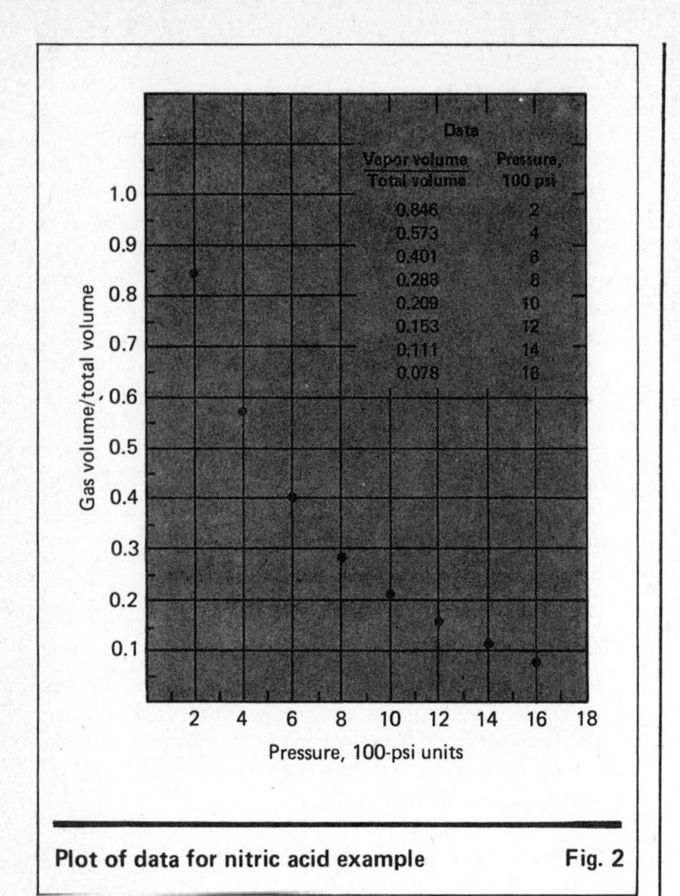


**Plot of data for nitric acid example** **Fig. 2**

## The author

**William Volk,** a private consultant, retired from Hydrocarbon Research, Inc., as Associate Laboratory Director in 1976 after 29 years with that organization. Prior to HRI, he worked with M. W. Kellogg Co. and on the Manhattan Project during WW II. He received a B.S. in Chemical Engineering from New York University and a Master's in Mathematical Statistics from Rutgers University. He is a registered professional engineer in New York and New Jersey and is a member of AIChE.

**Program for solving equations of the type $Y = a + b \cdot F(X)$** **Table I**

```
10 REM   CURVE FITTING FOR TWO VARIABLES
20 REM  FROM CHEMICAL ENGINEERING, APRIL 23, 1979
30 REM  BY WILLIAM VOLK
40 REM  TRANSLATED BY WILLIAM VOLK
50 REM   COPYRIGHT (C) 1984
60 REM  BY CHEMICAL ENGINEERING
70 HOME : VTAB 4
80 REM  SET DISPLAY TO FOUR DECIMALS
90 DEF  FN P(X) =  INT (1E4 * (X + .00005)) / 1E4
100  DIM X(100),Y(100)
110 N$(1) = "X"
120 N$(2) = "1/X"
130 N$(3) = "LOG(X)"
140 N$(4) = "SQR(X)"
150 N$(5) = "EXP(X)"
160 N$(6) = "X^2"
170  PRINT "PROGRAM CORRELATES A SET OF X-Y"
180  PRINT "DATA BY THE FOLLOWING EQUATIONS:"
190  PRINT
200  PRINT " Y = A + B * X"
210  PRINT " Y = A + B / X"
220  PRINT " Y = A + B * LOG(X)"
230  PRINT " Y = A + B * SQR(X)"
240  PRINT " Y = A + B * EXP(X)"
250  PRINT " Y = A + B * X^2"
260  PRINT
270  PRINT "THE REGRESSION CONSTANTS AND THE"
280  PRINT "CORRELATION COEFFICIENT FOR ALL THE"
290  PRINT "EQUATIONS ARE GIVEN IN THE ONE"
300  PRINT "CALCULATION."
310  PRINT
320  INPUT "PRESS RETURN TO ENTER DATA.  ";Q$
330  HOME : VTAB 8
340  PRINT "ENTER THE DATA IN PAIRS: X,Y."
350  PRINT
360  PRINT "WHEN ALL THE DATA ARE IN, ENTER"
370  PRINT "'END,0'."
380  PRINT
390  PRINT "THE DATA WILL BE SAVED IN THE COMPUTER"
400  REM  CHECK FOR IMPOSSIBLE CALCULATIONS
410  PRINT "MEMORY AS X(I),Y(I), FOR I GOING FROM"
420  PRINT "1 TO N."
430  PRINT
440  INPUT "ENTER FIRST PAIR NOW:         ";X$,Y
450  IF X$ = "END" GOTO 810
460 N = N + 1
470 X =  VAL (X$)
480 X(N) = X
490 Y(N) = Y
500  IF X < 0 THEN F2 = 1
```

**Program for solving equations of the type $Y = a + b \cdot F(X)$ (continued)** **Table I**

```
510  IF X = 0 THEN F1 = 1
520  IF X <  - 44 OR X > 44 THEN F3 = 1
530  REM  CALCULATE ALL SUMS AND SUMS OF SQUARES
540 S(1) = S(1) + X
550 SQ(1) = SQ(1) + X ^ 2
560 XY(1) = XY(1) + Y * X
570  IF F1 = 1 GOTO 680
580 S(2) = S(2) + 1 / X
590 SQ(2) = SQ(2) + (1 / X) ^ 2
600 XY(2) = XY(2) + Y / X
610  IF F2 = 1 GOTO 720
620 S(3) = S(3) +  LOG (X)
630 SQ(3) = SQ(3) + ( LOG (X)) ^ 2
640 XY(3) = XY(3) + Y *  LOG (X)
650 S(4) = S(4) +  SQR (X)
660 SQ(4) = SQ(4) + X
670 XY(4) = XY(4) + Y *  SQR (X)
680  IF F2 = 1 GOTO 720
690 S(6) = S(6) + X ^ 2
700 SQ(6) = SQ(6) + X ^ 4
710 XY(6) = XY(6) + Y * X ^ 2
720  IF F3 = 1 GOTO 760
730 S(5) = S(5) +  EXP (X)
740 SQ(5) = SQ(5) + ( EXP (X)) ^ 2
750 XY(5) = XY(5) + Y *  EXP (X)
760 SY = SY + Y
770 YQ = YQ + Y ^ 2
780  PRINT "ENTER NEXT PAIR OF VALUES"
790  INPUT "OR 'END,0'                    ";X$,Y
800  GOTO 450
810 PY = YQ - SY ^ 2 / N
820  HOME : PRINT
830  PRINT  TAB( 10);"THE RESULTS ARE:"
840  PRINT
850  PRINT "EQUATION INTERCEPT"; TAB( 20);"SLOPE"; TAB
     ( 28);"COEFFICIENT"
860  PRINT
870  FOR I = 1 TO 6
880  IF F1 = 1 AND I = 2 GOTO 1050
890  IF F1 = 1 AND I = 3 GOTO 1050
900  IF F1 = 1 AND I = 4 GOTO 1050
910  IF F2 = 1 AND I = 3 GOTO 1050
920  IF F2 = 1 AND I = 4 GOTO 1050
930  IF F2 = 1 AND I = 6 GOTO 1050
940  IF F3 = 1 AND I = 5 GOTO 1050
950  REM  CALCULATION OF REGRESSION COEFFICIENTS
960 X(I) = SQ(I) - S(I) ^ 2 / N
970 Y(I) = XY(I) - S(I) * SY / N
980 B(I) = Y(I) / X(I)
990 A(I) = SY / N - B(I) * S(I) / N
1000  REM  CALCULATION OF CORRELATION COEFFICIENT
1010 R(I) = B(I) * Y(I) / PY
1020  IF R(I) = 0 GOTO 1040
1030 R(I) =  SQR (R(I))
1040  PRINT N$(I); TAB( 10); FN P(A(I)); TAB( 20);
      FN P(B(I)); TAB( 30); FN P(R(I))
1050  NEXT I
1060  PRINT
1070  IF F1 = 1 OR F2 = 1 GOTO 1110
1080  IF F3 = 1 GOTO 1150
1090  GOTO 1190
1100  GOTO 1190
1110  PRINT "CORRELATIONS CAN NOT BE MADE WITH"
1120  PRINT "SOME OF THE FUNCTIONS AND NEGATIVE"
1130  PRINT "OR ZERO VALUES OF X."
1140  GOTO 1190
1150  PRINT "EXP(X) FUNCTION CAN NOT BE USED"
1160  PRINT "WITH X VALUES LARGER THAN 45, OR"
1170  PRINT "SMALLER THAN -45."
1180  GOTO 1190
1190  PRINT
1200  PRINT
1210  PRINT  TAB( 13);"END OF PROGRAM"
1220  END
```

**Example for: Curve fitting for two variables** **Table II**

**(Start of first display)**

```
PROGRAM CORRELATES A SET OF X-Y
DATA BY THE FOLLOWING EQUATIONS:

 Y = A + B * X
 Y = A + B / X
 Y = A + B * LOG(X)
 Y = A + B * SQR(X)
 Y = A + B * EXP(X)
 Y = A + B * X^2

THE REGRESSION CONSTANTS AND THE
CORRELATION COEFFICIENT FOR ALL THE
EQUATIONS ARE GIVEN IN THE ONE
CALCULATION.

PRESS RETURN TO ENTER DATA.
```

**(Start of next display)**

```
ENTER THE DATA IN PAIRS: X,Y.

WHEN ALL THE DATA ARE IN, ENTER
'END,0'.

THE DATA WILL BE SAVED IN THE COMPUTER
MEMORY AS X(I),Y(I), FOR I GOING FROM
1 TO N.
```

**Example for: Curve fitting for two variables (continued)** **Table II**

```
ENTER FIRST PAIR NOW:         2,.846
ENTER NEXT PAIR OF VALUES
OR 'END,0'                    4,.573
ENTER NEXT PAIR OF VALUES
OR 'END,0'                    6,.401
ENTER NEXT PAIR OF VALUES
OR 'END,0'                    8,.288
ENTER NEXT PAIR OF VALUES
OR 'END,0'                    10,.209
ENTER NEXT PAIR OF VALUES
OR 'END,0'                    12,.153
ENTER NEXT PAIR OF VALUES
OR 'END,0'                    14,.111
ENTER NEXT PAIR OF VALUES
OR 'END,0'                    16,.078
ENTER NEXT PAIR OF VALUES
OR 'END,0'                    END,0
```

**(Start of next display)**

```
          THE RESULTS ARE:

EQUATION INTERCEPT SLOPE    COEFFICIENT

X        .7882     -.0506    .9383
1/X      .0356     1.7472    .9705
LOG(X)   1.0896    -.3751    .9977
SQR(X)   1.1703    -.2907    .9769
EXP(X)   .3813     0         .446
X^2      .5844     -2.5E-03  .8441



             END OF PROGRAM
```

# Fitting data to a family of lines

MULTSLOPE, the program detailed in this section, is for a family of lines in which the slope is a function of a second independent variable.

☐ If data fall on a set of curves such as the straight lines in Fig. 1, or the curves in Fig. 2, or some other family of curves, this program will establish coefficients to fit the data with a relationship of the form:

$$Y = a + c \cdot \mathrm{F}_1(X_1) + d \cdot \mathrm{F}_1(X_1) \cdot \mathrm{F}_2(X_2)$$

The program, as written, is for a family of straight lines, with the slopes being a simple linear function of the second independent variable, and it includes the option of also "looking at" a log function of the second variable. If the family of curves is obviously not a series of straight lines, the program MULTSLOPE can be readily modified, following the procedure outlined in the preceding section (p. 68), with a function to account for the curvature.

If the relationship among the family of curves is also obviously not linear (e.g., if the second parameter varies to powers of ten), the program can be easily modified to use a more suitable function of the second parameter.

## A family of lines

It is quite common to have a situation in which data produce a series of curves that are of similar shape but are separated from each other by a second parameter. If the individual curves are straight lines, as illustrated in Fig. 1, they can be expressed by an equation:

$$Y = a + b \cdot X_1 \qquad (1)$$

where $b$ is the regression (or slope) of $Y$ upon $X_1$. In Fig. 1, the slope of the lines increases as $X_2$ goes from Case 1 to Case 4.

Family of curves to be correlated by program MULTSLOPE — Fig. 1

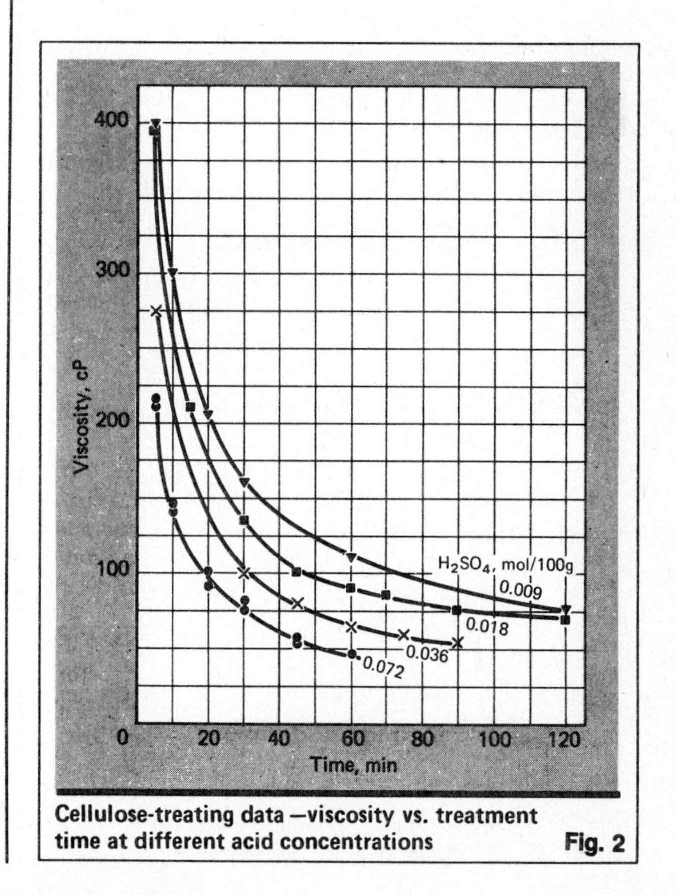


Cellulose-treating data —viscosity vs. treatment time at different acid concentrations — Fig. 2

Adapted from an article by William Volk, Consultant, originally published June 4, 1979.

If the slope is a linear function of the second variable, it may be expressed as:

$$b = c + d \cdot X_2 \qquad (2)$$

These two equations may be combined to yield an overall correlating equation:

$$Y = a + c \cdot X_1 + d \cdot X_1 \cdot X_2 \qquad (3)$$

If one or both of the relationships is *not* a simple linear function of the variable, but is a linear relationship of some *function* of the variable, as discussed for program MULTFIT [p. 68], Eq. (3) could be expanded as follows:

$$Y = a + c \cdot F_1(X_1) + d \cdot F_1(X_1) \cdot F_2(X_2) \qquad (4)$$

The correlation program presented here, MULTSLOPE, uses a linear relationship for the simple value of both independent variables. But the program is easily modified to use a different function of either variable, and the cellulose-reduction example that is given later in this article uses a modification of both ($F_1 = 1/X_1$ and $F_2 = \log X_2$).

A statistically significant correlation with MULTSLOPE indicates that the data may be represented by a family of curves. The shape of the curves is established by the function of $X_1$. The displacement of the curves in the family is set by the function of $X_2$.

The output from MULTSLOPE consists of the three equation constants $b_1$, $b_2$ and $b_0$, in that order, plus $r$, the correlation coefficient [p. 69]. If two functions of the second parameter are tested with one input of data, the output is two sets of equation constants and two correlation coefficients.

The program also has a routine for checking $\Sigma Y$, $\Sigma X_1$, $\Sigma X_2$, and also the number of data sets entered before running the main calculation. There is also provision for deleting data if errors are discovered.

## The program

Table I shows the program. It is written for an Apple II computer. Table II shows the computer displays.

## Cellulose-reduction example

For this example, some data with distinct curvature have been selected. A plot of the data is shown in Fig. 2. The data are for treating cellulose with sulfuric acid to reduce the cellulose chain structure. The data are viscosity (which measures the reduction in the cellulose chain), the time of treatment, and the acid concentration. In the figure, the ordinate is the viscosity, the abscissa is the time, and the acid concentration is the parameter that sets each group of data apart.

The curves in Fig. 2 show $Y$ decreasing with increases in $X$, and are concave upward. A function of $X$ suggested by the first article of this series might give a better correlation than $X$ itself. Therefore, before testing all the data with MULTSLOPE, one set of data (that at 0.072 acid concentration) was tested with MULTFIT. The correlation coefficients for the different functions of $X$ are:

| Function | Correlation coefficient |
|---|---|
| $X$ | 0.8828 |
| $1/X$ | 0.9872 |
| $\log(X)$ | 0.9852 |
| $X$ | 0.9440 |

The $1/X$ function gives the best correlation for one set of data, so this function is used in MULTSLOPE for $X_1$; and inasmuch as the $X_2$ parameter is doubled for each set of data, two functions of $X_2$ are tested, the simple $X_2$ and $\log(X_2)$. The output for the data of Table III is:

| | | |
|---|---|---|
| 1,997.0832 | ($b_1$ | For equation: $Y = b_0 + b_1 \cdot (1/X_1) + b_2 \cdot (1/X_1) \cdot X_2$ |
| −18,072.4876 | ($b_2$ | |
| 61.6300 | ($b_0$ | |
| 0.9691 | ($r$—correlation coefficent) | |
| −835.28 | ($b_1$ | For equation: $Y = b_0 + b_1 \cdot (1/X_1) + b_2(1/X_1) \log X_2$ |
| −598.98 | ($b_2$ | |
| 61.468 | ($b_0$ | |
| 0.9730 | ($r$—correlation coefficient) | |

The log function of $X_2$ is slightly better than the direct function—94.7% of the variation is accounted for, compared with 93.9%. The engineer would have to decide whether this slight improvement were worth the extra complication in the equation. MULTSLOPE, however, in one simple running, gives two highly significant correlations of the data.

**Program for fitting data to a family of lines** **Table I**

```
10  REM   FITTING DATA TO A FAMILY OF CURVES
20  REM  FROM CHEMICAL ENGINEERING, JUNE 4, 1979
30  REM  BY WILLIAM VOLK
40  REM  TRANSLATED BY WILLIAM VOLK
50  REM   COPYRIGHT (C) 1984
60  REM  BY CHEMICAL ENGINEERING
70  HOME : VTAB 4
80  DIM X(100),Y(100),Z(100)
90  REM  SET DISPLAY AT FOUR DECIMALS
100  DEF  FN P(X) =  INT (1E4 * (X + .00005)) / 1E4
110  PRINT "PROGRAM CORRELATES A GROUP OF X-Y DATA"
120  PRINT "WHICH MAY BE SEPARATED INTO A FAMILY"
130  PRINT "OF CURVES BY A SECOND PARAMETER."
140  PRINT
150  PRINT "EXAMPLES MIGHT BE:
160  PRINT
170  PRINT "VISCOSITY VS. TEMPERATURE AND DIFFERENT"
180  PRINT "CONCENTRATIONS OF SOME COMPONENT."
190  PRINT
200  PRINT "PRESSURE DROP IN SOLIDS TRANSPORT"
210  PRINT "AT DIFFERENT SOLIDS RATES, AND"
220  PRINT "AT DIFFERENT GAS VELOCITIES."
230  PRINT
240  PRINT "PROGRAM PROVIDES FOR SEVERAL FUNCTIONS"
250  PRINT "OF BOTH INDEPENDENT VARIABLES."
260  PRINT
```

**Program for fitting data to a family of lines (continued)** **Table I**

```
270  INPUT "PRESS RETURN TO CONTINUE.   ";Q$
280  HOME : VTAB 4
290  PRINT "LINES OF SIMILAR CURVATURE MAY  BE"
300  PRINT "REPRESENTED BY A GROUP OF FUNCTIONS:"
310  PRINT "X, LOG(X), 1/X, SQR(X)."
320  PRINT
330  PRINT "OR FOR THE OPPOSITE CURVATURE BY:
340  PRINT "EXP(X), OR X TO SOME POWER GREATER THAN ONE."
350  PRINT
360  PRINT "ONE OF THESE FUNCTIONS SHOULD BE "
370  PRINT "SELECTED AFTER INSPECTION OF THE DATA."
380  PRINT
390  PRINT "THE SPACING AMONG THE FAMILY OF"
400  PRINT "LINES MAY BE LINEAR OR A LOG FUNCTION."
410  PRINT
420  INPUT "PRESS RETURN TO CONTINUE.   ";Q$
430  HOME : VTAB 4
440  PRINT "THE PROGRAM WILL CORRELATE Y AGAINST"
450  PRINT "SOME FUNCTION, F, OF X:"
460  PRINT
470  PRINT "      Y = A + B * F(X)"
480  PRINT
490  PRINT "AND WILL CORRELATE THE RECRESSION COEF-"
500  PRINT "FICIENT, B, AGAINST A FUNCTION, G,"
510  PRINT "OF THE PARAMETER:"
520  PRINT
530  PRINT "      B = C + D * G(PARAMETER)"
540  PRINT
550  PRINT "YOUR DATA WILL BE SAVED IN THE COMPUTER"
560  PRINT "AS X(I),Y(I),Z(I); I GOING FROM 1 TO N,"
570  PRINT "AND WHERE Z IS THE PARAMETER."
580  PRINT "YOU MAY SELECT DIFFERENT COMBINATIONS"
590  PRINT "OF THE FUNCTIONS 'F' AND 'G' TO FIND"
600  PRINT "A SATISFACTORY CORRELATION."
610  PRINT
620  PRINT "PRESS RETURN TO SELECT THE CORRELATING"
630  INPUT "FUNCTIONS, AND TO ENTER THE DATA.   ";Q$
640  HOME : VTAB (4)
650  PRINT "SELECT ONE OF THE FOLLOWING GROUPS"
660  PRINT "OF CORRELATING FUNCTIONS:"
670  PRINT
680  PRINT " 1.  X.              (LINEAR RELATION)"
690  PRINT " 2.  1/X             (NO ZERO VALUES.)"
700  PRINT " 3.  NATURAL LOG X   (NO NEGATIVE OR"
710  PRINT " 4.  SQUARE ROOT X    ZERO VALUES.)
720  PRINT " 5.  EXP(X)          (COMPUTER LIMIT "
730  PRINT "                      +/- 45.)
740  PRINT " 6.  X TO A POWER    (YOU SELECT THE)
750  PRINT "                      THE POWER.)
760  PRINT
770  INPUT "MAKE YOUR SELECTION BY NUMBER.    ";SE
780  IF SE < 1 OR SE > 6 GOTO 800
790  GOTO 840
800  PRINT
810  PRINT "YOUR OPTIONS WERE 1 TO 6.  YOU ENTERED"
820  PRINT SE;".  TRY AGAIN."
830  GOTO 760
840  IF SE < 6 GOTO 880
850  PRINT
860  PRINT "ENTER YOUR POWER VALUE, GREATER THAN 1,"
870  INPUT "FOR THE FUNCTION OF X        ";PO
880  PRINT
890  PRINT "SELECT CORRELATION FOR THE FAMILY OF"
900  PRINT "CURVES: 1 FOR LINEAR, 2 FOR LOGS."
910  INPUT "NOTE: THAT'S NATURAL LOG.  ";FM
920  PRINT
930  IF FM < 1 OR FM > 2 GOTO 950
940  GOTO 980
950  PRINT
960  PRINT "YOUR OPTIONS WERE 1 OR 2.  TRY AGAIN.
970  GOTO 880
980  IF F1 = 1 GOTO 1180
990  HOME : VTAB 8
1000  PRINT "ENTER THE DATA: X, PARAMETER, Y."
1010  PRINT
1020  PRINT "WHEN ALL THE DATA ARE IN, ENTER"
1030  PRINT "'END', 0,0."
1040  PRINT
1050  PRINT "DATA WILL BE SAVED AS X(I),Z(I),Y(I),"
1060  PRINT "I GOING FROM 1 TO N."
1070  PRINT
1080  INPUT "ENTER THE FIRST DATA:  ";X$,Z,Y
1090 N = N + 1
1100 X(N) =  VAL (X$)
1110 Y(N) = Y
1120 Z(N) = Z
1130  GOTO 1150
1140 N = N + 1
1150  INPUT "NEXT OR:  'END,0,0'    ";X$,Z,Y
1160  IF X$ = "END" GOTO 1180
1170  GOTO 1090
1180  HOME : VTAB 8
1190 F1 = 0
1200  PRINT "IT TAKES A FEW SECONDS"
1210  PRINT "FOR THE CALCULATION."
1220  FOR I = 1 TO 9
1230 S(I) = 0
1240  NEXT I
1250  ONERR  GOTO 2060
1260  FOR I = 1 TO N
1270  ON SE GOTO 1280,1310,1340,1370,1400,1430
1280 X = X(I)
1290 F$ = "X"
1300  GOTO 1460
1310 X = 1 / X(I)
1320 F$ = "1/X"
1330  GOTO 1460
1340 X =  LOG (X(I))
1350 F$ = "LOG(X)"
1360  GOTO 1460
1370 X =  SQR (X(I))
1380 F$ = "SQR(X)"
1390  GOTO 1460
1400 X =  EXP (X(I))
```

**Program for fitting data to a family of lines (continued)** **Table I**

```
1410 F$ = "EXP(X)"
1420  GOTO 1460
1430 X = X(I) ^ PO
1440 PO$ =  STR$ (PO)
1450 F$ = "X^" + PO$
1460  IF FM = 2 GOTO 1500
1470 W = X * Z(I)
1480 G$ = "Z"
1490  GOTO 1520
1500 W = X *  LOG (Z(I))
1510 G$ = "LOG(Z)"
1520 S(1) = S(1) + X
1530 S(2) = S(2) + X ^ 2
1540 S(3) = S(3) + W
1550 S(4) = S(4) + W ^ 2
1560 S(5) = S(5) + X * W
1570 S(6) = S(6) + Y(I)
1580 S(7) = S(7) + (Y(I)) ^ 2
1590 S(8) = S(8) + X * Y(I)
1600 S(9) = S(9) + W * Y(I)
1610  NEXT I
1620 Q(1) = S(2) - S(1) ^ 2 / N
1630 Q(2) = S(4) - S(3) ^ 2 / N
1640 Q(3) = S(5) - S(1) * S(3) / N
1650 Q(4) = S(7) - S(6) ^ 2 / N
1660 Q(5) = S(8) - S(1) * S(6) / N
1670 Q(6) = S(9) - S(3) * S(6) / N
1680 TP = Q(1) * Q(2) - (Q(3)) ^ 2
1690 B1 = (Q(5) * Q(2) - Q(6) * Q(3)) / TP
1700 B2 = (Q(6) * Q(1) - Q(5) * Q(3)) / TP
1710 B0 = (S(6) - B1 * S(1) - B2 * S(3)) / N
1720 R =  SQR ((B1 * Q(5) + B2 * Q(6)) / Q(4))
1730  HOME : PRINT
1740  PRINT "THE CORRELATING EQUATIONS ARE:"
1750  PRINT
1760  PRINT " Y = "; FN P(B0);" + B * ";F$
1770  PRINT "
1780  PRINT "THE REGRESSION, B, IN THE ABOVE"
1790  PRINT "EQUATION HAS THE FOLLOWING CORRELATION:"
1800  PRINT
1810  PRINT " B = "; FN P(B1);" + "; FN P(B2);" * ";G$
1820  PRINT
1830  PRINT "WHERE 'Z' IS THE PARAMETER."
1840  PRINT
1850  PRINT "THE CORRELATION COEFFICIENT IS: "; FN P(R)
1860  PRINT
1870  PRINT
1880  PRINT "THE FOLLOWING OPTIONS ARE AVAILABLE:"
1890  PRINT
1900  PRINT "1. ANOTHER CORRELATION WITH SAME DATA."
1910  PRINT "2. ANOTHER CORRELATION WITH DIFFERENT"
1920  PRINT "   DATA."
1930  PRINT "3. END THE PROGRAM."
1940  PRINT
1950  INPUT "SELECT YOUR OPTION BY NUMBER.   ";OP
1960  ON OP GOTO 2010,640,2030
1970  PRINT
1980  PRINT "YOUR OPTIONS WERE 1 TO 3.  YOU ENTERED"
1990  PRINT OP".  TRY AGAIN."
2000  GOTO 1870
2010 F1 = 1
2020  GOTO 640
2030  PRINT
2040  PRINT  TAB( 15);"END OF PROGRAM"
2050  GOTO 2120
2060  PRINT
2070  PRINT "COMPUTER HAS FOUND AN ERROR AND"
2080  PRINT "CAN NOT CONTINUE.  THERE MAY BE"
2090  PRINT "FUNCTIONS THE COMPUTER CAN NOT HANDLE:"
2100  PRINT "SQR(NEGATIVES), OR DIVISION BY ZERO."
2110  PRINT "CHECK YOUR DATA."
2120  END
```

**Example for: Fitting data to a family of lines** **Table II**

**(Start of first display)**

```
PROGRAM CORRELATES A GROUP OF X-Y DATA
WHICH MAY BE SEPARATED INTO A FAMILY
OF CURVES BY A SECOND PARAMETER.

EXAMPLES MIGHT BE:

VISCOSITY VS. TEMPERATURE AND DIFFERENT
CONCENTRATIONS OF SOME COMPONENT.

PRESSURE DROP IN SOLIDS TRANSPORT
AT DIFFERENT SOLIDS RATES, AND
AT DIFFERENT GAS VELOCITIES.

PROGRAM PROVIDES FOR SEVERAL FUNCTIONS
OF BOTH INDEPENDENT VARIABLES.

PRESS RETURN TO CONTINUE.
```

**(Start of next display)**

```
LINES OF SIMILAR CURVATURE MAY  BE
REPRESENTED BY A GROUP OF FUNCTIONS:
X, LOG(X), 1/X, SQR(X).

OR FOR THE OPPOSITE CURVATURE BY:
EXP(X), OR X TO SOME POWER GREATER THAN ONE.

ONE OF THESE FUNCTIONS SHOULD BE
SELECTED AFTER INSPECTION OF THE DATA.
```

**Example for: Fitting data to a family of lines (continued)** **Table II**

```
THE SPACING AMONG THE FAMILY OF
LINES MAY BE LINEAR OR A LOG FUNCTION.

PRESS RETURN TO CONTINUE.
```

**(Start of next display)**

```
THE PROGRAM WILL CORRELATE Y AGAINST
SOME FUNCTION, F, OF X:

     Y = A + B * F(X)

AND WILL CORRELATE THE RECRESSION COEF-
FICIENT, B, AGAINST A FUNCTION, G,
OF THE PARAMETER:

     B = C + D * G(PARAMETER)

YOUR DATA WILL BE SAVED IN THE COMPUTER
AS X(I),Y(I),Z(I); I GOING FROM 1 TO N,
AND WHERE Z IS THE PARAMETER.
YOU MAY SELECT DIFFERENT COMBINATIONS
OF THE FUNCTIONS 'F' AND 'G' TO FIND
A SATISFACTORY CORRELATION.

PRESS RETURN TO SELECT THE CORRELATING
FUNCTIONS, AND TO ENTER THE DATA.
```

**(Start of next display)**

```
SELECT ONE OF THE FOLLOWING GROUPS
OF CORRELATING FUNCTIONS:

 1.  X.              (LINEAR RELATION)
 2.  1/X             (NO ZERO VALUES.)
 3.  NATURAL LOG X   (NO NEGATIVE OR
 4.  SQUARE ROOT X    ZERO VALUES.)
 5.  EXP(X)          (COMPUTER LIMIT
                      +/- 45.)
 6.  X TO A POWER    (YOU SELECT THE)
                      THE POWER.)

MAKE YOUR SELECTION BY NUMBER.    2

SELECT CORRELATION FOR THE FAMILY OF
CURVES: 1 FOR LINEAR, 2 FOR LOGS.
NOTE: THAT'S NATURAL LOG.  1
```

**(Start of next display)**

```
ENTER THE DATA: X, PARAMETER, Y.

WHEN ALL THE DATA ARE IN, ENTER
'END', 0,0.

DATA WILL BE SAVED AS X(I),Z(I),Y(I),
I GOING FROM 1 TO N.
```

```
ENTER THE FIRST DATA:  5,.072,215
NEXT OR:  'END,0,0'    5,.072,210
NEXT OR:  'END,0,0'    10,.072,145
NEXT OR:  'END,0,0'    10,.072,140
NEXT OR:  'END,0,0'    20,.072,100
NEXT OR:  'END,0,0'    20,.072,90
NEXT OR:  'END,0,0'    30,.072,80
NEXT OR:  'END,0,0'    30,.072,75
NEXT OR:  'END,0,0'    45,.072,55
NEXT OR:  'END,0,0'    45,.072,53
NEXT OR:  'END,0,0'    60,.072,45
NEXT OR:  'END,0,0'    5,.036,275
NEXT OR:  'END,0,0'    30,.036,100
NEXT OR:  'END,0,0'    45,.036,80
NEXT OR:  'END,0,0'    60,.036,65
NEXT OR:  'END,0,0'    75,.036,60
NEXT OR:  'END,0,0'    90,.036,55
NEXT OR:  'END,0,0'    5,.018,400
NEXT OR:  'END,0,0'    15,.018,210
NEXT OR:  'END,0,0'    30,.018,135
NEXT OR:  'END,0,0'    45,.018,100
NEXT OR:  'END,0,0'    60,.018,90
NEXT OR:  'END,0,0'    70,.018,85
NEXT OR:  'END,0,0'    90,.018,75
NEXT OR:  'END,0,0'    120,.018,70
NEXT OR:  'END,0,0'    5,.009,400
NEXT OR:  'END,0,0'    10,.009,300
NEXT OR:  'END,0,0'    20,.009,205
NEXT OR:  'END,0,0'    30,.009,160
NEXT OR:  'END,0,0'    60,.009,110
NEXT OR:  'END,0,0'    120,.009,75
NEXT OR:  'END,0,0'    END,0,0
```

**(Start of next display)**

```
IT TAKES A FEW SECONDS
FOR THE CALCULATION.
```

**(Start of next display)**

```
THE CORRELATING EQUATIONS ARE:

 Y = 61.63 + B * 1/X

THE REGRESSION, B, IN THE ABOVE
EQUATION HAS THE FOLLOWING CORRELATION:

 B = 1997.0832 + -18072.4876 * Z

WHERE 'Z' IS THE PARAMETER.

THE CORRELATION COEFFICIENT IS: .9691


THE FOLLOWING OPTIONS ARE AVAILABLE:

1. ANOTHER CORRELATION WITH SAME DATA.
2. ANOTHER CORRELATION WITH DIFFERENT
   DATA.
```

**Example (continued)** **Table II**

```
3. END THE PROGRAM.

SELECT YOUR OPTION BY NUMBER.   1
SELECT ONE OF THE FOLLOWING GROUPS
OF CORRELATING FUNCTIONS:

 1.  X.              (LINEAR RELATION)
 2.  1/X             (NO ZERO VALUES.)
 3.  NATURAL LOG X   (NO NEGATIVE OR
 4.  SQUARE ROOT X    ZERO VALUES.)
 5.  EXP(X)          (COMPUTER LIMIT
                      +/- 45.)
 6.  X TO A POWER    (YOU SELECT THE)
                      THE POWER.)

MAKE YOUR SELECTION BY NUMBER.   2

SELECT CORRELATION FOR THE FAMILY OF
CURVES: 1 FOR LINEAR, 2 FOR LOGS.
NOTE: THAT'S NATURAL LOG.  2

IT TAKES A FEW SECONDS
FOR THE CALCULATION.

THE CORRELATING EQUATIONS ARE:

 Y = 61.4679 + B * 1/X

THE REGRESSION, B, IN THE ABOVE
EQUATION HAS THE FOLLOWING CORRELATION:

B = -835.2785 + -598.9813 * LOG(Z)

WHERE 'Z' IS THE PARAMETER.

THE CORRELATION COEFFICIENT IS: .973


THE FOLLOWING OPTIONS ARE AVAILABLE:

1. ANOTHER CORRELATION WITH SAME DATA.
2. ANOTHER CORRELATION WITH DIFFERENT
   DATA.
3. END THE PROGRAM.

SELECT YOUR OPTION BY NUMBER.   3

              END OF PROGRAM
```

**Cellulose-treating data*** **Table III**

| Viscosity, cP | Time, min | $H_2SO_4$ mol/100g |
|---|---|---|
| 215 | 5 | 0.072 |
| 210 | 5 | 0.072 |
| 145 | 10 | 0.072 |
| 140 | 10 | 0.072 |
| 100 | 20 | 0.072 |
| 90 | 20 | 0.072 |
| 80 | 30 | 0.072 |
| 75 | 30 | 0.072 |
| 55 | 45 | 0.072 |
| 53 | 45 | 0.072 |
| 45 | 60 | 0.072 |
| 275 | 5 | 0.036 |
| 100 | 30 | 0.036 |
| 80 | 45 | 0.036 |
| 65 | 60 | 0.036 |
| 60 | 75 | 0.036 |
| 55 | 90 | 0.036 |
| 400 | 5 | 0.018 |
| 210 | 15 | 0.018 |
| 135 | 30 | 0.018 |
| 100 | 45 | 0.018 |
| 90 | 60 | 0.018 |
| 85 | 70 | 0.018 |
| 75 | 90 | 0.018 |
| 70 | 120 | 0.018 |
| 400 | 5 | 0.009 |
| 300 | 10 | 0.009 |
| 205 | 20 | 0.009 |
| 160 | 30 | 0.009 |
| 110 | 60 | 0.009 |
| 75 | 120 | 0.009 |
| 4,258 | 1,265 | 1,206 |

*From figures in Melm et al., *Ind. and Eng. Chem.*, Vol. 44, No. 12, p. 2,905.

## Reference

1. Volk, William, "Applied Statistics for Engineers," McGraw-Hill, New York, 2nd ed., 1969.

# Correlating one dependent variable with two independent variables

This Apple II computer program, named MULT-2, will produce an equation to fit a set of data when there are two independent variables, and will indicate the relative contribution of each variable to the correlation.

☐ The two preceding sections dealt with data that could be plotted and then tested, with some suitable relationships, as indicated by the plot. This section deals with the relationship between one dependent and two independent variables that would be represented by a three-dimensional model not readily illustrated in a simple graph. The equation, however, is common enough:

$$Y = b_0 + b_1X_1 + b_2X_2 \quad (1)$$

$Y$ is the dependent variable and $X_1$ and $X_2$ are the independent variables. This program calculates the equation constants $b_0$, $b_1$ and $b_2$ for a given set of data.

## The basis of the program

The "least squares" solution of Eq. (1) is fairly well known as the *multiple linear regression* solution for two independent variables. It appears in many statistics texts and is available in most computer "statistics packages." This equation is used when it is expected that the dependent variable is affected independently by the two independent variables: for example, growth as a function of time and nutrients; conversion as a function of temperature and catalyst concentration; yield as a function of pressure and concentration.

The novelty in the program MULT-2 is that, while it gives the constants for Eq. (1), it also gives the constants for both of the linear equations with one independent variable:

$$Y = c_0 + c_1X_1 \quad (2)$$

$$Y = d_0 + d_1X_2 \quad (3)$$

and provides a comparison among the correlation coefficients of all three equations. The user can then determine whether the correlation with the two independent variables is significantly better than the correlations with only one independent variable.

Eq. (1) will always give at least as good a fit to the data as either (2) or (3) in terms of accounting for the variation of the dependent variable. However, from a statistical or practical point of view, addition of the second independent variable to the correlating equation may not be justified for the amount of improvement obtained.

Calculation of the constants for the least-squares solution of Eq. (1) requires all the numbers that are also necessary for the calculation of the constants for Eq. (2) and (3). MULT-2 uses these values to calculate the constants for all three equations and also calculates the correlation coefficients for all three. It then provides a statistical comparison between the best two to determine whether the difference is significant.

The usual statistical test employed to determine whether a two-independent-variable correlation is better than a single-independent-variable correlation is the *variance F test.* This compares the additional variation in the dependent variable that is accounted for by the addition of the second term with the residual variation that is not accounted for.

The "*F*" statistic, the correlation coefficient ($r$), and the more common "*t*" statistic are all related in a manner that is beyond the scope of this article, but which is explained in some of the standard statistics texts (e.g., Volk, W., "Applied Statistics for Engineers," McGraw-Hill, New York, 1969). The relationship between the correlation coefficients of a two-independent-variable equation, and of an equation using only one of the independent variables, and the $t$ statistic, is shown in the following equation:

$$t_\nu = \sqrt{\nu[(r_2^2 - r_1^2)/(1 - r_2^2)]} \quad (4)$$

where $\nu$ is the number of degrees of freedom associated with $r_2$, the correlation coefficient of the two-independent-variable equation. $r_1$ is the correlation coefficient for the equation using only one of the independent variables. For cases covered by this article, $\nu = N - 3$, where $N$ is the number of sets of data.

For those familiar with the $t$ test, a precise statement

**Adapted from an article by William Volk, Consultant, originally published September 10, 1979.**

can be made about the two correlation coefficients from the value of $t$ and the number of degrees of freedom. For practical purposes, if the $t$ value is less than 2.0, the two-independent-variable correlation is not significantly better than the single-variable correlation with which it is compared.

MULT-2 calculates $t$ at the end of the computation, and compares the two-independent-variable equation with the better of the single-variable equations.

## Using the program

The program, listed in Table I, is interactive and self-explanatory for the most part. The first display screen explains what the program does, and then the user is invited to begin entering data.

Data are entered as sets, with each set having the order $X_1$, $X_2$, $Y$. At least four sets of data must be entered, so that there is at least one degree of freedom in the $t$ calculation. There is no means of correcting bad data after they are entered, so one must simply start over. After entering the last set of data, the user enters "END,0,0" and presses the return key to obtain the results.

The calculated results are the constants and correlation coefficients for the two one-variable equations ($Y = b_0 + b_1X_1$ and $Y = b_0 + b_1X_2$) and for the two-variable equation ($Y = b_0 + b_1X_1 + b_2X_2$). The program also calculates the $t$ statistic comparing the two-variable equation with the best of the one-variable equations. Finally, it evaluates the meaning of this t value for the given number of data, printing this out as the "probability of error in saying there is no improvement."

The program stores the data as X1(I), X2(I), and Y(I), so it can be retrieved by the usual BASIC commands.

To see how the program works, let us look at an example.

## Agricultural example

The data for this example are from some agricultural experiments on plant-available phosphorus as a possible function of the organic and inorganic phosphorus in the soil. The data are given in Table II.

A first assumption might be that the plant-available phosphorus, $Y$, is a function of both the inorganic phosphorus, $X_1$, and the organic phosphorus, $X_2$, in the soil. Thus:

$$Y = b_0 + b_1X_1 + b_2X_2.$$

This equation, tested with a multivariate least-squares solution, gives a correlation coefficient of 0.6945, which is statistically significant for the amount of data involved, and might be satisfactory.

However, if the data are tested with MULT-2, the results are as shown in Table III, the example displays.

1.8434 ($b_1$ for the equation $Y = b_0 + b_1X_1$)
59.2590 ($b_0$ for the same equation)
0.6934 ($r$—correlation coefficient, inorganic phosphorus)
0.7023 ($b_1$ for equation $Y = b_0 + b_1X_2$)
51.7013 ($b_0$ for the same equation)
0.3545 ($r$—correlation coefficient, organic phosphorus)
1.7898 ($b_1$ for the equation $Y = b_0 + b_1X_1 + b_2X_2$)
0.0866 ($b_2$ for the same equation)
56.2510 ($b_0$ for the same equation)
0.6945 ($r$—correlation coefficient, both phosphoruses)
0.2088 ($t$—comparison of third equation with first)

These results show that the correlation coefficient for the third equation is almost the same as for the first equation. The very low $t$ value shows the same thing. The correlation of plant-available phosphorus is almost entirely with the inorganic phosphorus, and the organic phosphorus contributes nothing significant to the correlation.

## Gasoline example

This example deals with some data for which the correlation using two independent variables is significantly better than with either one separately. The data are given in Table IV. They are a portion of some gasoline octane-number data correlated against catalyst purity and weight % carbon on the catalyst. These data, tested with MULT-2, give the following results, shown in Table V, ($X_1$ is catalyst purity, $X_2$ is carbon on catalyst, and $Y$ is octane number):

1.1274 ($b_1$ for the equation $Y = b_0 + b_1X_1$)
−24.8467 ($b_0$ for the same equation)
0.4703 ($r$—correlation coefficient)
−0.2030 ($b_1$ for the equation $Y = b_0 + b_1X_2$)
88.1152 ($b_0$ for the same equation)
0.6561 ($r$—correlation coefficient)
1.919 ($b_1$ for equation $Y = b_0 + b_1X_1 + b_2X_2$)
−0.2903 ($b_2$ for the same equation)
−102.009 ($b_0$ for the same equation)
0.996 ($r$—correlation coefficient)
23.6084 ($t$—comparing third equation with second)
0.0001 (Probability of error in saying third equation is no better than second)

The correlation coefficient of 0.6561 for the second equation indicates a significant correlation between octane number and carbon on the catalyst. However, the correlation coefficient of 0.996 for the third equation shows an obviously much better correlation with both variables. The $t$ value of 23.6 substantiates this statement—there is only a 1 in 10,000 chance that the third equation is not significantly better.

**Program for correlating one dependent variable with two independent variables** **Table I**

```
10  REM  CORRELATING ONE DEPENDENT VARIABLE WITH TWO
    INDEPENDENT VARIABLES.
20  REM  FROM CHEMICAL ENGINEERING, SEPTEMBER 10, 1979
30  REM  BY WILLIAM VOLK
40  REM  TRANSLATED BY WILLIAM VOLK
50  REM   COPYRIGHT (C) 1984
60  REM  BY CHEMICAL ENGINEERING
70  HOME : VTAB 4
80  REM  SET DISPLAY TO FOUR DECIMALS
90  DEF  FN P(X) =  INT (1E4 * (X + .00005)) / 1E4
100  DIM X1(100),X2(100),Y(100)
110  PRINT "PROGRAM CORRELATES ONE DEPENDENT"
120  PRINT "VARIABLE AGAINST TWO INDEPENDENT"
130  PRINT "VARIABLES."
140  PRINT
150  PRINT "IT CALCULATES TWO SINGLE-VARIABLE"
160  PRINT "CORRELATIONS, AND ONE TWO-VARIABLE"
170  PRINT "CORRELATION."
180  PRINT
190  PRINT "IT CALCULATES A COMPARISON"
200  PRINT "OF THE THREE RESULTS."
210  PRINT
220  PRINT "DATA ARE ENTERED AS: X1, X2, Y"
230  PRINT
240  PRINT "THE DATA INPUT IS TERMINATED WITH:"
250  PRINT
260  PRINT "             END, 0, 0"
270  PRINT
280  INPUT "PRESS RETURN TO ENTER DATA.  ";Q$
290  HOME : VTAB 4
300  PRINT "ENTER DATA: X1, X2, Y  SEPARATED BY"
310  PRINT "COMMAS.  WHEN ALL THE DATA ARE IN"
320  PRINT "ENTER: 'END, 0, 0'"
330  PRINT
340  PRINT "DATA WILL BE SAVED IN COMPUTER MEMORY"
350  PRINT "AS X1(I), X2(I), Y(I), I GOING FROM"
360  PRINT "1 TO N."
370  PRINT
380  INPUT "ENTER THE FIRST SET:    ";X$,X2,Y
390 N = N + 1
400 X1(N) =  VAL (X$)
410 X2(N) = X2
420 Y(N) = Y
430 X1 =  VAL (X$)
440 S1 = S1 + Y
450 S2 = S2 + Y ^ 2
460 S3 = S3 + X1
470 S4 = S4 + X1 ^ 2
480 S5 = S5 + X2
490 S6 = S6 + X2 ^ 2
500 S7 = S7 + X1 * Y
510 S8 = S8 + X2 * Y
520 S9 = S9 + X1 * X2
530  INPUT "NEXT SET OR: 'END,0,0'  ";X$,X2,Y
540  IF X$ = "END" GOTO 560
550  GOTO 390
560  HOME : VTAB 8
570  PRINT "CALCULATION TAKES "
580  PRINT "A FEW SECONDS."
590 P1 = S2 - S1 ^ 2 / N
600 P2 = S4 - S3 ^ 2 / N
610 P3 = S6 - S5 ^ 2 / N
620 P4 = S7 - S3 * S1 / N
630 P5 = S8 - S5 * S1 / N
640 P6 = S9 - S3 * S5 / N
650  REM  SINGLE-VARIABLE EQUATIONS
660 B1(1) = P4 / P2
670 B0(1) = (S1 - B1(1) * S3) / N
680 R(1) =  SQR (B1(1) * P4 / P1)
690 B1(2) = P5 / P3
700 B0(2) = (S1 - B1(2) * S5) / N
710 R(2) =  SQR (B1(2) * P5 / P1)
720 TP = P2 * P3 - P6 ^ 2
730  REM   TWO-VARIABLE EQUATIONS
740 B1(3) = (P4 * P3 - P5 * P6) / TP
750 B2(3) = (P5 * P2 - P4 * P6) / TP
760 B0(3) = (S1 - B1(3) * S3 - B2(3) * S5) / N
770 R(3) =  SQR ((B1(3) * P4 + B2(3) * P5) / P1)
780 NU = N - 3
790  IF R(1) > R(2) GOTO 820
800 R = R(2)
810  GOTO 830
820 R = R(1)
830 T =  SQR (NU * (R(3) ^ 2 - R ^ 2) / (1 - R(3) ^ 2))
840  GOTO 1160
850  IF PR < 1E - 5 THEN PR = 1E - 4
860  FOR I = 1 TO 3
870 B0(I) =  FN P(B0(I))
880 B1(I) =  FN P(B1(I))
890 B2(I) =  FN P(B2(I))
900 R(I) =  FN P(R(I))
910  NEXT I
920  HOME
930  PRINT  TAB( 10);"THE RESULTS ARE:"
940  PRINT
950  PRINT "Y = B0 + B1 * X1"; TAB( 22);"Y = B0 + B1 * X2"
960  PRINT
970  PRINT "B0 = "; FN P(B0(1)); TAB( 25); FN P(B0(2))
980  PRINT "B1 = "; FN P(B1(1)); TAB( 25); FN P(B1(2))
990  PRINT "R =  "; FN P(R(1)); TAB( 25); FN P(R(2))
1000  PRINT
1010  PRINT "    Y = B0 + B1 * X1 + B2 * X2"
1020  PRINT
1030  PRINT  TAB( 10);"B0 = "; FN P(B0(3))
1040  PRINT  TAB( 10);"B1 = "; FN P(B1(3))
1050  PRINT  TAB( 10);"B2 = "; FN P(B2(3))
1060  PRINT  TAB( 10);"R  = "; FN P(R(3))
1070  PRINT
1080  PRINT "T FOR COMPARISON OF TWO VARIABLE"
1090  PRINT "CORRELATION WITH BETTER OF THE SINGLE"
```

**Program listing (continued)** **Table I**

```
1100 PRINT "VARIABLE IS:   "; FN P(T)
1110 PRINT
1120 PRINT "PROBABILITY OF ERROR IN SAYING THERE"
1130 PRINT "IS NO IMPROVEMENT IS:  "; FN P(PR)
1140 GOTO 1340
1150 REM  ROUTINE FOR PROBABILITY CALCULATION
1160 TH =  ATN ( ABS (T) /  SQR (NU))
1170 PI = 3.14159265
1180 PA = 1
1190 IF NU / 2 -  INT (NU / 2) = 0 GOTO 1280
1200 IF NU > 1 GOTO 1230
1210 PR = 1 - (2 * TH / PI)
1220 GOTO 850
1230 FOR I = NU - 3 TO 2 STEP  - 2
1240 PA = 1 + PA * (I / (I + 1) * ( COS (TH)) ^ 2)
1250 NEXT I
1260 PR = 1 - (2 / PI) * (TH +  SIN (TH) *  COS (TH) * PA
1270 GOTO 850
1280 IF NU = 2 GOTO 1320
1290 FOR I = (NU - 3) TO 1 STEP  - 2
1300 PA = 1 + PA * (I / (I + 1) * ( COS (TH)) ^ 2)
1310 NEXT I
1320 PR = 1 -  SIN (TH) * PA
1330 GOTO 850
1340 PRINT
1350 PRINT  TAB( 13);"END OF PROGRAM"
1360 END
```

**Data used in agricultural example** **Table II**

| | Plant available phosphorus $Y$ | Inorganic phosphorus $X_1$ | Organic phosphorus $X_2$ | (parts per million) |
|---|---|---|---|---|
| | 64 | 0.4 | 53 | |
| | 60 | 0.4 | 23 | |
| | 71 | 3.1 | 19 | |
| | 61 | 0.6 | 34 | |
| | 54 | 4.7 | 24 | |
| | 77 | 1.7 | 65 | |
| | 81 | 9.4 | 44 | |
| | 93 | 10.1 | 31 | |
| | 93 | 11.6 | 29 | |
| | 51 | 12.6 | 58 | |
| | 76 | 10.9 | 37 | |
| | 96 | 23.1 | 46 | |
| | 77 | 23.1 | 50 | |
| | 93 | 21.6 | 44 | |
| | 95 | 23.1 | 56 | |
| | 54 | 1.9 | 36 | |
| | 168 | 26.8 | 58 | |
| | 99 | 29.9 | 51 | |
| **Totals** | 1463 | 215.0 | 758 | |

Data from: Eid, M.T., Black, C.A., Kempthorne, O., and Zoellner, J.A., Iowa Agricultural Experiment Station Research Bulletin 406.

**Example 1 for: Correlating one dependent variable with two independent variables** **Table III**

**(Start of first display)**

```
PROGRAM CORRELATES ONE DEPENDENT
VARIABLE AGAINST TWO INDEPENDENT
VARIABLES.

IT CALCULATES TWO SINGLE-VARIABLE
CORRELATIONS, AND ONE TWO-VARIABLE
CORRELATION.

IT CALCULATES A COMPARISON
OF THE THREE RESULTS.

DATA ARE ENTERED AS: X1, X2, Y

THE DATA INPUT IS TERMINATED WITH:

           END, 0, 0

PRESS RETURN TO ENTER DATA.
```

**(Start of next display)**

```
ENTER DATA: X1, X2, Y  SEPARATED BY
COMMAS.  WHEN ALL THE DATA ARE IN
ENTER: 'END, 0, 0'

DATA WILL BE SAVED IN COMPUTER MEMORY
AS X1(I), X2(I), Y(I), I GOING FROM
1 TO N.
```

```
ENTER THE FIRST SET:     .4,53,64
NEXT SET OR: 'END,0,0'  .4,23,60
NEXT SET OR: 'END,0,0'  3.1,19,71
NEXT SET OR: 'END,0,0'  .6,34,61
NEXT SET OR: 'END,0,0'  4.7,24,54
NEXT SET OR: 'END,0,0'  1.7,65,77
NEXT SET OR: 'END,0,0'  9.4,44,81
NEXT SET OR: 'END,0,0'  10.1,31,93
NEXT SET OR: 'END,0,0'  11.6,29,93
NEXT SET OR: 'END,0,0'  12.6,58,51
NEXT SET OR: 'END,0,0'  10.9,37,76
NEXT SET OR: 'END,0,0'  23.1,46,96
NEXT-SET OR: 'END,0,0'  23.1,50,77
NEXT SET OR: 'END,0,0'  21.6,44,93
NEXT SET OR: 'END,0,0'  23.1,56,95
NEXT SET OR: 'END,0,0'  1.9,36,54
NEXT SET OR: 'END,0,0'  26.8,58,168
NEXT SET OR: 'END,0,0'  29.9,51,99
NEXT SET OR: 'END,0,0'  END,0,0
```

**(Start of next display)**

```
CALCULATION TAKES
A FEW SECONDS.
```

**(Start of next display)**

```
          THE RESULTS ARE:

Y = B0 + B1 * X1      Y = B0 + B1 * X2

B0 = 59.259              51.7013
```

**Example 1 (continued)** **Table III**

```
B1 = 1.8434          .7023
R =  .6934           .3545

     Y = B0 + B1 * X1 + B2 * X2

          B0 = 56.251
          B1 = 1.7898
          B2 = .0866
          R  = .6945

T FOR COMPARISON OF TWO VARIABLE
CORRELATION WITH BETTER OF THE SINGLE
VARIABLE IS:   .2088

PROBABILITY OF ERROR IN SAYING THERE
IS NO IMPROVEMENT IS:  .8374

          END OF PROGRAM
```

**Data used in gasoline example** **Table IV**

| | Octane number | Catalyst purity, % | Carbon, weight % |
|---|---|---|---|
| | $Y$ | $X_1$ | $X_2$ |
| | 88.6 | 99.8 | 3.0 |
| | 86.4 | 99.7 | 10.0 |
| | 87.2 | 99.6 | 7.0 |
| | 88.4 | 99.5 | 2.0 |
| | 87.2 | 99.4 | 5.0 |
| | 86.8 | 99.3 | 6.0 |
| | 86.1 | 99.2 | 8.0 |
| | 87.3 | 99.1 | 3.0 |
| | 86.4 | 99.0 | 5.0 |
| | 86.6 | 98.9 | 4.0 |
| | 87.1 | 98.8 | 2.0 |
| Totals | 958.1 | 1092.3 | 55.0 |

**Example 2 for: Correlating one dependent variable with two independent variables** **Table V**

**(Start of first display)**

```
PROGRAM CORRELATES ONE DEPENDENT
VARIABLE AGAINST TWO INDEPENDENT
VARIABLES.

IT CALCULATES TWO SINGLE-VARIABLE
CORRELATIONS, AND ONE TWO-VARIABLE
CORRELATION.

IT CALCULATES A COMPARISON
OF THE THREE RESULTS.

DATA ARE ENTERED AS: X1, X2, Y

THE DATA INPUT IS TERMINATED WITH:

          END, 0, 0

PRESS RETURN TO ENTER DATA.
```

**(Start of next display)**

```
ENTER DATA: X1, X2, Y  SEPARATED BY
COMMAS.  WHEN ALL THE DATA ARE IN
ENTER: 'END, 0, 0'

DATA WILL BE SAVED IN COMPUTER MEMORY
AS X1(I), X2(I), Y(I), I GOING FROM
1 TO N.

ENTER THE FIRST SET:     99.8,3,88.6
NEXT SET OR: 'END,0,0'  99.7,10,86.4
NEXT SET OR: 'END,0,0'  99.6,7,87.2
NEXT SET OR: 'END,0,0'  99.5,2,88.4
NEXT SET OR: 'END,0,0'  99.4,5,87.2
NEXT SET OR: 'END,0,0'  99.3,6,86.8
NEXT SET OR: 'END,0,0'  99.2,8,86.1
NEXT SET OR: 'END,0,0'  99.1,3,87.3
NEXT SET OR: 'END,0,0'  99,5,86.4
NEXT SET OR: 'END,0,0'  98.9,4,86.6
NEXT SET OR: 'END,0,0'  98.8,2,87.1
NEXT SET OR: 'END,0,0'  END,0,0
```

**(Start of next display)**

```
CALCULATION TAKES
A FEW SECONDS.
```

**(Start of next display)**

```
        THE RESULTS ARE:

Y = B0 + B1 * X1      Y = B0 + B1 * X2

B0 = -24.8467           88.1152
B1 = 1.1274             -.203
R =  .4703              .6561

   Y = B0 + B1 * X1 + B2 * X2

        B0 = -102.009
        B1 = 1.919
        B2 = -.2903
        R  = .996

T FOR COMPARISON OF TWO VARIABLE
CORRELATION WITH BETTER OF THE SINGLE
VARIABLE IS:   23.6084

PROBABILITY OF ERROR IN SAYING THERE
IS NO IMPROVEMENT IS:  1E-04

          END OF PROGRAM
```

# Second-degree-polynomial computer program

If a second-degree-polynomial equation describes the relationship between two variables in a set of data, this Apple II program, MULTSQUARE, will define the equation coefficients and also provide data about the statistical significance of the relationship.

☐ MULTSQUARE tests to find out whether data may conform to the relationship $Y = b_0 + b_1X + b_2X^2$. The program also tests the data against the equations $Y = b_0 + b_1X^2$ and $Y = b_0 + b_1X$. Correlation coefficients are worked out for each of the three equations to test for proper fit. Finally, a $t$ test compares the best two coefficients for statistical significance.

## Second-degree polynomial

It was pointed out in the discussion of the program MULTFIT (p. 68) that data with $Y$ values that increase with an increase in $X$ and curve upward might be expressed with an equation:

$$Y = b_0 + b_1X^2 \quad (1)$$

If there is a change in the relationship between $Y$ and $X$, $Y$ first decreasing and later increasing as $X$ increases, the data might be fitted with the polynomial:

$$Y = b_0 + b_1X + b_2X^2 \quad (2)$$

where $b_1$ is negative and $b_2$ is positive.

These two relationships are illustrated in Fig. 1. If the curves are concave downward, the same relationship holds but the signs of the regression coefficients are reversed.

If a plot of the data shows an orderly curve, as illustrated in Fig. 1, the selection of a correlating equation is not too difficult. However, in the more usual case, we have data such as in Fig. 2, and it is not clear which, if any, relationship between the variables yields a satisfactory fit.

MULTSQUARE fits both Eq. 1 and 2 to the data, and also "tries" the simple linear equation:

$$Y = b_0 + b_1X \quad (3)$$

The least-squares solution to the second-degree polynomial, Eq. 2, is discussed in most statistics texts, and computer and calculator programs are readily available. The additional features of MULTSQUARE are that this program also examines the other equations and presents a quantitative measure for comparing all three.

The usual approach to data that may be correlated by either a simple linear equation (3) or by a polynomial (2) is first to try the linear equation and, if this does not appear satisfactory, to then try the second-degree polynomial. The lines shown in Fig. 2 are the best fits for Eq. 1, 2 and 3. It is certainly not obvious which is the most satisfactory, or in fact whether any of the three is statistically significant.

MULTSQUARE provides for testing all three equations. It gives the equation constants and also the correlation coefficients for all three. It also gives the $t$ value for the comparison of the best two. It is inherent in the arithmetic that the correlation coefficient for Eq. 2 will be at least as good (as large) as the correlation coefficients for Eq. 1 and 3. However, it may not be *statistically* larger. The $t$ test can be used as a measure of statistical significance.

A brief discussion of the relationship between $t$ and the correlation coefficients was given in the previous article (MULT-2). The number of degrees of freedom associated with $t$ is three less than the number of data points. The calculated $t$ value is compared with the tabulated values at the proper number of degrees of freedom to determine the significance level of the difference between the correlation coefficients.

The program provides the equation constants, the three correlation coefficients and the $t$ test value with one input of data. It also provides for checking the data input and for making corrections of erroneous data have been entered.

## Using the program

The program, listed in Table I, is interactive and self-explanatory for the most part. The first display screen explains what the program does, and then the user is invited to begin entering data.

Data are entered as sets, in the order $X$, $Y$. At least four sets of data must be entered, so that there is at least one degree of freedom in the $t$ calculation. There is no means of correcting bad data after they are entered, so one must simply start over. After entering the last set of data, the user enters "END,0" and presses the return key to obtain the results.

The calculated results are the constants and correlation coefficients for the linear equation ($Y = b_0 + b_1 X$) and the two second-degree equations ($Y = b_0 + b_1 X_2$ and

To meet the author see p. 70.

Adapted from an article by William Volk, Consultant, originally published December 17, 1979.

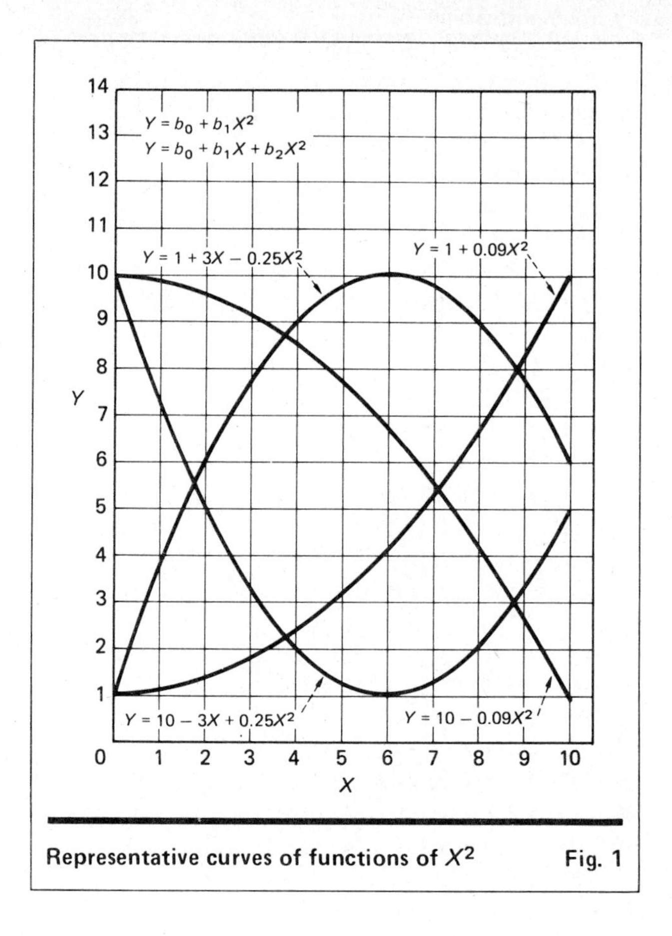


Representative curves of functions of $X^2$ Fig. 1

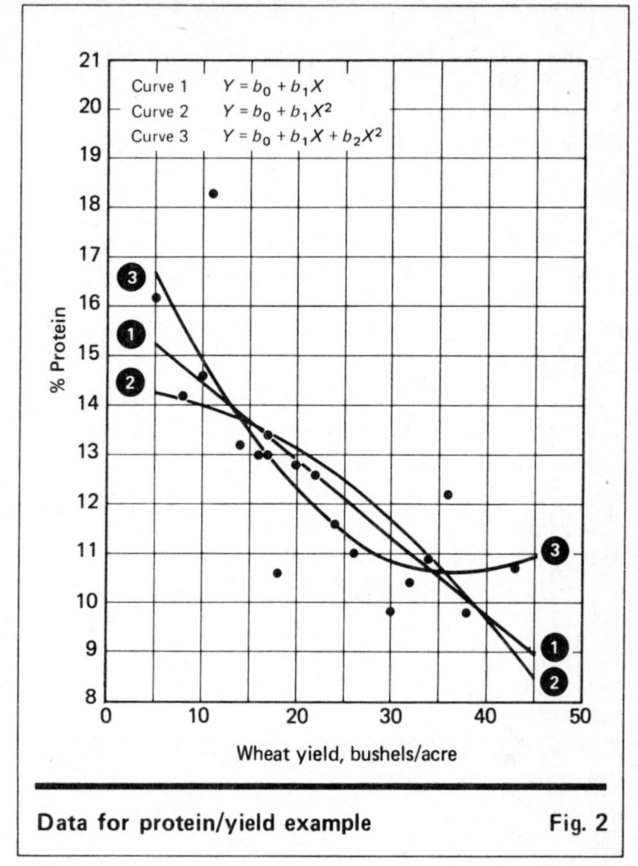


Data for protein/yield example Fig. 2

$Y = b_0 + b_1 X + b_2 X_2$). The program also calculates the $t$ statistic comparing the last equation with the better of the first two. Finally, it evaluates the meaning of this $t$ value for the given number of data, printing this out as the "probability of error in saying there is no improvement."

The program stores the data as X(I) and Y(I), so it can be retrieved by the usual BASIC commands.

## Protein/yield example

This example examines the data used for Fig. 2. The data are given in Table II, and Table III shows the computer-generated displays. The results are shown at the bottom of Table III.

The first-degree linear equation has a correlation coefficient of 0.7814, which is very significant statistically. This equation is plotted as the straight line in Fig. 2. The second-degree linear equation has a correlation coefficient of 0.6988, which is also significant but is not as good as the first-degree equation. This equation is also plotted in Fig. 2 and is concave downward. The downward curvature might not be expected from the general appearance of the data, but if you recall the first section in this series dealing with the program MULTFIT (p. 000), the shape of the $Y = b_0 + b_1X^2$ correlation, fitted to data with $Y$ decreasing with increase in $X$, is concave downward.

The polynomial second-degree equation has a correlation coefficient of 0.8374 that, when compared with the linear correlation coefficient of 0.7814, gives a $t$ value of 2.2021. This $t$ value is just barely significant, having a probability value of 0.04, where values of 0.05 or less are usually accepted as significant. This $t$ value is interpreted in the results (Table III) as the "probability of error in saying there is no improvement."

When a plot of data shows a definite form, the selection of a suitable correlation curve is usually simple. When data show a real trend, but with a considerable scatter, the best correlation form is not obvious. The data in Fig. 2 are of this type. There is a definite decrease in protein with an increase in yield, but the shape of this trend is not clear. All of the equations have a statistically significant correlation coefficient. With 19 data points, any value of $r$ greater than 0.6 is significant at the 0.01 level—i.e., there is less than 1 chance in 100 of having a correlation coefficient this large if the data were merely random points.

The selection of the best equation in this case is a choice between the linear and the second-degree polynomial. The latter is slightly better statistically, but the user must decide whether the improvement justifies the extra complication of a second-degree equation. Examination of the plot may help in making this decision.

**The MULTSQUARE program for second-degree-polynomial-type relationships** **Table I**

```
10  REM   SECOND-DEGREE POLYNOMIAL COMPUTER PROGRAM
20  REM   FROM CHEMICAL ENGINEERING, NOVEMBER 19, 1979
30  REM   BY WILLIAM VOLK
40  REM  TRANSLATED BY WILLIAM VOLK
50  REM   COPYRIGHT (C) 1984
60  REM  BY CHEMICAL ENGINEERING
70  HOME : VTAB 2
80  REM  SET DISPLAY TO FOUR DECIMALS AND DATA TO 100
    ITEMS
90  DEF  FN P(X) =  INT (1E4 * (X + .00005)) / 1E4
100  DIM X(100),Y(100)
110  PRINT "PROGRAM CORRELATES A DEPENDENT VARIABLE"
120  PRINT "AGAINST AN INDEPENDENT VARIABLE WITH"
130  PRINT "A SECOND DEGREE POLYNOMIAL:"
140  PRINT "
150  PRINT "        Y = B0 + B1 * X + B2 * X^2"
160  PRINT
170  PRINT "IT ALSO CALCULATES THE CORRELATIONS FOR:"
180  PRINT
190  PRINT "           Y = B0 + B1 * X"
200  PRINT "           Y = B0 + B1 * X^2"
210  PRINT
220  PRINT "AND IT GIVES A COMPARISON OF THE"
230  PRINT "THREE RESULTS."
240  PRINT
250  PRINT "DATA ARE ENTERED AS:   X,Y"
260  PRINT
270  PRINT "THE DATA INPUT IS"
280  PRINT "TERMINATED WITH:     END,0"
290  PRINT
300  INPUT "PRESS RETURN TO ENTER DATA.  ";Q$
310  HOME : VTAB 4
320  PRINT "ENTER DATA AS: X, Y  SEPARATED BY"
330  PRINT "A COMMA.  WHEN ALL THE DATA ARE IN"
340  PRINT "ENTER:  'END, 0'"
350  PRINT
360  PRINT "THE DATA WILL BE SAVED IN COMPUTER"
370  PRINT "MEMORY AS X(I), Y(I), FOR I GOING"
380  PRINT "FROM 1 TO N."
390  PRINT
400  INPUT "ENTER FIRST SET:     ";X$,Y
410 X2 =  VAL (X$) ^ 2
420 N = N + 1
430 X1 =  VAL (X$)
440 X(N) = X1
450 Y(N) = Y
460  REM  SUMS, SUMS OF SQUARES AND SUMS OF CROSS
     PRODUCTS
470 S1 = S1 + Y
480 S2 = S2 + Y ^ 2
490 S3 = S3 + X1
500 S4 = S4 + X1 ^ 2
510 S5 = S5 + X2
520 S6 = S6 + X2 ^ 2
530 S7 = S7 + X1 * Y
540 S8 = S8 + X2 * Y
550 S9 = S9 + X1 * X2
560  INPUT "NEXT SET OF: 'END,0  ";X$,Y
570  IF X$ = "END" GOTO 590
580  GOTO 410
590  HOME : VTAB 8
600  PRINT "CALCULATION TAKES A"
610  PRINT "FEW SECONDS."
620  REM  CALCULATION OF SUMS OF DEVIATION FROM MEANS
630 P1 = S2 - S1 ^ 2 / N
640 P2 = S4 - S3 ^ 2 / N
650 P3 = S6 - S5 ^ 2 / N
660 P4 = S7 - S3 * S1 / N
670 P5 = S8 - S5 * S1 / N
680 P6 = S9 - S3 * S5 / N
690 B1(1) = P4 / P2
700 B0(1) = (S1 - B1(1) * S3) / N
710 R(1) =  SQR (B1(1) * P4 / P1)
720 B1(2) = P5 / P3
730 B0(2) = (S1 - B1(2) * S5) / N
740 R(2) =  SQR (B1(2) * P5 / P1)
750 TP = P2 * P3 - P6 ^ 2
760 B1(3) = (P4 * P3 - P5 * P6) / TP
770 B2(3) = (P5 * P2 - P4 * P6) / TP
780 B0(3) = (S1 - B1(3) * S3 - B2(3) * S5) / N
790 R(3) =  SQR ((B1(3) * P4 + B2(3) * P5) / P1)
800  REM  DEGREES OF FREEDOM
810 NU = N - 3
820  IF R(1) > R(2) GOTO 850
830 R = R(2)
840  GOTO 860
850 R = R(1)
860 T =  SQR (NU * (R(3) ^ 2 - R ^ 2) / (1 - R(3) ^ 2))
870  GOTO 1190
880  IF PR < 1E - 5 THEN PR = 1E - 4
890  FOR I = 1 TO 3
900 B0(I) =  FN P(B0(I))
910 B1(I) =  FN P(B1(I))
920 B2(I) =  FN P(B2(I))
930 R(I) =  FN P(R(I))
940  NEXT I
950  HOME
960  PRINT  TAB( 10);"THE RESULTS ARE:"
970  PRINT
980  PRINT "Y =  B0 + B1 * X"; TAB( 22);"Y = B0 + B1 * X^2"
990  PRINT
1000  PRINT "B0 = "; FN P(B0(1)); TAB( 25); FN P(B0(2))
1010  PRINT "B1 = "; FN P(B1(1)); TAB( 25); FN P(B1(2))
1020  PRINT "R =  "; FN P(R(1)); TAB( 25); FN P(R(2))
1030  PRINT
1040  PRINT "   Y = B0 + B1 * X + B2 * X^2"
1050  PRINT
1060  PRINT  TAB( 10);"B0 = "; FN P(B0(3))
1070  PRINT  TAB( 10);"B1 = "; FN P(B1(3))
1080  PRINT  TAB( 10);"B2 = "; FN P(B2(3))
1090  PRINT  TAB( 10);"R  = "; FN P(R(3))
1100  PRINT
```

**Program listing (continued)** **Table I**

```
1110  PRINT "T FOR COMPARISON OF THE POLYNOMIAL"
1120  PRINT "WITH THE BETTER OF THE SINGLE"
1130  PRINT "VARIABLE EQUATIONS IS:  "; FN P(T)
1140  PRINT
1150  PRINT "PROBABILITY OF ERROR IN SAYING THERE"
1160  PRINT "IS NO IMPROVEMENT IS:  "; FN P(PR)
1170  GOTO 1370
1180  REM  PROBABILITY CALCULATION
1190 TH =  ATN ( ABS (T) /  SQR (NU))
1200 PI = 3.14159265
1210 PA = 1
1220  IF NU / 2 -  INT (NU / 2) = 0 GOTO 1310
1230  IF NU > 1 GOTO 1260
1240 PR = 1 - (2 * TH / PI)
1250  GOTO 880
1260  FOR I = NU - 3 TO 2 STEP  - 2
1270 PA = 1 + PA * (I / (I + 1) * ( COS (TH)) ^ 2)
1280  NEXT I
1290 PR = 1 - (2 / PI) * (TH +  SIN (TH) *  COS (TH)
     * PA)
1300  GOTO 880
1310  IF NU = 2 GOTO 1350
1320  FOR I = (NU - 3) TO 1 STEP  - 2
1330 PA = 1 + PA * (I / (I + 1) * ( COS (TH)) ^ 2)
1340  NEXT I
1350 PR = 1 -  SIN (TH) * PA
1360  GOTO 880
1370  PRINT  TAB( 13);"END OF PROGRAM"
1380  END
```

**Data for protein/yield example** **Table II**

| % Protein<br>Y | Wheat yield, bushels/acre<br>X |
|---|---|
| 10.7 | 43 |
| 9.8 | 38 |
| 12.2 | 36 |
| 10.9 | 34 |
| 10.4 | 32 |
| 9.8 | 30 |
| 11.0 | 26 |
| 11.6 | 24 |
| 12.6 | 22 |
| 12.8 | 20 |
| 10.6 | 18 |
| 13.4 | 17 |
| 13.0 | 17 |
| 13.0 | 16 |
| 13.2 | 14 |
| 18.3 | 11 |
| 14.6 | 10 |
| 14.2 | 8 |
| 16.2 | 5 |
| 238.3 | 421 |

These data are from Snedecor, George W., "Statistical Methods," Iowa State College Press. They are a random selection of one-fifth of the original data.

**Example for: Second-degree-polynomial computer program** **Table III**

**(Start of first display)**

```
PROGRAM CORRELATES A DEPENDENT VARIABLE
AGAINST AN INDEPENDENT VARIABLE WITH
A SECOND DEGREE POLYNOMIAL:

        Y = B0 + B1 * X + B2 * X^2

IT ALSO CALCULATES THE CORRELATIONS FOR:

          Y = B0 + B1 * X
          Y = B0 + B1 * X^2

AND IT GIVES A COMPARISON OF THE
THREE RESULTS.

DATA ARE ENTERED AS:    X,Y

THE DATA INPUT IS
TERMINATED WITH:      END.0

PRESS RETURN TO ENTER DATA.
```

**(Start of next display)**

```
ENTER DATA AS: X, Y  SEPARATED BY
A COMMA.  WHEN ALL THE DATA ARE IN
ENTER:  'END, 0'

THE DATA WILL BE SAVED IN COMPUTER
MEMORY AS X(I), Y(I), FOR I GOING
FROM 1 TO N.

ENTER FIRST SET:       43,10.7
NEXT SET OF: 'END,0  38,9.8
NEXT SET OF: 'END,0  36,12.2
NEXT SET OF: 'END,0  34,10.9
NEXT SET OF: 'END,0  32,10.4
NEXT SET OF: 'END,0  30,9.8
NEXT SET OF: 'END.0  26,11
NEXT SET OF: 'END,0  24,11.6
NEXT SET OF: 'END,0  20,12.8
NEXT SET OF: 'END,0  18,10.6
NEXT SET OF: 'END,0  17,13.4
NEXT SET OF: 'END,0  17.13
```

**Example for: Second-degree-polynomial computer program (continued)** **Table III**

```
NEXT SET OF: 'END,0  16,13
NEXT SET OF: 'END,0  14,13.2
NEXT SET OF: 'END,0  11,18.3
NEXT SET OF: 'END,0  10,14.6
NEXT SET OF: 'END,0  8,14.2
NEXT SET OF: 'END,0  5,16.2
NEXT SET OF: 'END,0  END,0
CALCULATION TAKES A
FEW SECONDS.
```

**(Start of next display)**

```
       THE RESULTS ARE:

Y =  B0 + B1 * X      Y = B0 + B1 * X^2

B0 = 16.0518            14.3251
B1 = -.1585             -2.9E-03
R =  .7814              .6995

    Y = B0 + B1 * X + B2 * X^2

          B0 = 18.8022
          B1 = -.4544
          B2 = 6.2E-03
          R  = .841

T FOR COMPARISON OF THE POLYNOMIAL
WITH THE BETTER OF THE SINGLE
VARIABLE EQUATIONS IS:  2.2251

PROBABILITY OF ERROR IN SAYING THERE
IS NO IMPROVEMENT IS:  .0418
            END OF PROGRAM
```

# Solution to a three-variable problem

This section presents MULT-3, which provides a correlation of one dependent variable with up to three independent variables—taken one, two or three at a time.

☐ Multivariable-correlation solutions by the least-squares method are a fairly common tool for chemical engineers. Statistics texts and computer statistics packages usually describe the technique.

In general, two approaches are employed: one is to determine the correlation by using all of the independent variables and then, by statistical analysis, to delete those variables that do not make a significant contribution to the correlation.

The other method, common in computer-statistics packages, is to test the independent variables one at a time, select the best one, hold that one and test the others along with the first, select the best pair that includes the best one, and so on. The first method is cumbersome and may involve more work than is necessary; the second may not give the best correlation.

The program presented here, MULT-3, looks at all the possible relationships between the dependent variable and the independent variables, first one at a time, then all pairs, and finally gives the correlation with all three independent variables. It provides the regression coefficients and the correlation coefficient for all the relationships. The reason for this total solution is that the independent variable that shows the best correlation in a simple linear equation may not be one of the pair that yields the best solution when the variables are taken two at a time. The example given later demonstrates this situation.

Even if the independent variable that provides the best correlation when the variables are taken one at a time is included with another for the best pair (which is most often the case), it should be of interest to the experimenter to see the degree of correlation that exists among all the variables. The program MULT-3 gives this information, by yielding the solution to the following equations:

$$Y = b_0 + b_1X_1 \quad (1) \qquad Y = b_0 + b_1X_1 + b_2X_2 \quad (4)$$

$$Y = b_0 + b_1X_2 \quad (2) \qquad Y = b_0 + b_1X_1 + b_2X_3 \quad (5)$$

$$Y = b_0 + b_1X_3 \quad (3) \qquad Y = b_0 + b_1X_2 + b_2X_3 \quad (6)$$

$$Y = b_0 + b_1X_1 + b_2X_2 + b_3X_3 \quad (7)$$

To meet the author see p. 70.

along with the correlation coefficients for each. It also tests the statistical significance of the correlation coefficients of the multivariable equations against the correlation coefficients of the equations having fewer variables.

## Using the program

The program, listed in Table I, is interactive and self-explanatory for the most part. The first display screen explains what the program does, and then the user is invited to begin entering data.

Data are entered as sets, in the order $X_1, X_2, X_3, Y$. At least five sets of data must be entered, so that there is at least one degree of freedom in the $t$ calculation. There is no means of correcting bad data after they are entered, so one must simply start over. After entering the last set of data, the user enters "END, 0, 0, 0" and presses the return key to obtain the results.

The first set of results is for the single-variable equations (Eq. 1–3): $Y = b_0 + b_1 X_1$, $Y = b_0 + b_1 X_2$, and $Y = b_0 + b_1 X_3$.

The next set of results, obtained by pressing the return key, is for the two-variable equations (Eq. 4–6): $Y = b_0 + b_1 X_1 + b_2 X_2$, $Y = b_0 + b_1 X_1 + b_2 X_3$, and $Y = b_0 + b_1 X_2 + b_2 X_3$.

The last set of results, obtained by pressing the return key, is for the three-variable equation (Eq. 7): $Y = b_0 + b_1 X_1 + b_2 X_2 + b_3 X_3$. The program calculates the regression coefficients and correlation coefficient for each of the equations. After presenting the results for the three-variable equation, it also calculates the $t$ statistic comparing the last equation with the best of the two-variable equations. Finally, the program evaluates the meaning of this $t$ value for the given number of data, printing this out as the "probability of error in saying there is no improvement."

The program stores the data as X1(I), X2(I), X3(I) and Y(I), so they can be retrieved by the usual BASIC commands.

Solutions of the single- and two-variable equations follow the routines discussed in popular statistics texts [1, 2, 3]. Solution of the three-variable equation involves the solution of three simultaneous equations, to get $b_1$,

Adapted from an article by William Volk, Consultant, originally published December 17, 1979.

$b_2$ and $b_3$, then solution for $b_0$ by the following equation:

$$b_0 = \bar{Y} - b_1 \bar{X}_1 - b_2 \bar{X}_2 - b_3 \bar{X}_3 \quad (8)$$

This program uses the Doolittle procedure [4] to solve the three equations simultaneously. This works by eliminating variables through division and subtraction until only one equation with one unknown is left. The value from this equation is then substituted into the others to get the rest of the unknowns.

## Crystal flowrate example

Table II gives some data on the flowrate of a certain crystal as dependent variable; and moisture, crystal dimension ratio, and impurity as independent variables. These data were selected from a larger quantity of data [5] to demonstrate MULT-3. The regression coefficients from the total data are within reasonable confidence limits of the coefficients calculated from the data of Table II, and the smaller amount of data simplifies the presentation. Table III shows the computer-generated displays, and the results, for this example.

The example shows that the best correlation of flowrate with one independent variable is with moisture level—the correlation coefficient equals 0.6457. However, the best correlation of flowrate with two independent variables is not with moisture and one of the others, but is with crystal dimension ratio and purity—a correlation coefficient of 0.8065. The correlation coefficient with all three independent variables is only 0.8119, not significantly better than with the best two. Just how this result is interpreted is up to the experimenter, but the use of MULT-3 permits the engineer to evaluate all of the correlations and does not get him locked in with the first best one.

To run the example, enter the data in Table II in $X_1$, $X_2$, $X_3$, $Y$ sets. After the last set, enter "END, 0, 0, 0." Then press the return key to get the correlation constants and coefficients for the three single-variable equations. Table III, the example display, shows these data and results.

Now press the return key again to get the correlation constants and coefficients for the three two-variable equations. The best correlation is with the combination $X_2$ and $X_3$. Press the return key once more to get the correlation constants and coefficient for the three-variable equation. The final result is the $t$ statistic comparing the three-variable correlation with the best of the two-variable ones.

The program interprets the $t$ statistic as a "probability of error in saying there is no improvement." In this case, the $t$ value of 0.5989 is definitely not significant—there is only a 0.5588 probability of error in saying the three-variable equation is no better than the best of the two-variable equations.

It is very unusual to find data in which the independent variable that gives the best single-variable correlation is not included with others in giving improved correlation. The data of the example were selected to show one such case. What MULT-3 does that most other multivariable correlations do not do is give an opportunity to determine whether or not this situation occurs.

**MULT-3 program for the correlation of three independent variables** **Table I**

```
10  REM  SOLUTION TO A THREE-VARIABLE PROBLEM
20  REM  FROM CHEMICAL ENGINEERING, DECEMBER 17, 1979
30  REM  BY WILLIAM VOLK
40  REM  TRANSLATED BY WILLIAM VOLK
50  REM  COPYRIGHT (C) 1984
60  REM  BY CHEMICAL ENGINEERING
70  HOME : VTAB 2
80  REM  SET DISPLAY TO FOUR DECIMALS AND DATA TO 100
    ENTRIES
90  DEF  FN P(X) =  INT (1E4 * (X + .00005)) / 1E4
100  DIM Y(100),X1(100),X2(100),X3(100)
110 PI = 3.14159265
120  PRINT "PROGRAM CALCULATES CORRELATION OF ONE"
130  PRINT "DEPENDENT VARIABLE AGAINST THREE"
140  PRINT "INDEPENDENT VARIABLES."
150  PRINT
160  PRINT "IT CALCULATES THE CORRELATIONS WITH ALL"
170  PRINT "COMBINATIONS OF INDEPENDENT VARIABLES."
180  PRINT
190  PRINT "IT GIVE A STATISTICAL COMPARISON"
200  PRINT "OF THE RESULTS."
210  PRINT
220  PRINT "DATA ARE ENTERED AS: X1, X2, X3, Y"
230  PRINT "SEPARATED BY COMMAS."
240  PRINT
250  PRINT "THE DATA INPUT IS"
260  PRINT "TERMINATED WITH:     END,  0,  0,  0"
270  PRINT
280  PRINT "DATA WILL BE SAVED IN COMPUTER MEMORY"
290  PRINT "AS X1(I), X2(I), X3(I), Y(I), FOR"
300  PRINT "I GOING FROM 1 TO N."
310  PRINT
320  INPUT "PRESS RETURN TO ENTER DATA.  ";Q$
330  HOME : VTAB 4
340  PRINT "ENTER DATA: X1, X2, X3, Y  SEPARATED"
350  PRINT "BY COMMAS.  WHEN ALL DATA ARE IN"
360  PRINT "ENTER: 'END, 0, 0, 0'"
370  PRINT
380  INPUT "ENTER FIRST SET:      ";X$,D(3),D(4),D(1)
390 D(2) =  VAL (X$)
400 N = N + 1
410  REM  CONVERT INPUT TO X & Y VARIABLES
420 X1(N) = D(2)
430 X2(N) = D(3)
440 X3(N) = D(4)
450 Y(N) = D(1)
460  REM  SUMS, SUMS OF SQUARES AND SUMS OF CROSS PRODUCTS
470  FOR I = 1 TO 4
480 S(I) = S(I) + D(I)
490  FOR J = 1 TO 4
```

**MULT-3 program for the correlation of three independent variables (continued)** **Table I**

```
500 Q(I,J) = Q(I,J) + D(I) * D(J)
510  NEXT J
520  NEXT I
530  INPUT "NEXT OR:'END,0,0,0' ";X$,D(3),D(4),D(1)
540  IF X$ = "END" GOTO 560
550  GOTO 390
560  FOR I = 1 TO 4
570  FOR J = 1 TO 4
580 P(I,J) = Q(I,J) - S(I) * S(J) / N
590  NEXT J
600  NEXT I
610  FOR I = 1 TO 3
620 B(1,I) = P(1,I + 1) / P(I + 1,I + 1)
630 B(0,I) = (S(1) - B(1,I) * S(I + 1)) / N
640 R(I) =  SQR (B(1,I) * P(1,I + 1) / P(1,1))
650 B(0,I) =  FN P(B(0,I))
660 B(1,I) =  FN P(B(1,I))
670 R(I) =  FN P(R(I))
680  NEXT I
690  HOME : VTAB 4
700  PRINT "FOR THE SINGLE-VARIABLE EQUATION:"
710  PRINT
720  PRINT "         Y = B0 + B1 * X"
730  PRINT
740  PRINT  TAB( 12);"X1"; TAB( 22);"X2"; TAB( 32);"X3"
750  PRINT
760  PRINT "B0 ="; TAB( 10);B(0,1); TAB( 20);B(0,2); TAB( 30);B(0,3)
770  PRINT
780  PRINT "B1 ="; TAB( 10);B(1,1); TAB( 20);B(1,2); TAB( 30);B(1,3)
790  PRINT
800  PRINT "R  ="; TAB( 10);R(1); TAB( 20);R(2); TAB( 30);R(3)
810  PRINT : PRINT : PRINT : PRINT
820  INPUT "PRESS RETURN FOR FURTHER RESULTS. ";Q$
830 TP(1) = P(2,2) * P(3,3) - P(2,3) ^ 2
840 TP(2) = P(2,2) * P(4,4) - P(2,4) ^ 2
850 TP(3) = P(3,3) * P(4,4) - P(3,4) ^ 2
860 B(1,4) = (P(1,2) * P(3,3) - P(1,3) * P(2,3)) / TP(1)
870 B(2,4) = (P(1,3) * P(2,2) - P(1,2) * P(2,3)) / TP(1)
880 B(1,5) = (P(1,2) * P(4,4) - P(1,4) * P(2,4)) / TP(2)
890 B(2,5) = (P(1,4) * P(2,2) - P(1,2) * P(2,4)) / TP(2)
900 B(1,6) = (P(1,3) * P(4,4) - P(1,4) * P(3,4)) / TP(3)
910 B(2,6) = (P(1,4) * P(3,3) - P(1,3) * P(3,4)) / TP(3)
920 B(0,4) = (S(1) - B(1,4) * S(2) - B(2,4) * S(3)) / N
930 B(0,5) = (S(1) - B(1,5) * S(2) - B(2,5) * S(4)) / N
940 B(0,6) = (S(1) - B(1,6) * S(3) - B(2,6) * S(4)) / N
950 R(4) =  SQR ((B(1,4) * P(1,2) + B(2,4) * P(1,3)) / P(1,1))
960 R(5) =  SQR ((B(1,5) * P(1,2) + B(2,5) * P(1,4)) / P(1,1))
970 R(6) =  SQR ((B(1,6) * P(1,3) + B(2,6) * P(1,4)) / P(1,1))
980 AW = B(1,4)
990  FOR I = 4 TO 6
1000 R(I) =  FN P(R(I))
1010  FOR J = 0 TO 2
1020 B(J,I) =  FN P(B(J,I))
1030  NEXT J
1040  NEXT I
1050  HOME : VTAB 4
1060  PRINT "      FOR THE TWO-VARIABLE EQUATION:"
```

**MULT-3 program for the correlation of three independent variables (continued)** **Table I**

```
1070  PRINT
1080  PRINT "         Y =  B0 + B1 * X1 + B2 * X2"
1090  PRINT
1100  PRINT  TAB( 11);"X1,X2"; TAB( 21);"X1,X3"; TAB( 31);"X2,X3"
1110  PRINT
1120  PRINT "B0 ="; TAB( 10);B(0,4); TAB( 20);B(0,5); TAB( 30);B(0,6)
1130  PRINT "B1 ="; TAB( 10);B(1,4); TAB( 20);B(1,5); TAB( 30);B(1,6)
1140  PRINT "B2 ="; TAB( 10);B(2,4); TAB( 20);B(2,5); TAB( 30);B(2,6)
1150  PRINT
1160  PRINT "R  ="; TAB( 10);R(4); TAB( 20);R(5); TAB( 30);R(6)
1170  PRINT : PRINT : PRINT
1180  INPUT "PRESS RETURN FOR ADDITIONAL RESULTS.  ";X$
1190  FOR J = 2 TO 4
1200  FOR I = 1 TO 3
1210 Q(J,I) = P(J,I) / P(J,4)
1220  NEXT I
1230  NEXT J
1240  FOR J = 2 TO 3
1250  FOR I = 1 TO 3
1260 Q(J,I) = Q(J,I) - Q(J + 1,I)
1270  NEXT I
1280  NEXT J
1290 B(2,7) = (Q(2,1) / Q(2,2) - Q(3,1) / Q(3,2)) / (Q(2,3) / Q(2,2) - Q(3,3) /
Q(3,2))
1300 B(1,7) = (Q(2,1) - B(2,7) * Q(2,3)) / Q(2,2)
1310 B(3,7) = (P(2,1) - B(1,7) * P(2,2) - B(2,7) * P(2,3)) / P(2,4)
1320 B(0,7) = (S(1) - B(1,7) * S(2) - B(2,7) * S(3) - B(3,7) * S(4)) / N
1330  FOR I = 0 TO 3
1340 B(I,7) =  FN P(B(I,7))
1350  NEXT I
1360  HOME : VTAB 4
1370  PRINT " FOR THE THREE-VARIABLE EQUATION:"
1380  PRINT
1390  PRINT "  Y = B0 + B1 * X1 + B2 * X2 + B3 * X3"
1400  PRINT
1410  PRINT "B0 = "; TAB( 20);B(0,7)
1420  PRINT "B1 = "; TAB( 20);B(1,7)
1430  PRINT "B2 ="; TAB( 20);B(2,7)
1440  PRINT "B3 ="; TAB( 20);B(3,7)
1450  PRINT
1460 R(7) =  SQR ((B(1,7) * P(1,2) + B(2,7) * P(1,3) + B(3,7) * P(1,4)) / P(1,1)
)
1470  PRINT "R  ="; TAB( 20); FN P(R(7))
1480  GOSUB 1590
1490  PRINT
1500  PRINT "T FOR COMPARISON OF THE THREE VARIABLE"
1510  PRINT "EQUATION WITH THE BEST OF THE TWO"
1520  PRINT "VARIABLE EQUATIONS IS:     "; FN P(T)
1530  PRINT
1540  PRINT "PROBABILITY OF ERROR IN SAYING THERE"
1550  PRINT "IS NO IMPROVEMENT IS: "; FN P(PR)
1560  PRINT
1570  GOTO 1810
1580 R = R(4)
1590  IF R(5) > R THEN R = R(5)
1600  IF R(6) > R THEN R = R(6)
1610  REM  DEGREES OF FREEDOM AND PROBABILITY CALCULATION
```

**Program listing (continued)** **Table I**

```
1620 NU = N - 4
1630 T =  SQR (NU * (R(7) ^ 2 - R ^ 2) / (1 - R(7) ^
     2))
1640 TH =  ATN ( ABS (T) /  SQR (NU))
1650 PA = 1
1660  IF NU / 2 -  INT (NU / 2) = 0 GOTO 1750
1670  IF NU > 1 GOTO 1700
1680 PR = 1 - (2 * TH / PI)
1690  RETURN
1700  FOR I = NU - 3 TO 2 STEP  - 2
1710 PA = 1 + PA * (I / (I + 1) * ( COS (TH)) ^ 2)
1720  NEXT I
1730 PR = 1 - (2 / PI) * (TH +  SIN (TH) *  COS (TH) *
     PA)
1740  RETURN
1750  IF NU = 2 GOTO 1790
1760  FOR I = (NU - 3) TO 1 STEP  - 2
1770 PA = 1 + PA * (I / (I + 1) * ( COS (TH)) ^ 2)
1780  NEXT I
1790 PR = 1 -  SIN (TH) * PA
1800  RETURN
1810  PRINT
1820  PRINT  TAB( 15);"END OF PROGRAM."
1830  END
```

**Data for crystal flowrate example** **Table II**

| | $Y$ (Flowrate, g/s) | $X_1$ (Moisture content, %) | $X_2$ (Dimension ratio, length/breadth) | $X_3$ (Impurity, %) |
|---|---|---|---|---|
| | 3.21 | 0.12 | 3.2 | 0.01 |
| | 3.25 | 0.12 | 2.7 | 0.00 |
| | 4.00 | 0.17 | 2.7 | 0.00 |
| | 3.62 | 0.24 | 2.8 | 0.00 |
| | 3.76 | 0.10 | 2.6 | 0.00 |
| | 4.55 | 0.11 | 2.0 | 0.02 |
| | 5.32 | 0.10 | 2.0 | 0.07 |
| | 4.39 | 0.10 | 2.0 | 0.02 |
| | 4.59 | 0.17 | 2.2 | 0.03 |
| | 5.00 | 0.17 | 2.4 | 0.04 |
| | 3.68 | 0.15 | 2.4 | 0.02 |
| | 3.18 | 0.23 | 2.2 | 0.10 |
| | 5.00 | 0.21 | 1.9 | 0.04 |
| | 0.00 | 0.37 | 2.3 | 0.14 |
| | 3.70 | 0.28 | 2.4 | 0.05 |
| | 3.40 | 0.32 | 3.3 | 0.08 |
| | 0.00 | 0.28 | 3.5 | 0.12 |
| | 2.33 | 0.22 | 3.0 | 0.06 |
| Totals | 62.98 | 3.46 | 45.6 | 0.80 |

**Example for: Solution to a three-variable problem** **Table III**

**(Start of first display)**

```
PROGRAM CALCULATES CORRELATION OF ONE
DEPENDENT VARIABLE AGAINST THREE
INDEPENDENT VARIABLES.

IT CALCULATES THE CORRELATIONS WITH ALL
COMBINATIONS OF INDEPENDENT VARIABLES.

IT GIVE A STATISTICAL COMPARISON
OF THE RESULTS.

DATA ARE ENTERED AS: X1, X2, X3, Y
SEPARATED BY COMMAS.

THE DATA INPUT IS
TERMINATED WITH:    END,  0,  0,  0

DATA WILL BE SAVED IN COMPUTER MEMORY
AS X1(I), X2(I), X3(I), Y(I), FOR
I GOING FROM 1 TO N.

PRESS RETURN TO ENTER DATA.
```

**(Start of next display)**

```
ENTER DATA: X1, X2, X3, Y  SEPARATED
BY COMMAS.  WHEN ALL DATA ARE IN
ENTER: 'END, 0, 0, 0'

ENTER FIRST SET:      .12,3.2,.01,3.21
NEXT OR:'END,0,0,0' .12,2.7,0,3.25
NEXT OR:'END,0,0,0' .17,2.7,0,4
NEXT OR:'END,0,0,0' .24,2.8,0,3.62
NEXT OR:'END,0,0,0' .1,2.6,0,3.76
NEXT OR:'END,0,0,0' .11,2,.02,4.55
NEXT OR:'END,0,0,0' .1,2,.07,5.32
NEXT OR:'END,0,0,0' .1,2,.02,4.39
NEXT OR:'END,0,0,0' .17,2.2,.03,4.59
NEXT OR:'END,0,0,0' .17,2.4,.04,5
NEXT OR:'END,0,0,0' .15,2.4,.02,3.68
NEXT OR:'END,0,0,0' .23,2.2,.1,3.18
NEXT OR:'END,0,0,0' .21,1.9,.04,5
NEXT OR:'END,0,0,0' .37,2.3,.14,0
NEXT OR:'END,0,0,0' .28,2.4,.05,3.7
NEXT OR:'END,0,0,0' .32,3.3,.08,3.4
NEXT OR:'END,0,0,0' .28,3.5,.12,0
NEXT OR:'END,0,0,0' .22,3,.06,2.33
NEXT OR:'END,0,0,0' END,0,0,0
```

**(Start of next display)**

```
FOR THE SINGLE-VARIABLE EQUATION:

        Y = B0 + B1 * X

             X1         X2         X3

B0 =     5.7466    7.8863     4.48
```

**Example for: Solution to a three-variable problem (continued)** **Table III**

```
B1 =      -11.6935  -1.7319   -22.0747

R  =      .6457     .5591     .6378




PRESS RETURN FOR FURTHER RESULTS.
```

**(Start of next display)**

```
      FOR THE TWO-VARIABLE EQUATION:

            Y =  B0 + B1 * X1 + B2 * X2

             X1,X2      X1,X3      X2,X3

B0 =       8.3507     5.3968     8.295
B1 =       -9.37620001-7.0051    -1.5382
B2 =       -1.2038    -12.4066   -20.2368

R  =       .7426      .6916      .8065



PRESS RETURN FOR ADDITIONAL RESULTS.
```

**(Start of next display)**

```
 FOR THE THREE-VARIABLE EQUATION:

  Y = B0 + B1 * X1 + B2 * X2 + B3 * X3

B0 =                 8.3545
B1 =                 -2.6198
B2 =                 -1.4239
B3 =                 -16.7576

R  =                 .8119

T FOR COMPARISON OF THE THREE VARIABLE
EQUATION WITH THE BEST OF THE TWO
VARIABLE EQUATIONS IS:     .5989

PROBABILITY OF ERROR IN SAYING THERE
IS NO IMPROVEMENT IS: .5588


                END OF PROGRAM.
```

## References


1. Brownlee, K. A., "Statistical Theory and Methodology," 2nd ed., Wiley, New York, 1960.
2. Snedecor, G. W., "Statistical Methods," Iowa State College Press, Ames, Iowa, 1956.
3. Volk, W.,' "Applied Statistics for Engineers," 2nd ed., McGraw-Hill, New York, 1969.
4. Anderson, R. L., & Bancroft, T. A., "Statistical Theory in Research," McGraw-Hill, New York, 1952.
5. Davies, O. L., "Statistical Methods in Research and Production," Hafner Pub. Co., New York, 1957.

# Correlating the hyperbolic function

A considerable amount of chemical engineering data present asymptotic properties characteristic of the hyperbolic correlation $Y = (a + bX)/(1 + cX)$. This program solves the equation by an approximate least-squares technique.

□ Very often, the engineer is faced with the task of fitting the best possible curve to a set of precise data taken in the laboratory for the correlation of physical properties such as density, specific heat, viscosity, thermal conductivity and vapor pressure. In such cases, the three-constant equation $Y = (a + bX)/(1 + cX)$ adapts itself quite well to the curvature of the data and usually yields excellent results (see Ref. 1).

## The program

**It is generally believed that representation of data by the methods of conventional regression analysis is confined to the "straight line" $Y = b_0 + b_1 F(X)$, or the parabola, or—more generally—to an equation linear with respect to the coefficients $b_0$, $b_1$, and so on (see Ref. 3, p 2–770).**

However, in the case of the hyperbola, the following equations satisfy the least-squares criterion $\Sigma(Y_i - Y_c)^2$ = minimum:

$$\Sigma Y_i = an + b\Sigma X_i - c\Sigma X_i Y_i \quad (1)$$

$$\Sigma X_i Y_i = a\Sigma X_i + b\Sigma X_i^2 - c\Sigma X_i^2 Y_i \quad (2)$$

$$\Sigma X_i Y_i^2 = a\Sigma X_i Y_i + b\Sigma X_i^2 Y_i - c\Sigma X_i^2 Y_i^2 \quad (3)$$

Solution for the constants $a$, $b$ and $c$ in terms of the data $X_i$, $Y_i$ is straightforward.

Table I lists the BASIC program that fits an equation of the form $Y = (a + bX)/(1 + cX)$ to a set of $X$, $Y$ data. The program is interactive, and self-explanatory for the most part. Simply enter the data in the order $X$, $Y$, and the program calculates the constants $a$, $b$ and $c$. It also calculates a correlation coefficient, $r$, that indicates the goodness of fit—a correlation of 1.0 is perfect.

When these calculations are complete, the program can also calculate values of $Y$ for given values of $X$. Finally, the data are stored as X(I) and Y(I), so they can be retrieved by the usual BASIC commands.

## Vapor pressure example

Suppose we wish to correlate the vapor pressure of 1,3 butadiene as a function of temperature, from 40 mm Hg abs. up to 30 atm absolute. Perry gives the values shown in Table II [3]. Table III shows the computer-generated displays and results.

The Clausius-Clapeyron equation predicts that ln $P$ versus $1/T$ should produce a straight line. However, some curvature always occurs, as taken into account by the Antoine equation (Ref. 3, p. 3–228).

Calling $X = 1/T \times 10^3$ and $Y = \ln P$, the least-squares fit to $Y = (a + bX)/(1 + cX)$ yields:

$$a = 15.42 \qquad b = -2.64 \qquad c = -0.0417$$

so the final equation of butadiene vapor pressure is:

$$\ln P = (15.42 - 2640/T)/(1 - 41.7/T)$$

where $T$ is temperature, K, and $p$ is pressure, mm Hg. (Note that $c \ll b$, as it should be according to the Clausius-Clapeyron prediction; that is, the curve approaches a straight line.)

The calculated pressures at each temperature are shown in the last column of Table II. Deviations from the observed values are on the order of 1% or less.

Adapted from an article by Jean R. Brosens, Proquigel, originally published April 7, 1980.

## Final remarks

Since the hyperbolic equation is nonlinear, a determination of the quality of fit of the estimate, such as the standard deviation

$$s = \sqrt{\frac{\Sigma(Y_i - Y_c)^2}{n - 3}}$$

cannot be expressed as a combination of terms of the type $\Sigma X^m Y^n$ where $m$ and $n$ may assume the values 0, 1, 2 . . . and so on. In the program:

$$r = \sqrt{1 - \left(\frac{n\ \Sigma(Y_i - Y_c)^2}{\Sigma Y^2 - (\Sigma Y)^2}\right)}$$

### Nomenclature

$a, b, c$ Coefficients in the hyperbolic equation

$$Y = (a + bX)/(1 + cX)$$

$n$ Number of data pairs $X_i$, $Y_i$

$s$ Standard deviation of the estimate (equal to 0 for a perfect fit)

$X_i$, $Y_i$ Experimental (data) values of $X$ and $Y$

$Y_c$ $Y$ value calculated from

$$Y_c = (a + bX_i)/(1 + cX_i)$$

**Program for correlating the hyperbolic function** **Table I**

```
10 REM  CORRELATING THE HYPERBOLIC FUNCTION
20 REM   FROM CHEMICAL ENGINEERING, APRIL 7, 1980
30 REM   BY JEAN R. BROSENS
40 REM  TRANSLATED BY WILLIAM VOLK
50 REM   COPYRIGHT (C) 1984
60 REM  BY CHEMICAL ENGINEERING
70 HOME : VTAB 4
80 REM  SET DATA INPUT AT 100 AND DISPLAY TO FOUR
   DECIMALS
90 DIM X(100),Y(100)
100 DEF FN P(X) = ( INT (1E4 * (X + .00005))) / 1E4
110 PRINT "   CORRELATION FOR THE RELATION:"
120 PRINT
130 PRINT "   Y = (A + B * X)/(1 + C * X)"
140 PRINT
150 PRINT "DATA ARE ENTERED AS: X,Y"
160 PRINT
170 PRINT "WHEN ALL THE DATA"
180 PRINT "ARE IN, ENTER:    END, 0"
190 PRINT
200 PRINT "DATA WILL BE SAVED IN COMPUTER MEMORY"
210 PRINT "AS X(I), Y(I), FOR I GOING FROM"
220 PRINT "1 TO N."
230 PRINT
240 INPUT "PRESS RETURN TO ENTER DATA.  ";Q$
250 HOME : VTAB 4
260 PRINT "ENTER THE DATA IN PAIRS: X, Y"
270 PRINT "SEPARATED BY A COMMA."
280 PRINT
290 PRINT "WHEN ALL THE DATA ARE IN, ENTER"
300 PRINT "'END, 0'."
310 PRINT
320 INPUT "ENTER FIRST DATA PAIR:  ";X$,Y
330 N = N + 1
340 X =  VAL (X$)
350 X(N) = X
360 Y(N) = Y
370 S1 = S1 + X
380 S2 = S2 + X ^ 2
390 S3 = S3 + Y
400 S4 = S4 + X * Y
410 S5 = S5 + X * Y ^ 2
420 S6 = S6 + X ^ 2 * Y
430 S7 = S7 + (X * Y) ^ 2
440 S8 = S8 + Y ^ 2
450 INPUT "NEXT PAIR OR: 'END, 0'  ";X$,Y
460 IF X$ = "END" GOTO 480
470 GOTO 330
480 A1 = S2 * S7 - S6 ^ 2
490 B1 = S1 * S6 - S2 * S4
500 A2 = (S3 * S6 - S4 ^ 2) * A1 - (S4 * S7 - S5 * S6)
   * B1
510 B2 = (N * S6 - S1 * S4) * A1 - (S1 * S7 - S4 * S6)
   * B1
520 A = A2 / B2
530 B = (S3 / S4 - S4 / S6) - A * (N / S4 - S1 / S6)
540 B = B / (S1 / S4 - S2 / S6)
550 C = (A * N + B * S1 - S3) / S4
560 YP = S8 - S3 ^ 2 / N
570 FOR I = 1 TO N
580 D = D + (Y(I) - ((A + B * X(I)) / (1 + C * X(I))))
   ^ 2
590 NEXT I
600 IF D > YP GOTO 620
610 GOTO 640
620 FL = 1
630 GOTO 650
640 R = SQR ((YP - D) / YP)
650 HOME : VTAB 4
660 IF FL = 0 GOTO 730
670 PRINT "HYPERBOLIC FUNCTION CORRELATION"
680 PRINT "IS NOT APPLICABLE TO YOUR DATA.
690 PRINT "DEVIATIONS FROM THE CORRELATION ARE"
700 PRINT "GREATER THAN FROM THE MEAN"
710 PRINT "HOWEVER, THE RESULTS ARE:"
720 PRINT
730 PRINT "CONSTANTS FOR CORRELATION:"
740 PRINT
```

**Program for correlating the hyperbolic function (continued)** **Table I**

```
750  PRINT "Y = (A + B * X)/(1 + C * X)"
760  PRINT
770  PRINT "ARE:"
780  PRINT "
790  PRINT "A ="; TAB( 25); FN P(A)
800  PRINT "B ="; TAB( 25); FN P(B)
810  PRINT "C ="; TAB( 25); FN P(C)
820  PRINT
830  PRINT "CORRELATION"
840 R =  FN P(R)
850  IF R = 1 THEN R = .9999
860  PRINT "   COEFFICIENT, R ="; TAB( 25);R
870  PRINT
880  PRINT "IF YOU WISH TO ESTIMATE A VALUE OF Y"
890  PRINT "FROM AN X VALUE, ENTER THE X VALUE."
900  INPUT "IF NOT, ENTER 'NO'.            ";Q$
910  IF Q$ = "NO" GOTO 1150
920  IF  ASC (Q$) > 59 GOTO 940
930  GOTO 980
940  PRINT
950  PRINT "YOU SHOULD ENTER A NUMBER OR 'NO'.  YOU"
960  PRINT "ENTERED ";Q$;".  TRY AGAIN."
970  GOTO 870
980 X =  VAL (Q$)
990 Y = (A + B * X) / (1 + C * X)
1000  PRINT
1010  PRINT "ESTIMATED Y WHEN X =";X;" IS:  "; FN P(Y)
1020  PRINT
1030  PRINT "DO YOU WANT ANOTHER ESTIMATE?  ANSWER"
1040  INPUT "Y OR N.  ";X$
1050  IF X$ = "N" GOTO 1150
1060  IF X$ = "Y" GOTO 1110
1070  PRINT
1080  PRINT "ENTER Y(ES) OR N(O).  YOU "
1090  PRINT "ENTERED ";X$;". TRY AGAIN.
1100  GOTO 1020
1110  PRINT
1120  INPUT "ENTER X   ";X
1130  PRINT
1140  GOTO 990
1150  PRINT
1160  PRINT  TAB( 13);"END OF PROGRAM"
1170  END
```

**Vapor pressure of butadiene** **Table II**

| $t$, °C | $T$, K | $1/T \times 10^3$ | $P$, mm Hg | $\ln P$ | $P_{calc.}$, mm Hg |
|---|---|---|---|---|---|
| -61.3 | 211.9 | 4.719 | 40 | 3.689 | 39.9 |
| -55.1 | 218.1 | 4.585 | 60 | 4.094 | 60.3 |
| -46.8 | 226.4 | 4.417 | 100 | 4.605 | 100.3 |
| -33.9 | 239.3 | 4.179 | 200 | 5.298 | 203.0 |
| -19.3 | 253.9 | 3.939 | 400 | 5.991 | 406.6 |
| -4.4 | 268.8 | 3.720 | 760 | 6.633 | 755.4 |
| 15.3 | 288.5 | 3.466 | 2 x 760 | 7.326 | 2 x 762 |
| 47.0 | 320.2 | 3.123 | 5 x 760 | 8.243 | 5 x 765 |
| 76.0 | 349.2 | 2.864 | 10 x 760 | 8.936 | 10 x 752 |
| 114.0 | 387.2 | 2.583 | 20 x 760 | 9.629 | 20 x 768 |
| 139.8 | 413.0 | 2.421 | 30 x 760 | 10.035 | 30 x 766 |

**Example for: Correlating the hyperbolic function** **Table III**

**(Start of first display)**

```
   CORRELATION FOR THE RELATION:

   Y = (A + B * X)/(1 + C * X)

DATA ARE ENTERED AS: X,Y

WHEN ALL THE DATA
ARE IN, ENTER:     END, 0

DATA WILL BE SAVED IN COMPUTER MEMORY
AS X(I), Y(I), FOR I GOING FROM
1 TO N.

PRESS RETURN TO ENTER DATA.
```

**(Start of next display)**

```
ENTER THE DATA IN PAIRS: X, Y
SEPARATED BY A COMMA.

WHEN ALL THE DATA ARE IN, ENTER
'END, 0'.

ENTER FIRST DATA PAIR:  4.179,3.689
NEXT PAIR OR: 'END, 0'  4.585,4.094
NEXT PAIR OR: 'END, 0'  4.417,4.605
NEXT PAIR OR: 'END, 0'  4.179,5.298
NEXT PAIR OR: 'END, 0'  3.939,5.991
NEXT PAIR OR: 'END, 0'  3.720,6.633
NEXT PAIR OR: 'END, 0'  3.466,7.326
NEXT PAIR OR: 'END, 0'  3.123,8.243
```

**Example for: Correlating the hyperbolic function (continued)** **Table III**

```
NEXT PAIR OR: 'END, 0'  2.864,8.936
NEXT PAIR OR: 'END, 0'  2.583,9.629
NEXT PAIR OR: 'END, 0'  2.421,10.035
NEXT PAIR OR: 'END, 0'  END,0
```

**(Start of next display)**

```
CONSTANTS FOR CORRELATION:

Y = (A + B * X)/(1 + C * X)

ARE:

A =                 11.2019
B =                 -2.356
C =                 -.1765

CORRELATION
  COEFFICIENT, R =     .9157

IF YOU WISH TO ESTIMATE A VALUE OF Y
FROM AN X VALUE, ENTER THE X VALUE.
IF NOT, ENTER 'NO'.          4

ESTIMATED Y WHEN X =4 IS:  6.0444

DO YOU WANT ANOTHER ESTIMATE? ANSWER
Y OR N.  N

            END OF PROGRAM
```

## References


1. Johnson, Eric E., Curve fitting easily done with hyperbolic equation, *Chem. Eng.*, Aug. 11, 1969, pp. 122–126.
2. Prahl, Walter H., Solving graphical problems on the calculator, *Chem. Eng.*, Oct. 22, 1979, pp. 118–126.
3. Perry, Robert H., and other eds., "Chemical Engineers' Handbook," 4th ed., McGraw-Hill, New York, 1963.

## The author


**Jean Robert Brosens** is planning director for Proquigel Ltda., a plastics manufacturer in São Paulo, Brazil. Previously, he was project manager for Rhodia, the Brazilian branch of the Rhône-Poulenc group. Mr. Brosens was born in Brussels, Belgium, and completed his B.S.Ch.E. and M.S.Ch.E. at Massachusetts Institute of Technology.


## Comments

Sir: This is a comment on "Correlating the hyperbolic function" by J. R. Brosens in your April 7, 1980 issue.

Conventional hyperbolic equations, such as $y = a + b/(c + x)$, the linearizable $y = c(a - x)/(b + x)$, and $y = (a + bx)/(1 + cx)$, have an $x$ in the denominator. Their residuals contain different $x_i$'s in the denominator so they cannot be summed directly.

The conventional least squares technique sums the squares of the residuals, so it is not applicable. This fact had been generally recognized, until Brosens claimed to have found least-squares equations for $Y = (a + bX)/(1 + cX)$.

His apparent solution is due to a mathematical error: he dropped the denominator from one side of the equation. The individual residual is $d_i = (a + bX_i)/(1 + cX_i) - Y_i$.

Multiplication by the denominator gives the correct equation $d_i(1 + cX_i) = a + bX_i - Y_i(1 + cX_i)$. Dropping the factor $(1 + cX_i)$ from the left side gives the incorrect equation $d_i = a + bX_i - Y_i(1 + cX_i)$. Applying the least-squares routine to it gives Brosen's equations.

Thus, the least-squares equations do not give the minimum of $\Sigma\, d_i^2$ as claimed, but that of $\Sigma(d_i(1 + cX_i))^2$. They are thus theoretically wrong.

How far are they practically wrong? $\Sigma\, d_i^2$ is identical with $\Sigma\,(d_i(1 + cX_i))^2$ only if both are zero, that is when *all* points lie *exactly* on the hyperbola; but in that case, no least-squares method is necessary. In all other cases, the deviation depends upon the size of the residuals. As long as they are small, the result may be acceptable as an approximation; but for larger ones, the results are too far from the correct value to be acceptable even as approximations.

WALTER H. PRAHL
*Technical Management Consultant*
*St. Petersburg Beach, Fla.*

**Author replies:**

Strictly speaking, Mr. Prahl is right. My correlation is, in fact, only an approximation of the exact solution.

Mr. Prahl correctly deduces that my equations minimize $\Sigma\,[d_i(1 + cX_i)]^2$ instead of $\Sigma\, d_i^2$; therefore, my approximate equations work only when the residuals $d_i$ are small. However, the above expressions show that they also work well for the (most frequent) case when the value of coefficient $c$ is relatively small; in other words, when the curvature of the hyperbola is low. In such cases, my method gets results quickly, avoiding tedious plotting and iterations.

JEAN R. BROSENS
*Rio de Janeiro, Brazil*

# Program for normal and log-normal distributions

## This program shows how well a given cumulative distribution corresponds with a normal or log-normal distribution.

☐ A normal distribution of values falls along a bell-shaped curve, as shown in Fig. 1. If the ordinate is converted from frequency to cumulative-percent-less-than (*CPLT*), the result is as shown in Fig. 2. And if this ordinate is plotted on a probability scale, the result is a straight line, as shown in Fig. 3.

A log normal distribution is shown in Fig. 4. If the $x$ values are plotted on a log scale (see Fig. 5), the log normal distribution looks like a normal distribution on an arithmetic plot. So if a log normal distribution is plotted with the ordinates on a probability scale, and the abscissas on a log scale, the result is again a straight line (Fig. 6).

A common example of a log normal distribution (for which this program is written) is the frequency of particle sizes obtained in size reduction operations. Fig. 7 presents a histogram of a particle size distribution, showing percent of ground material (frequency) versus particle size range for a common mineral. The data points, plotted on rectangular graph paper in Fig. 7, are shown in Fig. 8 on log probability paper, which plots the particle size (log scale) versus *CPLT* (probability scale).

### The program

The program (Table I) makes use of the fact that probability (or frequency), expressed as cumulative-percent-less-than, can conveniently be converted to "$z$" units—a $z$ unit being equivalent to one standard deviation from the mean of the normal curve. This conversion is easily made by using a table of normal curve areas, as found in practically all statistics texts and handbooks.

The program (Table I) takes the following data: cumulative-percent-less-than, and the corresponding values of $x$, starting from the lowest cumulative percent. The values of $x$ are assumed to be in increasing order, and *CPLT* is assumed to increase with increasing $x$, as in Fig. 2.

Once the *CPLT* vs. $x$ data are entered, the user chooses which distribution. The program calculates the mean and standard deviation of $x$, ignoring the two extreme values entered. It then calculates the values of $x$ that *would* correspond to the given *CPLTs* if the distribution were perfectly normal or log-normal and had the mean and standard deviation of the given $x$ data. It also calculates the correlation coefficient, $r$, which describes how closely the given *CPLT* vs. $x$ data match the chosen distribution.

### Example

Table II shows how the program analyzes the particle-size-distribution data of Fig. 8. The data are entered as (*CPLT*, $x$) pairs, beginning with the lowest *CPLT*. Note that *CPLT* must be entered as a percent, and not as a decimal fraction.

Since the log-probability graph in Fig. 8 looks more like a log-normal distribution than a normal one, the log-normal option was chosen. The program's results are: the original *CPLT* values; the original $x$ values; and the $x$ values that would correspond to the given *CPLTs* if the distribution were perfectly log-normal.

The final result is the correlation coefficient, $r$, which is 0.9977 in this case. Since the best possible coefficient is 1, this indicates a nearly perfect fit.

**Adapted from an article by Jack F. Chapin, Berks Campus, Pennsylvania State University, originally published December 15, 1980.**

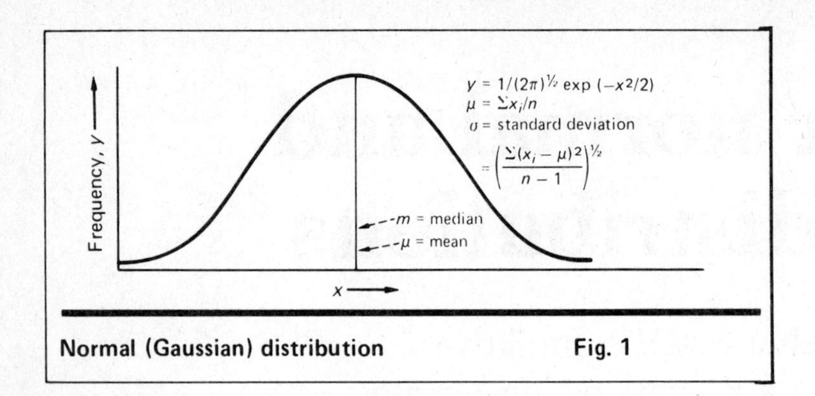


Normal (Gaussian) distribution Fig. 1

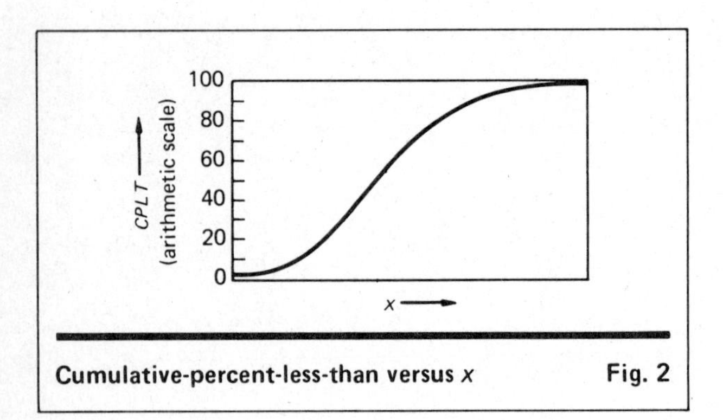


Cumulative-percent-less-than versus $x$ Fig. 2

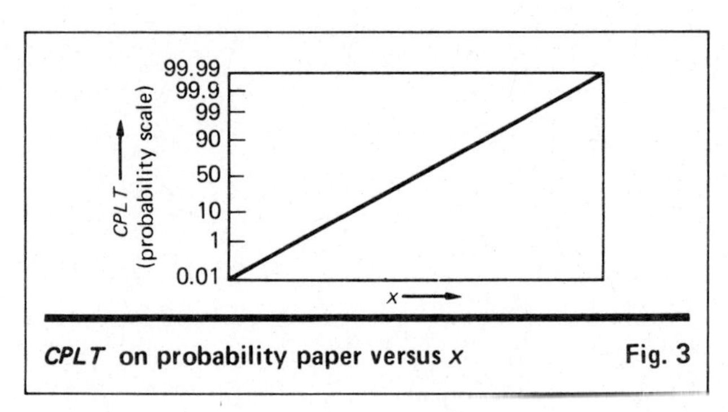


*CPLT* on probability paper versus $x$ Fig. 3

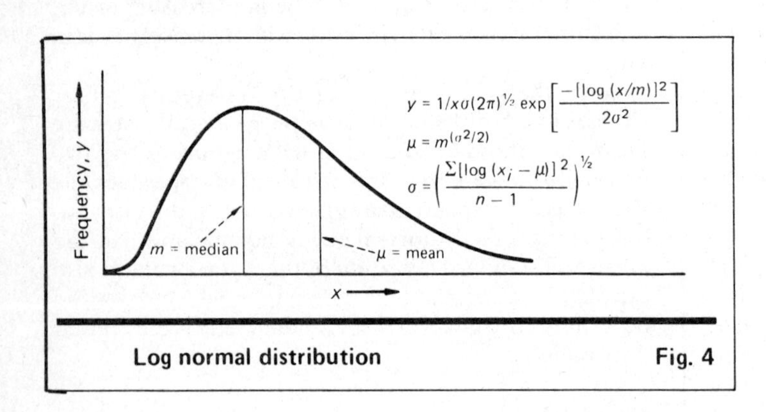


Log normal distribution Fig. 4

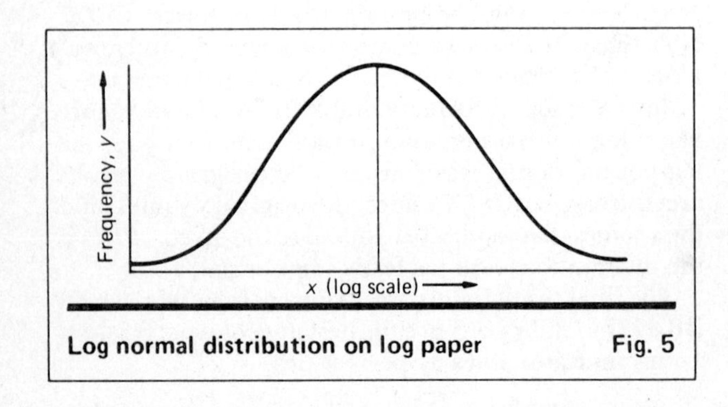


Log normal distribution on log paper Fig. 5

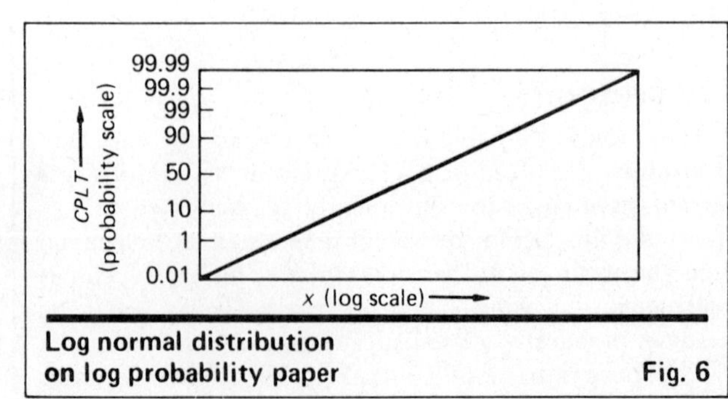


Log normal distribution on log probability paper Fig. 6

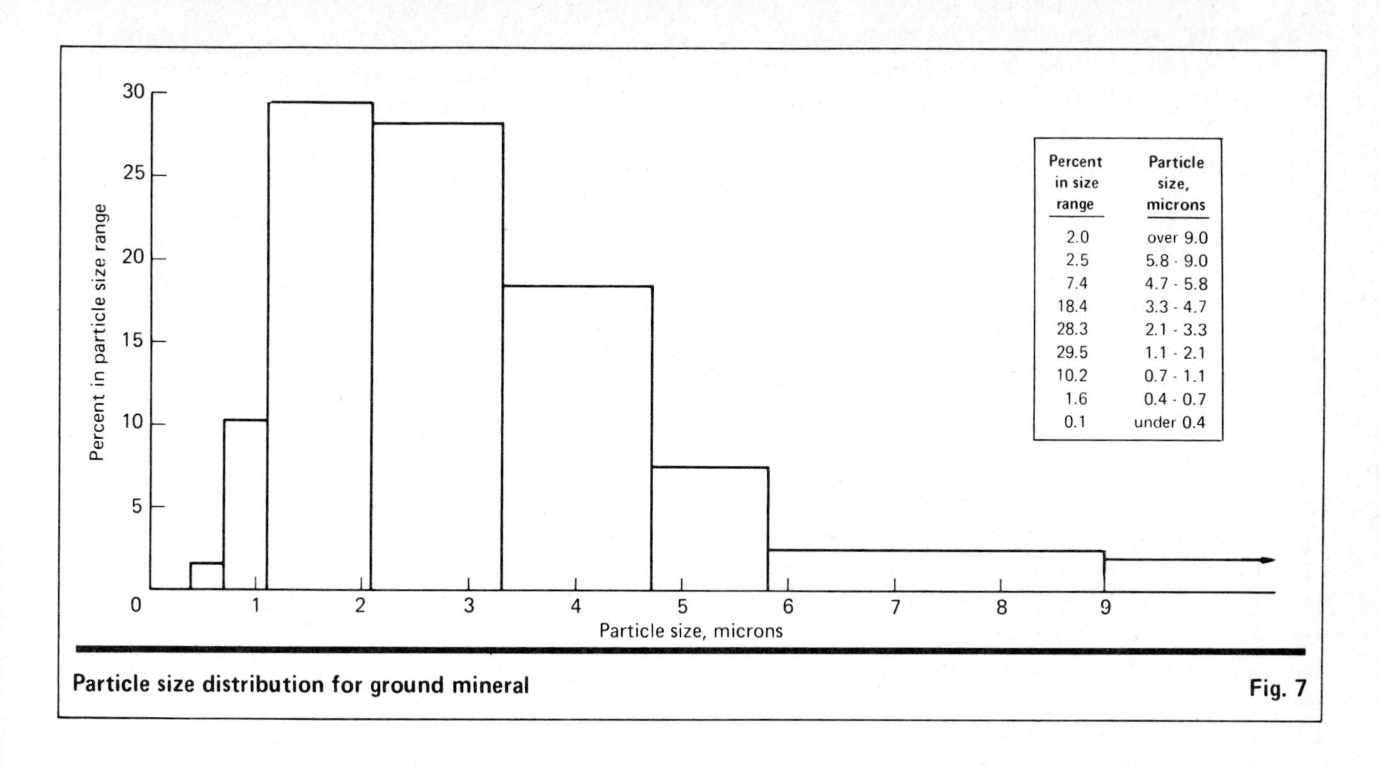


**Particle size distribution for ground mineral** **Fig. 7**

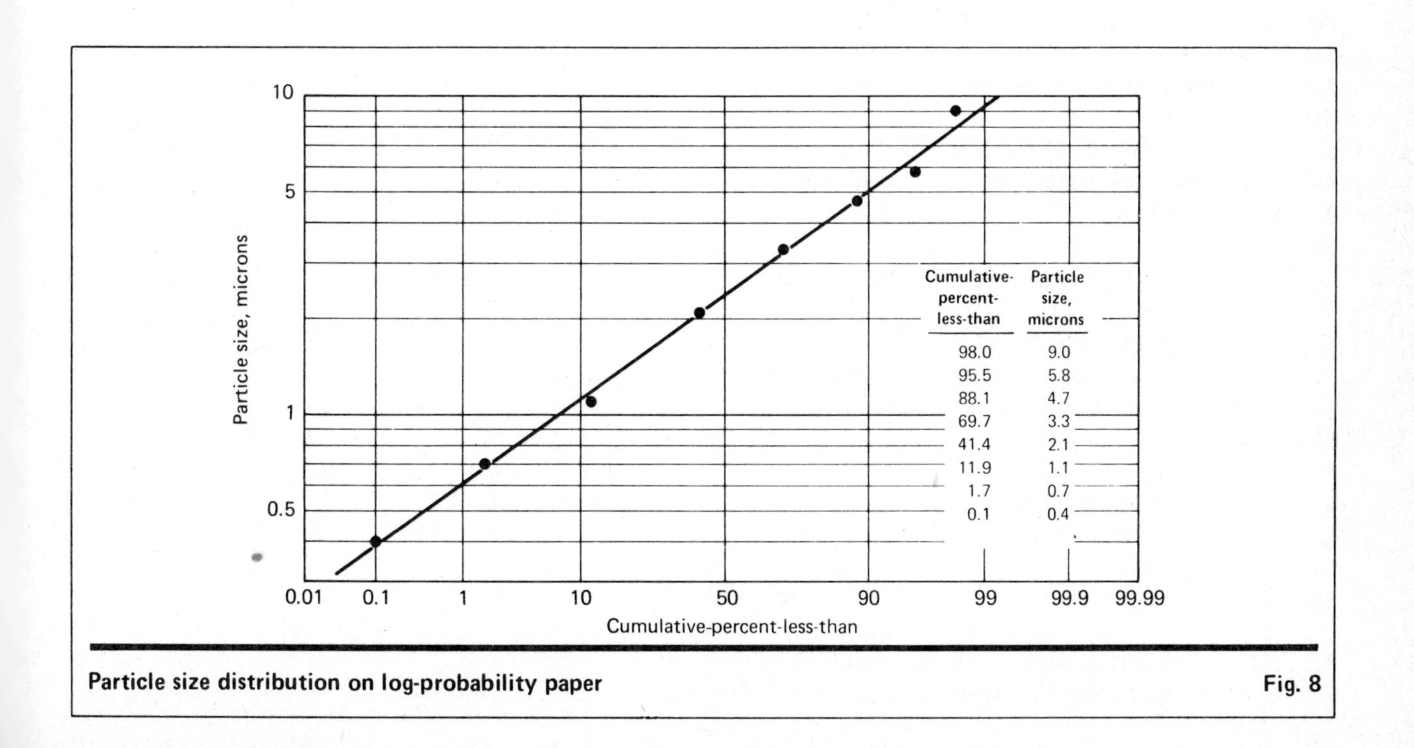


**Particle size distribution on log-probability paper** **Fig. 8**

**Program for normal and log-normal distributions** **Table I**

```
10  REM   PROGRAM FOR NORMAL AND LOG-NORMAL DISTRIBUTIONS
20  REM   FROM CHEMICAL ENGINEERING, DECEMBER 15, 1980
30  REM   BY JACK F. CHAPIN
40  REM  TRANSLATED BY WILLIAM VOLK
50  REM   COPYRIGHT (C) 1984
60  REM  BY CHEMICAL ENGINEERING
70  REM   THIS PROGRAM IS A MODIFICATION OF THE ORIGINAL PROCEDURE.  THE ORIGINA
L USED AN ITERATIVE TRAPEZOIDAL NUMERICAL INTEGRATION TO OBTAIN A NORMAL APPROXI
MATION.
80  REM  THIS PROGRAM CALCULATES THE MEAN AND STANDARD DEVIATION OF THE INPUT DA
TA, AND THEN CALCULATES PRECISE NORMAL PROBABILITIES FROM THESE PARAMETERS.
90  REM  THE TWO EXTREME VALUES, WHICH HAVE NO ACTUAL LIMITS TO THEIR RANGES, AR
E NOT INCLUDED IN THE CALCULATION OF THE MEAN AND STANDARD DEVIATION OF THE DATA

100  HOME : VTAB 2
110 PI = 3.14159265
120  REM  SET DISPLAY TO FOUR DECIMALS AND DATA TO 50 ENTRIES
130  DEF  FN P(X) =  INT (1E4 * (X + .00005)) / 1E4
140  DIM T(50),X(50),CP(50),YE(50)
150  PRINT "PROGRAM TAKES FREQUENCY DATA AND"
160  PRINT "CALCULATES EITHER PROBABILITY OR LOG"
170  PRINT "PROBABILITY CORRELATION OF X VS. "
180  PRINT "CPLT (CUMULATIVE PERCENT LESS THAN Y)."
190  PRINT
200  PRINT "IT FIRST CALCULATES THE MEAN AND"
210  PRINT "STANDARD DEVIATION OF THE INPUT DATA,"
220  PRINT "EXCLUDING THE TWO EXTREME VALUES."
230  PRINT
240  PRINT "IT THEN USES THE NORMAL PROBABILITY"
250  PRINT "TO CALCULATE THE X (OR LOG X) VALUE"
260  PRINT "ASSOCIATED WITH EACH CUMULATIVE"
270  PRINT "PERCENTAGE VALUE."
280  PRINT
290  PRINT "IT COMPARES THE X VALUES BASED ON"
300  PRINT "THE NORMAL DISTRIBUTION WITH THE INPUT"
310  PRINT "VALUES FOR THE CORRELATION COEFFICIENT."
320  PRINT
330  INPUT "PRESS RETURN TO CONTINUE.    ";Q$
340  HOME : VTAB 2
350  PRINT "DATA ARE THE CUMULATIVE PERCENT LESS"
360  PRINT "THAN SOME MEASUREMENT (CPLT), AND THAT"
370  PRINT "MEASUREMENT (X)."
380  PRINT
390  PRINT "THE PROGRAM WILL CALCULATE EITHER THE"
400  PRINT "PROBABILITIES ASSOCIATED WITH THE X"
410  PRINT "VARIABLE, OR WITH LOG (X) - THE"
420  PRINT "LOG PROBABILITY RELATIONSHIP.  (THE"
430  PRINT "LOGS ARE TO BASE 10.)"
440  PRINT
450  PRINT "IF YOU WANT A LOG PROBABILITY"
460  PRINT "CALCULATION, ENTER 'L'.  IF NOT"
470  INPUT "ENTER 'N'.                        ";A$
480  IF A$ = "L" GOTO 550
490  IF A$ = "N" GOTO 580
500  PRINT
510  PRINT "YOUR OPTIONS WERE 'L' OR 'N'.  YOU"
520  PRINT "ENTERED ";A$". TRY AGAIN."
```

**Program for normal and log-normal distributions (continued)** **Table I**

```
530  PRINT
540  GOTO 450
550 F1 = 1
560 N$ = "LOG NORMAL"
570  GOTO 590
580 N$ = "    NORMAL"
590  HOME : VTAB 4
600  PRINT "TO RUN THE PROGRAM:"
610  PRINT
620  PRINT "YOU ENTER THE PERCENT OF OBSERVA-"
630  PRINT "TIONS LESS THAN SOME MINIMUM VALUE,"
640  PRINT "AND THAT VALUE."
650  PRINT
660  PRINT "YOU THEN ENTER SUCCESSIVE CUMULATIVE"
670  PRINT "PERCENTS, AND THE MEASUREMENT UP TO"
680  PRINT "THAT CUMULATIVE PERCENT."
690  PRINT
700  PRINT "ENTER THE FIRST VALUES: CPLT (THE CUM-"
710  PRINT "LATIVE PERCENT LESS THAN SOME VALUE), "
720  PRINT "AND X (THE VALUE)."
730  PRINT
740  PRINT "WHEN ALL THE DATA ARE IN, ENTER 'END,0'"
750  PRINT
760  INPUT "ENTER THE VALUES NOW:    ";CP$,X
770 N = N + 1
780 T(N) =  VAL (CP$)
790 CP(N) = T(N) - T(N - 1)
800 X(N) = X
810  IF F1 = 0 GOTO 830
820 X(N) =  LOG (X(N)) /  LOG (10)
830 SX = SX + X(N)
840 SC = SC + CP(N)
850  IF N = 1 GOTO 880
860 SF = SF + CP(N) * (X(N) + X(N - 1)) / 2
870 QF = QF + CP(N) * ((X(N) + X(N - 1)) / 2) ^ 2
880  INPUT "NEXT VALUES OR: 'END,0' ";CP$,X
890  IF CP$ = "END" GOTO 910
900  GOTO 770
910 MN = SF / (SC - CP(1))
920 MY = SX / N
930 SG = (QF - SF ^ 2 / (SC - CP(1))) / (SC - CP(1))
940 SD =  SQR (SG)
950  HOME : PRINT
960  PRINT  TAB( 10);"CALCULATED RESULTS"
970  PRINT
980  PRINT "CUMULATIVE"; TAB( 15);"UPPER"; TAB( 30);"X"
990  PRINT "  % LESS "; TAB( 15);"RANGE"; TAB( 25);
     "CALCULATED"
1000  PRINT "  THAN"; TAB( 17);"OF"; TAB( 30);"BY"
1010  PRINT "    X"; TAB( 17);"X"; TAB( 25);N$
1020  PRINT
1030 SD =  SQR (SG)
1040  FOR I = 1 TO N
1050 PR = PR + CP(I) / 100
1060  IF PR < .5 GOTO 1100
1070 PR = 1 - PR
1080 F3 = 1
1090  REM  NORMAL PROBABILITY CALCULATION
1100 W3 =  SQR ( LOG (1 / PR ^ 2))
1110 ZP = (2.515517 + .802853 * W3 + .010328 * W3 ^ 2) /
(1 + 1.432788 * W3 + .189269 * W3 ^ 2 + .001308 * W3 ^ 3)
1120 ZP = W3 - ZP
1130  IF F3 = 1 GOTO 1160
1140 YE(I) = MN - ZP * SD
1150  GOTO 1180
1160 YE(I) = MN + ZP * SD
1170 PR = 1 - PR
1180 PC = PC + CP(I)
1190  IF F1 = 1 GOTO 1220
1200  PRINT PC; TAB( 15);X(I); TAB( 25); FN P(YE(I))
1210  GOTO 1230
1220  PRINT PC; TAB( 15);10 ^ X(I); TAB( 25); FN P(10 ^
      YE(I))
1230 D = D + (X(I) - YE(I)) ^ 2
1240 W = W + (X(I) - YM) ^ 2
1250  NEXT I
1260 R =  SQR ((W - D) / W)
1270  PRINT
1280  PRINT "CORRELATION COEFFICIENT:     "; FN P(R)
1290  PRINT
1300  PRINT  TAB( 13);"END OF PROGRAM"
1310  END
```

**Example for: Normal and log-normal distributions** **Table II**

**(Start of first display)**

```
PROGRAM TAKES FREQUENCY DATA AND
CALCULATES EITHER PROBABILITY OR LOG
PROBABILITY CORRELATION OF X VS.
CPLT (CUMULATIVE PERCENT LESS THAN Y).

IT FIRST CALCULATES THE MEAN AND
STANDARD DEVIATION OF THE INPUT DATA,
EXCLUDING THE TWO EXTREME VALUES.

IT THEN USES THE NORMAL PROBABILITY
TO CALCULATE THE X (OR LOG X) VALUE
ASSOCIATED WITH EACH CUMULATIVE
PERCENTAGE VALUE.

IT COMPARES THE X VALUES BASED ON
THE NORMAL DISTRIBUTION WITH THE INPUT
VALUES FOR THE CORRELATION COEFFICIENT.

PRESS RETURN TO CONTINUE.
```

**(Start of next display)**

```
DATA ARE THE CUMULATIVE PERCENT LESS
THAN SOME MEASUREMENT (CPLT), AND THAT
MEASUREMENT (X).

THE PROGRAM WILL CALCULATE EITHER THE
```

**Example for: Normal and log-normal distributions (continued)** **Table II**

```
PROBABILITIES ASSOCIATED WITH THE X
VARIABLE, OR WITH LOG (X) - THE
LOG PROBABILITY RELATIONSHIP.  (THE
LOGS ARE TO BASE 10.)

IF YOU WANT A LOG PROBABILITY
CALCULATION, ENTER 'L'.  IF NOT
ENTER 'N'.                        L
```

**(Start of next display)**

```
TO RUN THE PROGRAM:

YOU ENTER THE PERCENT OF OBSERVA-
TIONS LESS THAN SOME MINIMUM VALUE,
AND THAT VALUE.

YOU THEN ENTER SUCCESSIVE CUMULATIVE
PERCENTS, AND THE MEASUREMENT UP TO
THAT CUMULATIVE PERCENT.

ENTER THE FIRST VALUES: CPLT (THE CUM-
LATIVE PERCENT LESS THAN SOME VALUE),
AND X (THE VALUE).

WHEN ALL THE DATA ARE IN, ENTER 'END,0'

ENTER THE VALUES NOW:   .1,.4
NEXT VALUES OR: 'END,0' 1.7,.7
NEXT VALUES OR: 'END,0' 11.9,1.1
NEXT VALUES OR: 'END,0' 41.4,2.1
NEXT VALUES OR: 'END,0' 69.7,3.3
NEXT VALUES OR: 'END,0' 88.1,4.7
NEXT VALUES OR: 'END,0' 95.5,5.8
NEXT VALUES OR: 'END,0' 98,9
NEXT VALUES OR: 'END,0' END,0
```

**(Start of next display)**

CALCULATED RESULTS

| CUMULATIVE % LESS THAN X | UPPER RANGE OF X | X CALCULATED BY LOG NORMAL |
|---|---|---|
| .1 | .4 | .3855 |
| 1.7 | .7 | .6715 |
| 11.9 | 1.1 | 1.1501 |
| 41.4 | 2.1 | 1.9955 |
| 69.7 | 3.3 | 3.0339 |
| 88.1 | 4.7 | 4.4377 |
| 95.5 | 5.8 | 5.9605 |
| 98 | 9 | 7.3171 |

CORRELATION COEFFICIENT: .9977

END OF PROGRAM

## The author

**Jack F. Chapin** is in charge of the Associate Degree Program in Chemical Engineering Technology at Berks Campus, Pennsylvania State University, R.D. #5 Tulpehocken Rd., Reading, PA 19608, tel: 215-375-4211. He received his B.S. in chemical engineering and an M.S. in chemical engineering practice at M.I.T. and has worked with The Polymer Corp. and Consolidation Coal Co. Professor Chapin is a member of AIChE and the Reading [Pa.] Chemists' Club.

# Section III
# Physical Properties Correlation

Properties of chemical compounds
Volumes of pure liquids
Volumes of saturated liquid mixtures
Viscosities of pure gases
Viscosities of liquids and mixtures
Vapor pressure vs. temperature
Latent heat of vaporization
Gas equations
Equation-of-state variables
Properties of gas mixtures
Thermodynamic properties of pure gases and binary mixtures
Thermal conductivities of pure gases

# Properties of chemical compounds

With this program one can calculate a dozen physical, thermodynamic and transport properties of a broad range of chemical compounds.

□ Carl Yaws and others have collected lists of formulas and correlation constants to calculate twelve properties of industrially important compounds. These were published as articles in *Chemical Engineering* [*1–3*]. These have also been compiled in one book, which includes charts that let you read the properties directly [*4*].

There are twelve properties, six for the liquid state and six for the gas state: heat capacity of ideal gas, heat of formation of ideal gas, free energy of formation of ideal gas, thermal conductivity of gas, viscosity of gas, heat capacity of liquid, thermal conductivity of liquid, surface tension of liquid, viscosity of liquid, density of liquid, vapor pressure, and heat of vaporization. Table I lists the properties, the formulas, the required data, and where to get the required constants.

These formulas are of several different types—linear, quadratic, etc.—so they can be inconvenient to use unless collected in one place. This Apple II program does just that.

## The program

For each of the twelve properties, the program prompts the user to provide the necessary constants. At least two constants and one temperature are always required. In addition, the surface-tension and heat-of-vaporization calculations require a known value at some temperature. Once the constants are entered, the user can calculate a property again for different temperature points.

For once-only applications, these properties can be read from tables, estimated from graphs, or calculated on a scientific calculator. Indeed, the book that contains the constants also has the graphs [*4*]. But for repetitive work, such as determining the value of a property at ten temperatures, a program can save a lot of time.

Table II lists the program. The user simply selects the property and is prompted from that point to the end of the calculation. The program accepts either Fahrenheit or Centigrade temperatures: All temperatures entered have to be appended by an "F" or a "C" so that the program knows how to interpret them. Once the various constants are entered, the property can be calculated for any number of temperatures.

Note that the program has no means of checking whether the correlation is valid for the temperature entered. The references containing the constants and formulas [*1–4*] tell the valid ranges.

## Example

Table I lists example constants and results, and Table III shows the actual displays and results for one of the examples—heat capacity of methane at 25°C. The other examples can be worked basically the same way by selecting other properties when prompted by the program. Once again, note that two of the properties require a known value at some temperature.

## The author

**Erminio Santi** works for Monda S.p.A., 37058 Sanguinetto (VR), Viale Roma, 9, Italy, as production manager in a food plant. His experience has been in food engineering and food processing. He holds a degree in chemical engineering from the University of Padua, is a licensed professional engineer, and is registered in the Italian Ordine Professionale degli Ingegneri.


## References


1. Yaws, C. L., et al., Correlation Constants for Chemical Compounds, *Chem. Eng.*, Aug. 16, 1976, pp. 79-87.
2. Yaws, C. L., et al., Correlation Constants for Liquids, *Chem. Eng.*, Oct. 25, 1976, pp. 127-135.
3. Yaws, C. L., et al., Correlation Constants for Chemical Compounds, *Chem. Eng.*, Nov. 22, 1976, pp. 153-162.
4. Yaws, C. L., et al., "Physical Properties, A Guide to the Physical, Thermodynamic and Transport Property Data of Industrially Important Chemical Compounds," *Chemical Engineering*, McGraw-Hill, New York, 1977.


Adapted from an article by Erminio Santi, Monda S.p.A., originally published June 2, 1980.

**Correlation formulas and example results** **Table I**

### 1. Heat capacity of ideal gas

**The formula:** $C_p^\circ = A + BT + CT^2 + DT^3$

**The units:** $C_p^\circ$ = cal/g-mol K, $T$ = Kelvin

**The example:** Methane

See Table 22-I [1] for correlation constants

| | |
|---|---|
| Enter $A$ | 5.04 |
| Enter $B$ | 9.03 E-3 |
| Enter $C$ | 8.87 E-6 |
| Enter $D$ | −5.4 E-9 |
| Enter $T$ | 25C |
| $C_p^\circ$ = | 8.378 |

### 2. Heat of formation of ideal gas

**The formula:** $\Delta H_f^\circ = A + BT + CT^2$

**The units:** $\Delta H_f^\circ$ = kcal/g-mol, $T$ = Kelvin

**The example:** Ethylene

See Table 22-II [1] for correlation constants.

| | |
|---|---|
| Enter $A$ | 14.8 |
| Enter $B$ | −8.8 E-3 |
| Enter $C$ | 3.15 E-6 |
| Enter $T$ | 126.8C |
| $\Delta H_f^\circ$ = | 11.78 |

### 3. Free energy of formation of ideal gas

**The formula:** $\Delta G_f^\circ = A + BT$

**The units:** $\Delta G_f^\circ$ = kcal/g-mol, $T$ = Kelvin

**The example:** Methane

See Table 22-III [1] for correlation constants

| | |
|---|---|
| Enter $A$ | −20.1 |
| Enter $B$ | .0249 |
| Enter $T$ | 126.8C |
| $\Delta G_f^\circ$ | −10.14 |

### 4. Thermal conductivity of gas

**The formula:** $K_G = A + BT + CT^2 + DT^3$

**The units:** $K_G$ = kcal/s-cm-K, $T$ = Kelvin

**The example:** Hydrogen

See Table 23-I [3] for correlation constants

| | |
|---|---|
| Enter $A$ | 19.34 |
| Enter $B$ | 1.5974 |
| Enter $C$ | −9.93 E-4 |
| Enter $D$ | 37.29 E-8 |
| Enter $T$ | 500C |
| $K_G$ = | 883.1 |

### 5. Viscosity of gas

**The formula:** $\mu_G = A + BT + CT^2$

**The units:** $\mu_G$ = micropoise, $T$ = Kelvin

**The example:** Methane

See Table 23-II [3] for correlation constants

| | |
|---|---|
| Enter $A$ | 15.96 |
| Enter $B$ | .3439 |
| Enter $C$ | −8.14 E-5 |
| Enter $T$ | 403C |
| $\mu_G$ = | 211.3 |

### 6. Liquid heat capacity

**The formula:** $C_p = A + BT + CT^2 + DT^3$

**The units:** $C_p$ = cal/g-K, $T$ = Kelvin

**The example:** Benzene

See Table 24-II [2] for correlation constants

| | |
|---|---|
| Enter $A$ | −1.481 |
| Enter $B$ | .01546 |
| Enter $C$ | −4.37 E-5 |
| Enter $D$ | 4.409 E-8 |
| Enter $T$ | 20C |
| $C_p$ = | 4064 |

### 7. Thermal conductivity of liquid

**The formula:** $K_L = A + BT + CT^2$

**The units:** $K_L$ = micro-cal/s-cm-K

**The example:** Toluene

See Table 24-IV [2] for correlation constants

| | |
|---|---|
| Enter $A$ | 485.1 |
| Enter $B$ | −.5384 |
| Enter $C$ | −5.9 E-5 |
| Enter $T$ | 150C |
| $K_L$ = | 246.7 |

### 8. Surface tension of liquid

**The formula:** $\sigma = \sigma_1 \, [(T_c - T)/(T_c - T_1)]^n$

**The units:** $\sigma$ = dyne/cm, $T$ = Kelvin

**The example:** Benzene

See Table 24-I [2] for correlation constants

| | |
|---|---|
| Enter $T_c$ | 288.94C |
| Enter $\sigma_1$ | 28.88 |
| Enter $T_1$ | 20C |
| Enter $n$ | 1.2243 |
| Enter $T$ | 150C |
| $\sigma$ = | 21.20 |

### 9. Viscosity of liquid

**The formula:** $\log \mu_L = A + B/T + CT + DT^2$

**The units:** $\mu_L$ = centipoise, $T$ = Kelvin

**The example:** Water

See Table 23-III [3] for correlation constants

| | |
|---|---|
| Enter $A$ | −10.73 |
| Enter $B$ | 1,828 |
| Enter $C$ | .01966 |
| Enter $D$ | −1.466 E-5 |
| Enter $T$ | 250C |
| $\mu_L$ = | .1089 |

### 10. Density of liquid

**The formula:** $\rho_L = AB^{-(1-T_r)^{2/7}}$

**The units:** $\rho_L$ = g/cm³, $T$ = Kelvin, $T_r = T/T_c$

**The example:** Chlorobenzene

See Table 24-III [2] for correlation constants

| | |
|---|---|
| Enter $A$ | .3706 |
| Enter $B$ | .2708 |
| Enter $T_c$ | 359.2C |
| Enter $T$ | 50C |
| $\rho_L$ = | 1.075 |

### 11. Vapor pressure

**The formula:**

$\log P_V = A + B/T + C \log T + DT + ET^2$

**The units:** $P_V$ = mm Hg, $T$ = Kelvin

**The example:** Toluene

See Table 23-IV [3] for correlation constants

| | |
|---|---|
| Enter $A$ | 115.21 |
| Enter $B$ | −4,918.1 |
| Enter $C$ | −43.467 |
| Enter $D$ | .038548 |
| Enter $E$ | −1.3496 E-5 |
| Enter $T$ | OC |
| $P_V$ = | 6.676 |

### 12. Heat of vaporization

**The formula:** $\Delta H_V = \Delta H_{V1} \, [(T_c - T)/(T_c - T_1)]^n$

**The units:** $\Delta H_V$ = cal/g, $T$ = Kelvin

**The example:** Methyl chloride

See Table 22-IV [1] for correlation constants

| | |
|---|---|
| Enter $T_c$ | 143.1C |
| Enter $H_{V1}$ | 102.2 |
| Enter $T_1$ | −23.8C |
| Enter $n$ | 0.38 |
| Enter $T$ | 40C |
| $\Delta H_V$ = | 85.11 |

**Program for properties of chemical compounds** **Table II**

```
10 REM  PROPERTIES OF CHEMICAL COMPOUNDS
20 REM   FROM CHEMICAL ENGINEERING, JUNE 2, 1980
30 REM   BY ERMINIO SANTI
40 REM  TRANSLATED BY WILLIAM VOLK
50 REM   COPYRIGHT (C) 1984
60 REM  BY CHEMICAL ENGINEERING
70 DIM A$(12),B$(12),Q$(12)
80 REM  SET DISPLAY TO FOUR DECIMALS
90 DEF  FN P(X) =  INT (1E4 * (X + .00005)) / 1E4
100  HOME : VTAB 2
110  PRINT "THIS PROGRAM PROVIDES FOR THE SOLUTION"
120  PRINT "OF EMPIRICAL EQUATIONS TO DETERMINE"
130  PRINT "A NUMBER OF PROPERTIES OF CHEMICAL"
140  PRINT "COMPOUNDS."
150  PRINT
160  PRINT "THE FOLLOWING PROPERTIES CAN BE"
170  PRINT "CALCULATED:"
180  PRINT
190  PRINT "HEAT OF FORMATION,"; TAB( 24);"HEAT
     CAPACITY,"
200  PRINT "THERMAL CONDUCTIVITY,"; TAB( 24);"FREE
     ENERGY,"
210  PRINT "SURFACE TENSION,"; TAB( 24);"VISCOSITY,"
220  PRINT "LIQUID DENSITY,"; TAB( 24);"VAPOR PRESSURE,"
230  PRINT "HEAT OF VAPORIZATION."
240  PRINT
250  PRINT "TEMPERATURE AND EMPIRICAL CONSTANTS ARE"
260  PRINT "REQUIRED.  FOR SURFACE TENSION AND HEAT"
270  PRINT "OF VAPORIZATION, THE CRITICAL TEMP-"
280  PRINT "ERATURE AND A VALUE AT SOME OTHER"
290  PRINT "TEMPERATURE ARE REQUIRED."
300  PRINT
310  INPUT "PRESS RETURN TO CONTINUE.    ";Q$
320  HOME : VTAB 4
330  PRINT "INDICATE THE PROPERTY TO BE DETERMINED"
340  PRINT "BY NUMBER.  THE PROGRAM WILL SHOW"
350  PRINT "THE DATA REQUIRED."
360  PRINT
370  PRINT " 1.  HEAT CAPACITY OF IDEAL GAS."
380  PRINT " 2.  HEAT OF FORMATION OF IDEAL GAS."
390  PRINT " 3.  FREE ENERGY OF IDEAL GAS."
400  PRINT " 4.  THERMAL CONDUCTIVITY OF GAS."
410  PRINT " 5.  VISCOSITY OF GAS
420  PRINT " 6.  HEAT CAPACITY OF LIQUID."
430  PRINT " 7.  THERMAL CONDUCTIVITY OF LIQUID."
440  PRINT " 8.  SURFACE TENSION OF LIQUID."
450  PRINT " 9.  VISCOSITY OF LIQUID."
460  PRINT "10.  DENSITY OF LIQUID."
470  PRINT "11.  VAPOR PRESSURE."
480  PRINT "12.  HEAT OF VAPORIZATION."
490 A$(1) = "GAS HEAT CAPACITY"
500 B$(1) = "C(P), CAL/(G-MOL)(K)
510 A$(2) = "GAS HEAT OF FORMATION"
520 B$(2) = "H(F), K-CAL/G-MOL"
530 A$(3) = "GAS FREE ENERGY"
540 B$(3) = "G(F), K-CAL/G-MOL"
550 A$(4) = "GAS THERMAL CONDUCTIVITY"
560 B$(4) = "K(G), K-CAL/(S)(CM)(K)
570 A$(5) = "GAS VISCOSITY"
580 B$(5) = "MU(G), MICROPOISE"
590 A$(6) = "LIQUID HEAT CAPACITY"
600 B$(6) = "C(P), CAL/(G)(K)
610 A$(7) = "LIQUID THERMAL CONDUCTIVITY"
620 B$(7) = "K(L), MICRO-CAL/(S)(CM)(K)"
630 A$(8) = "LIQUID SURFACE TENSION"
640 B$(8) = "SIGMA, DYNES/CM"
650 A$(9) = "LIQUID VISCOSITY"
660 B$(9) = "MU(L), CENTIPOISE"
670 A$(10) = "LIQUID DENSITY"
680 B$(10) = "RHO(L), G/CU  CM"
690 A$(11) = "VAPOR PRESSURE"
700 B$(11) = "P(V), MM H2O"
710 A$(12) = "HEAT OF VAPORIZATION"
720 B$(12) = "H(V), CAL/G"
730  PRINT
740  INPUT "MAKE YOUR SELECTION BY NUMBER.    ";OP
750  HOME : VTAB 4
760 Q$(8) = "SURFACE TENSION"
770 Q$(12) = "HEAT OF VAPORIZATION"
780  PRINT "TO CALCULATE ";A$(OP)
790  PRINT
800  IF OP = 8 OR OP = 12 GOTO 850
810  PRINT "YOU NEED THE CONSTANTS TO SOLVE"
820  PRINT "THE FOLLOWING EQUATION, AND THE"
830  PRINT "TEMPERATURE IN DEGREES C OR F.
840  PRINT
850  ON OP GOTO 860,930,1000,860,930,860,930,1060,1430,
     1490,1650,1060
860  PRINT : PRINT
870  PRINT B$(OP);" = "
880  PRINT  TAB( 6);"A + B*T + C*T^2 + D*T^3"
890  PRINT : PRINT
900  GOSUB 1890
910 CP = A + B * T + C * T ^ 2 + D * T ^ 3
920  GOTO 2130
930  PRINT : PRINT
940  PRINT B$(OP);" = "
950  PRINT  TAB( 6);"A + B*T + C*T^2"
960  PRINT : PRINT
970  GOSUB 1890
980 CP = A + B * T + C * T ^ 2
990  GOTO 2130
1000  PRINT : PRINT
1010  PRINT B$(OP);" = A + B*T"
1020  PRINT : PRINT
1030  GOSUB 1890
1040 CP = A + B * T
1050  GOTO 2130
1060  PRINT "YOU NEED THE FOLLOWING DATA:"
1070  PRINT "CRITICAL TEMP.; ";Q$(OP)
1080  PRINT "AT SOME TEMPERATURE, AND THAT"
1090  PRINT "TEMPERATURE; AND THE CORRELATION"
```

**Program for properties of chemical compounds (continued)** **Table II**

```
1100  PRINT "PARAMETER, N, IN THE FOLLOWING"
1110  PRINT "EQUATION."
1120  PRINT
1130  IF OP = 12 GOTO 1720
1140  PRINT "SIGMA = SIGMA(T1)*((TC-T)/(TC-T1))^N
1150  PRINT
1160  PRINT "INDICATE WHETHER TEMERATURES ARE IN"
1170  PRINT "DEGREES C OR F BY FOLLOWING THE "
1180  PRINT "VALUES WITH C OR F: 140C OR 140F"
1190  PRINT
1200  INPUT "ENTER CRITICAL TEMPERATURE  ";T$(1)
1210  IF OP = 12 GOTO 1740
1220  INPUT "ENTER KNOWN VALUE OF SIGMA  ";SG
1230  INPUT "ENTER TEMPERATURE OF SIGMA  ";T$(2)
1240  INPUT "ENTER PARAMETER N           ";N1
1250  INPUT "ENTER TEST TEMPERATURE      ";T$(3)
1260 N = 3
1270  REM  DETERMINES WHETHER INPUT WAS DEGREES C OR F.
      THERE ARE SIMILAR ROUTINES FOR OTHER INPUT.
1280  FOR I = 1 TO N
1290 T(I) =  VAL ( LEFT$ (T$(I), LEN (T$(I)) - 1))
1300  IF  RIGHT$ (T$(I),1) = "C" GOTO 1380
1310  IF  RIGHT$ (T$(1),1) = "F" GOTO 1370
1320  HOME : PRINT
1330  PRINT "YOU DIDN'T INDICATE DEGREES C OR F."
1340  PRINT "YOU HAVE TO START OVER"
1350  PRINT
1360  GOTO 850
1370 T(I) = (T(I) - 32) / 1.8
1380  NEXT I
1390 T = T(3)
1400  IF OP = 10 GOTO 1600
1410 CP = SG * ((T(1) - T) / (T(1) - T(2))) ^ N1
1420  GOTO 2130
1430  PRINT "LOG(MU) = A + B/T + C*T + D*T^2"
1440  PRINT : PRINT
1450  GOSUB 1890
1460 CP = A + B / T + C * T + D * T ^ 2
1470 CP = 10 ^ CP
1480  GOTO 2130
1490  PRINT "RHO(L) = A/B^((1-T/TC)^(2/7))"
1500  PRINT : PRINT
1510  GOSUB 1890
1520  PRINT
1530  PRINT "FOLLOW TEMPERATURES WITH
1540  PRINT "C OR F: 140C OR 140F."
1550  PRINT
1560  INPUT "ENTER CRITICAL TEMPERATURE  ";T$(1)
1570  INPUT "ENTER TEST TEMPERATURE      ";T$(2)
1580 N = 2
1590  GOTO 1280
1600 T = T(2) + 273.16
1610 T(1) = T(1) + 273.16
1620 CP = (1 - T / T(1)) ^ (2 / 7)
1630 CP = A / B ^ CP
1640  GOTO 2130
1650  PRINT "LOG P(V) = A + B/T + C*LOG(T) + D*T + E*T^2"
1660  PRINT : PRINT
1670  GOSUB 1890
1680 CP = A + B / T + C *  LOG (T) /  LOG (10) + D * T +
     E * T ^ 2
1690 CP = 10 ^ CP
1700  GOTO 2130
1720  PRINT "H(V) = H(V(1))*((TC-T)/(TC-T1))^N"
1730  GOTO 1150
1740  INPUT "ENTER KNOWN H(V) VALUE      ";SG
1750  INPUT "ENTER TEMPERATURE OF H(V)   ";T$(2)
1760  GOTO 1240
1770  PRINT "VALUES WITH C OR F: 140C OR 140F"
1780  PRINT "TEMPERATURE, SEPARATED BY COMMAS."
1790  PRINT
1800  PRINT "ENTER THE TEST TEMPERATURE"
1810  PRINT "AND FOLLOW THE TEMPERATURE"
1820  PRINT "WITH AN F OR C:"
1830  PRINT "140F, OR 140C."
1840  PRINT
1850  INPUT "ENTER TEMPERATURE NOW       ";T$
1860  GOTO 2000
1870  PRINT
1880  RETURN
1890  INPUT "ENTER A                     ";A
1900  INPUT "ENTER B                     ";B
1910  IF OP = 3 GOTO 1790
1920  IF OP = 10 THEN  RETURN
1930  INPUT "ENTER C                     ";C
1940  IF OP = 2 OR OP = 5 OR OP = 7 GOTO 1790
1950  INPUT "ENTER D                     ";D
1960  IF OP = 11 GOTO 1980
1970  GOTO 1790
1980  INPUT "ENTER E                     ";E
1990  GOTO 1790
2000 T1 =  VAL ( LEFT$ (T$, LEN (T$) - 1))
2010  IF  RIGHT$ (T$,1) = "C" GOTO 2100
2020  IF  RIGHT$ (T$,1) = "F" GOTO 2080
2030  PRINT
2040  PRINT "YOU DID NOT INDICATE DEGREES C OR F"
2050  PRINT
2060  INPUT "ENTER TEMPERATURE AGAIN     ";T$
2070  GOTO 2000
2080 T = (T1 - 32) / 1.8 + 273.16
2090  GOTO 2110
2100 T = T1 + 273.16
2110  IF OP = 8 OR OP = 12 THEN T = T - 273.16
2120  RETURN
2130  PRINT
2140  PRINT B$(OP);" = "; FN P(CP)
2150  PRINT
2160  PRINT "FOR ANOTHER ";A$(OP)
2170  PRINT "CALCULATION AT A DIFFERENT TEMPERATURE"
2180  PRINT "ENTER THE TEMPERATURE, FOLLOWED BY"
2190  PRINT "F OR C.  IF YOU DON'T WANT ANOTHER"
2200  INPUT "CALCULATION, ENTER 'END'    ";T$
2210  IF T$ = "END" OR T$ = "E" GOTO 2310
2220  IF  ASC (T$) < 48 OR  ASC (T$) > 57 GOTO 2240
```

**Program for properties of chemical compounds (continued)** **Table II**

```
2230  GOTO 2290
2240  PRINT
2250  PRINT "INPUT SHOULD BE 'END' OR A TEMPERATURE."
2260  PRINT "YOU ENTERED ";T$;". TRY AGAIN."
2270  PRINT
2280  GOTO 2160
2290  GOSUB 2000
2300  ON OP GOTO 910,980,1040,910,980,910,980,1410,1460,
      1620,1680,1410
2310  PRINT
2320  PRINT  TAB( 10);"END OF PROGRAM."
2330  END
```

**Example for: Program for properties of chemical compounds** **Table III**

**(Start of first display)**

```
THIS PROGRAM PROVIDES FOR THE SOLUTION
OF EMPIRICAL EQUATIONS TO DETERMINE
A NUMBER OF PROPERTIES OF CHEMICAL
COMPOUNDS.

THE FOLLOWING PROPERTIES CAN BE
CALCULATED:

HEAT OF FORMATION,       HEAT CAPACITY,
THERMAL CONDUCTIVITY,    FREE ENERGY,
SURFACE TENSION,         VISCOSITY,
LIQUID DENSITY,          VAPOR PRESSURE,
HEAT OF VAPORIZATION.

TEMPERATURE AND EMPIRICAL CONSTANTS ARE
REQUIRED.  FOR SURFACE TENSION AND HEAT
OF VAPORIZATION, THE CRITICAL TEMP-
ERATURE AND A VALUE AT SOME OTHER
TEMPERATURE ARE REQUIRED.

PRESS RETURN TO CONTINUE.
```

**(Start of next display)**

```
INDICATE THE PROPERTY TO BE DETERMINED
BY NUMBER.  THE PROGRAM WILL SHOW
THE DATA REQUIRED.

 1.  HEAT CAPACITY OF IDEAL GAS.
 2.  HEAT OF FORMATION OF IDEAL GAS.
 3.  FREE ENERGY OF IDEAL GAS.
 4.  THERMAL CONDUCTIVITY OF GAS.
 5.  VISCOSITY OF GAS
 6.  HEAT CAPACITY OF LIQUID.
 7.  THERMAL CONDUCTIVITY OF LIQUID.
 8.  SURFACE TENSION OF LIQUID.
 9.  VISCOSITY OF LIQUID.
10.  DENSITY OF LIQUID.
11.  VAPOR PRESSURE.
12.  HEAT OF VAPORIZATION.

MAKE YOUR SELECTION BY NUMBER.     7
```

**(Start of next display)**

```
TO CALCULATE LIQUID THERMAL CONDUCTIVITY

YOU NEED THE CONSTANTS TO SOLVE
THE FOLLOWING EQUATION, AND THE
TEMPERATURE IN DEGREES C OR F.


K(L), MICRO-CAL/(S)(CM)(K) =
      A + B*T + C*T^2

ENTER A                         485.1
ENTER B                         -58.84E-2
ENTER C                         -.59E-4

ENTER THE TEST TEMPERATURE
AND FOLLOW THE TEMPERATURE
WITH AN F OR C:
140F, OR 140C.

ENTER TEMPERATURE NOW        150C

K(L), MICRO-CAL/(S)(CM)(K) = 225.5479

FOR ANOTHER LIQUID THERMAL CONDUCTIVITY
CALCULATION AT A DIFFERENT TEMPERATURE
ENTER THE TEMPERATURE, FOLLOWED BY
F OR C.  IF YOU DON'T WANT ANOTHER
CALCULATION, ENTER 'END'    302F

K(L), MICRO-CAL/(S)(CM)(K) = 225.5479

FOR ANOTHER LIQUID THERMAL CONDUCTIVITY
CALCULATION AT A DIFFERENT TEMPERATURE
ENTER THE TEMPERATURE, FOLLOWED BY
F OR C.  IF YOU DON'T WANT ANOTHER
CALCULATION, ENTER 'END'    END

             END OF PROGRAM.
```

# Volumes of pure liquids

## This program offers the user a choice among four correlations in predicting the molar volume of saturated and compressed pure liquids.

☐ There are a number of correlations available for predicting the volume of saturated pure liquids at various temperatures. Based on the principle of corresponding states, these require critical-property information plus a number of empirical constants.

This Apple II program is built around four such saturated-liquid correlations: Lyckman, Eckert and Prausnitz [1]; Rackett [3]; the Yamada-Gunn modification of Rackett's equation [4]; and Lu, Reuther, Hsi and Chiu [5]. The program also includes the Chueh and Prausnitz [2] equation, which predicts the effect of pressure on molar volume and thus can be used to correct the saturated-liquid volume for the higher pressures.

Table I lists the program, which is interactive and self-explanatory for the most part. However, the user is allowed to choose among the four saturated-liquid correlations, and these are not, of course, explained in depth within the program. Thus, before going over the program let us look at these equations and at the higher-pressure correction equation.

### Saturated-molar-volume correlations

The program offers four correlations for the molar volume of saturated pure liquids. In turn:

■ Lyckman, Eckert and Prausnitz's equation [1] for the molar density of a saturated pure liquid yields the following relationship for saturated molar volume ($V_S$, $cm^3/g\text{-mol}$):

$$V_s/V_c = V_{r1} + \omega V_{r2} + \omega^2 V_{r3} \quad (1)$$

$$V_{rj} = a_j + b_j T_r + c_j T_r^2 + d_j T_r^3 + (e_j/T_r) + f_j \ln(1 - T_r) \quad (2)$$

where $T_r$ is the reduced temperature ($T/T_c$).

There are six constants for each $V_{rj}$ term, thus 18 constants overall. These are included in the program, and need not be entered. The only required data are the pure liquid's critical temperature ($T_c$, K), acentric factor ($\omega$), critical volume ($V_c$, $cm^3/g\text{-mol}$), and temperature ($T$, °C).

■ Rackett [3] proposed the relationship:

$$\log V_s = [1 + (1 - T_r)^{2/7}] \times [\log Z_{RA} - \log (P_c/RT_c)] \quad (3)$$

The term $Z_{RA}$ is not a critical compressibility factor, but an adjustable parameter determined from available data. In 1978, Spencer and Adler reviewed the volumetric data and arrived at "best" values for $Z_{RA}$ for many substances [6].

■ Yamada and Gunn [4] altered the Rackett equation to:

$$V_s = V_{scr} Z^{(1-T_r)2/7} \quad (4)$$

Eq. (4) has two adjustable parameters—the compressibility factor ($Z$) and the scaling volume ($V_{scr}$). The Rackett value may be used for the compressibility factor, or a value may be calculated with the generalized relationship:

$$Z = 0.29056 - 0.08775\omega \quad (5)$$

The scaling volume can be determined with Eq. (4), using an experimental value for $V_s$ (referred to as $V_s'$). The program does this. The value for $V_s'$ should be chosen at conditions far removed from the critical point, because $V_{scr}$ differs somewhat from the critical volume.

■ Lu, Ruether, Hsi and Chiu [5] modified Guggenheim's correlation [7] into:

$$V_s = V_{scr}/[1 + \alpha(1 - T_r) + \beta(1 - T_r)^{1/3}] \quad (6)$$

Here, $\alpha = 0.73098 + 0.28908\omega$, and $\beta = 1.75238 + 0.74293\omega$.

The scaling volumes of Yamada and Gunn can be used, or they can be calculated by:

$$V_{scr} = V_{0.6}/(0.3862 - 0.0866\omega) \quad (7)$$

Here, $V_{0.6}$ is the molar volume when the reduced temperature is 0.6, well removed from the critical point. Lu, et al., give scaling volumes for many substances.

### Effect of pressure on volume

To estimate the effect of higher pressure on the volume of the liquid, Chueh and Prausnitz proposed the relationship:

$$V = V_s \left[1 + \frac{9Z_c N(P - P_s)}{P_c}\right]^{-1/9} \quad (8)$$

Here:

$$N = (1.0 - 0.89\omega)[\exp (6.9547 - 76.2853T_r + 191.3060T_r^2 - 203.5472T_r^3 + 82.7631T_r^4)] \quad (9)$$

Adapted from an article by James H. Weber, University of Nebraska, originally published March 26, 1979.

The additional data required are vapor pressure ($P_s$), critical pressure ($P_c$), and the constants necessary to calculate the value of $N$. The critical compressibility factor ($Z_c$) is calculated from the other critical properties, and gas law constant ($R$ = 82.07).

The value of $V_s$ is assumed to be known—i.e., at a given temperature, from one of the four correlations.

Though it should be obvious, please note that this equation is valid only for pressures above the saturation pressure for the given temperature. At pressures below the saturation pressure, the substance is a vapor and not a liquid.

## Using the program

Table II shows the displays the user would encounter in calculating the molar volume for saturated liquid propylene at 37.78°C (100°F) via the Lyckman, Eckert and Prausnitz correlation. It also shows the displays encountered in going further and calculating molar volume at a different pressure via the Chueh and Prausnitz correlation.

Table III shows the data required for calculating the molar volume of saturated, pure propylene by any of the four correlations, and compares the molar volumes they estimate for 37.78°C with the volume measured experimentally by Zordan and Henry [8]—the estimated volumes are within 0.2 $cm^3/g$-mol. Finally, it also shows the estimated volumes for compressed propylene at various pressures, and compares them with experimental ones—again, they are within 0.2 $cm^3/g$-mol.

## The author

**James H. Weber** is Regent's Professor of Chemical Engineering at the University of Nebraska-Lincoln (Lincoln, NE 68588), where he has been for the past 30 years, having previously been chairman of the Dept. of Chemical Engineering for 13 years. He has also served as a consultant to Gas Processors Assn., C. F. Braun & Co., and Phillips Petroleum Co. Holder of B.S., M.S. and Ph.D. degrees in chemical engineering from the University of Pittsburgh, he is a member of AIChE and ACS, a registered engineer in the State of Nebraska, and author or coauthor of 60 articles published in technical and scientific magazines.

## References


1. Lyckman, E. W., Eckert, C. A., and Prausnitz, J. M., *Chem. Eng. Sci.*, Vol. 20, 1965, p. 703.
2. Chueh, P. L., and Prausnitz, J. M., *AIChE J.*, Vol. 13, 1967, p. 1099, and Vol. 15, 1969, p. 471.
3. Rackett, H. G., *J. Chem. Eng. Data*, Vol. 15, 1970, p. 514.
4. Yamada, T., and Gunn, J., *J. Chem. Eng. Data*, Vol. 18, 1973, p. 234.
5. Lu, B. C.-Y., Ruether, J. A., Hsi, C., and Chiu, C.-H., *J. Chem. Eng. Data*, Vol. 18, 1973, p. 241.
6. Spencer, C. F., and Adler, S. B., *J. Chem. Eng. Data*, Vol. 23, 1978, p. 92.
7. Guggenheim, E. A., *J. Chem. Phys.*, Vol. 13, 1945, p. 253.
8. Zordan, T. A., and Henry, R. M., *J. Chem. Eng. Data*, Vol. 20, 1975, p. 343.


**Program for predicting volumes of pure liquids** **Table I**

```
10 REM   VOLUMES OF PURE LIQUIDS
20 REM   FROM CHEMICAL ENGINEERING, MARCH 26, 1979
30 REM   BY JAMES H. WEBER
40 REM  TRANSLATED BY WILLIAM VOLK
50 REM   COPYRIGHT (C) 1984
60 REM  BY CHEMICAL ENGINEERING
70 HOME : VTAB 2
80 REM  SET DISPLAYS TO TWO OR FOUR DECIMALS
90 DEF  FN P(X) =  INT (1E4 * (X + .00005)) / 1E4
100  DEF  FN Q(X) =  INT (100 * (X + .005)) / 100
110 A$(1) = "LYCKMAN, ECKERT & PRAUSNITZ EQUATION"
120 A$(2) = "RACKETT'S EQUATION"
130 A$(3) = "YAMADA & GUNN EQUATION"
140 A$(4) = "LU, RUETHER, HSI & CHIU EQUATION"
150  PRINT "PROGRAM OFFERS THE FOLLOWING PURE"
160  PRINT "LIQUID VOLUME CALCULATIONS:"
170  PRINT
180  PRINT "1. ";A$(1)
190  PRINT "   REQUIRING CRITICAL TEMPERATURE"
200  PRINT "   AND PRESSURE, AND ACENTRIC FACTOR."
210  PRINT
220  PRINT "2. ";A$(2);",REQUIRING"
230  PRINT "   CRITICAL TEMPERATURE AND PRESSURE,"
240  PRINT "  AND A COMPRESSIBILITY FACTOR."
250  PRINT
260  PRINT "3. ";A$(3);", WHICH HAS"
270  PRINT "   SEVERAL ALTERNATIVES:"
280  PRINT " A COMPRESSIBILITY AND SCALING FACTORS."
290  PRINT " B ACENTRIC AND SCALING FACTORS "
300  PRINT " C KNOWN VOLUME & TEMPERATURE IN PLACE"
310  PRINT "   OF SCALING FACTOR IN A & B."
320  PRINT
330  PRINT "4. ";A$(4)
340  PRINT "   REQUIREMENTS AS YAMADA & GUNN."
350  PRINT
360  INPUT "MAKE YOUR SELECTION BY NUMBER, 1 TO 4. ";OP
370  HOME : PRINT
380  PRINT A$(OP)
390  PRINT
400  ON OP GOTO 420,600,740,1220
410  REM   LYCKMAN, ECKERT & PRAUSNITZ
420  PRINT "ENTER THE FOLLOWING DATA:"
430  PRINT
440  INPUT "CRITICAL TEMPERATURE, DEG. K      ";TC
450  INPUT "CRITICAL VOLUME, CC/G-MOL         ";VC
460  INPUT "ACENTRIC FACTOR, OMEGA            ";OM
470  INPUT "CALCULATION TEMPERATURE, DEG. C   ";T
480 TR = (T + 273.16) / TC
490  IF TR < 1 GOTO 540
500  PRINT : PRINT
510  PRINT "REDUCED TEMPERATURE IS GREATER THAN 1."
520  PRINT "THESE CORRELATIONS CAN NOT BE USED."
530  GOTO 1870
540 V0 = .11917 + .009513 * TR + .21091 * TR ^ 2 - .069
   22 * TR ^ 3 + .0748 / TR - .084476 *  LOG (1 - TR)
550 V1 = .98465 - 1.60378 * TR + 1.82484 * TR ^ 2 - .61
   432 * TR ^ 3 - .34546 / TR + .087037 *  LOG (1 - TR)
560 V2 =  - .55314 - .15793 * TR - 1.01601 * TR ^ 2 + .34
```

**Program for predicting volumes of pure liquids (continued)** **Table I**

```
   095 * TR ^ 3 + .46795 / TR - .239938 *  LOG (1 - TR)
570 V = (V0 + OM * V1 + OM ^ 2 * V2) * VC
580  GOTO 1180
590  REM  RACKETT
600  PRINT "ENTER THE FOLLOWING DATA:'
610  PRINT
620  INPUT "CRITICAL TEMPERATURE, DEG. K       ";TC
630  INPUT "CRITICAL PRESSURE, ATM.            ";PC
640  INPUT "COMPRESSIBILITY FACTOR             ;ZR
650  INPUT "CALCULATION TEMPERATURE, DEG. C    ";T
660 TR = (T + 273.16) / TC
670  IF TR < 1 GOTO 690
680  GOTO 500
690 TP = ((1 + (1 - TR) ^ (2 / 7)) * ( LOG (ZR) /  LOG
    (10)) - ( LOG (PC / (82.07) / TC) /  LOG (10)))
700 V = 10 ^ TP
710 OM = (ZR - .29056) / ( - .08775)
720  GOTO 1180
730  REM  YAMADA AND GUNN
740  PRINT "YOU NEED THE CRITICAL TEMPERATURE,"
750  PRINT "YOUR TEST TEMPERATURE,"
760  PRINT "AND ONE OF THE FOLLOWING:"
770  PRINT
780  PRINT "1. ACENTRIC FACTOR AND SCALING FACTOR."
790  PRINT "2. ACENTRIC FACTOR AND A KNOWN VOLUME"
800  PRINT "   AND TEMPERATURE."
810  PRINT "3. COMPRESSIBILITY FACTOR AND SCALING"
820  PRINT "   FACTOR."
830  PRINT "4. COMPRESSIBILITY FACTOR AND KNOWN"
840  PRINT "   VOLUME AND TEMPERATURE."
850  PRINT
860  INPUT "MAKE YOUR CHOICE BY NUMBER: ";YP
870  HOME : VTAB 4
880  PRINT A$(3)
890  PRINT : PRINT
900  INPUT "CRITICAL TEMPERATURE, DEG. K       ";TC
910  INPUT "CALCULATION TEMPERATURE, DEG. C    ";T
920 TR = (T + 273.16) / TC
930  IF TR < 1 GOTO 950
940  GOTO 500
950  ON YP GOTO 1080,1130,960,1000
960  INPUT "COMPRESSIBILITY FACTOR             ";ZR
970  INPUT "SCALING FACTOR                     ";VS
980  PRINT
990  GOTO 1050
1000  INPUT "COMPRESSIBILITY FACTOR            ";ZR
1010  INPUT "KNOWN VOLUME, CC/G-MOL            ";V1
1020  INPUT "TEMPERATURE FOR VOLUME, DEG. C    ";T1
1030  PRINT
1040 VS = V1 / (ZR ^ ((1 - (T1 + 273.16) / TC) ^ (2 /
    7)))
1050 V = VS * ZR ^ ((1 - TR) ^ (2 / 7))
1060 OM = (ZR - .29056) / ( - .08775)
1070  GOTO 1180
1080  INPUT "ACENTRIC FACTOR, OMEGA            ";OM
1090  INPUT "SCALING FACTOR                    ";VS
1100 ZR = .29056 - .08775 * OM
1110  IF F1 = 1 GOTO 1040
1120  GOTO 1050
1130  INPUT "ACENTRIC FACTOR, OMEGA            ";OM
1140  INPUT "KNOWN VOLUME, CC/G-MOL            ";V1
1150  INPUT "TEMPERATURE FOR VOLUME, DEG. C    ";T1
1160 F1 = 1
1170  GOTO 1070
1180  PRINT
1190  PRINT "LIQUID VOLUME, CC/G-MOL IS "; FN P(V)
1200  GOTO 1540
1210  REM  LU, REUTHER
1220  PRINT "YOU NEED:"
1230  PRINT
1240  PRINT "THE CRITICAL TEMPERATURE, DEG. K.,"
1250  PRINT "THE TEST TEMPERATURE, DEG. C.",
1260  PRINT "THE ACENTRIC FACTOR, OMEGA, AND"
1270  PRINT
1280  PRINT "1. EITHER THE SCALING FACTOR, OR"
1290  PRINT "2. THE VOLUME AT REDUCED TEMPERATURE"
1300  PRINT "   OF 0.6."
1310  PRINT
1320  INPUT "ENTER 1, OR 2 FOR YOUR CHOICE.  ";CH
1330  HOME : PRINT
1340  PRINT A$(OP)
1350  PRINT
1360  PRINT "ENTER:"
1370  PRINT
1380  INPUT "CRITICAL TEMPERATURE, DEG. K      ";TC
1390  INPUT "TEST TEMPERATURE, DEG. C          ";T
1400 TR = (T + 273.16) / TC
1410  IF TR < 1 GOTO 1430
1420  GOTO 500
1430  INPUT "ACENTRIC FACTOR, OMEGA            ";OM
1440  ON CH GOTO 1450,1470
1450  INPUT "THE SCALING FACTOR                ";VS
1460  GOTO 1500
1470  INPUT "VOLUME, CC/G-MOL AT TR = .6      · ";V1
1480  PRINT "TEMPERATURE OF 0.6."
1490  ON CH GOTO 1500,1520
1500 V = VS / (1 + (.73098 + .28908 * OM) * (1 - TR) +
    (1.75238 + .74293 * OM) * (1 - TR) ^ (1 / 3))
1510  GOTO 1180
1520 VS = V1 / (.3862 - .0866 * OM)
1530  GOTO 1500
1540  PRINT
1550  PRINT "YOU CAN USE AN EQUATION OF CHUEH AND"
1560  PRINT "PRAUSNITZ TO ESTIMATE THE VOLUME AT"
1570  PRINT "ANOTHER PRESSURE.  YOU WILL NEED THE"
1580  PRINT "VAPOR PRESSURE AT ";T;" DEG. C."
1590  PRINT "AND THE CRITICAL VOLUME, CC/G-MOL."
1600  PRINT
1610  PRINT "DO YOU WANT THIS CALCULATION?"
1620  INPUT "ANSWER Y OR N.  ";AN$
1630  IF AN$ = "N" GOTO 1870
1640  IF AN$ = "Y" GOTO 1690
1650  PRINT
1660  PRINT "YOUR OPTIONS WERE Y OR N.  YOU ENTERED"
```

**Program for predicting volumes of pure liquids (continued)** **Table I**

```
1670  PRINT AN$;". TRY AGAIN."
1680  GOTO 1600
1690  HOME : VTAB 4
1700  PRINT "FOR ESTIMATED VOLUME AT A TEST PRESSURE"
1710  PRINT "DATA REQUIRED ARE:"
1720  PRINT
1730  PRINT "CRITICAL VOLUME, CC/G-MOL. AND"
1740  PRINT "VAPOR PRESSURE, ATM, AT ";T;" DEG. C."
1750  PRINT "TEST PRESSURE, ATM."
1760  PRINT
1770  PRINT "ENTER DATA NOW:"
1780  PRINT
1790  INPUT "CRITICAL VOLUME, CC/G-MOL  ";VC
1800  INPUT "VAPOR PRESSURE, ATM        ";PS
1810  INPUT "TEST PRESSURE, ATM         ";P1
1820 N = (1 - .89 * OM) *  EXP (6.9547 - 76.2853 * TR +
     191.3060 * TR ^ 2 - 203.5472 * TR ^ 3 + 82.7631 *
     TR ^ 4)
1830 V1 = V / (1 + 9 * VC / 82.07 / TC * N * (P1 - PS))
     ^ (1 / 9)
1840  PRINT
1850  PRINT "VOLUME AT ";P1;" ATM. "
1860  PRINT "       IS "; FN Q(V1);" CC/G-MOL."
1870  PRINT
1880  PRINT  TAB( 10);"END OF PROGRAM"
1890  END
```

**Example for: Volumes of pure liquids** **Table II**

**(Start of first display)**

```
PROGRAM OFFERS THE FOLLOWING PURE
LIQUID VOLUME CALCULATIONS:

1. LYCKMAN, ECKERT & PRAUSNITZ EQUATION
   REQUIRING CRITICAL TEMPERATURE
   AND PRESSURE, AND ACENTRIC FACTOR.

2. RACKETT'S EQUATION,REQUIRING
   CRITICAL TEMPERATURE AND PRESSURE,
  AND A COMPRESSIBILITY FACTOR.

3. YAMADA & GUNN EQUATION, WHICH HAS
   SEVERAL ALTERNATIVES:
 A COMPRESSIBILITY AND SCALING FACTORS.
 B ACENTRIC AND SCALING FACTORS.
 C KNOWN VOLUME & TEMPERATURE IN PLACE
   OF SCALING FACTOR IN A & B.

4. LU, RUETHER, HSI & CHIU EQUATION
   REQUIREMENTS AS YAMADA & GUNN.

MAKE YOUR SELECTION BY NUMBER, 1 TO 4. 1
```

**(Start of next display)**

```
LYCKMAN, ECKERT & PRAUSNITZ EQUATION

ENTER THE FOLLOWING DATA:

CRITICAL TEMPERATURE, DEG. K      365
CRITICAL VOLUME, CC/G-MOL         181
ACENTRIC FACTOR, OMEGA            .148
CALCULATION TEMPERATURE, DEG. C   37.78

LIQUID VOLUME, CC/G-MOL IS 87.0332

YOU CAN USE AN EQUATION OF CHUEH AND
PRAUSNITZ TO ESTIMATE THE VOLUME AT
ANOTHER PRESSURE.  YOU WILL NEED THE
VAPOR PRESSURE AT 37.78 DEG. C.
AND THE CRITICAL VOLUME, CC/G-MOL.

DO YOU WANT THIS CALCULATION?
ANSWER Y OR N.  Y
```

**(Start of next display)**

```
FOR ESTIMATED VOLUME AT A TEST PRESSURE
DATA REQUIRED ARE:

CRITICAL VOLUME, CC/G-MOL. AND
VAPOR PRESSURE, ATM, AT 37.78 DEG. C.
TEST PRESSURE, ATM.

ENTER DATA NOW:

CRITICAL VOLUME, CC/G-MOL  181
VAPOR PRESSURE, ATM        15
TEST PRESSURE, ATM         40.827

VOLUME AT 40.827 ATM.
       IS 84.63 CC/G-MOL.

          END OF PROGRAM
```

**Comparisons of calculated and experimental volumes of liquid propylene** **Table III**

**Properties of propylene:**

|  |  |
|---|---|
| Molecular weight | 42.081 |
| Critical temperature, $T_c$, °K | 365.0 |
| Critical pressure, $P_c$, atm | 45.6 |
| Critical volume, $V_c$, cm$^3$/g-mol | 181 |
| Critical compressibility factor, $Z_c$ | 0.275 |
| Acentric factor, $\omega$ | 0.148 |
| Compressibility factor, $Z_{RA}$ | 0.27821 |
| Scaling volume, $V_{scr}$, cm$^3$/g-mol | 183 |

**Saturated volumes at 37.78°C:**

| $V_s$, cm$^3$/g-mol | Method |
|---|---|
| 87.0 | Lyckman, Eckert and Prausnitz [1] |
| 87.1 | Rackett [3] |
| 87.2 | Yamada and Gunn [4] |
| 87.1 | Lu, Ruether, Hsi and Chiu [5] |
| 87.2 | Zordan and Henry [8]* |

**Volumes at 37.78 °C and different pressures:**

| Pressure, P, atm | Volume, V, cm$^3$/g-mol: Zordan and Henry [8] * | Volume, V, cm$^3$/g-mol: Chueh and Prausnitz [2] |
|---|---|---|
| 27.218 | 86.0 | 85.8 |
| 40.827 | 84.8 | 84.7 |
| 54.436 | 83.8 | 83.6 |
| 68.045 | 82.8 | 82.7 |
| 81.654 | 82.0 | 81.9 |
| 95.264 | 81.2 | 81.1 |
| 108.87 | 80.5 | 80.4 |

*Experimental

# Volumes of saturated liquid mixtures

Three correlations for determining the molal volumes of saturated liquid mixtures are programmed. The results compare favorably with experimental data.

☐ This Apple II program calculates the volumes of saturated liquid mixtures by the methods of Harmens [*1,2*], Rackett [*3*], and Spencer and Danner [*4*]. In the last case, the pseudocritical temperature may be determined by the methods proposed by Kay [*5*], Li [*6*], and Chueh and Prausnitz [*7*]. The calculated results are compared to the experimental values for the methane-*n*-pentane and the propane-benzene systems at 37.78°C (100°F) given by Sage and Lacey [*8*].

The earliest correlation, simple to use but very limited in scope, is that of Harmens. Later, Rackett offered a modification of his method for pure substances. Spencer and Danner proposed a modification of the Rackett equation and studied the effect of three different values of pseudocritical temperature. The pseudocritical temperature is important, because it is the main, and sometimes the only, parameter that is adjusted.

## Molal volume by Harmens' correlation

In 1965, Harmens offered the correlation:

$$1/V_b = \rho_b = C \times F(T_r) \qquad (1)$$

He tabulated values for $C$ for straight hydrocarbons up to $C_7$. These can be calculated by Eq. (2):

$$C = \rho_c(0.43875 - 0.625\ Z_c) \qquad (2)$$

The function $F(T_r)$ is calculated from:

$$F(T_r) = 15.81 - 17.71\ T_r + 22.67\ T_r^2 - 15.07\ T_r^3 \qquad (3)$$

Relationship (3) was determined by Spencer and Danner from Harmens' tabulated information. The $C$ value for a mixture is obtained from the pure component values by:

$$x_{w1}/C_1 + x_{w2}/C_2 = 1/C_m \qquad (4)$$

And the pseudocritical temperature is estimated by:

$$T_{cm} = \sum_{i=1}^{n} r_i T_{ci} \qquad (5)$$

Here:

$$r_i = (x_i + x_{wi})/2 \qquad (6)$$

See the "Nomenclature" for explanations of the variables used here. And note that saturation for a liquid mixture is here taken to mean liquid at the bubble point.

## Using the program

Table I lists the program, which includes the Harmens correlation and the two others. The program is interactive, and self-explanatory for the most part. Table II shows example displays for molal-volume calculations for mixtures of methane/*n*-pentane and propane/benzene. Note that the program lists the data required for each correlation before asking the user to choose one.

To calculate saturated-liquid molal volume ($V_c$) via the Harmens equation, select the Harmens equation by entering "1" when prompted by the program. Then enter the necessary data: mole fraction ($x_i$), molecular weight, critical temperature ($T_{ci}$), critical volume ($V_{ci}$), and critical compressibility ($Z_{ci}$), for each of the components. When data for the last component have been entered, input "END," and enter the temperature when prompted. The program then calculates and displays $V_s$, cm$^3$/g-mol, for the mixture.

Now let us look at the correlations of Rackett and of Spencer and Danner.

## Rackett's modification

In 1971, Rackett modified a correlation of his for pure components into:

$$V_b = V_{cm} Z_{cm}^{[(1-T/T_{cb})^{2/7}]} \qquad (7)$$

Here:

$$V_{cm} = \Sigma x_i V_{ci} \qquad (8)$$

And:

$$Z_{cm} = \Sigma x_i Z_{ci} \qquad (9)$$

Hence, the only adjustable parameter is $T_{cb}$, which, for a mixture, is calculated by means of:

$$T_{cb} = \sum_{i=1}^{n} x_i b_i T_{ci} \Big/ \sum_{i=1}^{n} x_i b_i \qquad (10)$$

Adapted from an article by James H. Weber, University of Nebraska, originally published May 7, 1979.

## Nomenclature

- $b$ — Coefficient of Rackett, defined by Eq. (11)
- $C$ — Parameter defined by Eq. (2)
- $k$ — Interaction coefficient proposed by Chueh and Prausnitz, defined by Eq. (19)
- $P$ — Pressure, atm
- $R$ — Gas law constant, (atm) (cm$^3$)/(g-mol) (K)
- $r$ — Arithmetic average of mole fraction and weight fraction of a component
- $T$ — Temperature, K
- $V$ — Molal volume, cm$^3$/g-mol
- $x$ — Mole fraction
- $x_w$ — Weight fraction
- $Z$ — Compressibility factor
- $Z_{RA}$ — Modified compressibility factor proposed by Rackett
- $\phi$ — Critical volume fraction, defined by Eq. (15)
- $\rho$ — Density, g-mol/cm$^3$

**Subscripts**

- $b$ — Bubble point
- $c$ — Critical point
- $i$ — Component $i$ in mixture
- $j$ — Component $j$ in mixture
- $m$ — Mean value
- $r$ — Reduced property

Here:

$$b_i = \exp\left[0.000633 \sum_{i=1}^{n} x_j \,(T_{ci} - T_{cj})^{9/7}\right] \quad (11)$$

For aromatic hydrocarbons and non-hydrocarbons, the right-hand side of Eq. (11) must be multiplied by a factor, $c_i$. Values for this coefficient are given in Table III.

The Rackett correlation is choice "2" on the program menu; the necessary data are as displayed in Table II. If the "ln $c_i$" factor is not available for some component or components, enter zero, and the program will assume that $c_i = 1$.

## Rackett modified by Spencer and Danner

Spencer and Danner modified the Rackett equation to:

$$V_b = \frac{1}{\rho_b} = R\left[\sum_{i=1}^{n} x_i \,(T_{ci}/P_{ci})\right] Z_{RAm}^{[1+(1-T_r)^{2/7}]} \quad (12)$$

Here:

$$Z_{RAm} = \sum_{i=1}^{n} x_i Z_{RAi} \quad (13)$$

Eq. (12) has two adjustable parameters—$Z_{RA}$, the Rackett compressibility factor; and $T_{cm}$, the pseudocritical temperature of the mixture, needed to calculate $T_r$ $(=T/T_{cm})$.

In the program, three options—those of Kay, Li, and Chueh and Prausnitz—are given for calculating the pseudocritical temperature.

Kay's method is:

$$T_{cm} = \sum_{i=1}^{n} x_i T_{ci} \quad (14)$$

If Kay's method is chosen, then the program will request that the critical temperatures ($T_{ci}$) be entered in the same order as the components were.

Li proposed using the critical-volume fraction to calculate the pseudocritical temperature:

$$\phi_i = (x_i V_{ci})/(\sum_i x_i V_{ci}) \quad (15)$$

And:

$$T_{cm} = \Sigma \phi_i T_{ci} \quad (16)$$

Note that Li's method requires both critical temperature ($T_{ci}$) and critical volume ($V_{ci}$) for each component.

Chueh and Prausnitz proposed the following set of equations for calculating the pseudocritical temperature:

$$T_{cm} = \sum_i \sum_j \phi_i \phi_j T_{cij} \quad (17)$$

$$T_{cij} = (T_{ci} T_{cj})^{1/2} \,(1 - k_{ij}) \quad (18)$$

$$k_{ij} = 1.0 - \left[\frac{(V_{ci}^{1/3}\, V_{cj}^{1/3})^{1/2}}{(V_{ci}^{1/3}\, V_{cj}^{1/3})/2}\right] \quad (19)$$

As in the case of Li's equation, the equations of Chueh and Prausnitz require both critical temperature ($T_{ci}$) and critical volume ($V_{ci}$) for each component.

## Example results

Table II shows example displays for a binary mixture of 90% (molal) $n$-pentane and 10% methane at 37.78°C (100°F). Table IV lists the saturated-liquid molal volume calculated for this system by each of the correlations, and compares the calculated result with the experimental one of Sage and Lacey.

Table IV also lists calculated and experimental molal volumes for other mixtures of $n$-pentane and methane; calculated and experimental molal volumes for mixtures of propane and benzene; and the physical properties needed for all of the calculations.

Note in Table IV that the Harmens equation yields unreliable answers for the $n$-pentane/methane system at 37.78°C. This is because the system is too close to its critical temperature. Finally, note that for the 30%-$n$-pentane mixture, the pseudocritical temperature is below 37.78°C by Rackett's and Kay's definitions, so no solution is possible at that temperature.

## References


1. Harmens, A., *Chem. Eng. Sci.*, Vol. 20, 1965, p. 813.
2. Harmens, A., *Chem. Eng. Sci.*, Vol. 21, 1966, p. 725.
3. Rackett, H. G., *J. Chem. Eng. Data*, Vol. 16, 1971, p. 308.
4. Spencer, C. F., and Danner, R. P., *J. Chem. Eng. Data*, Vol. 18, 1973, p. 230.
5. Kay, W. B., *Ind. Eng. Chem.*, Vol. 30, 1938, p. 459.
6. Li, C. C., *Canadian J. Chem. Eng.*, Vol. 19, 1971, p. 709.
7. Chueh, P. L., and Prausnitz, J. M., *AIChE J.*, Vol. 13, 1967, p. 1099.
8. Sage, B. H., and Lacey, W. N., "Thermodynamic Properties of the Lighter Paraffin Hydrocarbons and Nitrogen," Am. Petroleum Inst., New York, N.Y., 1950, pp. 137 and 176.

**Program for predicting volumes of saturated liquid mixtures** **Table I**

```
10  REM   VOLUMES OF SATURATED LIQUID MIXTURES
20  REM   FROM CHEMICAL ENGINEERING, MAY 7, 1979
30  REM   BY JAMES H. WEBER
40  REM  TRANSLATED BY WILLIAM VOLK
50  REM   COPYRIGHT (C) 1984
60  REM  BY CHEMICAL ENGINEERING
70  HOME : VTAB 2
80  REM  SET DISPLAY TO FOUR DECIMALS
90  DEF  FN P(X) =  INT (1E4 * (X + .00005)) / 1E4
100 A$(1) = "HARMENS' CORRELATION"
110 A$(2) = "RACKETT'S CORRELATION"
120 A$(3) = "SPENCER AND DANNER CORRELATION"
130  PRINT "PROGRAM OFFERS THE FOLLOWING LIQUID"
140  PRINT "MIXTURE VOLUME CALCULATIONS:"
150  PRINT
160  FOR I = 1 TO 3
170  PRINT I;".  ";A$(I)
180  NEXT I
190  PRINT
200  PRINT "THE FOLLOWING DATA ARE REQUIRED:"
210  PRINT
220  PRINT  TAB( 15);"                   SPENCER"
230  PRINT  TAB( 15);"HARMENS  RACKETT  DANNER"
240  PRINT
250  PRINT "MOL FRACTION"; TAB( 18);"X"; TAB( 27);"X";
     TAB( 36);"X"
260  PRINT "MOLECULAR WEIGHT"; TAB( 18);"X"
270  PRINT "CRITICAL"
280  PRINT "TEMPERATURE, K"; TAB( 18);"X"; TAB( 27);"X"
     ; TAB( 36);"X"
290  PRINT "VOLUME,CC/G-MOL"; TAB( 18);"X"; TAB( 27);
     "X"; TAB( 33);"MAYBE"
300  PRINT "PRESSURE, ATM"; TAB( 36);"X"
310  PRINT "COMPRESSIBILITY"; TAB( 18);"X"; TAB( 27);
     "X"; TAB( 32);"MODIFIED"
320  PRINT "LN(C) FACTOR"; TAB( 27);"X"
330  PRINT
340  INPUT "MAKE YOUR CHOICE BY NUMBER: 1,2,OR 3 ";OP
350  HOME : VTAB 4
360  PRINT "FOR ";A$(OP)
370  PRINT "ENTER THE FOLLOWING DATA FOR EACH"
380  PRINT "COMPONENT."
390  IF OP = 2 GOTO 410
400  PRINT "WHEN ALL DATA ARE IN, ENTER 'END'."
410  ON OP GOTO 420,830,1170
420  PRINT
430  REM  HARMENS' CORRELATION
440 N = N + 1
450  INPUT "MOL FRACTION                    ";A$
460  IF A$ = "END" GOTO 660
470 X(N) =  VAL (A$)
480  INPUT "MOLECULAR WEIGHT                ";MW(N)
490  INPUT "CRITICAL TEMPERATURE, DEG. K    ";TC(N)
500  INPUT "CRITICAL VOLUME, CC/G-MOL       ";VC(N)
510  INPUT "COMPRESSIBILITY FACTOR          ";Z(N)
520  PRINT
530  PRINT "ENTER THE NEXT SET OR 'END'"
540  GOTO 440
550  INPUT "ENTER DATA NOW: ";A$,MW,TC,VC,Z
560  PRINT "
570  IF A$ = "END" GOTO 660
580 N = N + 1
590 X(N) =  VAL (A$)
600 MW(N) = MW
610 TC(N) = TC
620 VC(N) = VC
630 Z(N) = Z
640  PRINT "ENTER THE NEXT SET OR 'END,0,0,0,0"
650  GOTO 510
660 N = N - 1
670  PRINT
680  INPUT "ENTER TEST TEMPERATURE, DEG. C    ";T
690  FOR I = 1 TO N
700 SW = SW + X(I) * MW(I)
710  NEXT I
720  FOR I = 1 TO N
730 XW(I) = X(I) * MW(I) / SW
740 R(I) = (X(I) + XW(I)) / 2
750 TM = TM + R(I) * TC(I)
760 C(I) = 1 / VC(I) * (.43875 - .625 * Z(I))
770 CV = CV + XW(I) / C(I)
780  NEXT I
790 TR = (T + 273.16) / TM
800 FT = 15.81 - 17.71 * TR + 22.67 * TR ^ 2 - 15.07 *
     TR ^ 3
810 V = 1 / (FT / CV)
820  GOTO 1880
830  PRINT
840  REM  RACKETT CORRELATION
850  PRINT "CORRELATION APPLIES ONLY TO BINARY"
860  PRINT "MIXTURES."
870  PRINT
880  PRINT "ENTER FOLLOWING DATA FOR EACH COMPONENT"
890  PRINT
900 N = N + 1
910  INPUT "MOL FRACTION                      ";X(N)
920  INPUT "CRITICAL TEMPERATURE, DEG. K      ";TC(N)
930  INPUT "CRITICAL VOLUME, CC/G-MOL         ";VC(N)
940  INPUT "CRITICAL COMPRESSIBILITY FACTOR   ";ZC(N)
950  PRINT "FOR AROMATICS OR NON-HYDROCARBONS"
960  PRINT "A 'LN(C)' FACTOR IS NEEDED.  IF THIS"
970  PRINT "IS NOT AVAILABLE, ENTER ZERO."
980  INPUT "ENTER VALUE NOW                   ";C(N)
990  IF N = 2 GOTO 1030
1000  PRINT
1010  PRINT "ENTER THE SECOND SET."
1020  GOTO 900
1030  PRINT
```

**Program for predicting volumes of saturated liquid mixtures (continued)** **Table I**

```
1040  INPUT "ENTER TEST TEMPERATURE, DEG. C  ";T
1050  FOR I = 1 TO 2
1060 B(I) =  EXP (.000633 * ( ABS (TC(1) - TC(2))) ^
    (9 / 7) * X(I) + C(I))
1070 DE = DE + B(I) * X(I)
1080 NU = NU + B(I) * X(I) * TC(I)
1090 ZC = ZC + X(I) * ZC(I)
1100 VC = VC + X(I) * VC(I)
1110  NEXT I
1120 TC = NU / DE
1130  IF (T + 273.16) / TC < 1 GOTO 1150
1140  GOTO 1780
1150 V = VC * ZC ^ ((1 - (T + 273.16) / TC) ^ (2 / 7))
1160  GOTO 1880
1170  PRINT
1180  REM  SPENCER AND DANNER
1190 N = N + 1
1200  INPUT "MOL FRACTION                         ";X$
1210  IF X$ = "END" GOTO 1290
1220 X(N) =  VAL (X$)
1230  INPUT "CRITICAL TEMPERATURE, DEG. K       ";TC(N)
1240  INPUT "CRITICAL PRESSURE, ATM             ";PC(N)
1250  INPUT "MODIFIED COMPRESSIBILITY FACTOR    ";ZR(N)
1260  PRINT
1270  PRINT "ENTER NEXT SET, OR 'END'."
1280  GOTO 1190
1290 N = N - 1
1300  HOME : VTAB 4
1310  PRINT "ENTER THE TEST TEMPERATURE,"
1320  INPUT "DEG. C.       ";T
1330  HOME : VTAB 4
1340  PRINT "CALCULATION OF THE PSEUDOCRITICAL"
1350  PRINT "TEMPERATURE IS REQUIRED.  THREE"
1360  PRINT "METHODS ARE AVAILABLE:"
1370  PRINT
1380  PRINT "1.  KAY'S - REQUIRING NO ADDITIONAL DATA."
1390  PRINT "2.  LI'S - REQUIRING THE CRITICAL"
1400  PRINT "    VOLUME OF EACH COMPONENT."
1410  IF N > 2 GOTO 1450
1420  PRINT
1430  PRINT "3.  CHUEH AND PRAUSNITZ METHOD WHICH"
1440  PRINT "    ALSO REQUIRES CRITICAL VOLUME."
1450  PRINT
1460  INPUT "MAKE YOUR CHOICE BY NUMBER  ";CH
1470  ON CH GOTO 1570,1480,1480
1480  PRINT
1490  PRINT "ENTER ";N;" VALUES OF CRITICAL VOLUME"
1500  PRINT "CC/G-MOL, IN THE SAME ORDER AS THE"
1510  PRINT "ORIGINAL DATA; ONE VALUE AT A TIME."
1520  PRINT
1530  FOR I = 1 TO N
1540  PRINT
1550  INPUT "ENTER VALUE NOW ";VC(I)
1560  NEXT I
1570  FOR I = 1 TO N
1580 ZM = ZM + X(I) * ZR(I)
1590 BR = BR + (X(I) * (TC(I) / PC(I)))
1600  IF CH > 1 GOTO 1630
1610 TM = TM + X(I) * TC(I)
1620  GOTO 1640
1630 VT = VT + X(I) * VC(I)
1640  NEXT I
1650  IF CH = 1 GOTO 1760
1660  FOR I = 1 TO N
1670 PH(I) = X(I) * VC(I) / VT
1680 TM = TM + PH(I) * TC(I)
1690  NEXT I
1700  IF CH = 2 GOTO 1760
1710 K = (VC(1) * VC(2)) ^ (1 / 6)
1720 K = K / ((VC(1) ^ (1 / 3) + VC(2) ^ (1 / 3)) / 2)
1730 K = 1 - K ^ 3
1740 TJ =  SQR (TC(1) * TC(2)) * (1 - K)
1750 TM = 2 * TJ * PH(1) * PH(2) + PH(1) ^ 2 * TC(1) +
    PH(2) ^ 2 * TC(2)
1760 TR = (T + 273.16) / TM
1770  IF TR < 1 GOTO 1870
1780  HOME : PRINT
1790  PRINT "PSEUDOCRITICAL TEMPERATURE IS LESS"
1800  PRINT "THAN TEST TEMPERATURE.  TRY ANOTHER"
1810  PRINT "METHOD."
1820 BR = 0:VT = 0:ZM = 0:TM = 0
1830  IF OP <  > 2 GOTO 1860
1840 DE = 0:NU = 0:ZC = 0:VC = 0:N = 0:CV = 0
1850  GOTO 150
1860  GOTO 1340
1870 V = 82.07 * BR * ZM ^ (1 + (1 - TR) ^ (2 / 7))
1880  HOME : VTAB 4
1890  PRINT A$(OP);" INDICATES THE"
1900  PRINT "VOLUME OF THE MIXTURE,"
1910  PRINT "IN CC/G-MOL IS: "; FN P(V)
1920  PRINT
1930  PRINT  TAB( 13);"END OF PROGRAM"
1940  END
```

**Example for: Volumes of saturated liquid mixtures** **Table II**

**(Start of first display)**

```
PROGRAM OFFERS THE FOLLOWING LIQUID
MIXTURE VOLUME CALCULATIONS:

1.  HARMENS' CORRELATION
2.  RACKETT'S CORRELATION
3.  SPENCER AND DANNER CORRELATION
```

```
THE FOLLOWING DATA ARE REQUIRED:

                                   SPENCER
                   HARMENS  RACKETT  DANNER

MOL FRACTION          X        X       X
MOLECULAR WEIGHT X
CRITICAL
```

**Example for: Volumes of saturated liquid mixtures (continued)** **Table II**

```
TEMPERATURE, K   X          X        X
VOLUME,CC/G-MOL  X          X      MAYBE
PRESSURE, ATM                        X
COMPRESSIBILITY  X          X     MODIFIED
LN(C) FACTOR                X

MAKE YOUR CHOICE BY NUMBER: 1,2,OR 3 1
```

**(Start of next display)**

```
FOR HARMENS' CORRELATION
ENTER THE FOLLOWING DATA FOR EACH
COMPONENT.
WHEN ALL DATA ARE IN, ENTER 'END'.

MOL FRACTION                    .9
MOLECULAR WEIGHT                72.15
CRITICAL TEMPERATURE, DEG. K    470.3
CRITICAL VOLUME, CC/G-MOL       304
COMPRESSIBILITY FACTOR          .262

ENTER THE NEXT SET OR 'END'
MOL FRACTION                    .1
MOLECULAR WEIGHT                16.04
CRITICAL TEMPERATURE, DEG. K    190.6
CRITICAL VOLUME, CC/G-MOL       99
COMPRESSIBILITY FACTOR          .288

ENTER THE NEXT SET OR 'END'
MOL FRACTION                    END

ENTER TEST TEMPERATURE, DEG. C     37.78
```

**(Start of next display)**

```
HARMENS' CORRELATION INDICATES THE
VOLUME OF THE MIXTURE,
IN CC/G-MOL IS: 115.0079

          END OF PROGRAM
```

**(Start of first display)**

```
PROGRAM OFFERS THE FOLLOWING LIQUID
MIXTURE VOLUME CALCULATIONS:

1.  HARMENS' CORRELATION
2.  RACKETT'S CORRELATION
3.  SPENCER AND DANNER CORRELATION

THE FOLLOWING DATA ARE REQUIRED:

                                SPENCER
            HARMENS   RACKETT   DANNER

MOL FRACTION     X          X        X
MOLECULAR WEIGHT X
CRITICAL
TEMPERATURE, K   X          X        X
VOLUME,CC/G-MOL  X          X      MAYBE
PRESSURE, ATM                        X
COMPRESSIBILITY  X          X     MODIFIED
LN(C) FACTOR                X

MAKE YOUR CHOICE BY NUMBER: 1,2,OR 3 2
```

**(Start of next display)**

```
FOR RACKETT'S CORRELATION
ENTER THE FOLLOWING DATA FOR EACH
COMPONENT.

CORRELATION APPLIES ONLY TO BINARY
MIXTURES.

ENTER FOLLOWING DATA FOR EACH COMPONENT

MOL FRACTION                        .5
CRITICAL TEMPERATURE, DEG. K        470.3
CRITICAL VOLUME, CC/G-MOL           304
CRITICAL COMPRESSIBILITY FACTOR     .262
FOR AROMATICS OR NON-HYDROCARBONS
A 'LN(C)' FACTOR IS NEEDED.  IF THIS
IS NOT AVAILABLE, ENTER ZERO.
ENTER VALUE NOW                     0

ENTER THE SECOND SET.
MOL FRACTION                        .5
CRITICAL TEMPERATURE, DEG. K        190.6
CRITICAL VOLUME, CC/G-MOL           99-
CRITICAL COMPRESSIBILITY FACTOR     .288
FOR AROMATICS OR NON-HYDROCARBONS
A 'LN(C)' FACTOR IS NEEDED.  IF THIS
IS NOT AVAILABLE, ENTER ZERO.
ENTER VALUE NOW                     0

ENTER TEST TEMPERATURE, DEG. C  37.78
```

**(Start of next display)**

```
RACKETT'S CORRELATION INDICATES THE
VOLUME OF THE MIXTURE,
IN CC/G-MOL IS: 113.362

          END OF PROGRAM
```

**(Start of first display)**

```
PROGRAM OFFERS THE FOLLOWING LIQUID
MIXTURE VOLUME CALCULATIONS:

1.  HARMENS' CORRELATION
2.  RACKETT'S CORRELATION
3.  SPENCER AND DANNER CORRELATION

THE FOLLOWING DATA ARE REQUIRED:
```

**Example (continued)** **Table II**

```
                                    SPENCER
                  HARMENS  RACKETT  DANNER

MOL FRACTION       X        X        X
MOLECULAR WEIGHT X
CRITICAL
TEMPERATURE, K     X        X        X
VOLUME,CC/G-MOL    X        X      MAYBE
PRESSURE, ATM                        X
COMPRESSIBILITY    X        X     MODIFIED
LN(C) FACTOR                X

MAKE YOUR CHOICE BY NUMBER: 1,2,OR 3 3
```

**(Start of next display)**

```
FOR SPENCER AND DANNER CORRELATION
ENTER THE FOLLOWING DATA FOR EACH
COMPONENT.
WHEN ALL DATA ARE IN, ENTER 'END'.

MOL FRACTION                        .7923
CRITICAL TEMPERATURE, DEG. K       369.82
CRITICAL PRESSURE, ATM              41.93
MODIFIED COMPRESSIBILITY FACTOR    .27664

ENTER NEXT SET, OR 'END'.
MOL FRACTION                        .2077
CRITICAL TEMPERATURE, DEG. K       562.15
CRITICAL PRESSURE, ATM              48.33
MODIFIED COMPRESSIBILITY FACTOR    .26907

ENTER NEXT SET, OR 'END'.
MOL FRACTION                        END
ENTER THE TEST TEMPERATURE,
DEG. C.        100
```

**(Start of next display)**

```
CALCULATION OF THE PSEUDOCRITICAL
TEMPERATURE IS REQUIRED.  THREE
METHODS ARE AVAILABLE:

1.  KAY'S - REQUIRING NO ADDITIONAL DATA.
2.  LI'S - REQUIRING THE CRITICAL
    VOLUME OF EACH COMPONENT.

3.  CHUEH AND PRAUSNITZ METHOD WHICH
    ALSO REQUIRES CRITICAL VOLUME.

MAKE YOUR CHOICE BY NUMBER  1
SPENCER AND DANNER CORRELATION INDICATES THE
VOLUME OF THE MIXTURE,
IN CC/G-MOL IS: 111.1211

              END OF PROGRAM
```

**Coefficients for aromatics and non-hydrocarbons suggested by Rackett for Eq. (11)** **Table III**

| Substance | ln $c_i$ |
|---|---|
| Cyclopentane | -0.3 |
| Cyclohexane | -0.25 |
| Benzene | -0.4 |
| Toluene | -0.35 |
| Xylenes | -0.3 |
| Nitrogen | -0.2 |
| Carbon monoxide | -0.2 |
| Carbon dioxide | +0.15 |
| Hydrogen sulfide | -0.3 |

**Comparison of calculated and experimental molal volumes** **Table IV**

**Methane-*n*-pentane system at 37.78°C**

| | Pentane mole fraction | | |
|---|---|---|---|
| | 0.9 | 0.5 | 0.3 |
| **Results of:** | Molal volume, $V_b$,cm$^3$/g-mol | | |
| Sage and Lacey | 112.63 | 92.03 | 91.09 |
| Harmens | 115.0* | 119.4* | 131.9* |
| Rackett | 108.7 | | ** |
| Danner and Spencer, with $T_{cm}$ by: | | | |
| Kay | 101.0 | 106.3 | ** |
| Li | 98.6 | 81.4 | 78.0 |
| Chueh and Prausnitz | 99.0 | 84.7 | 89.8 |

*The method is unreliable when $T_r$ is greater than 0.90; by Harmens' definition, $T_r$ is 0.95 at 37.78°C.

**Because the pseudocritical temperature is below 37.78°C, a solution is not possible.

**Propane-benzene system at 37.78°C**

| | Propane mole fraction | | |
|---|---|---|---|
| | 0.1066 | 0.3069 | 0.7923 |
| **Results of:** | Molal volume, $V_b$,cm$^3$/g-mol | | |
| Sage and Lacey | 90.99 | 88.97 | 90.03 |
| Harmens | 92.3 | 91.9 | 92.8 |
| Rackett | 91.2 | 91.1 | 95.2 |
| Danner and Spencer, with $T_{cm}$ by: | | | |
| Kay | 89.9 | 89.1 | 89.9 |
| Li | 89.7 | 88.3 | 88.5 |
| Chueh and Prausnitz | 89.8 | 88.7 | 89.2 |

**Physical properties used in calculations**

| | Methane | Propane | *n*-Pentane | Benzene |
|---|---|---|---|---|
| MW | 16.04 | 44.09 | 72.15 | 78.11 |
| $T_c$ | 190.6† | 369.82 | 470.3 | 562.15 |
| $P_c$ | 45.4 | 41.93 | 38.3 | 48.33 |
| $V_c$ | 99. | 203. | 304. | 259. |
| $Z_c$ | 0.288 | 0.281 | 0.262 | 0.271 |
| $Z_{RA}$ | 0.2876 | 0.27664 | 0.2685 | 0.26907 |

†Harmens recommends using 195.0 K in his proposed calculation.

Data sources: Reid, R. C., Prausnitz, J. M., and Sherwood, T. K., "The Properties of Gases and Liquids," 3rd ed., McGraw-Hill Book Co., 1977; Spencer, C. F. and Adler, S. B., *J. Chem. Eng. Data,* Vol. 23, 1978, p. 82; Spencer, C. F. and Danner, R. P., *J. Chem. Eng. Data,* Vol. 18, 1973, p. 230.

# Viscosities of pure gases

A number of correlations—some theoretical, many empirical—have been developed for predicting the transport property of viscosity. This program lets the user choose among four of them.

☐ This Apple II program offers four methods for predicting viscosities of pure gases—both polar and nonpolar—at both low and high pressures.

The two correlations for low pressure (i.e., 1 atm) are those of Chapman and Enskog and of Yoon and Thodos. Both of these require the temperature, molecular weight, critical temperature and critical pressure of the pure gas, and the Chapman-Enskog correlation also requires an acentric factor. Both can handle polar gases, but these require additional physical constants, and the Yoon-Thodos correlation also distinguishes between gases that are hydrogen-bonding and those that are not.

The correlations for nonpolar and polar gases at high pressure are those of Jossi, Stiel and Thodos, and of Reichenberg. These require the same basic data as the low-pressure correlations, but they also require a low-pressure viscosity figure and other data.

Table I lists the program, which is interactive and self-explanatory for the most part. The first screen display lists the correlations, and their required data, and the user picks a correlation from this menu. Then, data entry is all prompted. Table II shows example screen displays for viscosity calculations via two of the correlation methods.

To explain the program in more detail, let us now look at each of the correlations in turn. Then we shall show some example calculations for ethane (nonpolar) and ammonia (polar).

## Viscosity by Chapman-Enskog correlation

The Chapman-Enskog theoretical treatment of polar and nonpolar pure gases at low pressure resulted in the relationship [*1*]:

$$\eta = \frac{(5/16)(\pi MRT)^{1/2}}{(\pi\sigma^2)\Omega_v} = 26.69\,\frac{(MT)^{1/2}}{\sigma^2\Omega_v} \qquad (1)$$

For nonpolar gases, the collision integral, $\Omega_v$, using the Lennard-Jones 12-6 potential, has been represented mathematically by Neufeld, Janzen and Aziz [*2*]:

$$\Omega_v = \left(\frac{A}{T^{*B}}\right) + \frac{C}{\exp(DT^*)} + \frac{E}{\exp(FT^*)} \qquad (2)$$

Here, $T^* = kT/\epsilon$, and $A$, $B$, $C$, $D$, $E$ and $F$ are empirical constants.

If values of $\sigma$ and $\epsilon$ are not available, they may be calculated from relationships developed by Tee, Gotoh and Stewart [*3*]:

$$\sigma\,(P_c/T_c) = 2.3551 - 0.087\omega \qquad (3)$$

$$\epsilon/kT_c = 0.7915 + 0.1693\omega \qquad (4)$$

The viscosity of polar gases also may be calculated with Eq. (1). However, the collision integral, $\Omega_v$, must then be evaluated using the Stockmayer potential. The relationship between the two integrals is:

$$\Omega_v\,(\text{Stockmayer}) = \Omega_v\,(\text{Lennard-Jones}) + 0.2\delta^2/T^* \qquad (5)$$

Values of $\sigma$, $\epsilon/k$ and $\delta$ are not easily calculated, and data on a few common polar gases are given in Table III. The program asks whether the gas is polar, and if so it requests these constants.

## Viscosity by Yoon-Thodos correlation

Several correlations based on the theorem of corresponding states have been proposed for predicting viscosity data. One of the most useful is that offered by Yoon and Thodos [*4*]. This relationship, in varying forms, can be applied to both nonpolar and polar substances. Because none of these relationships include a pressure term, as was the case with Eq. (1), they are only valid at low pressure.

For nonpolar gases, Yoon and Thodos proposed:

$$\eta\xi = 4.610T_r^{0.618} - 2.04e^{-0.449T_r} + 1.94e^{-4.058T_r} + 0.1 \qquad (6)$$

For polar gases, Yoon and Thodos suggested:

1. For hydrogen bonding substances, $T_r < 2.0$:

$$\eta\xi = (0.755T_r - 0.055)\,Z_c^{-5/4} \qquad (7)$$

2. For nonhydrogen bonding substances, $T_r < 2.5$:

$$\eta\xi = (1.90T_r - 0.29)^{4/5}\,Z_c^{-2/3} \qquad (8)$$

**Adapted from an article by James H. Weber, University of Nebraska, originally published June 18, 1979.**

The program asks whether the gas is polar. If so, it also asks whether it is hydrogen bonding, and requests a value of either $Z_c$ or $V_c$—both $T_c$ and $P_c$ are already entered.

## Predicting viscosity at higher pressures

Eq. (1) through (8) apply to pure gases only at low pressure. As might be imagined, the prediction of viscosities at increased pressures becomes more difficult and the results less reliable. Researchers have, therefore, usually depended on the theorem of corresponding states and either a residual viscosity ($\eta - \eta^o$) or a ratio ($\eta/\eta^o$). In both cases, the viscosity, $\eta^o$, at low pressure must be known or estimated.

For nonpolar gases, Jossi, Stiel and Thodos recommended [7]:

$$[(\eta - \eta^o)\xi + 1]^{0.25} = 1.0230 + 0.23364\rho_r + 0.58533\rho_r^2 - 0.40758\rho_r^3 + 0.093324\rho_r^4 \quad (9)$$

This relationship applies over a wide range of conditions ($0.1 \leqslant \rho_r < 3$).

For polar gases, Stiel and Thodos have proposed [8]:

$$(\eta - \eta^o)\xi = 1.656\rho_r^{1.111} \quad (\text{for } \rho_r \leq 0.1) \quad (10)$$

$$(\eta - \eta^o)\xi = 0.0607(9.045\rho_r + 0.63)^{1.739} \quad (\text{for } 0.1 \leq \rho_r \leq 0.9) \quad (11)$$

$$\log\{4 - \log[(\eta - \eta^o)\xi]\} = 0.6439 - 0.1005\rho_r - \Delta \quad (\text{for } 0.9 \leq \rho_r < 2.6) \quad (12)$$

$$\Delta = \begin{cases} 0 & (\text{for } 0.9 < \rho_r < 2.2) \\ 0.000475(\rho_r^3 - 10.65)^2 & (\text{for } 2.2 < \rho_r < 2.6) \end{cases} \quad (13)$$

The program asks whether the gas is polar or not, in order to choose among the two approaches. In either case, it requires the gas's molal volume at the conditions of interest ($V = 1/\rho$; and $V_r = 1/\rho_r$) and its viscosity at low pressure but the *same temperature*. If this is not available as experimental data, then one of the low-pressure correlations can be used to calculate it. This low-pressure viscosity is also required in the Reichenberg correlation.

## Reichenberg correlation for high pressures

A method using viscosity ratio for predicting the viscosity of nonpolar and polar gases at high pressure has been developed by Reichenberg [9]:

$$\eta/\eta^o = 1 + (1 - 0.45q) \times \{A'P_r^{1.5}/[B'P_r + (1 + C'P_r^{D'})^{-1}]\} \quad (14)$$

Here,

$$q = [668(\mu_P)^2 P_c]/T_c^2 \quad (15)$$

The terms $A'$, $B'$, $C'$ and $D'$ are functions of reduced temperature:

$$A' = (\alpha_1/T_r)\exp(\alpha_2 T_r^{-\alpha_3})$$
$$B' = A'(\beta_1 T_r - \beta_2)$$
$$C' = (\gamma_1/T_r)\exp(\gamma_2 T_r^{-\gamma_3})$$
$$D' = (\delta_1/T_r)\exp(\delta_2 T_r^{-\delta_3})$$

The $\alpha$, $\beta$, $\gamma$ and $\delta$ terms are empirical constants.

From Eq. (15), it can be seen that a value for the dipole moment (in debyes) of the substance is required. Although this may be somewhat of a handicap, the fact that volumetric data are not required represents an advantage over the methods proposed by Thodos et al. [7,8].

For this correlation, required data again include the gas's viscosity at low pressure but the same temperature, and this may be calculated by one of the two low-pressure correlations. The program also requires a value of $\mu_P$, the molecular dipole moment. If the gas is nonpolar, $\mu_P = 0$. If it is polar, then the value must be entered. Dipole moments for several nonpolar gases are listed in Table III.

## Nomenclature

|  |  |
|---|---|
| $A$, $B$, $C$, $D$, $E$, $F$ | Constants of Eq. (2) |
| $A'$, $B'$, $C'$, $D'$ | Constants of Eq. (14) |
| exp | Natural number $e$ (i.e., $\exp DT = e^{DT}$) |
| $k$ | Boltzmann's constant |
| $M$ | Molecular weight |
| $P$ | Pressure, atm |
| $q$ | Defined by Eq. (15) |
| $R$ | Gas law constant, 82.07 (atm) (cm$^3$)/(g-mol) (K) |
| $T$ | Temperature, K |
| $T^*$ | $kT/\epsilon$ |
| $t$ | Temperature, °C |
| $V$ | Molal volume, cm$^3$/g-mol |
| $Z$ | Compressibility factor, $PV/RT$ |
| $\alpha$, $\beta$, $\gamma$, $\delta$ | Constants for Eq. (14) |
| $\delta$ | Polar parameter in Eq. (5) |
| $\epsilon$ | Energy-potential parameter, ergs |
| $\eta$ | Viscosity, micropoise |
| $\mu_P$ | Dipole moment, debyes |
| $\xi$ | $T_c^{1/6}/(M^{1/2}P_c^{2/3})$ |
| $\rho$ | Density, g/cm$^3$ or g-mol/cm$^3$ |
| $\sigma$ | Molecular diameter, Å |
| $\omega$ | Pitzer's acentric factor |
| $\Omega$ | Collision integral |

**Subscripts**

| | |
|---|---|
| $c$ | Critical |
| $r$ | Reduced |

**Superscript**

| | |
|---|---|
| o | Low pressure |

## Example calculations

Table II shows example displays for two viscosity estimates: ethane (nonpolar) at 71.11°C, by the Chapman-Enskog correlation (Eq. 1); and ammonia (polar and hydrogen-bonding) at 37.78°C by the Yoon-Thodos correlation (Eq. 6).

Table IV lists more estimates for ethane and ammonia calculated by these correlations, and compares them with experimental data obtained by Sage and others [5,6]. Table V does the same for viscosity at higher pressures calculated via the correlations of Jossi, Stiel and Thodos (Eq. 9–13) and of Reichenberg (Eq. 14). Note in the physical-properties part of Table IV that the low-pressure viscosities ($\eta^\circ$) later used in the higher-pressure calculations are for ethane at 26.67°C and for ammonia at 171.11°C.

## References


1. Chapman, S. and Cowling, T. G., "The Mathematical Theory of Nonuniform Gases," Cambridge U. Press, New York, 1939.
2. Neufeld, P. D., Janzen, A. R., and Aziz, R. A., *J. Chem. Phys.*, Vol. 57, 1972, p. 1105.
3. Tee, L. S., Gotoh, S. and Stewart, W. E., *Ind. Eng. Chem. Fundamentals*, Vol. 5, No. 356, 1966, p. 363.
4. Yoon, P. and Thodos, G., *AIChE J*, Vol. 16, 1970, p. 300.
5. Carmichael, L. T. and Sage, B. H., *J. Chem. Eng. Data*, Vol. 8, 1963, p. 94.
6. Carmichael, L. T., Reamer, H. H. and Sage, B. H., *J. Chem. Eng. Data*, Vol. 8, 1963, p. 400.
7. Jossi, J. A., Stiel, L. I. and Thodos, G., *AIChE J*, Vol. 8, 1962, p. 59.
8. Stiel, L. I. and Thodos, G., *AIChE J*, Vol. 10, 1964, p. 275.
9. Reichenberg, D., "The Viscosities of Pure Gases at High Pressure," NPL Rep. Chem 38, National Physical Laboratory, Teddington, England, Aug. 1975.
10. Monchick, L. and Mason, E. A., *J. Chem. Phys.*, Vol. 35, 1961, p. 1676.
11. Canjar, L. N. and Manning, F. S., "Thermodynamic Properties and Reduced Correlations for Gases," Gulf Pub. Co., Houston, Tex., 1967.


**Program for predicting viscosities of pure gases** **Table I**

```
10  REM  VISCOSITIES OF PURE GASES
20  REM  FROM CHEMICAL ENGINEERING, JUNE 18, 1979
30  REM  BY JAMES H. WEBER
40  REM  TRANSLATED BY WILLIAM VOLK
50  REM  COPYRIGHT (C) 1984
60  REM  BY CHEMICAL ENGINEERING
70  HOME
80 A$(1) = "CHAPMAN-ENSKOG"
90 A$(2) = "YOON-THODOS"
100 A$(3) = "JOSSI, STIEL & THODOS"
110 A$(4) = "REICHENBERG"
120  REM  SET DISPLAY TO FOUR DECIMALS
130  DEF  FN P(X) =  INT (1E4 * (X + .00005)) / 1E4
140 PI = 3.14159265
150  PRINT "VISCOCITY OF PURE GASES."
160  PRINT "FOUR CORRELATIONS ARE AVAILABLE:"
170  PRINT
180  PRINT "FOR LOW PRESSURE:"
190  PRINT "1. ";A$(1);"          2.";A$(2)
200  PRINT
210  PRINT "FOR HIGH PRESSURE:"
220  PRINT "3. ";A$(3);"  4.";A$(4)
230  PRINT
240  PRINT "ALL REQUIRE TEST TEMPERATURE, MOLECULAR"
250  PRINT "WEIGHT, CRITICAL TEMPERATURE AND"
260  PRINT "PRESSURE; AND:"
270  PRINT
280  PRINT "1 - ACENTRIC FACTOR,"
290  PRINT "    AND ADDITIONAL DATA FOR POLAR GASES."
300  PRINT "2 - CRITICAL VOLUME FOR POLAR GASES."
310  PRINT
320  PRINT "3 & 4 - VISCOSITY AT LOW PRESSURE, AND"
330  PRINT "3 - MOLAL VOLUME AT TEST PRESSURE."
340  PRINT "4 - DIPOLE MOMENT."
350  PRINT
360  INPUT "SELECT CORRELATION BY NUMBER: ";OP
370  HOME : PRINT
380  PRINT A$(OP);" VISCOSITY CORRELATION"
390  PRINT
400  PRINT "ENTER THE FOLLOWING DATA:"
410  PRINT
420  INPUT "TEST TEMPERATURE, DEG. C         ";T
430  INPUT "MOLECULAR WEIGHT                 ";MW
440  INPUT "CRITICAL TEMPERATURE, DEG. K     ";TC
450  INPUT "CRITICAL PRESSURE, ATM           ";PC
460 TR = (T + 273.16) / TC
470 XI = TC ^ (1 / 6) / ( SQR (MW) * PC ^ (2 / 3))
480  ON OP GOTO 500,840,1080,1340
490  REM  CHAPMAN-ENSKOG
500  INPUT "ACENTRIC FACTOR                  ";OM
510  PRINT
520 SI = (2.3551 - .087 * OM) / ((PC / TC) ^ (1 / 3))
530 EK = (.7915 + .1693 * OM) * TC
540  INPUT "IS GAS POLAR?  ANSWER Y OR N.   ";A$
550  IF A$ = "N" GOTO 790
560  HOME : PRINT
570  PRINT
580  PRINT "VISCOSITY CALCULATION INCLUDES A"
590  PRINT "CALCULATION OF THE MOLECULAR DIAMETER"
600  PRINT "AND ENERGY POTENTIAL OVER BOLTZMAN'S"
610  PRINT "CONSTANT."
620  PRINT
630  PRINT "ALSO REQUIRED FOR POLAR GASES IS THE"
640  PRINT "STOCKMAYER POLAR PARAMETER."
650  PRINT
660  PRINT "CALCULATIONS OF MOLECULAR DIAMETER,"
670  PRINT "SIGMA, AND EPSILON OVER"
680  PRINT "K, ARE NOT RELIABLE"
690  PRINT "FOR POLAR GASES."
700  PRINT
710  PRINT "IF YOU HAVE ANY OF THESE VALUES, ENTER"
720  PRINT "THEM.  IF NOT, ENTER ZEROS."
730  PRINT
740  INPUT "MOLECULAR DIAMETER, SIGMA        ";S
750  INPUT "ENERGY/BOLTZMAN, EPSILON/K       ";E
760  INPUT "STOCKMAYER POLAR PARAMETER, DEL ";DE
770  IF S < > 0 THEN SI = S
780  IF E < > 0 THEN EK = E
```

**Program for predicting viscosities of pure gases (continued)** **Table I**

```
790 TS = (T + 273.16) / EK
800 OK = (1.16145 / TS ^ .14874) + (.52487 /  EXP (.7732
    * TS)) + 2.16178 /  EXP (2.43787 * TS)
810 OK = OK + .2 * (DE ^ 2) / TS
820 ET = 26.69 *  SQR (MW * (T + 273.16)) / OK / SI ^ 2
830  GOTO 1480
840  REM  YOON-THODOS
850  PRINT
860  PRINT "FOR POLAR GASES EITHER CRITICAL VOLUME"
870  PRINT "OR CRITICAL COMPRESSIBILITY FACTOR"
880  PRINT "IS NEEDED.  ENTER ONE OF THESE VALUES"
890  PRINT "IF YOURS IS A POLAR GAS.  IF NOT"
900  PRINT "ENTER ZERO."
910  INPUT "CRITICAL VOLUME, CC/G-MOL       ";VC
920  INPUT "CRITICAL COMPRESSIBILITY FACTOR ";ZC
930  IF VC = 0 AND ZC = 0 GOTO 1050
940  PRINT
950  INPUT "IS GAS HYDROGEN BONDED, ANSWER Y OR N. ";AN$
960  IF VC = 0 GOTO 980
970 ZC = PC * VC / TC / 82.07
980  IF AN$ = "N" GOTO 1020
990 ET = (.755 * TR - .055) * ZC ^ ( - 5 / 4)
1000  IF TR > 2 THEN F1 = 1
1010  GOTO 1060
1020 ET = (1.90 * TR - .29) ^ (4 / 5) * ZC ^ ( - 2 / 3)
1030  IF TR > 2.5 THEN F1 = 1
1040  GOTO 1060
1050 ET = (4.61 * TR ^ .618) - 2.04 *  EXP ( - .449 *
    TR) + 1.94 *  EXP ( - 4.058 * TR) + .1
1060 ET = ET / XI
1070  GOTO 1480
1080  REM  JOSSI,STIEL, ET AL
1090  INPUT "CRITICAL VOLUME, CC/G-MOL       ";VC
1100  INPUT "LOW PRESSURE VISCOSITY, MICRO-P ";Z1
1110  INPUT "TEST PRESSURE, ATM              ";P1
1120  INPUT "MOLAL VOLUME AT TEST PRESSURE   ";V1
1130 DR = VC / V1
1140  PRINT
1150  INPUT "IS THIS POLAR GAS? Y OR N       ";AN$
1160 XI = TC ^ (1 / 6) / SQR (MW) / PC ^ (2 / 3)
1170  IF AN$ = "Y" GOTO 1210
1180 TP = 1.0230 + .23364 * DR + .58533 * DR ^ 2 - .407
    58 * DR ^ 3 + .093324 * DR ^ 4
1190 ET = ((TP ^ 4) - 1) / XI + Z1
1200  GOTO 1480
1210  IF DR > .9 GOTO 1280
1220  IF DR > .1 GOTO 1250
1230 TP = 1.646 * DR ^ 1.111
1240  GOTO 1260
1250 TP = .0607 * (9.045 * DR + .63) ^ 1.739
1260 ET = TP / XI + Z1
1270  GOTO 1480
1280 TP = .6439 - .1005 * DR
1290  IF DR < 2.2 GOTO 1320
1300 DT = .000475 * (DR ^ 3 - 10.65) ^ 2
1310 TP = TP - DT
1320 TP = 10 ^ (4 - 10 ^ TP)
1330  GOTO 1260
1340  REM  REICHENBERG CORRELATION
1350  INPUT "DIPOLE MOMENT, DEBYES           ";DP
1360  INPUT "LOW PRESSURE VISCOCITY, MICRO-P ";Z1
1370  INPUT "TEST PRESSURE, ATM              ";P1
1380 TR = (T + 273.16) / TC
1390 PR = P1 / PC
1400 QQ = 668 * DP ^ 2 * PC / TC ^ 2
1410 AA = (.0019824 / TR) *  EXP (5.2683 * TR ^ ( - .5767))
1420 BB = AA * (1.6553 * TR - 1.276)
1430 CC = (.1319 / TR) *  EXP (3.7035 * TR ^ ( - 79.8678))
1440 DD = (2.9496 / TR) *  EXP (2.919 * TR ^ ( - 16.6169))
1450 TP = 1 + (1 - .45 * QQ) * (AA * PR ^ 1.5 / (BB * PR
    + (1 / (1 + CC * PR ^ DD))))
1460 ET = Z1 * TP
1470  GOTO 1480
1480  HOME : VTAB 8
1490  PRINT A$(OP);" CALCULATION"
1500  PRINT
1510  PRINT "INDICATES A VISCOSITY"
1520  PRINT "OF "; FN P(ET);" MICROPOISE AT ";T;" DEG. C."
1530  IF OP < 3 GOTO 1550
1540  PRINT "AND AT ";P1;" ATM. PRESSURE."
1550  PRINT
1560  IF OP = 3 AND F1 = 1 GOTO 1580
1570  GOTO 1640
1580  PRINT "REDUCED VOLUME IS TOO LOW FOR"
1590  GOTO 1630
1600  IF OP = 2 AND F1 = 1 GOTO 1620
1610  GOTO 1640
1620  PRINT "REDUCED TEMPERAURE IS TOO HIGH FOR"
1630  PRINT "RELIABLE RESULTS."
1640  PRINT
1650  PRINT  TAB( 15);"END OF PROGRAM"
1660  END
```

**Example for: Viscosities of pure gases** **Table II**

**(Start of first display)**

```
VISCOCITY OF PURE GASES.
FOUR CORRELATIONS ARE AVAILABLE:

FOR LOW PRESSURE:
1. CHAPMAN-ENSKOG          2.YOON-THODOS

FOR HIGH PRESSURE:
3. JOSSI, STIEL & THODOS  4.REICHENBERG

ALL REQUIRE TEST TEMPERATURE, MOLECULAR
WEIGHT, CRITICAL TEMPERATURE AND
PRESSURE; AND:

1 - ACENTRIC FACTOR,
    AND ADDITIONAL DATA FOR POLAR GASES.
```

**Example for: Viscosities of pure gases (continued)** **Table II**

```
2 - CRITICAL VOLUME FOR POLAR GASES.

3 & 4 - VISCOSITY AT LOW PRESSURE, AND
3 - MOLAL VOLUME AT TEST PRESSURE.
4 - DIPOLE MOMENT.

SELECT CORRELATION BY NUMBER: 1
```

**(Start of next display)**

```
CHAPMAN-ENSKOG VISCOSITY CORRELATION

ENTER THE FOLLOWING DATA:

TEST TEMPERATURE, DEG. C          71.11
MOLECULAR WEIGHT                  30.068
CRITICAL TEMPERATURE, DEG. K      305.4
CRITICAL PRESSURE, ATM            48.2
ACENTRIC FACTOR                   .098

IS GAS POLAR?  ANSWER Y OR N.     N
```

**(Start of next display)**

```
CHAPMAN-ENSKOG CALCULATION

INDICATES A VISCOSITY
OF 106.2181 MICROPOISE AT 71.11 DEG. C.


            END OF PROGRAM

VISCOCITY OF PURE GASES.
FOUR CORRELATIONS ARE AVAILABLE:

FOR LOW PRESSURE:
1. CHAPMAN-ENSKOG          2.YOON-THODOS

FOR HIGH PRESSURE:
3. JOSSI, STIEL & THODOS   4.REICHENBERG

ALL REQUIRE TEST TEMPERATURE, MOLECULAR
WEIGHT, CRITICAL TEMPERATURE AND
PRESSURE; AND:

1 - ACENTRIC FACTOR,
    AND ADDITIONAL DATA FOR POLAR GASES.
2 - CRITICAL VOLUME FOR POLAR GASES.

3 & 4 - VISCOSITY AT LOW PRESSURE, AND
3 - MOLAL VOLUME AT TEST PRESSURE.
4 - DIPOLE MOMENT.

SELECT CORRELATION BY NUMBER: 2
```

**(Start of next display)**

```
YOON-THODOS VISCOSITY CORRELATION

ENTER THE FOLLOWING DATA:

TEST TEMPERATURE, DEG. C          37.78
MOLECULAR WEIGHT                  17.032
CRITICAL TEMPERATURE, DEG. K      405.6
CRITICAL PRESSURE, ATM            111.3

FOR POLAR GASES EITHER CRITICAL VOLUME
OR CRITICAL COMPRESSIBILITY FACTOR
IS NEEDED.  ENTER ONE OF THESE VALUES
IF YOURS IS A POLAR GAS.  IF NOT
ENTER ZERO.
CRITICAL VOLUME, CC/G-MOL       72.5
CRITICAL COMPRESSIBILITY FACTOR 0

IS GAS HYDROGEN BONDED, ANSWER Y OR N. Y
```

**(Start of next display)**

```
YOON-THODOS CALCULATION

INDICATES A VISCOSITY
OF 108.0821 MICROPOISE AT 37.78 DEG. C.


            END OF PROGRAM
```

**Stockmayer potential parameters** **Table III**

| | Dipole moment, $\mu_p$, debyes | $\sigma$, Å | $\epsilon/k$, K | $\delta$ |
|---|---|---|---|---|
| $H_2O$ | 1.85 | 2.52 | 775 | 1.0 |
| $NH_3$ | 1.47 | 3.15 | 358 | 0.7 |
| HCl | 1.08 | 3.36 | 328 | 0.34 |
| $SO_2$ | 1.63 | 4.04 | 347 | 0.42 |
| $H_2S$ | 0.92 | 3.49 | 343 | 0.21 |
| $CH_3OH$ | 1.70 | 3.69 | 417 | 0.5 |
| $C_2H_5OH$ | 1.69 | 4.31 | 431 | 0.3 |
| $(CH_3)_2CO$ | 1.20 | 4.50 | 549 | 0.11 |

**Comparison of calculated and experimental viscosities of ethane and ammonia at 1 atm pressure** **Table IV**

| | Temperature, $t$, °C | Molal volume, $V$, $cm^3$/g-mol [11] | Viscosity, $\eta$, micropoise: Experimental | Calculated: Via Eq.(1) | Calculated: Via Eq.(6) |
|---|---|---|---|---|---|
| Ethane [5] | 26.67 | 24,442 | 93.84 | 93.00 | 94.63 |
| | 37.78 | 25,362 | 97.05 | 96.35 | 97.98 |
| | 71.11 | 28,141 | 106.91 | 106.22 | 107.87 |
| | 104.44 | 30,957 | 116.49 | 115.77 | 117.48 |
| | 137.78 | 33,641 | 125.78 | 125.00 | 126.82 |
| | 171.11 | 36,401 | 135.40 | 133.94 | 135.88 |
| | 204.44 | 39,141 | 144.88 | 142.56 | 144.68 |
| Ammonia [6] | 37.78 | 25,315 | 106.53 | 107.13 | 108.31 |
| | 71.11 | 28,080 | 119.60 | 119.27 | 121.14 |
| | 104.44 | 30,850 | 132.72 | 131.32 | 133.97 |
| | 137.78 | 33,614 | 145.70 | 143.24 | 146.80 |
| | 171.11 | 36,369 | 158.72 | 154.98 | 159.63 |
| | 204.44 | 39,120 | 171.65 | 166.52 | 172.46 |

**Physical properties used in calculations**

| | Ethane | Ammonia |
|---|---|---|
| Mol wgt | 30.068 | 17.032 |
| $T_c$ (K) | 305.4 | 405.6 |
| $P_c$ (atm) | 48.2 | 111.3 |
| $Z_c$ | 0.285 | 0.242 |
| $V_c$ ($cm^3$/g-mol) | 148.0 | 72.5 |
| $\omega$ | 0.098 | 0.250 |
| $\eta^o$ ($\mu$P) | 93.84 | 158.72 |

Data sources:
Reid, R.C., Prausnitz, J.M., and Sherwood, T.K., "The Properties of Gases and Liquids," 3rd ed., McGraw-Hill Book Co., 1977; Carmichael, L.T., and Sage, B.H., *J. Chem. Eng. Data,* Vol. 8, 1963; Carmichael, L.T., Reamer, H.H., and Sage, B.H., *J. Chem. Eng. Data,* Vol. 8, 1963; and Canjar, L.N., and Manning, F.S., "Thermodynamic Properties and Reduced Correlations for Gases," Gulf Pub. Co.

**Comparison of calculated and experimental viscosities of ethane and ammonia at elevated pressures** **Table V**

| | Pressure, $P$,atm | Molal volume, $V$, $cm^3$/g-mol [11] | Viscosity, $\eta$, micropoise: Experimental | Calculated: Via Eq. (9) | Calculated: Via Eq. (14) |
|---|---|---|---|---|---|
| Ethane at 26.67 °C [5] | 27.218 | 682.4 | 104.62 | 106.42 | 109.10 |
| | 42.916* | 270.7 | 136.22 | 138.07 | 123.16 |
| | | | | Via Eq. (10-13) | Via Eq. (14) |
| Ammonia at 171.11 °C [6] | 27.218 | 1,250 | 158.8 | 161.18 | 161.51 |
| | 68.046 | 440.9 | 161.59 | 166.57 | 169.30 |
| | 136.09 | 156.9 | 197.26 | 191.42 | 193.66 |
| | 204.14 | 71.99 | 316.96 | 272.58 | 269.59 |
| | 272.18 | 49.77 | 432.85 | 410.56 | 370.68 |
| | 340.23 | 43.71 | 499.86 | 509.22 | 432.00 |

*Saturation pressure

# Viscosities of liquids and mixtures

Correlations for calculating the viscosities of binary gaseous mixtures, pure liquids, and binary-liquid mixtures are programmed.

☐ This Apple II program estimates viscosities via six different correlations:

- For binary gaseous mixtures, the Wilke [1] correlation for low pressure, and the correlation of Dean and Stiel [2] for high pressure.
- For pure liquids, the correlations of Orrick and Erbar [5] and of Thomas [6] for moderate temperatures, and the correlation of Letsou and Stiel [8] for higher temperatures.
- For binary-liquid mixtures, the Lobe [9] correlation is used.

The program is listed in Table I, and displays for example cases are shown in Table II. Now let us look at each of the six correlations, in turn.

## Viscosity of gas mixtures

For gaseous mixtures at low pressure, Wilke used the general relationship:

$$\eta_m = \sum_{i=1}^{n} \left( y_i \eta_i \Big/ \sum_{j=1}^{n} y_i \phi_{ij} \right) \quad (1)$$

He proposed evaluating $\phi_{12}$ (for a binary mixture) by:

$$\phi_{12} = \frac{[1 + (\eta_1/\eta_2)^{1/2}(M_2/M_1)^{1/4}]^2}{[8(1 + M_1/M_2)]^{1/2}} \quad (2)$$

And $\phi_{21}$ by:

$$\phi_{21} = (\eta_2/\eta_1)(M_1/M_2)\phi_{12} \quad (3)$$

Since $\phi_{11} = \phi_{22} = 1$, Eq. (1) reduces to:

$$\eta_m = \frac{y_1\eta_1}{y_1 + y_2\phi_{12}} + \frac{y_2\eta_2}{y_2 + y_1\phi_{21}} \quad (4)$$

To use this correlation, enter the mole fraction, molecular weight, and pure-component viscosity (at the relevant temperature) for each of the two gases.

For mixtures of non-polar gases under pressure, Dean and Stiel used the theorem of corresponding states and correlated the residual viscosity as:

$$(\eta_m - \eta_m{}^o)\xi_m = 1.08[\exp(1.439\rho_{rm}) - \exp(-1.111\rho_{rm}{}^{1.858})] \quad (5)$$

Using Eq. (5) requires determining critical properties and density. To do this, Dean and Stiel followed the suggestions of Prausnitz and Gunn [3]:

$$T_{cm} = \sum_i y_i T_{ci} \quad (6)$$

$$Z_{cm} = \sum_i y_i Z_{ci} \quad (7)$$

$$V_{cm} = \sum_i y_i V_{ci} \quad (8)$$

$$P_{cm} = (Z_{cm} R T_{cm})/V_{cm} \quad (9)$$

Thus, this correlation requires critical data for both components. It also requires the viscosity of the mixture at low pressure and the relevant temperature ($\eta_m^0$), and the molar volume of the mixture at the higher pressure ($V_m$). Then, $V_{rm} = V_m/V_{cm}$, and $\rho_{rm} = 1/V_{rm}$, and the program has all the data needed to use Eq. (5).

Table III shows examples of viscosities calculated by the Wilke and Dean-Stiel correlations—Eq. (1) and (5)—for a gaseous mixture of methane and $n$-butane (0.384:0.616 ratio). It also provides all the data needed for these calculations, and shows how the estimates compare with experimental results.

## Viscosity of pure liquids

For pure liquids at moderate temperature, Orrick and Erbar proposed:

$$\ln \eta_L/\rho_L M = A + B/T \quad (10)$$

Here:

$$A = (6.95 + 0.21n) + GC(A) \quad (11)$$

$$B = 275 + 99n + GC(B) \quad (12)$$

This correlation is basically for hydrocarbons. Group contributions (GC) for $A$ and $B$ are the sums of the contributions of individual groups as listed in Table IV. The program asks for the molecular weight of the

Adapted from an article by James H. Weber, University of Nebraska, originally published July 30, 1979.

liquid, its density at the relevant temperature (g/cm$^3$), and the number of carbon atoms. It then goes through the groups listed in Table IV and asks how often they occur in the liquid molecule, and finally asks the temperature. Viscosity is reported in centipoise.

Similarly, Thomas suggested:

$$\log (8.569\, \eta_L/\rho_L^{1/2}) = \theta\,(1/T_r - 1) \quad (13)$$

This correlation, too, is built around group contributions (listed in Table V) and requires a density value. It also requires critical temperature, but not molecular weight.

For evaluating pure liquids at high temperatures ($0.76 \leq T_r \leq 0.98$), Letsou and Stiel suggested a corresponding-states approach that uses Pitzer's acentric factor:

$$\eta_L \xi = (\eta_L \xi)^0 + \omega\,(\eta_L \xi)^1 \quad (14)$$

Here:

$$(\eta_L \xi)^0 = 0.015174 - 0.02135 T_r + 0.0075 T_r^2 \quad (15)$$

$$(\eta_L \xi)^1 = 0.042552 - 0.07674 T_r + 0.0340 T_r^2 \quad (16)$$

Use of this correlation requires only $T_c$, $P_c$, $M$, $\omega$ and temperature.

Table VI shows examples of viscosities calculated for *n*-decane ($C_{10}H_{22}$) by each the Orrick-Erbar, Thomas, and Letsou-Stiel correlations—Eq. (10), (13) and (14). Note that *n*-decane has none of the groups listed in Table IV–V, except for C and H.

For liquid mixtures at moderate temperature, Lobe proposed:

$$\nu_m = \phi_1 \nu_1 e^{\phi_2 \alpha_2^*} + \phi_2 \nu_2 e^{\phi_1 \alpha_1} \quad (17)$$

$$\alpha_1^* = -1.7 \ln \nu_2/\nu_1 \quad (18)$$

$$\alpha_2^* = 0.27 \ln \nu_2/\nu_1 + (1.3 \ln \nu_2/\nu_1)^{1/2} \quad (19)$$

The $\phi$ terms are volume fractions on a mole basis, e.g.,

$$\phi_2 = \chi_2 V_{L2}/(\chi_2 V_{L2} + \chi_1 V_{L1}) \quad (20)$$

To use this correlation, input the mole fraction, molal volume, and pure-component viscosity at the relevant temperature for both components. The viscosity estimate is in centipoise.

## Nomenclature

- $M$ Molecular weight
- $n$ No. of carbon atoms, exclusive of those in groups listed in Table IV
- $P$ Pressure, atm
- $R$ Gas law constant, 82.07 (atm)(cm$^3$)/(g-mol)(K)
- $T$ Temperature, K
- $t$ Temperature, °C
- $V$ Volume
- $x$ Mole fraction, liquid phase
- $y$ Mole fraction, vapor phase
- $Z$ Compressibility factor
- $\alpha$ Parameter of Eq. (18) and (19)
- $\theta$ Thomas viscosity function, Table V
- $\eta$ Viscosity (micropoise, gas; centipoise, liquid)
- $\nu$ Kinematic viscosity, $\eta/\rho$
- $\xi$ $T_c^{1/6}/(M^{1/2}P_c^{2/3})$
- $\rho$ **Density, g/cm$^3$ or g-mol/cm$^3$**
- $\phi$ Parameter in Wilke correlation, Eq. (2) and (3), and molal volume fraction, Eq. (20)
- $\omega$ Pitzer's acentric factor

### Subscripts

- $c$ Critical
- $i$ Component $i$ in mixture
- $L$ Liquid
- $m$ Mixture
- $r$ Property at reduced conditions
- 1 Component 1 in a mixture
- 2 Component 2 in a mixture

### Superscript

- o Refers to low pressure

## References


1. Wilke, C. R., *J. Chem. Phys.*, Vol. 18, 1950, p. 517.
2. Dean, D. E., and Stiel, L. I., *AIChE J.*, Vol. 11, 1965, p. 526.
3. Prausnitz, J. M., and Gunn, R. D., *AIChE J.*, Vol. 4, 1958, pp. 420 and 494.
4. Carmichael, L. T., Berry, V., and Sage, B. H., *J. Chem. Eng. Data*, Vol. 12, 1967, p. 44.
5. Orrick, C., and Erbar, J. H., Reported in "Properties of Gases and Liquids" (3rd ed.) by Ried, R. C., Prausnitz, J. M., and Sherwood, T. K., McGraw-Hill Book Co., N.Y.
6. Thomas, L. H., *J. Chem. Soc.*, 1946, p. 573.
7. Carmichael, L. T., Berry, V., and Sage, B. H., *J. Chem. Eng. Data*, Vol. 14, 1969, p. 27.
8. Letsou, A., and Stiel, L. I., *AIChE, J.*, Vol. 19, 1973, p. 409.
9. Lobe, V. M., M.S. thesis, University of Rochester, Rochester, N.Y., 1973.


**Program for predicting viscosities of liquids and mixtures** **Table I**

```
10 REM  VISCOSITIES OF LIQUIDS AND MIXTURES
20 REM  FROM CHEMICAL ENGINEERING, JULY 30, 1979
30 REM  BY JAMES H. WEBER
40 REM  TRANSLATED BY WILLIAM VOLK
50 REM  COPYRIGHT (C) 1984
60 REM  BY CHEMICAL ENGINEERING
70 HOME : PRINT
80 REM  SET DISPLAY TO FOUR DECIMALS
90 DEF  FN P(X) =  INT (1E4 * (X + .00005)) / 1E4
100 PRINT "CALCULATE VISCOSITIES OF BINARY MIX-"
110 PRINT "TURES OF GASES AND LIQUIDS; AND VIS-"
120 PRINT "COSITIES OF PURE LIQUIDS."
130 PRINT
140 A$(1) = "WILKE CORRELATION"
150 A$(2) = "DEAN & STIEL CORRELATION"
160 A$(3) = "ORRICK & ERBAR CORRELATION"
180 A$(4) = "THOMAS CORRELATION"
190 A$(5) = "LETSOU & STIEL CORRELATION"
200 A$(6) = "LOBE CORRELATION"
210 PRINT "1.  FOR GAS MIXTURES AT LOW PRESSURES,"
220 PRINT "    THE ";A$(1)
230 PRINT
240 PRINT "2.  NONPOLAR GAS MIXTURES AT HIGH PRES-"
250 PRINT "    SURES, THE ";A$(2)
260 PRINT
270 PRINT "    FOR PURE LIQUIDS:"
280 PRINT "3.  ";A$(3);" OR"
290 PRINT "4.  ";A$(4)
```

**Program for predicting viscosities of liquids and mixtures (continued)** **Table I**

```
300  PRINT
310  PRINT "    LIQUIDS AT HIGH TEMPERATURE:"
320  PRINT "5.  ";A$(5)
330  PRINT
340  PRINT "6.  FOR BINARY LIQUID MIXTURES, THE"
350  PRINT "    ";A$(6)
360  PRINT
370  INPUT "MAKE YOUR CHOICE BY NUMBER:  ";OP
380  HOME : VTAB 4
390  PRINT A$(OP)
400  PRINT
410  ON OP GOTO 430,640,1070,1480,1750,1950
420  REM  WILKE CORRELATION
430  PRINT "FOR EACH COMPONENT ENTER"
440  PRINT
450  PRINT "THE MOL FRACTION,"
460  PRINT "THE MOLECULAR WEIGHT, AND"
470  PRINT "THE VISCOSITY, IN MICROPOISE, FOR THE"
480  PRINT "PURE GAS."
490  PRINT
500 N = N + 1
510  INPUT "ENTER MOL FRACTION            ";Y(N)
520  INPUT "ENTER MOLECULAR WEIGHT        ";MW(N)
530  INPUT "ENTER VISCOCITY, MICROPOISE  ";Z(N)
540  IF N = 2 GOTO 590
550  PRINT
560  PRINT "DO THE SAME FOR THE OTHER COMPONENT."
570  PRINT
580  GOTO 500
590 PH(1) = (1 +  SQR (Z(1) / Z(2)) * (MW(2) / MW(1)) ^
    (1 / 4)) ^ 2 /  SQR (8 * (1 + MW(1) / MW(2)))
600 PH(2) = Z(2) / Z(1) * MW(1) / MW(2) * PH(1)
610 ET = Y(1) * Z(1) / (Y(1) + Y(2) * PH(1)) + Y(2) *
    Z(2) / (Y(2) + Y(1) * PH(2))
620  GOTO 2240
630  REM  DEAN & STIEL CORRELATION
640  PRINT "THIS CORRELATION IS FOR A GAS MIXTURE"
650  PRINT "UNDER PRESSURE."
660  PRINT
670  PRINT "THE VISCOSITY OF THE MIXTURE AT LOW"
680  PRINT "PRESSURE, AND THE MOLAL VOLUME OF"
690  PRINT "THE MIXTURE ARE REQUIRED, ALONG WITH"
700  PRINT "CRITICAL PROPERTIES OF THE GAS"
710  PRINT "COMPONENTS."
720  PRINT
730  PRINT "FOR THE FIRST COMPONENT ENTER:"
740  PRINT
750 N = N + 1
760  INPUT "MOLECULAR WEIGHT                 ";MW(N)
770  INPUT "CRITICAL TEMPERATURE, DEG. K    ";TC(N)
780  INPUT "CRITICAL VOLUME, CC/G-MOL       ";VC(N)
790  INPUT "COMPRESSIBILITY FACTOR          ";ZR(N)
800  INPUT "MOL FRACTION                    ";Y(N)
810  IF N = 2 GOTO 860
820  PRINT
830  PRINT "THE SAME FOR THE SECOND COMPONENT:"
840  PRINT
850  GOTO 750
860  PRINT
870  PRINT "ENTER VISCOCITY OF THE MIXTURE"
880  INPUT "AT LOW PRESSURE, MICROPOISE     ";Z1
890  PRINT
900  PRINT "ENTER MOLAL VOLUME OF THE "
910  PRINT "MIXTURE, CC/G-MOL, AT THE "
920  INPUT "HIGH PRESSURE.                  ";V1
930  PRINT
940  FOR I = 1 TO 2
950 TM = TM + TC(I) * Y(I)
960 ZM = ZM + ZR(I) * Y(I)
970 MM = MM + MW(I) * Y(I)
980 VM = VM + VC(I) * Y(I)
990  NEXT I
1000 PM = ZM * TM * 82.07 / VM
1010 PS = TM ^ (1 / 6) /  SQR (MM) / PM ^ (2 / 3)
1020 RH = VM / V1
1030 TP = 1.08 * ( EXP (1.439 * RH) -  EXP ( - 1.111 *
     RH ^ 1.858))
1040 ET = TP / PS + Z1
1050  GOTO 2240
1060  REM  ORRICK & ERBON CORRELATION
1070  PRINT "DATA REQUIRED ARE:"
1080  PRINT
1090  PRINT "MOLECULAR WEIGHT,"
1100  PRINT "LIQUID DENSITY, G/CC, AND"
1110  PRINT "DETAILS OF MOLECULAR STRUCTURE."
1120  PRINT
1130  PRINT "ENTER DATA NOW:"
1140  PRINT
1150  INPUT "MOLECULAR WEIGHT                 ";MW
1160  INPUT "LIQUID DENSITY, G/CC             ";RO
1170  INPUT "NUMBER OF CARBON ATOMS           ";C
1180  INPUT "NUMBER OF C-FOUR RADICAL BONDS   ";R4
1190  INPUT "NUMBER OF C-THREE RADICAL BONDS  ";R3
1200  INPUT "NUMBER OF DOUBLE BONDS           ";DB
1210  INPUT "NUMBER OF FIVE MEMBER RINGS      ";R5
1220  INPUT "NUMBER OF SIX MEMBER RINGS       ";R6
1230  INPUT "NUMBER OF AROMATIC RINGS         ";AR
1240  INPUT "NUMBER OF ORTHO SUBSTITUTIONS    ";OT
1250  INPUT "NUMBER OF META SUBSTITUTIONS     ";MT
1260  INPUT "NUMBER OF PARA SUBSTITUTIONS     ";PA
1270  INPUT "NUMBER OF CHLORINE ATOMS         ";CL
1280  INPUT "NUMBER OF BROMINE ATOMS          ";BI
1290  INPUT "NUMBER OF IODINE ATOMS           ";IO
1300  INPUT "NUMBER OF -OH GROUPS             ";OH
1310  INPUT "NUMBER OF -COO- GROUPS           ";CO(2)
1320  INPUT "NUMBER OF -O- GROUPS             ";OO
1330  INPUT "NUMBER OF -COOH GROUPS           ";AC
1340  INPUT "NUMBER OF (-C-)=O GROUPS         ";CO(1)
1350  INPUT "TEST TEMPERATURE, DEG. C         ";T
1360 RA = RA - .15 * R3 - 1.2 * R4 + .24 * DB + .1 * R5
     - .45 * R6
1370 RB = RB + 35 * R3 + 400 * R4 - 90 * DB + 32 * R5 +
     250 * R6 + 20 * AR
1380 AR = AR - .12 * OT + .05 * MT - .01 * PA - .61 * CL
```

**Program for predicting viscosities of liquids and mixtures (continued)** **Table I**

```
     - 1.25 * BI - 1.75 * IO - 3 * OH - 1 * CO(2)
1390 BR = BR + 100 * OT - 34 * MT - 5 * PA + 200 * CL
     + 365 * BI + 400 * IO + 1600 * OH + 420 * CO(2)
1400 AR = AR - .38 * OO - .5 * CO(1) - .9 * AC
1410 BR = BR + 140 * O + 350 * CO(1) + 770 * AC
1420 A = AR - 6.95 - .21 * C
1430 B = 275 + 99 * C + BR
1440 TP = A + B / (T + 273.16)
1450 ET =  EXP (TP) * RO * MW
1460  GOTO 2240
1470  REM  THOMAS CORRELATION
1480  PRINT "DATA REQUIRED ARE:"
1490  PRINT "CRITICAL TEMPERATURE, DEG. K,"
1500  PRINT "LIQUID DENSITY, G/CC,"
1510  PRINT "AND DATA ON THE MOLECULAR STRUCTURE."
1520  PRINT
1530  PRINT "ENTER THE DATA NOW:"
1540  PRINT
1550  INPUT "CRITICAL TEMPERATURE, DEG. K      ";TC
1560  INPUT "LIQUID DENSITY, G/CC              ";RO
1570  INPUT "NUMBER OF CARBON ATOMS            ";C
1580  INPUT "NUMBER OF HYDROGEN ATOMS          ";H
1590  INPUT "NUMBER OF OXYGEN ATOMS            ";O
1600  INPUT "NUMBER OF CHLORINE ATOMS          ";CL
1610  INPUT "NUMBER OF BROMINE ATOMS           ";BR
1620  INPUT "NUMBER OF IODINE ATOMS            ";IO
1630  INPUT "NUMBER OF DOUBLE BONDS            ";DB
1640  INPUT "NUMBER OF C6H5 RINGS              ";C6
1650  INPUT "NUMBER OF SULFUR ATOMS            ";SU
1660  INPUT "NUMBER OF KETONES & ESTERS.       ";CO
1670  INPUT "NUMBER OF NITRILES                ";CN
1680 TH =  - .462 * C + .249 * H + .054 * O + .340 * CL
     + .326 * BR + .355 * IO + .478 * DB + .385 * C6 +
     .043 * SU + .105 * CO + .381 * CN
1690  INPUT "TEST TEMPERATURE, DEG. C          ";T
1700 TR = (T + 273.16) / TC
1710 TP = TH * (1 / TR - 1)
1720 ET =  SQR (RO) / 8.569 * 10 ^ TP
1730  GOTO 2240
1740  REM  LETSOU & STIEL CORRELATION
1750  PRINT "DATA REQUIRED ARE THE CRITICAL"
1760  PRINT "TEMPERATURE AND PRESSURE, THE MOLECULAR"
1770  PRINT "WEIGHT AND THE ACENTRIC FACTOR."
1780  PRINT "ENTER THE DATA NOW:"
1790  PRINT
1800  INPUT "CRITICAL TEMPERATURE, DEG. K      ";TC
1810  INPUT "CRITICAL PRESSURE, ATM            ";PC
1820  INPUT "MOLECULAR WEIGHT                  ";MW
1830  INPUT "ACENTRIC FACTOR, OMEGA            ";OM
1840  INPUT "TEST TEMPERATURE, DEG. C          ";T
1850 TR = (T + 273.16) / TC
1860  IF TR =  > .76 AND TR =  < .98 GOTO 1880
1870 F1 = 1
1880 PS = TC ^ (1 / 6) /  SQR (MW) / PC ^ (2 / 3)
1890 TP(0) = .015174 - .02135 * TR + .0075 * TR ^ 2
1900 TP(1) = .042552 - .07674 * TR + .0340 * TR ^ 2
1910 TP = TP(0) + OM * TP(1)
1920 ET = TP / PS
1930  GOTO 2240
1940  REM  LOBE - MIXTURES
1950  PRINT "DATA REQUIRED ARE THE MOL FRACTION,"
1960  PRINT "MOLAL VOLUME, CC/G-MOL, AND"
1970  PRINT "THE VISCOSITY IN CENTIPOISE"
1980  PRINT "OF EACH COMPONENT."
1990  PRINT
2000  PRINT "ENTER THE DATA NOW."
2010  PRINT
2020 N = N + 1
2030  INPUT "MOL FRACTION                   ";X(N)
2040  INPUT "MOLAL VOLUME, CC-G/MOL         ";V(N)
2050  INPUT "VISCOSITY, CENTIPOISE          ";Z(N)
2060  IF N = 2 GOTO 2110
2070  PRINT
2080  PRINT "SAME FOR SECOND COMPONENT"
2090  PRINT
2100  GOTO 2020
2110 VT = X(1) * V(1) + X(2) * V(2)
2120  FOR I = 1 TO 2
2130 PH(I) = X(I) * V(I) / VT
2140 NU(I) = Z(I) * V(I)
2150  NEXT I
2160 FR = NU(2) / NU(1)
2170  IF FR < 1 THEN FR = 1 / FR
2180 AL(2) = .27 *  LOG (FR) +  SQR (1.3 *  LOG (FR))
2190 AL(1) =  - 1.7 *  LOG (FR)
2200 NU = PH(1) * NU(1) *  EXP (PH(2) * AL(2))
2210 NU = NU + PH(2) * NU(2) *  EXP (PH(1) * AL(1))
2220 ET = NU / VT
2230  GOTO 2240
2240  PRINT
2250  PRINT "BASED ON ";A$(OP);" THE"
2260 MU$ = "MICROPOISE."
2270  IF OP > 2 THEN MU$ = "CENTIPOISE"
2280  PRINT "VISCOSITY IS "; FN P(ET);" ";MU$
2290  IF F1 = 0 GOTO 2330
2300  PRINT
2310  PRINT "REDUCED TEMPERATURE IS OUTSIDE THE"
2320  PRINT "RANGE FOR ";A$(5);" ."
2330  PRINT
2340  PRINT  TAB( 13);"END OF PROGRAM"
2350  END
```

**Example for: Viscosities of liquids and mixtures** **Table II**

**(Start of first display)**

```
CALCULATE VISCOSITIES OF BINARY MIX-
TURES OF GASES AND LIQUIDS; AND VIS-
COSITIES OF PURE LIQUIDS.

1.  FOR GAS MIXTURES AT LOW PRESSURES,
    THE WILKE CORRELATION

2.  NONPOLAR GAS MIXTURES AT HIGH PRES-
    SURES, THE DEAN & STIEL CORRELATION

    FOR PURE LIQUIDS:
3.  ORRICK & ERBAR CORRELATION OR
4.  THOMAS CORRELATION

    LIQUIDS AT HIGH TEMPERATURE:
5.  LETSOU & STIEL CORRELATION

6.  FOR BINARY LIQUID MIXTURES, THE
    LOBE CORRELATION

MAKE YOUR CHOICE BY NUMBER:  1
```

**(Start of next display)**

```
WILKE CORRELATION

FOR EACH COMPONENT ENTER

THE MOL FRACTION,
THE MOLECULAR WEIGHT, AND
THE VISCOSITY, IN MICROPOISE, FOR THE
PURE GAS.

ENTER MOL FRACTION           .384
ENTER MOLECULAR WEIGHT       16.043
ENTER VISCOCITY, MICROPOISE  116.1

DO THE SAME FOR THE OTHER COMPONENT.

ENTER MOL FRACTION           .616
ENTER MOLECULAR WEIGHT       58.124
ENTER VISCOCITY, MICROPOISE  77.09
```

**(Start of next display)**

```
BASED ON WILKE CORRELATION THE
VISCOSITY IS 86.2275 MICROPOISE.

            END OF PROGRAM
```

**(Start of first display)**

```
CALCULATE VISCOSITIES OF BINARY MIX-
TURES OF GASES AND LIQUIDS; AND VIS-
COSITIES OF PURE LIQUIDS.

1.  FOR GAS MIXTURES AT LOW PRESSURES,
    THE WILKE CORRELATION

2.  NONPOLAR GAS MIXTURES AT HIGH PRES-
    SURES, THE DEAN & STIEL CORRELATION

    FOR PURE LIQUIDS:
3.  ORRICK & ERBAR CORRELATION OR
4.  THOMAS CORRELATION

    LIQUIDS AT HIGH TEMPERATURE:
5.  LETSOU & STIEL CORRELATION

6.  FOR BINARY LIQUID MIXTURES, THE
    LOBE CORRELATION

MAKE YOUR CHOICE BY NUMBER:  3
```

**(Start of next display)**

```
ORRICK & ERBAR CORRELATION

DATA REQUIRED ARE:

MOLECULAR WEIGHT,
LIQUID DENSITY, G/CC, AND
DETAILS OF MOLECULAR STRUCTURE.

ENTER DATA NOW:

MOLECULAR WEIGHT                142.286
LIQUID DENSITY, G/CC            .719
NUMBER OF CARBON ATOMS          10
NUMBER OF C-FOUR RADICAL BONDS  0
NUMBER OF C-THREE RADICAL BONDS 0
NUMBER OF DOUBLE BONDS          0
NUMBER OF FIVE MEMBER RINGS     0
NUMBER OF SIX MEMBER RINGS      0
NUMBER OF AROMATIC RINGS        0
NUMBER OF ORTHO SUBSTITUTIONS   0
NUMBER OF META SUBSTITUTIONS    0
NUMBER OF PARA SUBSTITUTIONS    0
NUMBER OF CHLORINE ATOMS        0
NUMBER OF BROMINE ATOMS         0
NUMBER OF IODINE ATOMS          0
NUMBER OF -OH GROUPS            0
NUMBER OF -COO- GROUPS          0
NUMBER OF -O- GROUPS            0
NUMBER OF -COOH GROUPS          0
NUMBER OF (-C-)=O GROUPS        0
TEST TEMPERATURE, DEG. C        37.78
```

**Example for: Viscosities of liquids and mixtures (continued)** **Table II**

**(Start of next display)**

```
BASED ON ORRICK & ERBAR CORRELATION THE
VISCOSITY IS .7021 CENTIPOISE

          END OF PROGRAM
```

**(Start of first display)**

```
CALCULATE VISCOSITIES OF BINARY MIX-
TURES OF GASES AND LIQUIDS; AND VIS-
COSITIES OF PURE LIQUIDS.

1.  FOR GAS MIXTURES AT LOW PRESSURES,
    THE WILKE CORRELATION

2.  NONPOLAR GAS MIXTURES AT HIGH PRES-
    SURES, THE DEAN & STIEL CORRELATION

    FOR PURE LIQUIDS:
3.  ORRICK & ERBAR CORRELATION OR
4.  THOMAS CORRELATION

    LIQUIDS AT HIGH TEMPERATURE:
5.  LETSOU & STIEL CORRELATION

6.  FOR BINARY LIQUID MIXTURES, THE
    LOBE CORRELATION

MAKE YOUR CHOICE BY NUMBER:  5
```

**(Start of next display)**

```
LETSOU & STIEL CORRELATION

DATA REQUIRED ARE THE CRITICAL
TEMPERATURE AND PRESSURE, THE MOLECULAR
WEIGHT AND THE ACENTRIC FACTOR.
ENTER THE DATA NOW:

CRITICAL TEMPERATURE, DEG. K    617.6
CRITICAL PRESSURE, ATM          20.8
MOLECULAR WEIGHT                142.286
ACENTRIC FACTOR, OMEGA          .490
TEST TEMPERATURE, DEG. C        137.8
```

**(Start of next display)**

```
BASED ON LETSOU & STIEL CORRELATION THE
VISCOSITY IS .2317 CENTIPOISE

REDUCED TEMPERATURE IS OUTSIDE THE
RANGE FOR LETSOU & STIEL CORRELATION .

          END OF PROGRAM
```

**Calculated and experimental viscosities of gaseous mixture of methane and *n*-butane ($y_{c_1}$ = 0.384)** **Table III**

**Saturated vapor**

| $P$ (atm) | $t$ (°C) | Viscosity, $\eta$, micropoise: Experimental | Via Eq. (1) | Via Eq. (5) |
|---|---|---|---|---|
| 1.021 | 4.44 | 83 | 77 | |
| 6.124 | 37.78 | 88 | 87 | |
| 14.698 | 71.11 | 102 | 95 | |
| 31.64 | 104.4 | 121 | 104 | |

**Superheated vapor, $t$ = 237.8°C**

| $P$ (atm) | $t$ (°C) | Experimental | Via Eq. (1) | Via Eq. (5) |
|---|---|---|---|---|
| 68.05 | | 174 | | 188 |
| 136.1 | | 228 | | 239 |
| 204.1 | | 300 | | 305 |
| 272.2 | | 381 | | 365 |
| 340.2 | | 466 | | 425 |

**Physical properties used in the calculations**

| | Methane | | *n*-Butane |
|---|---|---|---|
| Mol wt | 16.043 | | 58.124 |
| $T_c$ (K) | 190.6 | | 425.2 |
| $P_c$ (atm) | 45.4 | | 37.5 |
| $V_c$ (cm$^3$/g-mol) | 99.0 | | 255. |
| $Z_c$ | 0.288 | | 0.274 |
| $\omega$ | 0.008 | | 0.193 |
| $\eta$ ($\mu$P), at P=1 atm | 106.2 | (4.44°C) | 68.75 |
| | 116.1 | (37.78°C) | 77.09 |
| | 125.5 | (71.11°C) | 85.32 |
| | 135.9 | (104.4°C) | 93.90 |

| | Mixture |
|---|---|
| $\eta^o_m$ ($\mu$P) (mixture) | 162 (237.8°C) |
| $V_m$ (cm$^3$/g-mol)* (mixture) | 536 (68.05 atm) |
| | 254 (136.1 atm) |
| | 176 (204.1 atm) |
| | 147 (272.2 atm) |
| | 130 (340.2 atm) |

Data sources: Reid, R. C., Prausnitz, J. M., and Sherwood, T. K., "Properties of Gases and Liquids," 3rd ed., McGraw-Hill Book Co., N.Y., 1977.

Carmichael, L. T., Berry, V., and Sage, B. H., *J. Chem. Eng. Data*, Vol. 10, 1965, p. 57, and Vol. 12, 1967, p. 44.

Carmichael, L. T., and Sage, B. H., *J. Chem. Eng. Data*, Vol. 8, 1963, p. 612.

*Estimated by generalized *P-V-T* relationship.

**Group contributions for $A$ and $B$ in Eq. (11) and (12) [5]** **Table IV**

| Group | GC($A$) | GC($B$) |
|---|---|---|
| $R-\overset{\vert}{\underset{R}{C}}-R$ | -0.15 | 35 |
| $R-\overset{R}{\underset{R}{C}}-R$ | -1.20 | 400 |
| Double bond | 0.24 | -90 |
| Five-member ring | 0.10 | 32 |
| Six-member ring | -0.45 | 250 |
| Aromatic ring | 0 | 20 |
| Ortho substitution | -0.12 | 100 |
| Meta substitution | 0.05 | -34 |
| Para substitution | -0.01 | - 5 |
| Chlorine | -0.61 | 220 |
| Bromine | -1.25 | 365 |
| Iodine | -1.75 | 400 |
| —OH | -3.00 | 1600 |
| —COO— | -1.00 | 420 |
| —O— | -0.38 | 140 |
| $-\overset{\vert}{C}=O$ | -0.50 | 350 |
| —COOH | -0.90 | 770 |

**Structural contributions to calculation of $\theta$ in Eq. (13) [6]** **Table V**

| | | | |
|---|---|---|---|
| C | -0.462 | Double bond | 0.478 |
| H | 0.249 | $C_6H_5$ | 0.385 |
| O | 0.054 | S | 0.043 |
| Cl | 0.340 | CO | 0.105* |
| Br | 0.326 | CN | 0.381** |
| I | 0.355 | | |

*Ketones and esters **Nitriles

**Comparison of calculated and experimental viscosities of $n$-decane in saturated liquid state** **Table VI**

| | | Viscosity, $\eta$, centipoise | | | |
|---|---|---|---|---|---|
| $P$(atm) | $t$(°C) | Experimental | Via Eq. (10) | Via Eq. (13) | Via Eq. (14) |
| 0.0000272 | 4.44 | 1.225 | 1.162 | 1.132 | * |
| 0.00497 | 37.78 | 0.730 | 0.713 | 0.649 | * |
| 0.0272 | 71.11 | 0.491 | 0.418 | 0.466 | * |
| 0.1082 | 104.4 | 0.355 | 0.347 | 0.334 | * |
| 0.3457 | 137.8 | 0.271 | 0.262 | 0.251 | * |
| 0.9179 | 171.1 | 0.209 | 0.210 | 0.197 | * |
| 2.122 | 204.4 | 0.160 | 0.172 | 0.158 | 0.151 |

*Below the recommended scope of correlation

**Physical properities used in calculation**

| | n-Decane |
|---|---|
| Mol wt | 142.286 |
| $T_c$ (K) | 617.6 |
| $P_c$ (atm) | 20.8 |
| $V_c$ (cm$^3$/g-mol) | 603 |
| $Z_c$ | 0.247 |
| $\omega$ | 0.490 |
| $\rho_L$ (g/cm$^3$)* | 0.744 (4.44°C) |
| | 0.719 (37.78°C) |
| | 0.691 (71.11°C) |
| | 0.663 (104.4°C) |
| | 0.636 (137.8°C) |
| | 0.609 (171.1°C) |
| | 0.576 (204.4°C) |

*From Sage, B. H.,and Lacy, W. N., "Thermodynamic of the Lighter Paraffin Hydrocarbons and Nitrogen," API, New York, 1950.

Other sources: Reid, R. C., Prausnitz, J. M., and Sherwood, T. K., "Properties of Gases and Liquids," 3rd ed., McGraw-Hill Book Co., N.Y., 1977.

# Vapor pressure vs. temperature

Because of the importance of these relationships in process calculations, they have been the object of many investigations. Five correlations are here programmed.

☐ This Apple II program offers five ways to calculate the vapor pressure of pure substances: the Antoine equation [*1*]; the Lee and Kesler [*2*], Gomez-Nieto and Thodos [*3,4*] and Riedel, Plank and Miller [*5*] expressions for the reduced vapor pressure; and the Frost and Kalkwarf [*6*] equation.

The program permits calculating vapor pressure if the temperature is known, and calculating temperature if the pressure is known. The latter usually involves a trial-and-error solution. Here, the Newton-Raphson convergence technique is used when possible. Results obtained by means of the program are compared with experimental data for propane.

Table I lists the program, and Table II shows example displays for calculations of vapor pressure and temperature. The program is interactive, and self-explanatory for the most part. For each correlation, the user has a choice of solving for $P^s$, given $t$, or of solving for $t$, given $P^s$. (The variables used here are explained in the nomenclature table.)

## Antoine equation for lower pressures

Perhaps the most accurate expression for the vapor-pressure-temperature relationship up to a pressure of 1,500 mm Hg is the Antoine equation [*1*]:

$$\log P^s = A - [B/(t + C)] \tag{1}$$

Here, $A$, $B$ and $C$ are empirical constants, $t$ is expressed in degrees C, and log is to base 10.

The Antoine equation is choice "1" on the program menu. The next choice is whether to solve for pressure or temperature. Then, simply enter the three constants, and $t$ or $P^s$. Note that pressure must be expressed in mm Hg. The program immediately tells the desired result: either pressure, in both mm Hg and atm, or temperature. Table III gives the Antoine constants for propane, from API 44 [7].

## Correlations for higher pressures

Vapor pressure data up to 1,500 mm Hg are accurately correlated by the Antoine equation. At higher pressures, other relationships must be resorted to; these usually employ the theorem of corresponding states and relate reduced vapor pressure to reduced temperature. One such reduced relationship is that developed by Lee and Kesler [*2*]:

$$\ln P^s_r = f^{(o)}T_r + \omega f^{(1)}T_r \tag{2}$$

Here,
$$f^{(o)} = 5.92714 - (6.09648/T_r) - 1.28862 \ln T_r + 0.169347T_r^6 \tag{3}$$

$$f^{(1)} = 15.2518 - (15.6875/T_r) - 13.4721 \ln T_r + 0.43577T_r^6 \tag{4}$$

The acentric factor, $\omega$, is calculated by:

$$\omega = [-\ln P_c - (f_b^{(0)})]/f_b^{(1)} \tag{5}$$

In Eq. (5), $f_b^{(0)} = f^{(0)}$ of Eq. (3), with $T_{rb}$ substituted for $T_r$; and $f_b^{(1)} = f^{(1)}$ of Eq. (4), with $T_{rb}$ substituted for $T_r$.

This correlation is choice "3" on the menu. This, and the two following correlations, require the same input data: critical temperature and pressure, normal (1-atm) boiling point (°C), and either $t$ or $P^s$. Note that critical pressure must be expressed in atm, while a given vapor pressure must be expressed in mm Hg. In the case of estimating temperature, given pressure, the calculation takes a few seconds because it involves a trial-and-error procedure. The convergence method used is the Newton-Raphson method. The acceptable error in vapor pressure is set at 0.01 mm Hg, in statement 1520 of the program.

Table III compares results calculated via this correlation (Eq. (2)) with experimental ones [*8*] for propane. The required data are also listed.

## Gomez-Nieto and Thodos correlation

Another reduced-vapor-pressure-vs.-reduced-temperature expression has been derived by Gomez-Nieto and Thodos [*3,4*]:

$$\ln P_r^s = \beta[(1/T_r^m) - 1] + \gamma(T_r^n - 1) \tag{6}$$

Here, $\beta$, $m$ and $\gamma$ are empirical constants, and $n$ has been set equal to 7. Gomez-Nieto and Thodos reported values for these constants for a number of inorganic and hydrocarbon compounds. However, values for these constants can also be determined from:

$$m = 0.78425e^{0.089315s} - 8.5217/e^{0.74826s} \tag{7}$$

$$\beta = -4.26700 - \frac{221.79}{s^{2.5}e^{0.03848s^{2.5}}} + \frac{3.8126}{e^{2272.44/s^3}} \tag{8}$$

Adapted from an article by James H. Weber, University of Nebraska, originally published November 5, 1979.

$$\gamma = as + b\beta \qquad (9)$$

$$a = [(1/T_{rb}) - 1]/(1 - T_{rb}^{7}) \qquad (10)$$

$$b = [(1/T_{rb}^{m}) - 1]/(1 - T_{rb}^{7}) \qquad (11)$$

$$s = (T_b \ln P_c)/(T_c - T_b) \qquad (12)$$

These equations are part of the program; this correlation is choice "4" on the program menu. The only input data required are critical temperature and pressure, normal (1-atm) boiling point, and either $t$ or $P^s$. In the case of estimating $t$, given $P^s$, the calculation takes a few seconds. Note that critical pressure must be expressed in atm, while a given vapor pressure must be expressed in mm Hg. The acceptable error in the trial-and-error procedure is set at 0.01 mm Hg, again at statement 1520 of the program.

Table III compares results calculated via this correlation (Eq. (6)) with experimental ones for propane. Table IV compares the values of $m$, $\beta$ and $\gamma$ calculated via Eq. (7)–(12) with experimental ones reported by Gomez-Nieto and Thodos [4].

## Riedel, Plank and Miller correlation

The last relationship involving reduced vapor pressure and temperature that has been programmed is the Riedel, Plank and Miller correlation [5]:

$$\ln P_r^s = -G/T_r[1 - T_r^2 + k(3 + T_r)(1 - T_r)^3] \qquad (13)$$

$$G = 0.4835 + 0.4605h \qquad (14)$$

$$k = [h/G - (1 + T_{rb})]/[(3 + T_{rb})(1 - T_{rb})^2] \qquad (15)$$

$$h = T_{rb}(\ln P_c/1 - T_{rb}) \qquad (16)$$

This correlation is choice "5" on the program menu. The only input data required are critical temperature and pressure, normal (1-atm) boiling point, and either $t$ or $P^s$. In the case of estimating $t$, given $P^s$, the calculation takes a few seconds. Note that critical pressure must be expressed in atm, while a given vapor pressure must be expressed in mm Hg. The acceptable error in the trial-and-error procedure is set at 0.01 mm Hg, again at statement 1520 of the program.

Table III compares results calculated via this correlation (Eq. (13)) with experimental ones for propane.

## Frost and Kalkwarf correlation

The last correlation programmed is the Frost and Kalkwarf relationship [6]. Although this equation can be written in reduced form, the expression used is:

$$\ln P^s = A' + B'/T + C' \ln T + D'P^s/T^2 \qquad (17)$$

Here, $A'$, $B'$, $C'$ and $D'$ are empirical constants. The values used, based on the work of Harlacher and Braun [10], were determined by Reid, Prausnitz and Sherwood [9]. This equation is somewhat inconvenient because $P^s$ appears on both sides, and temperature appears as $\ln T$ and $T$. Hence, trial and error is required to calculate $t$ or $P^s$.

The program does the trial-and-error solution automatically. This correlation is choice "2" on the program menu. The input data required are the four constants—$A'$, $B'$, $C'$ and $D'$—and either $t$ or $P^s$. The acceptable error in the trial-and-error procedure is set at 0.01 mm Hg for pressure, at statement 1080 of the program, and at 0.01°C for temperature, at statement 1130 of the program.

Table III compares results calculated via this correlation (Eq. (17)) with experimental ones for propane, and lists the needed constants.

## Nomenclature

| | |
|---|---|
| $A$, $B$, $C$ | Constants of Antoine, Eq. (1) |
| $a$, $b$ | Constants of Gomez-Nieto and Thodos, Eq. (10) and (11) |
| $A'$, $B'$, $C'$, $D'$ | Constants of Frost and Kalkwarf, Eq. (17) |
| $f^{(0)}$, $f^{(1)}$ | Functions of Lee and Kesler, Eq. (3) and (4) |
| $G$, $h$, $k$ | Constants of Riedel, Plank and Miller, Eq. (14) (15) (16) |
| $m$ | Constant of Gomez-Nieto and Thodos, Eq. (6) |
| $P$ | Pressure, mm Hg or atm |
| $s$ | Constant of Gomez-Nieto and Thodos, Eq. (12) |
| $T$ | Temperature, K |
| $t$ | Temperature, °C |
| $\beta$, $\gamma$ | Constants of Gomez-Nieto and Thodos, Eq. (6) |
| $\omega$ | Pitzer's acentric factor |

Subscripts

| | |
|---|---|
| $b$ | Normal boiling point |
| $c$ | Critical |
| $r$ | Reduced |

Superscript

| | |
|---|---|
| $s$ | Saturated |

## References


1. Antoine, C., *Compte Rendus*, Vol. 107, No. 681, 1888, p. 836.
2. Lee, B. I., and Kesler, M. G., A.I.Ch.E. J., Vol. 21, 1975, p. 510.
3. Gomez-Nieto, M., and Thodos, G., *Ind. Eng. Chem. Fundam.*, Vol. 17, 1978, p. 45.
4. Ibid, Vol. 16, 1977, p. 254.
5. Miller, D. G., reported in Reid, Prausnitz and Sherwood, "The Properties of Gases and Liquids," 3rd ed., 1977.
6. Frost, A. A., and Kalkwarf, D. R., *J. Chem. Phys.*, Vol. 21, 1953, p. 264.
7. American Petrol. Inst., Research Project 44, Carnegie Press, 1953, p. 336.
8. Canjar, L. N., and Manning, F. S., "Thermodynamic Properties and Reduced Correlations for Gases," Gulf Pub. Co., Houston, Tex., 1967.
9. Reid, R. C., Prausnitz, J. M., and Sherwood, T. K., "The Properties of Gases and Liquids," 3rd ed., McGraw-Hill, New York, 1977, p. 189.
10. Harlacher, E. A., and Braun, W. G., *Ind. Eng. Chem. Proc. Design Devel.*, Vol. 9, 1970, p. 479.

**Program for predicting vapor pressure vs. temperature for pure substances** **Table I**

```
10  REM   VAPOR PRESSURE VS. TEMPERATURE
20  REM   FROM CHEMICAL ENGINEERING, NOVEMBER 5, 1979
30  REM   BY JAMES H. WEBER
40  REM  TRANSLATED BY WILLIAM VOLK
50  REM   COPYRIGHT (C) 1984
60  REM  BY CHEMICAL ENGINEERING
70  HOME : VTAB 2
80  REM  SET DISPLAY AT FOUR DECIMALS
90  DEF  FN P(X) =  INT (1E4 * (X + .00005)) / 1E4
100 A$(1) = "ANTOINE CORRELATION"
110 A$(2) = "FROST AND KALKWARF CORRELATION"
120 A$(3) = "LEE AND KESLER CORRELATION"
130 A$(4) = "GOMEZ-NIETO AND THODOS"
140 A$(5) = "RIEDEL, PLANK AND MILLER"
150 B$(0) = "VAPOR PRESSURE"
160 B$(1) = "TEMPERATURE"
170 C$(0) = "MM HG"
180 C$(1) = "DEG. C"
190  PRINT "FIVE CORRELATIONS FOR CALCULATING"
200  PRINT "VAPOR PRESSURE FOR A GIVEN TEMPERATURE"
210  PRINT "ARE GIVEN.  THEY WILL ALSO CALCULATE"
220  PRINT "THE TEMPERATURE FOR A GIVEN PRESSURE."
230  PRINT
240  PRINT "THE FIRST TWO REQUIRE ARBITRARY"
250  PRINT "CONSTANTS.   THE OTHERS REQUIRE THE"
260  PRINT "CRITICAL TEMPERATURE AND PRESSURE, AND"
270  PRINT "THE NORMAL (1 ATM) BOILING POINT "
280  PRINT
290  PRINT "THE CORRELATIONS ARE:"
300  PRINT
310  FOR I = 1 TO 5
320  PRINT I;".  ";A$(I)
330  NEXT I
340  PRINT
350  INPUT "CHOOSE BY NUMBER.                ";OP
360  HOME : VTAB 8
370  PRINT "IF YOU WANT TO ESTIMATE VAPOR PRESSURE,"
380  PRINT "ENTER P.  IF YOU WANT TO ESTIMATE"
390  PRINT "TEMPERATURE, ENTER T."
400  PRINT
410  INPUT "ENTER P OR T.       ";CH$
420  IF CH$ = "P" GOTO 520
430  IF CH$ = "T" GOTO 450
440  GOTO 470
450 F1 = 1
460  GOTO 520
470  PRINT
480  PRINT "YOUR CHOICES WERE P OR T.  YOU ENTERED"
490  PRINT CH$;".  TRY AGAIN."
500  PRINT
510  GOTO 370
520  HOME : VTAB 4
530  PRINT A$(OP)
540  PRINT
550  ON OP GOTO 570,840,1210,1210,1210
560  REM  ANTOINE CORRELATION
570  PRINT "THREE ARBITRARY ANTOINE CONSTANTS,"
580  PRINT "A, B, AND C ARE REQUIRED TO SOLVE"
590  PRINT
600  PRINT "           THE EQUATION:"
610  PRINT
620  PRINT "  LOG(P) = A - (B/(T + C))"
630  PRINT
640  PRINT "ENTER THESE AND THE"
650  IF F1 = 1 GOTO 680
660  PRINT "TEMPERATURE IN DEG. C"
670  GOTO 690
680  PRINT "PRESSURE IN MM HG"
690  PRINT
700  INPUT "ENTER THE CONSTANT A      ";A
710  INPUT "ENTER THE CONSTANT B      ";B
720  INPUT "ENTER THE CONSTANT C      ";C
730  IF F1 = 1 GOTO 770
740  PRINT "ENTER THE TEMPERATURE"
750  INPUT "IN DEG. C                 ";T
760  GOTO 810
770  PRINT "ENTER THE PRESSURE"
780  INPUT "IN MM HG                  ";P
790 AN =  - B / ( LOG (P) /  LOG (10) - A) - C
800  GOTO 1850
810 AN = 10 ^ (A - (B / (T + C)))
820  GOTO 1850
830  REM  FROST & KALKWARF CORRELATION
840  PRINT "FOUR ARBITRARY CONSTANTS ARE REQUIRED"
850  PRINT "TO SOLVE THE EQUATION:"
860  PRINT
870  PRINT "    LN(P) = A + B/T + C*LN(T) + D*P/T"
880  PRINT
890  PRINT "ENTER THESE CONSTANTS AND THE"
900  IF F1 = 1 GOTO 930
910  PRINT "TEMPERATURE IN DEG. C"
920  GOTO 940
930  PRINT "PRESSURE IN MM HG"
940  PRINT
950  PRINT "ENTER THE DATA NOW:"
960  PRINT
970  INPUT "ENTER A     ";A
980  INPUT "ENTER B     ";B
990  INPUT "ENTER C     ";C
1000  INPUT "ENTER D     ";D
1010  IF F1 = 1 GOTO 1040
1020  INPUT "ENTER TEMPERATURE, DEG. C    ";T
1030  GOTO 1050
1040  INPUT "ENTER PRESSURE, MM HG        ";P
1050 T1 = T + 273.16
1060  IF F1 = 1 GOTO 1110
1070 P1 =  EXP (A + B / T1 + C *  LOG (T1) + P * D /
     (T1 ^ 2))
1080  IF  ABS (P1 - P) < .01 GOTO 1160
1090 P = (P1 + P) / 2
1100  GOTO 1070
1110 TP =  LOG (P) - A
```

**Program for predicting vapor pressure vs. temperature for pure substances (continued)** **Table I**

```
1120 TQ = B / T1 + C *  LOG (T1) + P * D / (T1 ^ 2)
1130  IF  ABS (TP - TQ) < .01 GOTO 1160
1140 T1 = T1 + 4 * (TP - TQ)
1150  GOTO 1120
1160  IF F1 = 1 GOTO 1190
1170 AN = P1
1180  GOTO 1850
1190 AN = T1 - 273.16
1200  GOTO 1850
1210  PRINT "DATA REQUIRED ARE:"
1220  PRINT "CRITICAL TEMPERATURE, DEG. K,"
1230  PRINT "CRITICAL PRESSURE, ATM, AND"
1240  PRINT "NORMAL BOILING POINT, DEG. C."
1250  PRINT
1260  PRINT "ENTER THE DATA NOW:"
1270  PRINT
1280  INPUT "CRITICAL TEMPERATURE, DEG. K   ";TC
1290  INPUT "CRITICAL PRESSURE, ATM        ";PC
1300  INPUT "NORMAL BOILING POINT, DEG. C   ";TB
1310  IF F1 = 1 GOTO 1340
1320  INPUT "TEST TEMPERATURE, DEG. C       ";T
1330  GOTO 1370
1340  INPUT "TEST PRESSURE, MM HG           ";P
1350  PRINT
1360  PRINT "CALCULATION MAY TAKE A FEW SECONDS."
1370 TI = T + 273.16
1380 T1 = TI
1390 TR(1) = TI / TC
1400 TR(2) = (TB + 273.16) / TC
1410  IF OP = 4 GOTO 1670
1420  IF OP = 5 GOTO 1780
1430  REM  LEE & KESLER CORRELATION
1440  FOR I = 1 TO 2
1450 F0(I) = 5.92714 - (6.09648 / TR(I)) - 1.28862 *
     LOG (TR(I)) + .169347 * TR(I) ^ 6
1460 F1(I) = 15.2518 - (15.6875 / TR(I)) - 13.4721 *
     LOG (TR(I)) + .43577 * TR(I) ^ 6
1470  NEXT I
1480 OM = ( -  LOG (PC) - F0(2)) / F1(2)
1490 TP = F0(1) + OM * F1(1)
1500 AN =  EXP (TP) * PC * 760
1510  IF F1 = 0 GOTO 1850
1520  IF  ABS (AN - P) < .01 GOTO 1640
1530  IF F2 = 1 GOTO 1590
1540 P1 = AN
1550 TI = TI + 100
1560 T2 = TI
1570 F2 = 1
1580  GOTO 1390
1590 TI = TI - (TI - T1) / (AN - P1) * (AN - P)
1600 T1 = T2
1610 P1 = AN
1620 T2 = TI
1630  GOTO 1390
1640 AN = TI - 273.16
1650  GOTO 1850
1660  REM  GOMEZ-NIETO & THODOS CORRELATION
1670 S = (TB + 273.16) *  LOG (PC) / (TC - TB - 273.16)
1680 A = (1 / TR(2) - 1) / (1 - TR(2) ^ 7)
1690 M = .78425 *  EXP (.089315 * S) - 8.5217 /  EXP (.7
     4826 * S)
1700 B = ((1 / TR(2) ^ M) - 1) / (1 - TR(2) ^ 7)
1710 BT =  - 4.267 - 221.79 / (S ^ 2.5 *  EXP (.03848 *
     S ^ 2.5)) + 3.8126 /  EXP (2272.44 / S ^ 3)
1720 GM = A * S + B * BT
1730 TP = BT * (1 / TR(1) ^ M - 1) + GM * (TR(1) ^ 7 - 1)
1740 AN =  EXP (TP) * PC * 760
1750  IF F1 = 0 GOTO 1850
1760  GOTO 1520
1770  REM  RIEDEL, PLANK ET AL.
1780 H = TR(2) *  LOG (PC) / (1 - TR(2))
1790 G = .4835 + .4605 * H
1800 K = (H / G - 1 - TR(2)) / ((3 + TR(2)) * (1 - TR(2))
     ^ 2)
1810 TP = - G / TR(1) * (1 - TR(1) ^ 2 + K * (3 + TR(1))
     * (1 - TR(1)) ^ 3)
1820 AN =  EXP (TP) * PC * 760
1830  IF F1 = 0 GOTO 1850
1840  GOTO 1520
1850  HOME : VTAB (4)
1860  PRINT "THE ";A$(OP)
1870  IF OP < 4 GOTO 1890
1880  PRINT "CORRELATION"
1890  PRINT
1900  PRINT "GIVES A ";B$(F1);" OF"
1910  PRINT  FN P(AN);" ";C$(F1)
1920  IF F1 = 1 GOTO 1940
1930  PRINT  FN P(AN / 760);" ATM."
1940  PRINT
1950  PRINT  TAB( 13);"END OF PROGRAM"
1960  END
```

**Examples for: Vapor pressure vs. temperature** **Table II**

**(Start of first display)**

```
FIVE CORRELATIONS FOR CALCULATING
VAPOR PRESSURE FOR A GIVEN TEMPERATURE
ARE GIVEN.  THEY WILL ALSO CALCULATE
THE TEMPERATURE FOR A GIVEN PRESSURE.

THE FIRST TWO REQUIRE ARBITRARY
CONSTANTS.  THE OTHERS REQUIRE THE
CRITICAL TEMPERATURE AND PRESSURE, AND
THE NORMAL (1 ATM) BOILING POINT.

THE CORRELATIONS ARE:

1.  ANTOINE CORRELATION
2.  FROST AND KALKWARF CORRELATION
```

**Examples for: Vapor pressure vs. temperature (continued)** **Table II**

```
3.  LEE AND KESLER CORRELATION
4.  GOMEZ-NIETO AND THODOS
5.  RIEDEL, PLANK AND MILLER

CHOOSE BY NUMBER.              1
```

**(Start of next display)**

```
IF YOU WANT TO ESTIMATE VAPOR PRESSURE,
ENTER P.  IF YOU WANT TO ESTIMATE
TEMPERATURE, ENTER T.

ENTER P OR T.      P
```

**(Start of next display)**

```
ANTOINE CORRELATION

THREE ARBITRARY ANTOINE CONSTANTS,
A, B, AND C ARE REQUIRED TO SOLVE

        THE EQUATION:

  LOG(P) = A - (B/(T + C))

ENTER THESE AND THE
TEMPERATURE IN DEG. C

ENTER THE CONSTANT A      6.82973
ENTER THE CONSTANT B      813.2
ENTER THE CONSTANT C      248
ENTER THE TEMPERATURE
IN DEG. C                 15.56
```

**(Start of next display)**

```
THE ANTOINE CORRELATION

GIVES A VAPOR PRESSURE OF
5549.8924 MM HG
7.3025 ATM.

          END OF PROGRAM
```

**(Start of first display)**

```
FIVE CORRELATIONS FOR CALCULATING
VAPOR PRESSURE FOR A GIVEN TEMPERATURE
ARE GIVEN.  THEY WILL ALSO CALCULATE
THE TEMPERATURE FOR A GIVEN PRESSURE.

THE FIRST TWO REQUIRE ARBITRARY
CONSTANTS.   THE OTHERS REQUIRE THE
CRITICAL TEMPERATURE AND PRESSURE, AND
THE NORMAL (1 ATM) BOILING POINT.

THE CORRELATIONS ARE:

1.  ANTOINE CORRELATION
2.  FROST AND KALKWARF CORRELATION
3.  LEE AND KESLER CORRELATION
4.  GOMEZ-NIETO AND THODOS
5.  RIEDEL, PLANK AND MILLER

CHOOSE BY NUMBER.              3
```

**(Start of next display)**

```
IF YOU WANT TO ESTIMATE VAPOR PRESSURE,
ENTER P.  IF YOU WANT TO ESTIMATE
TEMPERATURE, ENTER T.

ENTER P OR T.      P
```

**(Start of next display)**

```
LEE AND KESLER CORRELATION

DATA REQUIRED ARE:
CRITICAL TEMPERATURE, DEG. K,
CRITICAL PRESSURE, ATM, AND
NORMAL BOILING POINT, DEG. C.

ENTER THE DATA NOW:

CRITICAL TEMPERATURE, DEG. K   369.97
CRITICAL PRESSURE, ATM         42.02
NORMAL BOILING POINT, DEG. C   -42.06
TEST TEMPERATURE, DEG. C       48.89
```

**(Start of next display)**

```
THE LEE AND KESLER CORRELATION

GIVES A VAPOR PRESSURE OF
12597.2004 MM HG
16.5753 ATM.

          END OF PROGRAM
```

**(Start of first display)**

```
FIVE CORRELATIONS FOR CALCULATING
VAPOR PRESSURE FOR A GIVEN TEMPERATURE
ARE GIVEN.  THEY WILL ALSO CALCULATE
THE TEMPERATURE FOR A GIVEN PRESSURE.

THE FIRST TWO REQUIRE ARBITRARY
CONSTANTS.   THE OTHERS REQUIRE THE
CRITICAL TEMPERATURE AND PRESSURE, AND
THE NORMAL (1 ATM) BOILING POINT.

THE CORRELATIONS ARE:

1.  ANTOINE CORRELATION
```

**Examples (continued)** **Table II**

```
2.  FROST AND KALKWARF CORRELATION
3.  LEE AND KESLER CORRELATION
4.  GOMEZ-NIETO AND THODOS
5.  RIEDEL, PLANK AND MILLER

CHOOSE BY NUMBER.            2
```

**(Start of next display)**

```
IF YOU WANT TO ESTIMATE VAPOR PRESSURE,
ENTER P.  IF YOU WANT TO ESTIMATE
TEMPERATURE, ENTER T.

ENTER P OR T.        T
```

**(Start of next display)**

```
FROST AND KALKWARF CORRELATION

FOUR ARBITRARY CONSTANTS ARE REQUIRED
TO SOLVE THE EQUATION:

   LN(P) = A + B/T + C*LN(T) + D*P/T

ENTER THESE CONSTANTS AND THE
PRESSURE IN MM HG

ENTER THE DATA NOW:

ENTER A      43.492
ENTER B      -3266.92
ENTER C      -4.179
ENTER D      1.81
ENTER PRESSURE, MM HG          5550
```

**(Start of next display)**

```
THE FROST AND KALKWARF CORRELATION

GIVES A TEMPERATURE OF
15.1781 DEG. C

          END OF PROGRAM
```

**Comparison of published and calculated results for propane** **Table III**

| Temperature, $t$(°C) | Vapor pressure [8] | Vapor pressure, $P^s$ (atm) Calculated Eq. (2) | Eq. (6) | Eq. (13) | Eq. (17) |
|---|---|---|---|---|---|
| −42.07 | 1.000 | 1.000 | 1.004 | 1.000 | 1.002 |
| −17.78 | 2.611 | 2.628 | 2.601 | 2.601 | 2.606 |
| 15.56 | 7.321 | 7.391 | 7.351 | 7.258 | 7.294 |
| 48.89 | 16.480 | 16.575 | 16.512 | 16.306 | 16.410 |
| 82.22 | 32.258 | 32.275 | 32.245 | 32.085 | 32.070 |
| 96.81 | 42.02 | 42.03 | 42.03 | 41. | 41.83 |
| | | Temperature, $t$ (°C), calculated | | | |
| | | −42.24 | −42.26 | −42.16 | −41.86 |
| | | −18.04 | −17.97 | −17.75 | −17.45 |
| | | 15.20 | 15.41 | 15.88 | 15.26 |
| | | 48.63 | 48.81 | 49.37 | 48.43 |
| | | 82.18 | 82.24 | 82.50 | 81.45 |
| | | 96.64 | 96.64 | 96.49* | 96.51** |

* Unstable near critical point
**Δ (calculated $t$ − estimated $t$) set equal to 0.001. In all other cases, Δ = 0.01

**Physical properties of propane used in calculations**

|  |  | |
|---|---|---|
| Mol wt | 44.097 | |
| $T_c$ | 369.97 | |
| $P_c$ | 42.02 | |
| $\omega$ | 0.152 | |
| $T_b$ | −42.06°C | |
| $A$ | 6.82973 | Antoine constants |
| $B$ | 813.20 | Antoine constants |
| $C$ | 248.00 | Antoine constants |
| $A'$ | 43.492 | Frost and Kalkwarf constants |
| $B'$ | −3,266.92 | Frost and Kalkwarf constants |
| $C'$ | −4.179 | Frost and Kalkwarf constants |
| $D'$ | 1.81 | Frost and Kalkwarf constants |

Data sources: references 3, 7, 8, 9 and 10.

**Constants of Gomez-Nieto and Thodos correlation for propane** **Table IV**

| | Calculated via Eq. (7) – (12) | Reported, ref. [4] |
|---|---|---|
| $\beta$ | −4.31582 | −4.36761 |
| m | 1.28568 | 1.27762 |
| $\gamma$ | 0.15575 | 0.14092 |

# Latent heat of vaporization

## This program, based on empirical and thermodynamically rigorous correlations, calculates the latent heat of vaporization of pure substances.

☐ Correlations for predicting latent heat of vaporization (or condensation) can be divided roughly into two groups: those that are largely empirical, and those based on the Clausius-Clapeyron equation, which uses vapor-pressure temperature relationships.

This Apple II program offers a choice of five correlations for latent heat of vaporization:

1. Riedel [*1*], which applies only at the normal boiling point.
2. Pitzer *et al.* [2], valid to the critical point.
3. Watson [*4*], requiring a known heat of vaporization at some temperature.
4. Clausius-Clapeyron, which applies only when vapor pressure is below 1,500 mm Hg.
5. A general thermodynamic correlation, which yields not only heat of vaporization but also vapor pressure, by the Lee and Kesler correlation [*6*]; vapor volume, by the Peng-Robinson correlation [*9*]; and liquid volume, by the Spencer and Danner correlation [*10*].

The program, listed in Table I, is interactive and self-explanatory for the most part. Table II shows example displays.

### Riedel's correlation

Although many correlations for predicting latent heat of vaporization at the normal boiling point are available, only Riedel's is included here:

$$\Delta H_{vb} = 1.093RT_c\left(T_{rb}\frac{(\ln P_c - 1)}{0.930 - T_{rb}}\right) \quad (1)$$

This correlation is choice "1" on the program menu. To use it, enter the critical temperature (K), critical pressure (atm), and normal (1-atm) boiling point (°C). The answer is reported in L-atm/g-mol. Note that 1 L-atm/g-mol is equal to 24.12 cal/g-mol.

In Table III, the latent heat value for methane calculated by this program is compared to that reported by Canjar and Manning [7]. Required data for methane are also shown in Table III.

If the normal boiling point is not known, it may be calculated by a correlation such as the Antoine equation [*5*]. Simply set vapor pressure equal to 760 mm Hg, and solve for *t:*

$$\log P^s = A - [B/(t + C)] \quad (2)$$

### Relationship of Pitzer et al.

Latent heat values over the temperature span from the normal boiling point to the critical point can be calculated with the relationship proposed by Pitzer et al. [*2*], extended by Carruth and Kobayashi [*3*] and put in mathematical form by Reid, Prausnitz and Sherwood [*8*]:

$$\Delta H_v/RT_c = 7.08(1 - T_r)^{0.354} + 10.95\omega(1 - T_r)^{0.456} \quad (3)$$

This correlation is choice "2" on the program menu. Required data are $T_c$, $\omega$, and $t$. Table III compares results from this correlation with experimental ones for methane.

### Watson correlation

If the latent heat at one temperature is known, Watson's correlation will prove useful:

$$\Delta H_{v(t_2)} = \Delta H_{v(t_1)}\left(\frac{1 - T_{r2}}{1 - T_{r1}}\right)^{0.378} \quad (4)$$

The relationship reflects the effect of temperature on $\Delta H_v$.

This correlation is choice "3" on the program menu. Using it requires a known value of $\Delta H_v$ at some temperature, plus $T_c$ and the temperature of interest. Table III compares results from this correlation with experimental ones for methane. The experimental value of $\Delta H_v$ at −161.49°C was taken as the known value.

### Clausius-Clapeyron equation

Another method for determining latent heat is with the Clausius-Clapeyron equation:

$$dP/dT = \Delta H_v/T\,\Delta V_v \quad (5)$$

The slope of the vapor-pressure/temperature relationship, $dP/dT$, can be determined from a number of correlations, including Antoine's, Eq. (2). For this particular case:

$$dP/dt = (2.303BP)/(t + C)^2 \quad (6)$$

Because the limit of the Antoine equation is usually 1,500 mm Hg, no large error is introduced by using the ideal gas law to find the volume of the saturated vapor and neglecting liquid volumes. With these assumptions,

Adapted from an article by James H. Weber, University of Nebraska, originally published January 14, 1980.

combining Eq. (5) and (6), and solving for the latent heat, gives:

$$\Delta H_v = (2.303BRT^2)/(t + C)^2 \quad (7)$$

This correlation is choice "4" on the program menu. The only data required are the Antoine constants $B$ and $C$, plus $t$. Table III compares results from this correlation with experimental ones, for methane. Note that this correlation should not be used when the vapor pressure is beyond 1,500 mm Hg (1.97 atm), because it relies on the Antoine equation.

## Generalized relationship

Choice "5" on the program menu is a generalized relationship based on Eq. (8), where the necessary data are calculated by other correlations. Any general vapor pressure vs. reduced temperature ($P^s$ vs. $T$) relationship can be related to heat of vaporization by:

$$-\frac{d \ln P_r^s}{d(1/T_r)} = \frac{\Delta H_v T_r}{\Delta V_v P^s} \quad (8)$$

Clearly, Eq. (8) can be solved for $\Delta H_v$; however, in addition to the left-hand side of the equation, the volume change, $\Delta V_v$, must also be determined. This can be done by using an equation of state to calculate both the saturated vapor and liquid values, or by using different equations to evaluate each of the two terms. Also, the vapor pressure, $P^s$, must be known.

In this program, the left side of Eq. (8) is evaluated with the Lee and Kesler relationship [6]:

$$-\frac{d \ln P_r^s}{d(1/T_r)} = 6.09648 - 1.28862T_r + 1.016T_r^7 + \omega(15.6875 - 13.4721T_r + 2.615T_r^7) = \psi \quad (9)$$

The saturated-vapor volume is evaluated by the Peng-Robinson correlation [9]:

$$P = \frac{RT}{V - b} - \frac{a(T)}{V(V + b) + b(V - b)} \quad (10)$$

Here:
$$a(T_c) = 0.45724(R^2T_c^2/P_c) \quad (11)$$
$$b(T_c) = 0.07780(RT_c/P_c) \quad (12)$$
$$a(T) = a(T_c)\alpha(T_r\omega) \quad (13)$$
$$b(T) = bT_c \quad (14)$$
$$\alpha^{1/2} = 1 + \kappa(1 - T_r^{1/2}) \quad (15)$$
$$\kappa = 0.37464 + 1.54226\omega - 0.26992\omega^2 \quad (16)$$

The saturated-liquid volume is calculated by means of the Spencer and Danner [10] modification of the Rackett equation [11]:

$$1/\rho_s = (RT_c/P_c)Z_{RA}^{[1+(1-T_r)^{2/7}]} \quad (17)$$

For completeness, the program also calculates vapor pressure, via the Lee and Kesler equation:

$$\ln P_r^s = f^{(0)}(T_r) + \omega f^{(1)}(T_r) \quad (18)$$

Here:
$$f^{(0)} = 5.92714 - (6.09648/T_r) - 1.28862 \ln T_r + 0.169347T_r^6 \quad (19)$$

$$f^{(1)} = 15.2518 - (15.6875/T_r) - 13.4721 \ln T_r + 0.43577T_r^6 \quad (20)$$

This generalized relationship requires the following data: critical temperature and pressure, Rackett compressibility factor, acentric factor, and temperature. The result is the latent heat of vaporization, plus the vapor volume, liquid volume, and vapor pressure.

Table III lists all these needed data, and compares the calculated latent heat of vaporization (Eq. (8)) with an experimental one, for methane. The vapor pressure of methane calculated by the Lee and Kesler relationship (Eq. (18)) is also compared with experimental values.

## Nomenclature

| | |
|---|---|
| $A$, $B$, $C$ | Antoine constants, Eq. (2) |
| $a$, $b$ | Peng-Robinson Constants, Eq. (10) |
| $f^{(0)}$, $f^{(1)}$ | Functions in Lee and Kesler correlation, Eq. (18) |
| $\Delta H_v$ | Latent heat of vaporization, (L)(atm)/g-mol |
| $M$ | Molecular weight |
| $P$ | Pressure, atm or mm Hg |
| $R$ | Gas Law Constant, 0.08207 (L)(atm)/(g-mol)(K) |
| $T$ | Temperature, K |
| $t$ | Temperature, °C |
| $V$ | Molal volume, cm$^3$/g-mol |
| $\Delta V_v$ | Volume change accompanying vaporization |
| $Z$ | Compressibility factor, $PV/RT$ |
| $Z_{RA}$ | Modified compressibility factor for Spencer and Danner relationship |
| $\alpha$, $\kappa$ | Parameters in Peng-Robinson correlation, Eq. (15) and (16) |
| $\rho$ | Density, g-mol/cm$^3$ |
| $\psi$ | $-d \ln P_r^s/d(1/T_r)$ |
| $\omega$ | Pitzer's acentric factor |

**Subscripts**

| | |
|---|---|
| $b$ | Normal boiling point |
| $c$ | critical point |
| $g$ | gas |
| $l$ | liquid |
| $r$ | reduced |

**Superscript**

| | |
|---|---|
| $s$ | saturation |

## References


1. Riedel, L., *Chem. Ing. Tech.,* Vol. 26, 1954, p. 679.
2. Pitzer, K. S., Lippmann, D. A., Curl, R. F., Huggins, C. M., and Peterson, D. E., *J. Am. Chem. Soc.,* Vol. 77, 1955, p. 3433.
3. Carruth, G. F., and Kobayashi, *Ind. Eng. Chem. Fund.,* Vol. 11, 1972, p. 509.
4. Watson, K. M., *Ind. Eng. Chem.,* Vol. 25, 1943, p. 398.
5. Antoine, C., *Comptes Rendus,* Vol. 107, No. 681, 1888, p. 836.
6. Lee, B. I., and Kesler, M. G., *AIChe J.,* Vol. 21, 1975, p. 510.
7. Canjar, L. N., and Manning, F. S., "Thermodynamic Properties and Reduced Correlations of Gases," Gulf Pub. Co., Houston, Tex., 1967.
8. Reid, R. C., Prausnitz, J. M., and Sherwood, T. K., "The Properties of Gases and Liquids," 3rd ed., McGraw-Hill Book Co., New York, 1977, p. 200.
9. Peng, D-Y., Robinson, D. B., *Ind. Eng. Chem. Fund.,* Vol. 15, 1976, p. 59.
10. Spencer, C. F., and Danner, R. P., *J. Chem. Eng. Data,* Vol. 17, 1972, p. 236.
11. Rackett, H. G., *J. Chem. Eng. Data,* Vol. 15, 1970, p. 514.

**Program for predicting latent heat of vaporization** **Table I**

```
10  REM   LATENT HEAT OF VAPORIZATION
20  REM   FROM CHEMICAL ENGINEERING, JANUARY 14, 1980
30  REM   BY JAMES H. WEBER
40  REM  TRANSLATED BY WILLIAM VOLK
50  REM   COPYRIGHT (C) 1984
60  REM  BY CHEMICAL ENGINEERING
70  HOME : VTAB 4
80 A$(1) = "RIEDEL'S"
90 A$(2) = "PITZER ET AL."
100 A$(3) = "WATSON'S"
110 A$(4) = "CLAUSIUS-CLAPEYRON"
120 A$(5) = "GENERAL"
130  REM  SET DISPLAY TO FOUR DECIMALS
140  DEF  FN P(X) =  INT (1E4 * (X + .00005)) / 1E4
150  REM  GAS CONSTANT
160 R = .08207
170  PRINT "PROGRAM CALCULATES THE LATENT HEAT OF"
180  PRINT "VAPORIZATION IN LITER-ATMOSPHERES/G-MOL"
190  PRINT "  (24.12 * L-ATM/G-MOL = CAL/G-MOL)"
200  PRINT
210  PRINT "FOLLOWING CORRELATIONS ARE AVAILABLE:"
220  PRINT
230  PRINT "1. RIEDEL'S CORRELATION FOR THE HEAT OF"
240  PRINT "   VAPORIZATION AT THE BOILING POINT."
250  PRINT "
260  PRINT "7. THE PITZER ET AL. CORRELATION FOR"
270  PRINT "   THE HEAT OF VAPORIZATION AT ANY"
280  PRINT "   TEMPERATURE BELOW THE CRITICAL."
290  PRINT
300  PRINT "3. WATSON CORRELATION WHICH REQUIRES"
310  PRINT "   A KNOWN HEAT OF VAPORIZATION."
320  PRINT
330  PRINT "4. CLAUSIUS-CLAPEYRON EQUATION WHICH"
340  PRINT "   REQUIRES THE ANTOINE CONSTANTS."
350  PRINT
360  PRINT "5. A GENERAL SOLUTION INVOLVING SEVERAL"
370  PRINT "   EMPIRICAL CORRELATIONS."
380  PRINT
390  INPUT "SELECT YOUR OPTION BY NUMBER      ";OP
400  HOME : VTAB 4
410  PRINT A$(OP);" CORRELATION:"
420  PRINT
430  IF OP = 5 GOTO 450
440  PRINT "THE DATA REQUIRED ARE:"
450  PRINT
460  ON OP GOTO 480,600,760,1000,1140
470  REM  RIEDEL CORRELATION
480  PRINT "CRITICAL TEMPERATURE, DEG. K"
490  PRINT "CRITICAL PRESSURE, ATM"
500  PRINT "BOILING POINT, DEG. C"
510  PRINT
520  PRINT "ENTER THE DATA NOW."
530  PRINT
540  INPUT "CRITICAL TEMPERATURE, DEG. K    ";TC
550  INPUT "CRITICAL PRESSURE, ATM          ";PC
560  INPUT "BOILING POINT, DEG. C           ";TB
570 HV = 1.093 * R * TC * ((TB + 273.16) / TC * ( LOG
    (PC) - 1) / (.930 - (TB + 273.16) / TC))
580  GOTO 1590
590  REM   PITZER, ET AL. CORRELATION
600  PRINT "THE CRITICAL TEMPERATURE, DEG. K"
610  PRINT "THE ACENTRIC FACTOR, OMEGA, AND"
620  PRINT "THE TEST TEMPERATURE, DEG. C"
630  PRINT
640  PRINT "ENTER THE DATA NOW:"
650  PRINT
660  INPUT "CRITICAL TEMPERATURE, DEG. K    ";TC
670  INPUT "ACENTRIC FACTOR, OMEGA          ";OM
680  INPUT "TEST TEMPERATURE, DEG. C        ";T
690 TR = (T + 273.16) / TC
700 TP = 7.08 * (1 - TR) ^ .354
710  PRINT "VALUE."
720 TP = TP + 10.95 * OM * (1 - TR) ^ .456
730 HV = R * TC * TP
740  GOTO 1590
750  REM  WATSON CORRELATION
760  PRINT "CRITICAL TEMPERATURE, DEG. K"
770  PRINT "HEAT OF VAPORIZATION AT SOME DEFINITE"
780  PRINT "TEMPERATURE, THAT TEMPERATURE IN"
790  PRINT "DEGREES C, AND A TEST TEMPERATURE"
800  PRINT "IN DEGREES C."
010  PRINT
820  PRINT "THE CALCULATED HEAT OF VAPORIZATION"
830  PRINT "WILL BE IN THE SAME UNITS AS THE INPUT"
840  PRINT "VALUE."
850  PRINT
860  PRINT "ENTER THE DATA NOW:"
870  INPUT "CRITICAL TEMPERATURE, DEG. K    ";TC
880  INPUT "KNOWN HEAT OF VAPORIZATION      ";HW
890  INPUT "TEMPERATURE, DEG. C OF ABOVE    ";TW
900  INPUT "TEST TEMPERATURE, DEG. C        ";T
910 TR = (T + 273.16) / TC
920 TS = (TW + 273.16) / TC
930  IF TR =  > 1 OR TX =  > 1 GOTO 950
940  GOTO 970
950 F1 = 1
960  GOTO 1590
970 HV = HW * ((1 - TR) / (1 - TS)) ^ .378
980  GOTO 1590
990  REM  CLAUSIUS-CLAPEYRON CALCULATION
1000  PRINT "ANTOINE'S CONSTANTS, B AND C FOR THE"
1010  PRINT "LIQUID IN QUESTION."
1020  PRINT
1030  PRINT "(NOTE:THIS CORRELATION HAS A LIMIT OF"
1040  PRINT "ABOUT 1500 MM HG VAPOR PRESSURE.)
1050  PRINT
1060  PRINT "ENTER THE DATA NOW:"
1070  PRINT
1080  INPUT "ANTOINE CONSTANT B             ";B
1090  INPUT "ANTOINE CONSTANT C             ";C
1100  INPUT "TEMPERATURE, DEG. C            ";T
1110 HV = 2.303 * B * R * (T + 273.16) ^ 2 / (T + C) ^ 2
```

**Program for predicting latent heat of vaporization (continued)** **Table I**

```
1120  GOTO 1590
1130  REM  GENERAL SOLUTION
1140  PRINT "THIS CALCULATION USES A LEE AND"
1150  PRINT "KESLER CORRELATION TO CALCULATE THE"
1160  PRINT "SATURATION VAPOR PRESSURE, AND A PENG-"
1170  PRINT "ROBINSON CORRELATION TO DETERMINE THE"
1180  PRINT "VAPOR VOLUME, AND A SPENCER AND DANNER"
1190  PRINT "CORRELATION FOR THE LIQUID VOLUME."
1200  PRINT "FROM THESE THE HEAT OF VAPORIZATION"
1210  PRINT "IS CALCULATED BY ANOTHER LEE AND"
1220  PRINT "KESLER RELATIONSHIP."
1230  PRINT
1240  PRINT "THE FOLLOWING DATA ARE REQUIRED:"
1250  PRINT
1260  PRINT "CRITICAL TEMPERATURE, DEG. K"
1270  PRINT "CRITICAL PRESSURE, ATM"
1280  PRINT "MODIFIED COMPRESSIBILITY FACTOR"
1290  PRINT "ACENTRIC FACTOR"
1300  PRINT "TEST TEMPERATURE, DEG. C"
1310  PRINT
1320  INPUT "PRESS RETURN TO ENTER DATA.          ";X$
1330  HOME : VTAB (4)
1340  INPUT "CRITICAL TEMPERATURE, DEG. K         ";TC
1350  INPUT "CRITICAL PRESSURE, ATM               ";PC
1360  INPUT "MODIFIED COMPRESSIBILITY FACTOR      ";ZR
1370  INPUT "ACENTRIC FACTOR, OMEGA               ";OM
1380  INPUT "TEST TEMPERATURE, DEG. C             ";T
1390 TK = T + 273.16
1400 TR = TK / TC
1410  IF TR < 1 GOTO 1440
1420 F1 = 1
1430  GOTO 1590
1440 FO = 5.92714 - (6.09648 / TR) - 1.28862 *  LOG (TR)
     + .169347 * TR ^ 6
1450 F9 = 15.2518 - (15.6875 / TR) - 13.4721 *  LOG (TR)
     + .43577 * TR ^ 6
1460 GT = 6.09648 - 1.28862 * TR + 1.016 * TR ^ 7 + OM
     * (15.6875 - 13.4721 * TR  + 2.615 * TR ^ 7)
1470 PS =  EXP (FO + OM * F9) * PC
1480 AL = ((.37464 + 1.54226 * OM - .26992 * OM ^ 2) *
     (1 -  SQR (TR)) + 1) ^ 2
1490 AF = .45724 * R ^ 2 * TC ^ 2 / PC * AL
1500 BT = .0778 * R * TC / PC
1510 E = 1
1520 VS = E * R * TK / PS
1530 TP = VS / (VS - BT) - (VS * AF / (VS * (VS + BT) +
     (BT * (VS - BT))) / (R * TK))
1540  IF  ABS (TP - E) < .0001 GOTO 1570
1550 E = (TP + E) / 2
1560  GOTO 1520
1570 LS = ZR ^ (1 + (1 - TR) ^ (2 / 7)) * (R * TC / PC)
1580 HV = (VS - LS) * PS * GT / TR
1590  HOME : VTAB 4
1600  PRINT "USING ";A$(OP);" CORRELATION"
1610  PRINT
1620  IF F1 = 0 GOTO 1660
1630  PRINT "REDUCED TEMPERATURE IS OUT OF RANGE."
1640  PRINT "TRY ANOTHER METHOD."
1650  GOTO 1840
1660  PRINT "THE HEAT OF VAPORIZATION IS"
1670  IF OP = 3 GOTO 1800
1680  PRINT  FN P(HV);" LITER-ATM/G-MOL."
1690  IF OP <  > 5 GOTO 1840
1700  PRINT
1710  PRINT "VAPOR PRESSURE BY LEE AND KESLER"
1720  PRINT "CORRELATION IS "; FN P(PS);" ATM."
1730  PRINT
1740  PRINT "VAPOR VOLUME BY PENG-ROBINSON CORRE-"
1750  PRINT "LATION IS "; FN P(VS);" CC/G-MOL."
1760  PRINT
1770  PRINT "LIQUID VOLUME BY SPENCER-DANNER"
1780  PRINT "CORRELATION IS "; FN P(LS);" CC/G-MOL."
1790  GOTO 1840
1800  PRINT  FN P(HV);" SAME UNITS AS INPUT VALUE."
1810  GOTO 1840
1820  PRINT
1830  PRINT "SAT. PRESSURE BY LEE & KESLER
1840  PRINT
1850  PRINT  TAB( 15);"END OF PROGRAM"
1860  END
```

**Examples for: Latent heat of vaporization** **Table II**

**(Start of first display)**

```
PROGRAM CALCULATES THE LATENT HEAT OF
VAPORIZATION IN LITER-ATMOSPHERES/G-MOL
  (24.12 * L-ATM/G-MOL = CAL/G-MOL)

FOLLOWING CORRELATIONS ARE AVAILABLE:

1. RIEDEL'S CORRELATION FOR THE HEAT OF
   VAPORIZATION AT THE BOILING POINT.

2. THE PITZER ET AL. CORRELATION FOR
   THE HEAT OF VAPORIZATION AT ANY
   TEMPERATURE BELOW THE CRITICAL.

3. WATSON CORRELATION WHICH REQUIRES
   A KNOWN HEAT OF VAPORIZATION.

4. CLAUSIUS-CLAPEYRON EQUATION WHICH
   REQUIRES THE ANTOINE CONSTANTS.

5. A GENERAL SOLUTION INVOLVING SEVERAL
   EMPIRICAL CORRELATIONS.

SELECT YOUR OPTION BY NUMBER      1
```

**Examples for: Latent heat of vaporization (continued)** **Table II**

**(Start of next display)**

```
RIEDEL'S CORRELATION:

THE DATA REQUIRED ARE:

CRITICAL TEMPERATURE, DEG. K
CRITICAL PRESSURE, ATM
BOILING POINT, DEG. C

ENTER THE DATA NOW.

CRITICAL TEMPERATURE, DEG. K     191.04
CRITICAL PRESSURE, ATM           46.06
BOILING POINT, DEG. C            -161.49
```

**(Start of next display)**

```
USING RIEDEL'S CORRELATION

THE HEAT OF VAPORIZATION IS
82.0574 LITER-ATM/G-MOL.

              END OF PROGRAM
```

**(Start of first display)**

```
PROGRAM CALCULATES THE LAtENI HEAT OF
VAPORIZATION IN LITER-ATMOSPHERES/G-MOL
  (24.12 * L-ATM/G-MOL = CAL/G-MOL)

FOLLOWING CORRELATIONS ARE AVAILABLE:

1. RIEDEL'S CORRELATION FOR THE HEAT OF
   VAPORIZATION AT THE BOILING POINT.

2. THE PITZER ET AL. CORRELATION FOR
   THE HEAT OF VAPORIZATION AT ANY
   TEMPERATURE BELOW THE CRITICAL.

3. WATSON CORRELATION WHICH REQUIRES
   A KNOWN HEAT OF VAPORIZATION.

4. CLAUSIUS-CLAPEYRON EQUATION WHICH
   REQUIRES THE ANTOINE CONSTANTS.

5. A GENERAL SOLUTION INVOLVING SEVERAL
   EMPIRICAL CORRELATIONS.

SELECT YOUR OPTION BY NUMBER      3
```

**(Start of next display)**

```
WATSON'S CORRELATION:

THE DATA REQUIRED ARE:

CRITICAL TEMPERATURE, DEG. K
HEAT OF VAPORIZATION AT SOME DEFINITE
TEMPERATURE, THAT TEMPERATURE IN
DEGREES C, AND A TEST TEMPERATURE
IN DEGREES C.

THE CALCULATED HEAT OF VAPORIZATION
WILL BE IN THE SAME UNITS AS THE INPUT
VALUE.

ENTER THE DATA NOW:
CRITICAL TEMPERATURE, DEG. K     191.04
KNOWN HEAT OF VAPORIZATION       82.4328
TEMPERATURE, DEG. C OF ABOVE     -161.49
TEST TEMPERATURE, DEG. C         -153.94
```

**(Start of next display)**

```
USING WATSON'S CORRELATION

THE HEAT OF VAPORIZATION IS
79.3763 SAME UNITS AS INPUT VALUE.

              END OF PROGRAM
```

**Comparison of published and calculated results for methane** **Table III**

| Temperature, $t$, °C | Saturation pressure, $P^s$, atm | | Latent heat of vaporization, $\Delta H_v$, (L) (atm)/(g-mol) | | | | | |
|---|---|---|---|---|---|---|---|---|
| | Ref. [7] | Eq. (18) | Ref. [7] | Eq. (1) | Eq. (3) | Eq. (4) | Eq. (7) | Eq. (8) |
| −161.49 ($t_b$) | 1.000 | 0.991 | 80.69 | 82.06 | 82.26 | * | 84.14 | 82.05 |
| −159.49 | 1.184 | 1.167 | 79.85 | | 81.52 | 79.92 | 83.94 | 81.46 |
| −156.71 | 1.477 | 1.450 | 79.00 | | 80.47 | 78.82 | 83.67 | 80.58 |
| −153.94 | 1.810 | 1.782 | 78.11 | | 79.39 | 77.70 | 83.42 | 79.66 |
| −140.05 | 4.389 | 4.365 | 73.11 | Not Applicable | 73.56 | 71.64 | | 74.08 |
| −123.38 | 10.207 | 10.207 | 64.97 | | 65.21 | 63.01 | Not Applicable | 65.22 |
| −106.72 | 20.210 | 20.198 | 54.11 | | 54.27 | 51.82 | | 53.25 |
| − 90.05 | 35.860 | 35.905 | 36.08 | | 36.31 | 33.78 | | 33.83 |
| − 82.12 ($t_c$) | 46.06 | 46.06 | 0 | | 0 | 0 | | – |

*Published value at $t_b$ used for $\Delta H_{v1}$

**Physical properties of methane used in the calculations:**

| | | | | |
|---|---|---|---|---|
| Mol wgt | 16.043 | $\omega$ | 0.008 | |
| $P_c$ | 46.06 atm | $A$ | 6.61184 | Antoine constants, Eq. (2) |
| $T_b$ | 111.67 K | $B$ | 389.93 | Antoine constants, Eq. (2) |
| $T_c$ | 191.04 K | $C$ | 266.00 | Antoine constants, Eq. (2) |
| $Z_{RA}$ | 0.2876 | | | |

Data Sources: Ref. [7], [8] and [10], and API Project 44, Carnegie Press, Pittsburgh, Pa., 1953

# Gas equations

This program solves for any variable in the ideal and van der Waals equations of state. Accelerated iteration speeds up the computation of number of gas moles and volume.

☐ Chemical engineers and chemists are aware that the ideal gas equation $PV = nRT$—relating the temperature, pressure and volume of a gas—is a combination of Boyle's, Charles' and the "pressure" laws. They also know that the ideal law holds exactly only for a hypothetical "ideal" gas, and approximately for *all* real gases, if the pressure is sufficiently low.

Unfortunately, many industrial processes operate at pressures well above atmospheric and at high temperatures. Under these conditions, almost all gases deviate markedly from ideality. Therefore, many other equations of state have been proposed to describe the behavior of real gases at high temperatures and pressures.

One of the best known of these equations is that of van der Waals:

$$[P + (n^2a/V^2)][V - nb] = nRT \qquad (1)$$

Here, $a$ and $b$ are constants for a particular gas.

This equation represents one modification of the ideal equation, with the "$n^2a$" term being a correction for the interatomic (or intermolecular) attractive forces, and the "$nb$" term for the volume of the atoms (or molecules) themselves.

As expected, the van der Waals equation provides a more accurate model for real gases at medium and high pressures. Unfortunately, it is somewhat difficult to use, because it can be seen to be cubic in both $n$ and $V$.

## The van der Waals constants

An imporant consideration is the derivation of the formulas for the two constants $a$ and $b$ for a gas. It is necessary that the van der Waals curve pass through the critical point for the gas concerned. It can easily be shown that the conditions for this to occur are the following at the critical point:

$$(\partial P/\partial V)_T = 0 = (\partial^2 P/\partial^2 V)_T \qquad (2)$$

Substituting the partial derivatives and solving the resulting equations leads to:

$$T_c = 8a/27bR \qquad (3)$$

$$P_c = a/27b^2 \qquad (4)$$

$$V_c = 3b \qquad (5)$$

The van der Waals constants are calculated from values of the critical parameters $P_c$, $T_c$ and $V_c$, which are tabulated in data books. In general, the tabulated values of $V_c$ are less precise than those for $T_c$ and $P_c$, so the program calculates $a$ and $b$ by means of Eq. (6) and (7), obtained by substitution in Eq. (3) and (4):

$$a = 27b^2P_c \qquad (6)$$

$$b = RT_c/8P_c \qquad (7)$$

## Solving the equations

In all the possible cases, the ideal equation is easily solved by simple rearrangements, as is the van der Waals equation for either $P$ or $T$.

As mentioned, because the van der Waals equation is cubic in both $n$ and $V$, solving for either of these variables is difficult. The simplest method of solution was found to be by iteration, using the sequences:

For $V$,

$$V_{m+1} = n\left[\frac{RT}{P + (n^2a/V_m^2)} + b\right] \qquad (8)$$

Here, $V_o = V_{ideal}$.

For $n$,

$$n_{m+1} = (1/b)\left[V - \frac{n_mRT}{P + (n_m^2/V^2)}\right] \qquad (9)$$

Here, $n_O = n_{ideal}$.

These iterative sequences converge rather slowly. However, the rate of convergence is accelerated by a technique known as the Steffensen iteration,* which uses Aitken's $\delta^2$-process to make the rate of convergence roughly quadratic. In practice, four Steffensen cycles are required to determine $n$ and $V$ values of maximum accuracy.

## The program

Table I lists the program, which is interactive and self-explanatory for the most part. Table II shows example displays.

The program solves both the ideal and the van der Waals equations; there is no need to specify which you want. The first input is the variable you want to solve

Adapted from an article by Phillip R. Rowley, James M. Brown Ltd., originally published February 11, 1980.

for: $V$, $P$, $T$, or $n$. Then, input the critical temperature ($T_c$) and the critical pressure ($P_c$). Finally, input the remaining three variables. The program then solves for the variable you want, by both equations of state, and displays the results.

*All the input and output units are SI.* That is: temperature in K (Kelvins), pressure in Pa (pascals, or $N/m^2$), volume in $m^3$ (cubic meters), and the amount of gas in mol (once called the gram-mole, and defined as the amount having the same number of atoms or molecules as there are atoms in 0.012 kg of carbon 12.) In SI units, the value of the gas constant $R$ is 8.3143 (Pa)($m^3$)/(mol)(K).

## Example

We know that 10 mol of methane occupies 1.756 liters at 100 atm and 0°C. What would the pressure be according to the ideal and the van der Waals equations of state?

First, change the units. In SI, the volume is 0.001756 $m^3$, and the temperature is 273.15 K (the given pressure is $10.13 \times 10^6$ Pa). The critical constants for methane are: $T_c = 190.6$ K; $P_c = 4.60 \times 10^6$ Pa.

Pressure is the unknown here, so select "P" in the program. Enter the data, as shown in Table II, and read the result: $12.93 \times 10^6$ Pa by the ideal equation of state, and $9.67 \times 10^6$ Pa by the van der Waals equation. Converting back to atm, this is 127.6 atm by the ideal equation, and 95.4 atm by the van der Waals equation. In any units, the van der Waals equation yields the more accurate result.

Using the above data, if you solve for any of the other variables instead of pressure the answers are:

Volume: 0.00224 $m^3$ by the ideal equation of state, 0.00168 $m^3$ by van der Waals'.

Temperature: 213.9 K by the ideal, 280.5 K by van der Waals'.

Number of mols: 7.83 mol by the ideal, 10.60 mol by van der Waals'.

## The author

**Phillip R. Rowley**, a Development Chemist in the Pigment Div. of James M. Brown Ltd. (Napier St., Fenton, Stoke-on-Trent, Staffordshire ST4 4NX, Great Britain), is responsible for R&D work from laboratory to plant scale. He holds a B.Sc. (Hons.) degree from the University of Manchester, specializing in inorganic chemistry. His main interests are inorganic and physical chemistry, the modeling of chemical processes, and the application of calculator programs to the solution of chemical problems. He runs a large independent TI-calculator users' group in the U.K.

**Program for solving gas equations** **Table I**

```
10 REM   GAS EQUATIONS
20 REM   FROM CHEMICAL ENGINEERING, FEBRUARY 11, 1980
30 REM   BY PHILIP R. ROWLEY
40 REM  TRANSLATED BY WILLIAM VOLK
50 REM   COPYRIGHT (C) 1984
60 REM  BY CHEMICAL ENGINEERING
70 HOME : VTAB 4
80 REM  SET DISPLAY AT FOUR DECIMALS.  NEXT LINE IS
    GAS CONSTANT
90 DEF  FN P(X) =  INT (1E4 * (X + .00005)) / 1E4
100 R = 8.3143
110 PRINT "PROGRAM USES VAN DER WAALS EQUATION TO"
120 PRINT "CALCULATE ONE OF THE TERMS: N,P,V, OR T"
130 PRINT "FROM THE OTHER THREE IN THE FOLLOWING"
140 PRINT "RELATION:"
150 PRINT
160 PRINT "   (P + A*(N/V)^2)*(V-NB) = NRT
170 PRINT
180 PRINT
190 PRINT "DATA REQUIRED ARE THE CRITICAL"
200 PRINT "TEMPERATURE AND PRESSURE,"
210 PRINT "AND THREE OF THE VARIABLES IN THE"
220 PRINT "EQUATION.
230 PRINT
240 PRINT "PROGRAM TAKES INPUT IN SYSTEME"
250 PRINT "INTERNATIONAL (SI) UNITS:"
260 PRINT
270 PRINT "MOL, PASCAL, CU. METER, K"
280 PRINT
290 INPUT "PRESS RETURN TO CONTINUE       ";Q$
300 HOME : VTAB 4
310 PRINT "PROGRAM GIVES BOTH THE IDEAL EQUATION"
320 PRINT "RESULT AND THE VAN DER WAALS RESULT."
330 PRINT
340 PRINT "ENTER SYMBOL OF VARIABLE TO BE"
350 INPUT "CALCULATED:  V, P, T, OR N.  ";X$
360 PRINT
370 PRINT "ENTER THE FOLLOWING DATA"
380 PRINT
390 INPUT "CRITICAL TEMPERATURE, K ";TC
400 INPUT "CRITICAL PRESSURE, PA   ";PC
410 PRINT
420 IF X$ = "T" GOTO 440
430 INPUT "TEMPERATURE, K          ";T
440 IF X$ = "P" GOTO 460
450 INPUT "PRESSURE, PA            ";P
460 IF X$ = "V" GOTO 490
470 INPUT "VOLUME, CU. M.              ";V
480 IF X$ = "N" GOTO 500
490 INPUT "NUMBER OF MOLS, MOL      ";N
500 PRINT
510 B = R * TC / 8 / PC
520 A = 27 * B ^ 2 * PC
530 IF X$ = "T" GOTO 570
540 IF X$ = "V" GOTO 660
550 IF X$ = "P" GOTO 870
560 GOTO 940
570 T = P + (N / V) ^ 2 * A
580 T = T * (V - N * B)
590 T = T / N / R
600 PRINT "TEMPERATURE, K"
610 TI = P * V / N / R
620 PRINT
630 PRINT "IDEAL EQUATION RESULT          "; FN P(TI)
640 PRINT "VAN DER WAALS' RESULT          "; FN P(T)
650 GOTO 1130
```

**Program for solving gas equations (continued)** **Table I**

```
660 V = N * R * P / T
670 V0 = V
680 C = C + 1
690  IF C > 4 GOTO 800
700 V2 = (N * (R * T / (P + ((N / V) ^ 2 * A)) + B))
710  IF F1 = 1 GOTO 760
720 V1 = V2
730 F1 = 1
740 V = V2
750  GOTO 700
760 F1 = 0
770  IF V2 - 2 * V1 + V0 = 0 GOTO 800
780 V = V0 - (V1 - V0) ^ 2 / (V2 - 2 * V1 + V0)
790  GOTO 670
800  PRINT "VOLUME, CU. M * 10^4"
810 VI = N * R * T / P * 1E4
820  PRINT
830  PRINT "BY IDEAL EQUATION IS          "; FN P(VI)
840 V = V * 1E4
850  PRINT "BY VAN DER WAALS' IS          "; FN P(V)
860  GOTO 1130
870 P = N * R * T / (V - N * B) - ((N / V) ^ 2 * A)
880  PRINT "PRESSURE, PA"
890  PRINT
900 PI = N * R * T / V
910  PRINT "BY IDEAL EQUATION IS          "; FN P(PI)
920  PRINT "BY VAN DER WAALS' IS          "; FN P(P)
930  GOTO 1130
940 N = P * V / R / T
950 N0 = N
960 C = C + 1
970  IF C > 4 GOTO 1080
980 N2 = (1 / B) * (V - (N * R * T) / (P + (N ^ 2 / V ^
    2 * A)))
990  IF F1 = 1 GOTO 1040
1000 N1 = N2
1010 F1 = 1
1020 N = N2
1030  GOTO 980
1040 F1 = 0
1050  IF (N2 - 2 * N1 + N0) = 0 GOTO 1080
1060 N = N0 - (N1 - N0) ^ 2 / (N2 - 2 * N1 + N0)
1070  GOTO 950
1080  PRINT "NUMBER OF MOLS"
1090 NI = P * V / R / T
1100  PRINT
1110  PRINT "BY IDEAL EQUATION IS          "; FN P(NI)
1120  PRINT "BY VAN DER WAALS' IS          "; FN P(N2)
1130  PRINT
1140  PRINT  TAB( 13);"END OF PROGRAM"
1150  END
```

**Example for: Gas equations** **Table II**

**(Start of first display)**

```
PROGRAM USES VAN DER WAALS EQUATION TO
CALCULATE ONE OF THE TERMS: N,P,V, OR T
FROM THE OTHER THREE IN THE FOLLOWING
RELATION:

  (P + A*(N/V)^2)*(V-NB) = NRT


DATA REQUIRED ARE THE CRITICAL
TEMPERATURE AND PRESSURE,
AND THREE OF THE VARIABLES IN THE
EQUATION.

PROGRAM TAKES INPUT IN SYSTEME
INTERNATIONAL (SI) UNITS:

MOL, PASCAL, CU. METER, K

PRESS RETURN TO CONTINUE
```

**(Start of next display)**

```
PROGRAM GIVES BOTH THE IDEAL EQUATION
RESULT AND THE VAN DER WAALS RESULT.

ENTER SYMBOL OF VARIABLE TO BE
CALCULATED:  V, P, T, OR N.  V

ENTER THE FOLLOWING DATA

CRITICAL TEMPERATURE, K 191
CRITICAL PRESSURE, PA   4.62E6

TEMPERATURE, K          273
PRESSURE, PA            1.293E7
NUMBER OF MOLS, MOL     10

VOLUME, CU. M * 10^4

BY IDEAL EQUATION IS         17.5546
BY VAN DER WAALS' IS         12.5196

             END OF PROGRAM
```

# Equation-of-state variables

## Compute compressibility factor, volume, fugacity, fugacity coefficient, enthalpy change, and the second and third virial coefficients for pure substances on the Apple II.

☐ A program for two equations of state, the Peng-Robinson [1] and the Benedict-Webb-Rubin [2], is presented. The former was selected as representative of the two constant $P$-$V$-$T$ relationships and is the latest modification of the well-known Redlich-Kwong equation [3]. The two constants in the Peng-Robinson are general; i.e., they are related to critical properties and the acentric factor.

The Benedict-Webb-Rubin equation relies on eight empirical constants that are particular to a given substance. Attempts to generalize them have not met with great success.

Table I lists the program, which for a given temperature and pressure, calculates these properties for the vapor phase of a pure substance: compressibility factor, volume, fugacity, fugacity coefficient, isothermal enthalpy change, and virial coefficients for a generalized equation of state. The program is self-explanatory; Table II shows displays for example calculations.

The user has a choice of the Peng-Robinson or the Benedict-Webb-Rubin equations of state. For the Peng-Robinson, the data required are: critical temperature and pressure, and acentric factor. The Benedict-Webb-Rubin equation requires the eight empirical constants, plus critical temperature and pressure.

Let us now go through the equations on which the program is based.

### Calculations with Peng-Robinson equation

The Peng-Robinson equation is:

$$P = \frac{RT}{V\text{-}b} - \frac{a(T)}{V(V+b) + b(V-b)} \qquad (1)$$

The constants $a$ and $b$ are generalized as follows:

$$a(T) = a(T_c) \times \alpha(T_r\omega) \qquad (2)$$

$$b(T) = b(T_c) \qquad (3)$$

Here:

$$a(T_c) = 0.45724(R^2T_c^2/P_c) \qquad (4)$$

$$b(T_c) = 0.07780(RT_c/P_c) \qquad (5)$$

$$\alpha^{1/2} = 1 + \kappa(1 - T_r^{1/2}) \qquad (6)$$

$$\kappa = 0.37464 + 1.54226\omega - 0.26992\omega^2 \qquad (7)$$

Hence, $b$ is a function of $T_c$ and $P_c$, while $\alpha$ is a function of those two quantities, as well as of $\omega$ and the reduced temperature.

Eq. (1) differs from the Redlich-Kwong equation in two important aspects: The $a$ term that accounts for the attractive forces between molecules is a more complicated function of temperature than it is in the Redlich-Kwong equation, in which the comparable term is $a_{RK}/T^{1/2}$; and the term $b(V\text{-}b)$ has been added to the denominator of the second term on the right side of Eq. (1).

### Compressibility factor and volume

Eq. (1) is cubic with respect to volume, and can be rewritten in terms of the compressibility factor, $Z$, as follows:

$$Z^3 - (1-B)Z^2 + (A - 3B^2 - 2B)Z - (AB - B^2 - B^3) = 0 \qquad (8)$$

Here:

$$A = aP/R^2T^2 \qquad (9)$$

$$B = bP/RT \qquad (10)$$

Eq. (8) is solved iteratively by the Newton method. The Z value is assumed to be correct when the left side of Eq. (8) is less than ±0.000001. This quantity was selected arbitrarily and can be changed by the user. This tolerance is defined in Line 690 of the program.

The program also calculates fugacity and fugacity coefficient. The Peng-Robinson equation for the fugacity coefficient is:

$$\ln\left(\frac{f}{p}\right) = Z - 1 - \ln(Z - B) - \left[\frac{A}{2(2^{1/2})B}\right]\ln\left(\frac{Z + 2.414B}{Z - 0.414B}\right) \qquad (11)$$

The program also calculates how enthalpy changes with an isothermal change in pressure—that is, the difference between the ideal-gas (low-pressure) enthalpy ($H^*$) and the enthalpy at the actual pressure ($H$). By the Peng-Robinson equation, this is:

$$(H - H^*)_T = RT(Z-1) + \left[\frac{T(d\alpha/dT) - a}{2(2^{1/2})b}\right]\ln\left(\frac{Z + 2.414B}{Z - 0.414B}\right) \qquad (12)$$

Adapted from an article by James H. Weber, University of Nebraska, originally published February 25, 1980.

## Virial coefficients

Frequently, knowledge of the second and third virial coefficients is desirable. If the virial equation is written in the form:

$$Z = 1 + \beta/V + \gamma/V^2 \quad (13)$$

then the second virial coefficient, $\beta$, by the Peng-Robinson equation, is:

$$\beta = b - [a(T)/RT] \quad (14)$$

and the third virial coefficient is:

$$\gamma = b^2 + [2a(T)b/RT] \quad (15)$$

Table III shows the program's Peng-Robinson calculations of $V$, $f/P$, and $(H - H^*)$ for methane at 37.73°C and various pressures. The Peng-Robinson values are those of Eq. (6), Eq. (8) and Eq. (11); the experimental values are those of Canjar and Manning [4]. Table III also shows calculated values of the virial coefficients, from Eq. (14) and Eq. (15), and compares them with experimental ones of Douslin et al. [5].

Also in Table III are the required data for methane, and the program's calculated values based on the Benedict-Webb-Rubin equation. Let us now go through the details of the Benedict-Webb-Rubin calculations.

## Benedict-Webb-Rubin equation

The Benedict-Webb-Rubin equation is:

$$Z = 1 + [B_0 - (A_0/RT) - (C_0/RT^3)](1/V) + [b - (a/RT)](1/V^2) + (a\alpha/RT)(1/V^5) + (c/RT^3)[(1 + \gamma V^{-2})/V^{-2}]e^{-\gamma V^{-2}} \quad (16)$$

Because volume is raised to the fifth power in Eq. (16), a trial-and-error solution is necessary, and the Newton convergence technique was used. The original estimate for $Z$ (1.0) is in the program, and the second estimate is the $Z$ value back-calculated from the volume determined using the original estimate of $Z$.

Once the volume is known, the fugacity and fugacity coefficient can be calculated. For the former:

$$RT \ln f = RT \ln(RT/V) + 2[B_0RT - A_0 - (C_0/T^2)]/V + 3/2(bRT - a)/V^2 + 6/5(a\alpha/V^5) + (c/T^2V^2) \times [(1 - e^{-\gamma/V^2})/(\gamma/V^2) + (1/2 + \gamma/V^2)e^{-\gamma/V^2}] \quad (17)$$

The fugacity coefficient is then simply $f/P$. Table III shows Eq. (17) values for methane.

To find the change in enthalpy with pressure, under isothermal conditions, the Benedict-Webb-Rubin equation gives:

$$(H - H^*)_T = [B_0RT - 2A_0 - (4C_0/T^2)]/V + 1/2(2bRT - 3a)/V^2 + 6/5(a\alpha/V^5) + (c/V^2T^2) \times \{3[(1 - e^{-\gamma/V^2})/(\gamma/V^2)] - (1/2 - \gamma/V^2)(e^{-\gamma/V^2})\} \quad (18)$$

Table III shows the calculated values for this isothermal enthalpy change.

Lastly, the program also calculates the second and third virial coefficients. For the Benedict-Webb-Rubin equation of state, the two equations are:

$$\beta = B_0 - (A_0/RT) - (C_0/RT^3) \quad (19)$$

$$\gamma = b - (a/RT) + (c\gamma/RT^3) \quad (20)$$

### Nomenclature

|  |  |
|---|---|
| $A$, $a$, $B$, $b$ | Constants of Peng-Robinson equation |
| $A_0$, $B_0$, $C_0$, $a$, $b$, $c$, $\alpha$, $\gamma$ | Constants of Benedict-Webb-Rubin equation |
| $f$ | Fugacity, atm |
| $H$ | Enthalpy, (L)(atm)/(g-mol) |
| $P$ | Pressure, atm |
| $R$ | Gas law constant, 0.08207 (L)(atm)/(g-mol)(K) |
| $T$ | Temperature, K |
| $t$ | Temperature, °C |
| $V$ | Volume, L/(g-mol) |
| $\alpha$ | Defined by Eq. (6) |
| $\beta$ | Second virial coefficient, L/(g-mol) |
| $\gamma$ | Third virial coefficient, $(L^2)/(g\text{-}mol^2)$ |
| $\kappa$ | Defined by Eq. (7) |
| $\omega$ | Pitzer's acentric factor |

**Subscripts**

|  |  |
|---|---|
| $c$ | Critical |
| $r$ | Reduced |

**Superscripts**

|  |  |
|---|---|
| * | Refers to ideal gas state |

## References

1. Peng, D.-Y., and Robinson, D. R., *Ind Eng. Chem. Fundamentals,* Vol. 15, 1976, p. 59.
2. Benedict, M., Webb, G. B., and Rubin, L. C., *J. Chem. Phys.,* Vol. 8, 1940, p. 334.
3. Redlich, O., and Kwong, J. N. S., *Chem. Rev.,* Vol. 44, 1948, p. 233.
4. Canjar, L. N., and Manning, F. S., "Thermodynamic Properties and Reduced Correlations for Gases," Gulf Publishing Co., Houston, Tex., 1967.
5. Douslin, D. R., Harrison, R. H., Moore, R. T., and McCullough, J. P., *J. Chem. Eng. Data,* Vol. 9, 1964, p. 358.

## The author

**James H. Weber** is Regent's Professor of Chemical Engineering at the University of Nebraska-Lincoln (Lincoln, NE 68588), where he has been for the past 30 years, having previously been chairman of the Dept. of Chemical Engineering for 13 years. He has also served as a consultant to Gas Processors Assn., C. F. Braun & Co., and Phillips Petroleum Co.
Holder of B.S., M.S. and Ph.D. degrees in chemical engineering from the University of Pittsburgh, he is a member of AIChE and ACS, a registered engineer in the State of Nebraska, and author or coauthor of 60 articles published in technical and scientific magazines.

**Program for equation-of-state variables** **Table I**

```
10  REM   EQUATION-OF-STATE VARIABLES
20  REM   FROM CHEMICAL ENGINEERING, FEBRUARY 25, 1980
30  REM   BY JAMES H. WEBER
40  REM  TRANSLATED BY WILLIAM VOLK
50  REM   COPYRIGHT (C) 1984
60  REM  BY CHEMICAL ENGINEERING
70  HOME : VTAB 2
80  REM  GAS CONSTANT AND SET DISPLAY TO FOUR PLACES
90 R = .08207
100  DEF  FN P(X) =  INT (1E4 * (X + .00005)) / 1E4
110  PRINT "PROGRAM CALCULATES TWO EQUATIONS OF"
120  PRINT "STATE:"
130  PRINT
140  PRINT  TAB( 6);"PENG-ROBINSON, AND"
150  PRINT  TAB( 6);"BENEDICT-WEBB-RUBIN"
160  PRINT
170  PRINT "WITH INPUT OF PRESSURE AND TEMPERATURE;"
180  PRINT "THE VOLUME, COMPRESSIBILITY FACTOR,"
190  PRINT "FUGACITY, ENTHALPY CHANGE, AND THE"
200  PRINT "SECOND AND THIRD VIRIAL COEFFICIENTS"
210  PRINT "ARE CALCULATED."
220  PRINT
230  PRINT "DATA REQUIRED ARE THE CRITICAL TEMPERA-"
240  PRINT "TURE AND PRESSURE AND:
250  PRINT
260  PRINT "FOR PENG-ROBINSON, THE ACENTRIC FACTOR;"
270  PRINT
280  PRINT "FOR B-W-R, EIGHT EMPIRICAL CONSTANTS"
290  PRINT "FOR THE PARTICULAR COMPOUND."
300  PRINT
310  INPUT "PRESS RETURN TO ENTER DATA       ";Q$
320  HOME : VTAB (4)
330  PRINT "SELECT EQUATION OF STATE"
340  PRINT "BY NUMBER, AND THEN ENTER THE REQUIRED"
350  PRINT "DATA."
360  PRINT
370  PRINT "1. FOR PENG-ROBINSON"
380  PRINT "2. FOR BENEDICT-WEBB-RUBIN"
390  PRINT
400  INPUT "MAKE A SELECTION BY NUMBER    ";OP
410  PRINT
420  PRINT "ENTER THE FOLLOWING DATA:"
430  PRINT
440  INPUT "CRITICAL TEMPERATURE, DEG. K     ";TC
450  INPUT "CRITICAL PRESSURE, ATM           ";PC
460  IF OP = 2 GOTO 480
470  INPUT "ACENTRIC FACTOR, OMEGA           ";OM
480  INPUT "GAS TEMPERATURE, DEG. C          ";T1
490 T = T1 + 273.16
500  INPUT "GAS PRESSURE, ATM                ";P
510  IF OP = 2 GOTO 860
520  REM  PENG-ROBINSON CALCULATION
530 K = .37464 + 1.54226 * OM - .26992 * OM ^ 2
540 TR = T / TC
550 AL = (1 + K * (1 -  SQR (TR))) ^ 2
560 AQ = .45724 * (R ^ 2 * TC ^ 2 / PC)
570 B = .0778 * R * TC / PC
580 A = AQ * AL
590 AU = A * P / R ^ 2 / T ^ 2
600 BU = B * P / R / T
610 Z = 1
620  GOSUB 680
630 F0 = FZ
640 Z = .8
650 F1 = 1
660  GOSUB 680
670  GOTO 660
680 FZ = Z ^ 3 - (1 - BU) * Z ^ 2 + (AU - 3 * BU ^ 2 -
    2 * BU) * Z - (AU * BU - BU ^ 2 - BU ^ 3)
690  IF  ABS (FZ) > 1E - 6 GOTO 710
700  GOTO 740
710  IF F1 = 0 GOTO 730
720 Z = Z - (1 - Z) / (F0 - FZ) * FZ
730  RETURN
```

**Program for equation-of-state variables (continued)** **Table I**

```
740 V = Z * R * T / P
750 TP = Z - 1 -  LOG (Z - BU) - (AU / 2 ^ (3 / 2) / BU) *  LOG ((Z + 2.414 * BU
) / (Z - .414 * BU))
760 FU =  EXP (TP)
770 EN = R * T * (Z - 1)
780 TP = AQ * (K ^ 2 * TR -  SQR (TR) * (K ^ 2 + K)) - A
790 TP = TP / (2 ^ (3 / 2)) / B *  LOG ((Z + 2.414 * BU) / (Z - .414 * BU))
800 EN = EN + TP
810 BA = B - (A / R / T)
820 GA = B ^ 2 + (2 * A * B / R / T)
830  PRINT
840  GOTO 1250
850  REM  BENEDICT-WEBB-RUBIN SOLUTION
860  PRINT
870  PRINT "SOLUTION OF THE BENEDICT-WEBB-RUBIN"
880  PRINT "EQUATION NEEDS THE VALUE OF EIGHT"
890  PRINT "CONSTANTS FOR THE COMPOUND INVOLVED."
900  PRINT "ENTER THESE CONSTANTS NOW:"
910  PRINT
920  INPUT "ENTER A0        ";A0
930  INPUT "ENTER B0        ";B0
940  INPUT "ENTER C0        ";C0
950  INPUT "ENTER A         ";A1
960  INPUT "ENTER B         ";B1
970  INPUT "ENTER C         ";C1
980  INPUT "ENTER ALPHA     ";A2
990  INPUT "ENTER GAMMA     ";B2
1000 Z0 = 1
1010 ZC = Z0
1020  GOSUB 1090
1030 Z1 = Z
1040 ZC = Z
1050  GOSUB 1090
1060  IF  ABS (ZC - Z) < .00005 GOTO 1150
1070 ZC = Z0 + (Z0 - Z1) * (Z0 - ZC) / (ZC - Z - Z0 + Z1)
1080  GOTO 1050
1090 V = ZC * R * T / P
1100 Z = 1 + (B0 - A0 / R / T - (C0 / R / T ^ 3)) / V
1110 Z = Z + (B1 - (A1 / R / T)) / V ^ 2
1120 Z = Z + (A1 * A2 / R / T / V ^ 5)
1130 Z = Z + (C1 / R / T ^ 3) * (1 + B2 / V ^ 2) / V ^ 2 *  EXP ( - B2 / V ^ 2)
1140  RETURN
1150 TP = R * T *  LOG (R * T / V) + 2 * (B0 * R * T - A0 - C0 / T ^ 2) / V
1160 TP = TP + 3 / 2 * (B1 * R * T - A1) / V ^ 2 + 6 / 5 * A1 * A2 / V ^ 5
1170 TP = TP + C / T ^ 2 / V ^ 2 * ((1 -  EXP ( - B2 / V ^ 2)) / B2 * V ^ 2 + (1
 / 2 + B2 / V ^ 2) *  EXP ( - B2 / V ^ 2))
1180 FV =  EXP (TP / R / T)
1190 FU = FV / P
1200 EN = (B0 * R * T - 2 * A0 - 4 * C0 / T ^ 2) / V
1210 EN = EN + 1 / 2 * (2 * B1 * R * T - 3 * A1) / V ^ 2 + 6 / 5 * A1 * A2 / V ^
 5
1220 EN = EN + (C1 / V ^ 2 / T ^ 2) * (3 * ((1 -  EXP ( - B2 / V ^ 2)) / (B2 / V
 ^ 2)) - (1 / 2 - B2 / V ^ 2) *  EXP ( - B2 / V ^ 2))
1230 BA = B0 - (A0 / R / T) - (C0 / R / T ^ 3)
1240 GA = B1 - (A1 / R / T) + (C1 * B2 / R / T ^ 3)
1250  HOME : VTAB 2
1260  PRINT "FOR THE FOLLOWING DATA:"
```

**Program for equation-of-state variables (continued)** **Table I**

```
1270  PRINT
1280  PRINT "CRITICAL TEMPERATURE, DEG. K    ";TC
1290  PRINT "CRITICAL PRESSURE, ATM          ";PC
1300  IF OP = 2 GOTO 1320
1310  PRINT "ACENTRIC FACTOR, OMEGA          ";OM
1320  PRINT "GAS TEMPERATURE, DEG. C         ";T1
1330  PRINT "GAS PRESSURE, ATM               ";P
1340  PRINT
1350  IF OP = 1 GOTO 1380
1360  PRINT "THE BENEDICT-WEBB-RUBIN EQUATION"
1370  GOTO 1390
1380  PRINT "THE PENG-ROBINSON EQUATION"
1390  PRINT
1400  PRINT "GIVES THE FOLLOWING RESULTS:"
1410  PRINT
1420  PRINT "COMPRESSIBILITY FACTOR, Z, IS   "; FN P(Z)
1430  PRINT "VOLUME, L/G-MOL IS              "; FN P(V)
1440  PRINT "FUGACITY COEFFICIENT IS         "; FN P(FU)
1450 FV = FU * P
1460  PRINT "FUGACITY, ATM IS                "; FN P(FV)
1470  PRINT "ENTHALPY CHANGE, L-ATM/G-MOL IS "; FN P(EN)
1480  PRINT "SECOND VIRIAL COEFFICIENT IS    "; FN P(BA)
1490  PRINT "THIRD VIRIAL COEFFICIENT IS    "; FN P(GA)
1500  PRINT
1510  PRINT  TAB( 13);"END OF PROGRAM"
1520  END
```

**Examples for: Equation-of-state variables** **Table II**

**(Start of first display)**

```
PROGRAM CALCULATES TWO EQUATIONS OF
STATE:

     PENG-ROBINSON, AND
     BENEDICT-WEBB-RUBIN

WITH INPUT OF PRESSURE AND TEMPERATURE;
THE VOLUME, COMPRESSIBILITY FACTOR,
FUGACITY, ENTHALPY CHANGE, AND THE
SECOND AND THIRD VIRIAL COEFFICIENTS
ARE CALCULATED.

DATA REQUIRED ARE THE CRITICAL TEMPERA-
TURE AND PRESSURE AND:

FOR PENG-ROBINSON, THE ACENTRIC FACTOR;

FOR B-W-R, EIGHT EMPIRICAL CONSTANTS
FOR THE PARTICULAR COMPOUND.

PRESS RETURN TO ENTER DATA
```

**(Start of next display)**

```
SELECT EQUATION OF STATE
BY NUMBER, AND THEN ENTER THE REQUIRED
DATA.

1. FOR PENG-ROBINSON
2. FOR BENEDICT-WEBB-RUBIN

MAKE A SELECTION BY NUMBER   1
```

**(Start of next display)**

```
ENTER THE FOLLOWING DATA:

CRITICAL TEMPERATURE, DEG. K     191.04
CRITICAL PRESSURE, ATM           46.06
ACENTRIC FACTOR, OMEGA           .008
GAS TEMPERATURE, DEG. C          37.73
GAS PRESSURE, ATM                6.804
```

**(Start of next display)**

```
FOR THE FOLLOWING DATA:

CRITICAL TEMPERATURE, DEG. K     191.04
CRITICAL PRESSURE, ATM           46.06
ACENTRIC FACTOR, OMEGA           8E-03
GAS TEMPERATURE, DEG. C          37.73
GAS PRESSURE, ATM                6.804

THE PENG-ROBINSON EQUATION

GIVES THE FOLLOWING RESULTS:

COMPRESSIBILITY FACTOR, Z, IS   .9869
VOLUME, L/G-MOL IS              3.7007
FUGACITY COEFFICIENT IS         .9869
FUGACITY, ATM IS                6.7147
ENTHALPY CHANGE, L-ATM/G-MOL IS -1.146
SECOND VIRIAL COEFFICIENT IS    -.0498
THIRD VIRIAL COEFFICIENT IS    4.7E-03

            END OF PROGRAM
```

**Examples for: Equation-of-state variables (continued)** **Table II**

**(Start of first display)**

```
PROGRAM CALCULATES TWO EQUATIONS OF
STATE:

    PENG-ROBINSON, AND
    BENEDICT-WEBB-RUBIN

WITH INPUT OF PRESSURE AND TEMPERATURE;
THE VOLUME, COMPRESSIBILITY FACTOR,
FUGACITY, ENTHALPY CHANGE, AND THE
SECOND AND THIRD VIRIAL COEFFICIENTS
ARE CALCULATED.

DATA REQUIRED ARE THE CRITICAL TEMPERA-
TURE AND PRESSURE AND:

FOR PENG-ROBINSON, THE ACENTRIC FACTOR;

FOR B-W-R, EIGHT EMPIRICAL CONSTANTS
FOR THE PARTICULAR COMPOUND.

PRESS RETURN TO ENTER DATA
```

**(Start of next display)**

```
SELECT EQUATION OF STATE
BY NUMBER, AND THEN ENTER THE REQUIRED
DATA.

1. FOR PENG-ROBINSON
2. FOR BENEDICT-WEBB-RUBIN

MAKE A SELECTION BY NUMBER   2
```

**(Start of next display)**

```
ENTER THE FOLLOWING DATA:

CRITICAL TEMPERATURE, DEG. K     191.04
CRITICAL PRESSURE, ATM           46.06
GAS TEMPERATURE, DEG. C          37.73
GAS PRESSURE, ATM                54.437

SOLUTION OF THE BENEDICT-WEBB-RUBIN
EQUATION NEEDS THE VALUE OF EIGHT
CONSTANTS FOR THE COMPOUND INVOLVED.
ENTER THESE CONSTANTS NOW:

ENTER AO        1.855
ENTER BO        .0426
ENTER CO        22570
ENTER A         .0494
ENTER B         .00338004
ENTER C         2545
ENTER ALPHA     .000124359
ENTER GAMMA     .006
```

**(Start of next display)**

```
FOR THE FOLLOWING DATA:
CRITICAL TEMPERATURE, DEG. K     191.04
CRITICAL PRESSURE, ATM           46.06
GAS TEMPERATURE, DEG. C          37.73
GAS PRESSURE, ATM                54.437

THE BENEDICT-WEBB-RUBIN EQUATION

GIVES THE FOLLOWING RESULTS:

COMPRESSIBILITY FACTOR, Z, IS   .9225
VOLUME, L/G-MOL IS              .4324
FUGACITY COEFFICIENT IS         .9146
FUGACITY, ATM IS                49.7867
ENTHALPY CHANGE, L-ATM/G-MOL IS -7.8096
SECOND VIRIAL COEFFICIENT IS    -.0393
THIRD VIRIAL COEFFICIENT IS     1.5E-03

          END OF PROGRAM
```

**Comparison of calculated and published equation-of-state values for methane at 37.73°C** **Table III**

| Pressure, $P$, atm | Volume, $V$, L/g-mol | | | Fugacity coefficient, $f/P$ | | | Enthalpy change, $(H-H^*)_T$, (L) (atm)/(g-mol) | | |
|---|---|---|---|---|---|---|---|---|---|
| | Ref. [4] | Eq. (8) | Eq. (16) | Ref. [4] | Eq. (11) | Eq. (17) | Ref. [4] | Eq. (12) | Eq. (18) |
| 1 | 25.46 | 25.47 | 25.48 | 0.9985 | 0.9981 | 0.9985 | −0.0743 | −0.1683 | −0.1395 |
| 3.402 | 7.482 | 7.450 | 7.461 | 0.9951 | 0.9934 | 0.9947 | −0.4050 | −0.5729 | −0.4754 |
| 6.804 | 3.709 | 3.701 | 3.711 | 0.9905 | 0.9869 | 0.9895 | −0.8836 | −1.146 | −0.9529 |
| 13.609 | 1.836 | 1.826 | 1.836 | 0.9807 | 0.9741 | 0.9790 | −1.876 | −2.293 | −1.914 |
| 27.218 | 0.9007 | 0.8902 | 0.8994 | 0.9619 | 0.9495 | 0.9577 | −3.830 | −4.584 | −3.859 |
| 40.83 | 0.5891 | 0.5793 | 0.5876 | 0.9438 | 0.9263 | 0.9363 | −5.817 | −6.861 | −5.827 |
| 54.437 | 0.4338 | 0.4247 | 0.4324 | 0.9264 | 0.9044 | 0.9146 | −7.916 | −9.110 | −7.810 |
| 68.046 | 0.3410 | 0.3328 | 0.3397 | 0.9097 | 0.8839 | 0.8927 | −9.867 | −11.315 | −9.794 |
| 102.07 | 0.2191 | 0.2128 | 0.2181 | 0.8714 | 0.8383 | 0.8374 | −14.764 | −16.530 | −14.663 |
| 170.11 | 0.1279 | 0.1238 | 0.1272 | 0.8114 | 0.7704 | 0.7334 | −23.195 | −24.984 | −23.056 |

**Virial coefficients**

| $\beta$, L/g-mol | | | $\gamma$, (L)²/(g-mol)² | | |
|---|---|---|---|---|---|
| Ref. [5] | (Eq. (14) | Eq. (19) | Ref. [5] | Eq. (15) | Eq. (20) |
| −0.04080 | −0.0498 | −0.0393 | 0.002254 | 0.0047 | 0.0015 |

**Physical properties of methane used in calculations**

|  |  | |  |  | |
|---|---|---|---|---|---|
| Mol Wgt | 16.043 | | $C_0$ | 22570. | Constants of Benedict-Webb-Rubin equation |
| $T_c$, K | 191.04 | | $a$ | 0.049400 | |
| $P_c$, atm | 46.06 | | $b$ | 0.00338004 | |
| $\omega$ | 0.008 | | $c$ | 2545. | |
| $A_0$ | 1.85500 | (cont'd) | $\alpha$ | 0.000124359 | |
| $B_0$ | 0.042600 | | $\gamma$ | 0.006 | |

Data sources: Ref. [4], and Reid, R. C., Prausnitz, J. M., and Sherwood, T. K., "Properties of Gases and Liquids," 2nd/3rd ed., McGraw-Hill Inc., New York, 1966/1977.

# Properties of gas mixtures

## Via equations of state, this program computes the volume, compressibility factor and isothermal enthalpy change of binary gaseous mixtures, and fugacites and fugacity coefficients of components.

☐ The previous program calculated the volume, compressibility factor, fugacity and isothermal enthalpy change of pure gases, using the Peng-Robinson [1] and Benedict-Webb-Rubin [2] equations of state. This program does the same, but for two-component gaseous mixtures. The equation constants are now functions of the mixture composition, but they are still calculated from pure-component data.

Table I lists the Apple II program, and Table II shows example displays. The program is self-explanatory for the most part, but let us go through the equations used.

### Volume via Peng-Robinson equation

The Peng-Robinson equation is:

$$P = \frac{RT}{V-b} - \frac{a}{V(V+b)+b(V-b)} \qquad (1)$$

Here, $a$ as a function of $T$, and $b$ as a function of $T_c$, are defined, respectively, as:

$$a(T) = 0.45724(R^2T_c^2/P_c)\alpha \qquad (2)$$

$$b(T_c) = 0.07780(RT_c/P_c) \qquad (3)$$

Eq. (4) and (4a) define $\alpha$ in terms of $T_r$ and $\omega$:

$$\alpha^{1/2} = 1 + \kappa(1 - T_r^{1/2}) \qquad (4)$$

$$\kappa = 0.37565 + 1.54226\omega - 0.26992\omega^2 \qquad (4a)$$

Eq. (5) and (6) make the handling of the constants $a$ and $b$ easier:

$$A = aP/R^2T^2 \qquad (5)$$

$$B = bP/RT \qquad (6)$$

The mixing rules for $a$ and $b$, the constants of the Peng-Robinson equation of state, are defined in Eq. (7) and (8) in terms of the mole fraction of $x$, any component:

$$a = \sum_i \sum_j x_i x_j a_{ij} \qquad (7)$$

$$b = \sum_i x_i b_i \qquad (8)$$

In Eq. (7):

$$a_{ij} = (1 - \delta_{ij})a_i^{1/2}a_j^{1/2} \qquad (9)$$

In Eq. (9), $\delta_{ij}$ is a binary interaction coefficient, which is determined empirically. It is the only term in the foregoing equations that is not evaluated from pure-substance data.

The Peng-Robinson equation is choice "1" on the program menu. Required data are: mole fraction, each component's critical temperature and pressure, and acentric factor, and the temperature and pressure of interest. Volume is one of the outputs, as is the related compressibility factor.

Table III shows compressibility factors calculated by this program (Eq. (1)) for an equimolar mixture of methane and ethane at 37.73°C, over a pressure range of 1 to 68.045 atm. Table III also shows the experimental data of Sage and Lacey [3], and the properties of methane and ethane used in the calculations.

### Fugacity and enthalpy

The fugacity of Component 1 in a binary mixture is calculated by the relationship:

$$\ln\frac{\hat{f}_1}{y_1P} = \frac{b_1}{b}(Z-1) - \ln(Z-B) - \frac{A}{2(2)^{1/2}B} \times \left(\frac{2\Sigma y_1 a_{1,2}}{a} - \frac{b_1}{b}\right)\ln\left(\frac{Z+2.414B}{Z-0.414B}\right) \qquad (10)$$

The isothermal change of enthalpy with pressure, using the Peng-Robinson expression, is calculated by:

$$(H-H^*)_T = RT(Z-1) + \left[\frac{T(da/dT)-a}{2(2)^{1/2}B}\right]\ln\left(\frac{Z+2.414B}{Z-0.414B}\right) \qquad (11)$$

Table III lists fugacity coefficients for ethane and methane in an equimolar mixture, as calculated by the Peng-Robinson equation (Eq. (10)), and compares them with published results. It also lists calculated values (Eq. (11)) for isothermal enthalpy change (i.e., actual enthalpy vs. ideal-gas enthalpy) for a mixture of 47.7% methane and 52.3% ethane at 38.61°C, and compares these with the published data of Powers et al. [4].

### Benedict-Webb-Rubin equation

The Benedict-Webb-Rubin equation of state [2] calculates the compressibility factor for a binary mixture the same way as it does a pure-component compressibility:

Adapted from an article by James H. Weber, University of Nebraska, originally published May 19, 1980.

$$Z = 1 + [B_o - (A_o/RT) - (C_o/RT^3)](V^{-1}) + [b - (a/RT)](V^{-2}) + (a\alpha/RT)(V^{-5}) + (c/RT^3)[(1 + \gamma V^{-2})(V^{-2})]e^{-\gamma V^{-2}} \quad (12)$$

In the case of a mixture, the values for the eight constants in Eq. (12) are determined via the following equations:

$$A_{om} = \left(\sum_i y_i A_{oi}^{1/2}\right)^2 \quad (13)$$

$$B_{om} = \frac{1}{4}\left(\sum_i y_i B_{oi}\right) + \frac{3}{4}\left(\sum_i y_i B_{oi}^{1/3}\right)\left(\sum_i y_i B_{oi}^{2/3}\right) \quad (14)$$

$$C_{om} = \left(\sum_i y_i C_{oi}^{1/2}\right)^2 \quad (15)$$

$$a_m = \left(\sum_i y_i a_i^{1/3}\right)^3 \quad (16)$$

$$b_m = \left(\sum_i y_i b_i^{1/3}\right)^3 \quad (17)$$

$$c_m = \left(\sum_i y_i c_i^{1/3}\right)^3 \quad (18)$$

$$\alpha_m = \left(\sum_i y_i \alpha_i^{1/3}\right)^3 \quad (19)$$

$$\gamma_m = \left(\sum_i y_i \gamma_i^{1/2}\right)^2 \quad (20)$$

Only pure-substance data and component compositions are used in the foregoing equations.

The Benedict-Webb-Rubin equation is choice "2" on the program menu. Required data are, for each component: critical temperature and pressure, and the eight constants for Eq. (13)–(20). The system temperature and pressure are also required. Table III compares the program's calculated compressibility factors (Eq. (12)) with published ones for an equimolar mixture of methane and ethane at 37.73°C. It also lists the constants for these components.

The effect of pressure on enthalpy under isothermal conditions for a mixture of constant composition is determined via the Benedict-Webb-Rubin equation:

$$(H - H^*)_T = [B_o RT - 2A_o - (4C_o/T^2)]/V + 1/2[(2bRT - 3a)/V^2] + 6/5(a\alpha/V^5) + (c/V^2T^2) \times \{3[(1 - e^{-\gamma/V^2})/(\gamma/V^2)] - (1/2 - \gamma/V^2)e^{-\gamma/V^2}\} \quad (21)$$

The Benedict-Webb-Rubin expression for calculating the fugacity of a component in a mixture is:

$$RT \ln \hat{f}_i = RT \ln (RTy_i/V) + [(B_{om} + B_{oi})RT - 2(A_{om}A_{oi})^{1/2} - 2(C_{om}C_{oi})^{1/2}T^{-2}]/V + (3/2)[RT(b_m^2 b_i)^{1/3} - (a_m^2 a_i)^{1/3}]/V^2 + (3/5)[a_m(\alpha_m^2 \alpha_i)^{1/3} + \alpha_m(a_m^2 a_i)^{1/3}]/V^5 + \{(3V^{-2})[(c_m^2 c_i)^{1/3}/T^2] [(1 - e^{-\gamma_m V^{-2}})/(\gamma_m V^{-2}) - [(e^{-\gamma_m V^{-2}})/2]\} - \{[(2V^{-2})(c_m T^{-2})(\gamma_i/\gamma_m)^{1/2}] [(1 - e^{-\gamma_m V^{-2}})/(\gamma_m V^{-2})] - [(e^{-\gamma_m V^{-2}})(1 + (\gamma_m V^{-2}/2))]\} \quad (22)$$

Table III lists fugacity coefficients for ethane and methane in an equimolar mixture, as calculated by the Benedict-Webb-Rubin equation (Eq. (22)), and compares them with published results. It also lists calculated values (Eq. (21)) for isothermal enthalpy change (i.e., actual enthalpy vs. ideal-gas enthalpy) for a mixture of 47.7% methane and 52.3% ethane at 38.61°C, and compares these with the published data of Powers et al. [4].

## Nomenclature

| | |
|---|---|
| $A$, $B$, $a$, $b$ | Constants of Peng-Robinson equation |
| $A_o$, $B_o$, $C_o$, $a$, $b$, $c$, $\alpha$, $\gamma$ | Constants of Benedict-Webb-Rubin equation |
| $H$ | Enthalpy, (L)(atm)/g-mol |
| $P$ | Pressure, atm |
| $R$ | Gas law constant, 0.08207(L)(atm)/(g-mol)(K) |
| $T$ | Temperature, K |
| $V$ | Volume, L/g-mol |
| $Z$ | Compressibility factor |
| $t$ | Temperature, °C |
| $\alpha$ | Defined by Eq. (4) |
| $\delta$ | Interaction coefficient, Eq. (9) |
| $\kappa$ | Defined by Eq. (4a) |
| $\phi$ | Fugacity coefficient, $f/yP$ |
| $\omega$ | Pitzer's acentric factor |

**Subscripts**

| | |
|---|---|
| $c$ | Critical |
| $i$ | Component $i$ in a mixture |
| $j$ | Component $j$ in a mixture |
| $m$ | Mixture |
| $r$ | Reduced |
| 1,2 | Numbers of components in a mixture |

**Superscripts**

| | |
|---|---|
| * | Property in ideal gas state |
| ^ | Property in a mixture |

## References


1. Peng, D.-Y., and Robinson, D. B., *Ind. Eng. Chem. Fundamentals,* Vol. 15, 1976, p. 59.
2. Benedict, M., Webb, G. B., and Rubin, L. C., *J. Chem. Phys.,* Vol. 8, 1940, p. 334.
3. Sage, B. H., and Lacey, W. N., "Thermodynamic Properties of the Lighter Paraffin Hydrocarbons and Nitrogen," Amer. Petrol. Inst., New York, N.Y., 1950.
4. Powers, J. E., Furtado, A. W., Kant, R., and Kwan, A., *J. Chem. Eng. Data,* Vol. 24, 1979, p. 46.

**Program for predicting properties of gas mixtures** **Table I**

```
10  REM   PROPERTIES OF GAS MIXTURES
20  REM   FROM CHEMICAL ENGINEERING, MAY 19, 1980
30  REM   BY JAMES H. WEBER
40  REM  TRANSLATED BY WILLIAM VOLK
50  REM   COPYRIGHT (C) 1984
60  REM  BY CHEMICAL ENGINEERING
70  HOME : VTAB 4
80  PRINT "PROGRAM DETERMINES THE PROPERTIES OF"
90  PRINT "BINARY GAS MIXTURES FROM TWO EQUATIONS"
100  PRINT "OF STATE:"
110  PRINT
120  PRINT "THE PENG-ROBINSON EQUATION, AND"
130  PRINT "THE BENEDICT-WEBB-RUBIN EQUATION."
140  PRINT
150  PRINT "CRITICAL TEMPERATURE AND PRESSURE FOR"
160  PRINT "THE TWO GASES ARE REQUIRED, AND FOR"
170  PRINT "THE B-W-R EQUATION, THE EQUATION"
180  PRINT "CONSTANTS FOR BOTH COMPONENTS."
190  PRINT
200  INPUT "PRESS RETURN TO ENTER DATA.     ";Q$
210  HOME : VTAB 4
220  REM  GAS CONSTANT AND SET DISPLAY TO FOUR DECIMALS
230 R = .08207
240  DEF  FN P(X) =  INT (1E4 * (X + .00005)) / 1E4
250  PRINT "SELECT THE EQUATION TO BE USED BY"
260  PRINT "NUMBER:"
270  PRINT
280  PRINT "1.  PENG-ROBINSON EQUATION."
290  PRINT "2.  BENEDICT-WEBB-RUBIN EQUATION."
300  PRINT
310  INPUT "MAKE YOUR SELECTION NOW:        ";OP
320  PRINT
330  PRINT "MOL FRACTION OF"
340  INPUT "  FIRST COMPONENT:  ";Y(1)
350 Y(2) = 1 - Y(1)
360  PRINT
370  PRINT "ENTER THE FOLLOWING DATA FOR BOTH"
380  PRINT "COMPONENTS.  SEPARATE THE VALUES"
390  PRINT "WITH A COMMA."
400  PRINT
410  PRINT "CRITICAL TEMPERATURE"
420  INPUT "  DEG. K          ";TC(1),TC(2)
430  PRINT "CRITICAL PRESSURE,"
440  INPUT "  ATM             ";PC(1),PC(2)
450  IF OP = 2 GOTO 520
460  PRINT "ACENTRIC FACTOR,"
470  INPUT "  OMEGA           ";OM(1),OM(2)
480  FOR I = 1 TO 2
490 KA(I) = .37565 + 1.54226 * OM(I) - .26992 * OM(I) ^ 2
500  NEXT I
510  IF OP = 1 GOTO 860
520  PRINT
530  PRINT "FOR THE BENEDICT-WEBB-RUBIN EQUATION"
540  PRINT "EIGHT EMPIRICAL CONSTANTS FOR EACH"
550  PRINT "COMPONENT ARE REQUIRED."
560  PRINT "ENTER THESE NOW, SEPARATED BY A COMMA."
570  PRINT
580  INPUT "A0            ";A0(1),A0(2)
590  INPUT "B0            ";B0(1),B0(2)
600  INPUT "C0            ";C0(1),C0(2)
610  INPUT "A             ";A(1),A(2)
620  INPUT "B             ";B(1),B(2)
630  INPUT "C             ";C(1),C(2)
640  INPUT "ALPHA         ";AP(1),AP(2)
650  INPUT "GAMMA         ";GM(1),GM(2)
660  FOR I = 1 TO 2
670 A0 = A0 + Y(I) *  SQR (A0(I))
680 B0 = B0 + Y(I) * B0(I)
690 BX = BX + Y(I) * B0(I) ^ (1 / 3)
700 BY = BY + Y(I) * B0(I) ^ (2 / 3)
710 C0 = C0 + Y(I) *  SQR (C0(I))
720 A1 = A1 + Y(I) * A(I) ^ (1 / 3)
730 B1 = B1 + Y(I) * B(I) ^ (1 / 3)
740 C1 = C1 + Y(I) * C(I) ^ (1 / 3)
750 AP = AP + Y(I) * AP(I) ^ (1 / 3)
760 GM = GM + Y(I) *  SQR (GM(I))
770  NEXT I
780 A0 = A0 ^ 2
790 B0 = B0 / 4 + (BX * BY) * 3 / 4
800 C0 = C0 ^ 2
810 A1 = A1 ^ 3
820 B1 = B1 ^ 3
830 C1 = C1 ^ 3
840 AP = AP ^ 3
850 GM = GM ^ 2
860  PRINT
870  PRINT "ENTER GAS TEMPERATURE"
880  INPUT "  DEG. C             ";TI
890 T = TI + 273.16
900  PRINT "ENTER GAS PRESSURE,"
910  INPUT "  ATM                ";P
920  IF OP = 2 GOTO 1400
930  REM  PENG-ROBINSON CALCULATION
940  FOR I = 1 TO 2
950 TR(I) = T / TC(I)
960 AP(I) = (1 + KA(I) * (1 -  SQR (TR(I)))) ^ 2
970 AQ(I) = .45724 * AP(I) * (TC(I) * R) ^ 2 / PC(I)
980 B(I) = .0778 * R * TC(I) / PC(I)
990  NEXT I
1000 AJ = (1 - DE) *  SQR (AQ(1) * AQ(2))
1010 A = 2 * AJ * Y(1) * Y(2) + Y(1) ^ 2 * AQ(1) + Y(2)
     ^ 2 * AQ(2)
1020 B = B(2) * Y(2) + B(1) * Y(1)
1030 AU = A * P / (R * T) ^ 2
1040 BU = B * P / R / T
1050 Z = 1
1060  GOSUB 1130
1070 F0 = FZ
1080 Z = .8
1090 F1 = 1
1100  GOSUB 1130
1110  GOTO 1100
```

**Program for predicting properties of gas mixtures (continued)** **Table I**

```
1120  REM  FUGACITY CALCULATION
1130 FZ = Z ^ 3 - (1 - BU) * Z ^ 2 + (AU - 3 * BU ^ 2 - 2 * BU) * Z - (AU * BU -
 BU ^ 2 - BU ^ 3)
1140  IF  ABS (FZ) > 1E - 6 GOTO 1160
1150  GOTO 1190
1160  IF F1 = 0 GOTO 1180
1170 Z = Z - (1 - Z) / (F0 - FZ) * FZ
1180  RETURN
1190 V = Z * R * T / P
1200  FOR I = 1 TO 2
1210 F(I) = B(I) / B * (Z - 1)
1220 F(I) = F(I) -  LOG (Z - BU)
1230 A9 = (AU / (2 ^ (3 / 2)) / BU)
1240 A8 = (2 * (Y(I) * AQ(I) + (1 - Y(1)) * AJ)) / A - B(I) / B
1250 A7 =  LOG ((Z + 2.414 * BU) / (Z - .414 * BU))
1260 F(I) = F(I) - A9 * A8 * A7
1270 F(I) =  EXP (F(I))
1280  NEXT I
1290  REM  ENTHALPY CALCULATION
1300 EN = R * T * (Z - 1)
1310 TP = KA(1) ^ 2 / TC(1) - KA(1) ^ 2 /  SQR (TC(1) * T) - KA(1) /  SQR (TC(1)
 * T)
1320 TP = TP * AQ(1) / AP(1) * Y(1) ^ 2
1330 TQ = (1 - DE) * Y(1) * Y(2) *  SQR (AQ(1) * AQ(2) / AP(1) / AP(2))
1340 TQ = TQ * (2 * KA(1) * KA(2) /  SQR (TC(1) * TC(2)) - (KA(1) * KA(2) + KA(1
 )) /  SQR (TC(1) * T) - (KA(1) * KA(2) + KA(2)) /  SQR (TC(2) * T))
1350 TP = TP + TQ + (KA(2) ^ 2 / TC(2) - KA(2) ^ 2 /  SQR (TC(2) * T) - KA(2) /
 SQR (TC(2) * T)) * Y(2) ^ 2 * AQ(2) / AP(2)
1360 TP = (TP * T - A) / (2 ^ (3 / 2) * B)
1370 EN = EN + TP *  LOG ((Z + 2.414 * BU) / (Z - .414 * BU))
1380  GOTO 1720
1390  REM  BENEDICT-WEBB-RUBIN CALCULATION
1400 Z0 = 1
1410 ZC = Z0
1420  GOSUB 1490
1430 Z1 = Z
1440 ZC = Z
1450  GOSUB 1490
1460  IF  ABS (ZC - Z) < .00005 GOTO 1560
1470 ZC = Z0 + (Z0 - Z1) * (Z0 - ZC) / (ZC - Z - Z0 + Z1)
1480  GOTO 1450
1490 V = ZC * R * T / P
1500 Z = 1 + (B0 - A0 / R / T - (C0 / R / T ^ 3)) / V
1510 Z = Z + (B1 - (A1 / R / T)) / V ^ 2
1520 Z = Z + (A1 * AP / R / T / V ^ 5)
1530 Z = Z + (C1 / R / T ^ 3) * (1 + GM / V ^ 2) / V ^ 2 *  EXP ( - GM / V ^ 2)
1540  RETURN
1550  REM  ENTHALPY CALCULATION
1560 EN = (B0 * R * T - 2 * A0 - 4 * C0 / T ^ 2) / V
1570 EN = EN + (2 * B1 * R * T - 3 * A1) / V ^ 2 / 2
1580 EN = EN + 6 / 5 * (A1 * AP / V ^ 5)
1590 EN = EN + (C1 / (V * T) ^ 2) * (3 * ((1 -  EXP ( - GM / V ^ 2)) / (GM / V ^
 2)) - (1 / 2 - GM / V ^ 2) *  EXP ( - GM / V ^ 2))
1600  REM  FUGACITY CALCULATION
1610 GD = GM / V ^ 2
1620 EX =  EXP ( - GD)
1630  FOR I = 1 TO 2
```

**Program for predicting properties of gas mixtures (continued)** **Table I**

```
1640 F(I) = R * T *  LOG (R * T * Y(I) / V)
1650 F(I) = F(I) + ((B0 + B0(I)) * R * T - 2 *  SQR (A0 * A0(I)) - 2 *  SQR (C0
* C0(I)) / T ^ 2) / V
1660 F(I) = F(I) + 3 / 2 * (R * T * (B1 ^ 2 * B(I)) ^ (1 / 3) - (A1 ^ 2 * A(I))
^ (1 / 3)) / V ^ 2
1670 F(I) = F(I) + 3 / 5 * (A1 * (AP ^ 2 * AP(I)) ^ (1 / 3) + AP * (A1 ^ 2 * A(I
)) ^ (1 / 3)) / V ^ 5
1680 F(I) = F(I) + 3 / V ^ 2 * (C1 ^ 2 * C(I)) ^ (1 / 3) / T ^ 2 * ((1 - EX) / G
D - EX / 2)
1690 F(I) = F(I) - (2 / V ^ 2 * C / T ^ 2 *  SQR (GM(I) / GM) * ((1 - EX) / GD -
 EX - GD * EX / 2))
1700 F(I) =  EXP (F(I) / R / T) / Y(I) / P
1710  NEXT I
1720  HOME : VTAB 4
1730  PRINT "THE MIXTURE VALUES, CALCULATED BY"
1740  PRINT
1750  IF OP = 2 GOTO 1780
1760  PRINT "THE PENG-ROBINSON EQUATION ARE:"
1770  GOTO 1790
1780  PRINT "THE BENEDICT-WEBB-RUBIN EQUATION ARE:"
1790  PRINT
1800  PRINT "COMPRESSIBILITY FACTOR, Z       "; FN P(Z)
1810  PRINT "GAS VOLUME, L/G-MOL             "; FN P(V)
1820  PRINT "FUGACITY COEFFICIENT:"
1830  PRINT "       FIRST COMPONENT          "; FN P(F(1))
1840  PRINT "       SECOND COMPONENT:        "; FN P(F(2))
1850  PRINT "ENTHALPY CHANGE IS              "; FN P(EN)
1860  PRINT
1870  PRINT  TAB( 13);"END OF PROGRAM"
1880  END
```

**Examples for: Properties of gas mixtures** **Table II**

**(Start of first display)**

```
PROGRAM DETERMINES THE PROPERTIES OF
BINARY GAS MIXTURES FROM TWO EQUATIONS
OF STATE:

THE PENG-ROBINSON EQUATION, AND
THE BENEDICT-WEBB-RUBIN EQUATION.

CRITICAL TEMPERATURE AND PRESSURE FOR
THE TWO GASES ARE REQUIRED, AND FOR
THE B-W-R EQUATION, THE EQUATION
CONSTANTS FOR BOTH COMPONENTS.

PRESS RETURN TO ENTER DATA.
```

**(Start of next display)**

```
SELECT THE EQUATION TO BE USED BY
NUMBER:

1.  PENG-ROBINSON EQUATION.
2.  BENEDICT-WEBB-RUBIN EQUATION.

MAKE YOUR SELECTION NOW:        1

MOL FRACTION OF
  FIRST COMPONENT:  .5

ENTER THE FOLLOWING DATA FOR BOTH
COMPONENTS.  SEPARATE THE VALUES
WITH A COMMA.

CRITICAL TEMPERATURE
  DEG. K            191.04,305.4
CRITICAL PRESSURE,
  ATM               46.06,48.2
ACENTRIC FACTOR,
  OMEGA             .008,.098

ENTER GAS TEMPERATURE
  DEG. C            37.73
ENTER GAS PRESSURE,
  ATM               40.827
```

**Examples for: Properties of gas mixtures (continued)** **Table II**

**(Start of next display)**

```
THE MIXTURE VALUES, CALCULATED BY

THE PENG-ROBINSON EQUATION ARE:

COMPRESSIBILITY FACTOR, Z        .8219
GAS VOLUME, L/G-MOL              .5136
FUGACITY COEFFICIENT:
       FIRST COMPONENT           .9486
       SECOND COMPONENT:         .7396
ENTHALPY CHANGE IS               -14.8616

          END OF PROGRAM
```

**(Start of first display)**

```
PROGRAM DETERMINES THE PROPERTIES OF
BINARY GAS MIXTURES FROM TWO EQUATIONS
OF STATE:

THE PENG-ROBINSON EQUATION, AND
THE BENEDICT-WEBB-RUBIN EQUATION.

CRITICAL TEMPERATURE AND PRESSURE FOR
THE TWO GASES ARE REQUIRED, AND FOR
THE B-W-R EQUATION, THE EQUATION
CONSTANTS FOR BOTH COMPONENTS.

PRESS RETURN TO ENTER DATA.
```

**(Start of next display)**

```
SELECT THE EQUATION TO BE USED BY
NUMBER:

1.  PENG-ROBINSON EQUATION.
2.  BENEDICT-WEBB-RUBIN EQUATION.

MAKE YOUR SELECTION NOW:        2

MOL FRACTION OF
  FIRST COMPONENT:  .477

ENTER THE FOLLOWING DATA FOR BOTH
COMPONENTS.  SEPARATE THE VALUES
WITH A COMMA.

CRITICAL TEMPERATURE
  DEG. K              191.04,305.4
CRITICAL PRESSURE,
  ATM                 46.06,48.2
```

**(Start of next display)**

```
FOR THE BENEDICT-WEBB-RUBIN EQUATION
EIGHT EMPIRICAL CONSTANTS FOR EACH
COMPONENT ARE REQUIRED.
ENTER THESE NOW, SEPARATED BY A COMMA.

A0          1.855,4.15556
B0          .0426,.0627724
C0          22570,179592
A           .0494,.34516
B           .00338004,.011122
C           2545,32767
ALPHA       .000124359,.000243389
GAMMA       .006,.0118

ENTER GAS TEMPERATURE
  DEG. C              38.61
ENTER GAS PRESSURE,
  ATM                 27.218
```

**(Start of next display)**

```
THE MIXTURE VALUES, CALCULATED BY


THE BENEDICT-WEBB-RUBIN EQUATION ARE:

COMPRESSIBILITY FACTOR, Z        .8924
GAS VOLUME, L/G-MOL              .8389
FUGACITY COEFFICIENT:
       FIRST COMPONENT           .9766
       SECOND COMPONENT:         .8362
ENTHALPY CHANGE IS               -9.2639

          END OF PROGRAM
```

**Comparison of calculated and published values** **Table III**

**Compressibility factors and component fugacity coefficients for an equimolal mixture of methane and ethane at 37.73°C**

| Pressure, | Compressibility factor, $Z$ | | | Component 1 coefficient, $\phi_1$ | | | Component 2 coefficient, $\phi_2$ | | |
|---|---|---|---|---|---|---|---|---|---|
| $P$, atm | Ref. [3] | Eq. (1) | Eq. (12) | Ref. [3] | Eq. (10) | Eq. (22) | Ref. [3] | Eq. (10) | Eq. (22) |
| 13.609 | 0.9475 | 0.9410 | 0.9486 | 0.9714 | 0.9802 | 0.9864 | 0.9297 | 0.9070 | 0.9161 |
| 27.218 | 0.8930 | 0.8815 | 0.8952 | 0.9484 | 0.9630 | 0.9748 | 0.8574 | 0.8203 | 0.8360 |
| 40.827 | 0.8350 | 0.8219 | 0.8397 | 0.9301 | 0.9486 | 0.9656 | 0.7844 | 0.7396 | 0.7595 |
| 54.436 | 0.7780 | 0.7632 | 0.7827 | 0.9165 | 0.9373 | 0.9594 | 0.7129 | 0.6648 | 0.6867 |
| 68.045 | 0.7230 | 0.7073 | 0.7259 | 0.9072 | 0.9291 | 0.9565 | 0.6446 | 0.5961 | 0.6182 |

**Isothermal enthalpy change for a 47.7% methane mixture with ethane at 38.61°C**

| Pressure, | Enthalpy change, $(H-H^*)_T$ | | |
|---|---|---|---|
| $P$, atm | Ref. [4] | Eq. (11) | Eq. (21) |
| 13.609 | −4.422 | −4.745 | −4.454 |
| 27.218 | −9.192 | −9.816 | −9.264 |
| 40.827 | −14.414 | −15.259 | −14.497 |
| 54.436 | −20.134 | −21.102 | −20.218 |
| 68.045 | −26.337 | −27.315 | −26.433 |

Data source: Reid, R. C., Prausnitz, J. M., and Sherwood, T. K., "Properties of Gases and Liquids," 2nd/3rd ed., McGraw-Hill, New York, 1966/1977.

**Physical properties used in calculations**

| | | Methane | Ethane |
|---|---|---|---|
| Mol wt | | 16.043 | 30.070 |
| $T_c$ | | 191.04 | 305.4 |
| $P_c$ | | 46.06 | 48.2 |
| $\omega$ | | 0.008 | 0.098 |
| $A_o$ | Constants of Benedict-Webb-Rubin equation | 1.85500 | 4.15556 |
| $B_o$ | | 0.042600 | 0.0627724 |
| $C_o$ | | 22,570 | 179,592 |
| $a$ | | 0.049400 | 0.345160 |
| $b$ | | 0.00338004 | 0.0111220 |
| $c$ | | 2,545 | 32,767 |
| $\alpha$ | | 0.000124359 | 0.000243389 |
| $\gamma$ | | 0.006 | 0.011800 |

# Thermodynamic properties of pure gases and binary mixtures

This program calculates enthalpy and entropy changes when the initial and final temperatures are given; or final temperatures when the other thermal variables are known.

☐ Isobaric heat-capacity expressions for pure gases usually take one of two forms:

$$C_p = a + bT + cT^2 + dT^3 \tag{1}$$

$$C_p = a + bT + (c/T^2) \tag{2}$$

By means of these expressions, $\Delta H$ (or $Q$) and $\Delta S$ can be calculated.

If mixing is ideal, the heat capacities of binary gaseous mixtures may be determined from the heat capacities of the pure components by simply combining the coefficients, i.e., $a_m = y_1a_1 + y_2a_2 + \cdots$. Hence, the program presented for pure gases can also be applied to mixtures if composition is taken into account.

Table I lists the Apple II program, which is menu-driven and mostly self-explanatory. Tables II and IV show example displays. Now, let us go through the features and equations.

## Enthalpy, temperature, entropy

To calculate $\Delta H$ (or $Q$) via Eq. (1), the required expression is:

$$Q = \Delta H = \int_{T_1}^{T_2} (a + bT + cT^2 + dT^3)dT \tag{3}$$

$$= a(T_2 - T_1) + \frac{b}{2}(T_2^2 - T_1^2) + \frac{c}{3}(T_2^3 - T_1^3) + \frac{d}{4}(T_2^4 - T_1^4) \tag{4}$$

Eq. (1) is choice 1 on the program menu. The required data are: mole fraction of the first component (= 1 if only one component), and the four constants (*a–d*) for each component. Then there is a choice of what to solve the equation for: enthalpy change ($\Delta H$), which requires two temperatures; or final temperature, which requires one temperature and the enthalpy change. Note that temperature is entered in °C; the program converts to K, as Eq. (1) and (2) require.

Solving for temperature, given enthalpy change, is a trial-and-error procedure. This program uses Newton's convergence technique. The first estimate of the second temperature is $T$ + 100 K. The acceptable tolerance between estimated temperature and calculated temperature is set at 0.01 in line 1300 of the program.

The final output includes heat capacity for both components, calculated by Eq. (1), and the entropy change. For heat capacity as expressed in Eq. (1), the entropy change in going from one temperature to another is expressed as:

$$\Delta S = a \ln\left(\frac{T_2}{T_1}\right) + b(T_2 - T_1) + \frac{c}{2}(T_2^2 - T_1^2) + \frac{d}{3}(T_2^3 - T_1^3) \tag{5}$$

Table III shows calculated values of enthalpy change, final temperature and entropy change for nitrogen going from 50°C to 100°C, 150°C and 200°C. The values obtained via the Eq. (1) form of heat capacity are denoted as such. The footnote indicates the constants for nitrogen; these are from Reid, Prausnitz and Sherwood [1].

## Properties via Eq. (2)

Eq. (2) can be used for the same type of calculations as described. For the enthalpy change, given the initial and final temperatures:

$$\Delta H = a(T_2 - T_1) + b(T_2^2 - T_1^2) - c\left(\frac{1}{T_2} - \frac{1}{T_1}\right) \tag{6}$$

Eq. (2) is choice "2" on the program menu. The required data are the same as for Eq. (1), except that there are only three constants (*a–c*). There is again a choice of

Adapted from an article by James H. Weber, University of Nebraska, originally published September 22, 1980.

what to solve the equation for: enthalpy change ($\Delta H$), which requires two temperatures; or final temperature, which requires one temperature and the enthalpy change. In solving for final temperature, the program again uses Newton's trial-and-error method, with a tolerated error of 0.01 (program line 1300). Note that temperature input and output are in °C.

Table III shows calculated values of enthalpy change and final temperature for nitrogen going from 50°C to 100°C, 150°C and 200°C. The footnote indicates the Eq. (2) constants for nitrogen; these are from Kelly [2].

The results include the heat capacity, from Eq. 2, and the entropy difference between the two temperatures:

$$\Delta S = a \ln\left(\frac{T_2}{T_1}\right) + b(T_2 - T_1) - \frac{c}{2}\left(\frac{1}{T_2^2} - \frac{1}{T_1^2}\right) \quad (7)$$

## Nomenclature

$a$, $b$, $c$, $d$ Constants of Eq. (1)

$a$, $b$, $c$ Constants of Eq. (2)

$H$ Enthalpy, cal/g-mol

$Q$ Heat, cal/g-mol

$S$ Entropy, cal/(g-mol)(K)

$T$ Temperature, K

$t$ Temperature, °C

**Subscripts**

$m$ Mixture

1 Component 1 in mixture

2 Component 2 in mixture

## References

1. Reid, R. C., Prausnitz, J. M., and Sherwood, T. K., "The Properties of Gases and Liquids," 3rd ed., McGraw-Hill, New York, 1977.
2. Kelly, K. K., U.S. Bur. Mines Bull. 584, 1960.

**Program predicts thermodynamic properties of pure gases and binary mixtures** **Table I**

```
10 REM   THERMODYNAMIC PROPERTIES OF PURE GASES AND
   BINARY MIXTURES
20 REM   FROM CHEMICAL ENGINEERING, SEPTEMBER 22, 1980
30 REM   BY JAMES H. WEBER
40 REM   TRANSLATED BY WILLIAM VOLK
50 REM   COPYRIGHT (C) 1984
60 REM  BY CHEMICAL ENGINEERING
70 HOME : VTAB 2
80 PRINT "PROGRAM CALCULATES THE ENTHALPY CHANGE"
90 PRINT "AND THE ENTROPY CHANGE FOR PURE GASES"
100 PRINT "OR BINARY GAS MIXTURES WHEN THE HEAT"
110 PRINT "CAPACITY IS CALCULATED FROM ONE OF THE"
120 PRINT "FOLLOWING EMPIRICAL EQUATIONS"
130 PRINT
140 PRINT "1: CP = A + BT + CT^2 + DT^3
150 PRINT
160 PRINT "2: CP = A + BT + C/T^2
170 PRINT
180 PRINT "YOU SELECT THE EQUATION TO BE USED,"
190 PRINT "(NOTE: YOU WILL NEED THE CONSTANTS FOR"
200 PRINT "THE GAS COMPONENTS), THEN YOU ENTER"
210 PRINT "THE MOL FRACTION AND THE CONSTANTS FOR"
220 PRINT "THE COMPONENTS."
230 PRINT
240 PRINT "IF YOU HAVE THE TWO TEMPERATURES, THE"
250 PRINT "PROGRAM WILL CALCULATE THE ENTHALPY"
260 PRINT "CHANGE.  IF YOU HAVE THE ENTHALPY"
270 PRINT "CHANGE, THE PROGRAM WILL CALCULATE"
280 PRINT "THE SECOND TEMPERATURE."
290 PRINT
300 INPUT "PRESS RETURN TO CONTINUE.";Q$
310 HOME : VTAB 4
320 DEF  FN P(X) =  INT (1E4 * (X + .00005)) / 1E4
330 P(X) = 7
340 PRINT "SELECT THE HEAT CAPACITY EQUATION FORM"
350 PRINT "BY NUMBER:"
360 PRINT
370 PRINT "1. CP = A + B*T + C*T^2 + D*T^3
380 PRINT
390 PRINT "2. CP = A + B*T + C/T^2"
400 PRINT
410 INPUT "MAKE YOUR SELECTION NOW  ";OP
420 IF OP = 1 OR OP = 2 GOTO 470
430 PRINT
440 PRINT "YOUR OPTIONS WERE '1' OR '2'."
450 PRINT "YOU ENTERED "OP". TRY AGAIN."
460 GOTO 400
470 PRINT
480 PRINT "ENTER THE MOL FRACTION OF THE FIRST"
490 PRINT "COMPONENT.  IF THE CALCULATION IS FOR "
500 PRINT "A PURE GAS, ENTER 1, OTHERWISE ENTER"
510 INPUT "A DECIMAL FRACTION. ";Y1
520 IF Y1 > 1 OR Y1 =  < 0 GOTO 540
530 GOTO 580
540 HOME : VTAB 4
550 PRINT "YOU SHOULD ENTER 1 OR A DECIMAL "
560 PRINT "FRACTION. YOU ENTERED ";Y1;".  TRY AGAIN."
570 GOTO 470
580 PRINT
590 PRINT "ENTER THE CONSTANTS FOR"
600 PRINT "THE HEAT CAPACITY"
610 IF Y1 < 1 GOTO 640
620 PRINT "CALCULATION:"
630 GOTO 650
640 PRINT "CALCULATION FOR THE FIRST COMPONENT:"
650 Y(1) = Y1
660 Y(2) = 1 - Y1
670 I = 1
680 INPUT " A                    ";A(I)
690 INPUT " B                    ";B(I)
700 INPUT " C                    ";C(I)
710 IF OP = 2 GOTO 730
720 INPUT " D                    ";D(I)
730 IF I = 2 GOTO 800
740 IF Y1 = 1 GOTO 800
750 I = 2
760 PRINT
770 PRINT "ENTER THE VALUES FOR THE SECOND"
```

**Program predicts thermodynamic properties of pure gases and binary mixtures (continued)** **Table I**

```
780  PRINT "COMPONENT:"
790  GOTO 680
800  FOR I = 1 TO 2
810 A = A + Y(I) * A(I)
820 B = B + Y(I) * B(I)
830 C = C + Y(I) * C(I)
840 D = D + Y(I) * D(I)
850  NEXT I
860  HOME : VTAB 5
870  PRINT "YOU MAY MAKE TWO CALCULATIONS:"
880  PRINT
890  PRINT "1. FIND THE ENTHALPY CHANGE BETWEEN"
900  PRINT "   TWO TEMPERATURES, OR"
910  PRINT
920  PRINT "2. FIND THE SECOND TEMPERATURE FOR A"
930  PRINT "   GIVEN ENTHALPY CHANGE."
940  PRINT
950  PRINT "BOTH WILL GIVE THE ENTROPY CHANGE."
960  PRINT : PRINT : PRINT
970  INPUT "MAKE YOUR SELECTION BY NUMBER     ";OQ
980  IF OQ = 1 OR OQ = 2 GOTO 1030
990  PRINT
1000  PRINT "YOUR OPTIONS WERE 1 OR 2.  YOU ENTERED"
1010  PRINT OQ;". TRY AGAIN."
1020  GOTO 960
1030  PRINT : PRINT : PRINT : PRINT
1040  PRINT "ENTER EITHER TWO TEMPERATURES IN"
1050  PRINT "DEGREES CENTIGRADE, OR ONE TEMPERATURE"
1060  PRINT "AND THE ENTHALPY CHANGE IN CAL/G-MOL"
1070  PRINT
1080  INPUT "ENTER THE DATA NOW  ";TI,TJ
1090 T1 = TI + 273.16
1100  IF OQ = 2 GOTO 1220
1110 T2 = TJ + 273.16
1120  IF OP = 2 GOTO 1150
1130 EN = A * (T2 - T1) + B / 2 * (T2 ^ 2 - T1 ^ 2) + C
     / 3 * (T2 ^ 3 - T1 ^ 3) + D / 4 * (T2 ^ 4 - T1 ^ 4)
1140  GOTO 1160
1150 EN = A * (T2 - T1) + B / 2 * (T2 ^ 2 - T1 ^ 2) - C
     / 2 * (1 / T2 - 1 / T1)
1160  IF OQ = 2 GOTO 1250
1170  IF OP = 2 GOTO 1200
1180 DS = A * LOG (T2 / T1) + B * (T2 - T1) + C / 2 *
     (T2 ^ 2 - T1 ^ 2) + D / 3 * (T2 ^ 3 - T1 ^ 3)
1190  GOTO 1340
1200 DS = A * LOG (T2 / T1) + B * (T2 - T1) - C / 2 *
     (1 / T2 ^ 2 - 1 / T1 ^ 2)
1210  GOTO 1340
1220 T2 = T1 + 100
1230 TX = T2
1240  ON OP GOTO 1130,1150
1250  IF F1 = 1 GOTO 1300
1260 F1 = 1
1270 EP = EN
1280 T2 = T2 - 10
1290  ON OP GOTO 1130,1150
1300  IF  ABS (TJ - EN) < .01 GOTO 1170
1310 T2 = T2 - (TX - T2) / (EP - EN) * (EN - TJ)
1320 CX = CX + 1
1330  ON OP GOTO 1130,1150
1340  HOME : VTAB 4
1350  FOR I = 1 TO 2
1360  IF OP = 2 GOTO 1400
1370 X(I) = A(I) + B(I) * T1 + C(I) * T1 ^ 2 + D(I) * T1
     ^ 3
1380 Y(I) = A(I) + B(I) * T2 + C(I) * T2 ^ 2 + D(I) * T2
     ^ 3
1390  GOTO 1420
1400 X(I) = A(I) + B(I) * T1 + C(I) / T1 ^ 2
1410 Y(I) = A(I) + B(I) * T2 + C(I) / T2 ^ 2
1420  NEXT I
1430 TJ =  INT (100 * (T2 - 273.15995)) / 100
1440  PRINT "HEAT CAPACITY, BTU/(G-MOL)(K)"
1450  IF Y1 = 1 GOTO 1470
1460  PRINT "FOR FIRST COMPONENT"
1470  PRINT "AT ";TI;" DEG. C IS "; FN P(X(1))
1480  PRINT "AT ";TJ;" DEG. C IS "; FN P(Y(1))"
1490  PRINT
1500  IF Y1 = 1 GOTO 1540
1510  PRINT "FOR SECOND COMPONENT
1520  PRINT "AT ";TI;" DEG. C IS "; FN P(X(2))
1530  PRINT "AT ";TJ;" DEG. C IS "; FN P(Y(2))
1540  PRINT
1550  PRINT "ENTHALPY CHANGE IS  "; FN P(EN);" CAL/G-MOL"
1560  PRINT
1570  IF OQ = 1 GOTO 1600
1580  PRINT "TEMPERATURE IS    "; FN P(TJ);" DEG. C."
1590  PRINT
1600  PRINT "ENTROPY CHANGE IS  "; FN P(DS)
1610  PRINT "               CAL/(G-MOL)(K)
1620  PRINT
1630  PRINT  TAB( 13);"END OF PROGRAM"
1640  END
```

**Example 1 for: Thermodynamic properties of pure gases and binary mixtures** **Table II**

**(Start of first display)**

```
PROGRAM CALCULATES THE ENTHALPY CHANGE
AND THE ENTROPY CHANGE FOR PURE GASES
OR BINARY GAS MIXTURES WHEN THE HEAT
CAPACITY IS CALCULATED FROM ONE OF THE
FOLLOWING EMPIRICAL EQUATIONS

1: CP = A + BT + CT^2 + DT^3

2: CP = A + BT + C/T^2

YOU SELECT THE EQUATION TO BE USED,
(NOTE: YOU WILL NEED THE CONSTANTS FOR
THE GAS COMPONENTS), THEN YOU ENTER
```

**Example 1 for: Thermodynamic properties of pure gases and binary mixtures (continued)** **Table II**

```
THE MOL FRACTION AND THE CONSTANTS FOR
THE COMPONENTS.

IF YOU HAVE THE TWO TEMPERATURES, THE
PROGRAM WILL CALCULATE THE ENTHALPY
CHANGE.  IF YOU HAVE THE ENTHALPY
CHANGE, THE PROGRAM WILL CALCULATE
THE SECOND TEMPERATURE.

PRESS RETURN TO CONTINUE.
```

**(Start of next display)**

```
SELECT THE HEAT CAPACITY EQUATION FORM
BY NUMBER:

1. CP = A + B*T + C*T^2 + D*T^3

2. CP = A + B*T + C/T^2

MAKE YOUR SELECTION NOW  1

ENTER THE MOL FRACTION OF THE FIRST
COMPONENT.  IF THE CALCULATION IS FOR
A PURE GAS, ENTER 1, OTHERWISE ENTER
A DECIMAL FRACTION. 1

ENTER THE CONSTANTS FOR
THE HEAT CAPACITY
CALCULATION:
A                    7.44
B                    -3.24E-3
C                    6.4E-6
D                    -2.79E-9
```

**(Start of next display)**

```
YOU MAY MAKE TWO CALCULATIONS:

1. FIND THE ENTHALPY CHANGE BETWEEN
   TWO TEMPERATURES, OR

2. FIND THE SECOND TEMPERATURE FOR A
   GIVEN ENTHALPY CHANGE.

BOTH WILL GIVE THE ENTROPY CHANGE.


MAKE YOUR SELECTION BY NUMBER     1


ENTER EITHER TWO TEMPERATURES IN
DEGREES CENTIGRADE, OR ONE TEMPERATURE
AND THE ENTHALPY CHANGE IN CAL/G-MOL

ENTER THE DATA NOW  50.150
```

**(Start of next display)**

```
HEAT CAPACITY, BTU/(G-MOL)(K)
AT 50 DEG. C IS 6.9672
AT 150 DEG. C IS 7.0036


ENTHALPY CHANGE IS  697.9908 CAL/G-MOL

ENTROPY CHANGE IS  1.8816
                CAL/(G-MOL)(K)

              END OF PROGRAM
```

**Comparison of calculated results of $\Delta H$, final temperature, and $\Delta S$ for nitrogen** **Table III**

| Temperature, °C | $\Delta H$, cal/g-mol | | $t_f$, °C | | $\Delta S$, cal/(g-mol) ($K$) | |
|---|---|---|---|---|---|---|
| $t_f$ | $C_p$† | $C_p$** | $C_p$† | $C_p$** | $C_p$† | $C_p$** |
| 100 | 348.54 | 352.19 | 100.00 | 100.06 | 1.0028 | 1.0132 |
| 150 | 697.99 | 707.81 | 150.04 | 149.70 | 1.8816 | 1.9074 |
| 200 | 1049.13 | 1066.48 | 200.06 | 199.76 | 2.6659 | 2.7085 |

† $C_p = 7.44 - (3.24 \times 10^{-3})\,T + (6.4 \times 10^{-6})\,T^2 - (2.79 \times 10^{-9})\,T^3$

** $C_p = 6.83 + (9 \times 10^{-4})\,T - (0.12 \times 10^{5})\,T^{-2}$

Initial temperature in all cases is 50°C.

**Example 2 for: Thermodynamic properties of pure gases and binary mixtures** **Table IV**

**(Start of first display)**

```
PROGRAM CALCULATES THE ENTHALPY CHANGE
AND THE ENTROPY CHANGE FOR PURE GASES
OR BINARY GAS MIXTURES WHEN THE HEAT
CAPACITY IS CALCULATED FROM ONE OF THE
FOLLOWING EMPIRICAL EQUATIONS

1: CP = A + BT + CT^2 + DT^3

2: CP = A + BT + C/T^2

YOU SELECT THE EQUATION TO BE USED,
(NOTE: YOU WILL NEED THE CONSTANTS FOR
THE GAS COMPONENTS), THEN YOU ENTER
THE MOL FRACTION AND THE CONSTANTS FOR
THE COMPONENTS.

IF YOU HAVE THE TWO TEMPERATURES, THE
PROGRAM WILL CALCULATE THE ENTHALPY
CHANGE.  IF YOU HAVE THE ENTHALPY
CHANGE, THE PROGRAM WILL CALCULATE
THE SECOND TEMPERATURE.

PRESS RETURN TO CONTINUE.
```

**(Start of next display)**

```
SELECT THE HEAT CAPACITY EQUATION FORM
BY NUMBER:

1. CP = A + B*T + C*T^2 + D*T^3

2. CP = A + B*T + C/T^2

MAKE YOUR SELECTION NOW  1

ENTER THE MOL FRACTION OF THE FIRST
COMPONENT.  IF THE CALCULATION IS FOR
A PURE GAS, ENTER 1, OTHERWISE ENTER
A DECIMAL FRACTION. .4

ENTER THE CONSTANTS FOR
THE HEAT CAPACITY
CALCULATION FOR THE FIRST COMPONENT:
 A                 5.04
 B                 9.3E-3
 C                 8.87E-6
 D                 -5.4E-9

ENTER THE VALUES FOR THE SECOND
COMPONENT:
 A                 7.44
 B                 -3.24E-3
 C                 6.4E-6
 D                 -2.79E-9
```

**(Start of next display)**

```
YOU MAY MAKE TWO CALCULATIONS:

1. FIND THE ENTHALPY CHANGE BETWEEN
   TWO TEMPERATURES, OR

2. FIND THE SECOND TEMPERATURE FOR A
   GIVEN ENTHALPY CHANGE.

BOTH WILL GIVE THE ENTROPY CHANGE.


MAKE YOUR SELECTION BY NUMBER     1


ENTER EITHER TWO TEMPERATURES IN
DEGREES CENTIGRADE, OR ONE TEMPERATURE
AND THE ENTHALPY CHANGE IN CAL/G-MOL

ENTER THE DATA NOW  50,150
```

**(Start of next display)**

```
HEAT CAPACITY, BTU/(G-MOL)(K)
FOR FIRST COMPONENT
AT 50 DEG. C IS 8.7895
AT 150 DEG. C IS 10.1545

FOR SECOND COMPONENT
AT 50 DEG. C IS 6.9672
AT 150 DEG. C IS 7.0036

ENTHALPY CHANGE IS  797.4857 CAL/G-MOL

ENTROPY CHANGE IS  2.1466
               CAL/(G-MOL)(K)

             END OF PROGRAM
```

# Thermal conductivities of pure gases

This Apple II program calculates the thermal conductivity of pure gases, and of gases at low pressure. It also calculates the effect of pressure on conductivity.

□ This program calculates the thermal conductivity ($\lambda$) of pure gases at atmospheric pressure by four equations: Euken, modified Euken, Stiel-Thodos, and Misic-Thodos (for hydrocarbons). It also offers the Roy-Thodos equation, for low pressure, which requires special constants. And, it can also calculate the thermal conductivity at higher pressures, given the 1-atm value.

Table I lists the program, and Table II shows example displays. Now let us look at the capabilities in detail.

## Thermal conductivity

The first decision in the program is whether to calculate thermal conductivity at 1 atm. The alternative is to calculate it at higher pressure. Then there is a choice between the four general correlations and the Roy-Thodos correlation.

The general equations require the following data: critical temperature and pressure; molecular weight; either heat capacity or the heat-capacity correlation constants; and temperature.

For determining the thermal conductivity of polyatomic gases at low pressure, the simplest correlations are the Eucken, Eq. (1), the modified Eucken, Eq. (2) [1], and the Stiel-Thodos, Eq. (3) [2]:

$$\lambda M/\eta = C_v + 4.47 \qquad (1)$$

$$\lambda M/\eta = 1.32C_v + 3.52 \qquad (2)$$

$$\lambda M/\eta = 1.15C_v + 4.04 \qquad (3)$$

If $C_p$ is given, $C_v$ is calculated by Eq. (4). Or $C_p$ may be calculated first, by Eq. (5):

$$C_v = C_p - R \qquad (4)$$

$$C_p = a + bT + cT^2 + dT^3 \qquad (5)$$

Viscosity is determined with the Yoon-Thodos correlation [3]:

$$\eta\xi = 4.610T_r^{0.618} - 2.04e^{-0.449T_r} + 1.94e^{-4.058T_r} + 0.1 \qquad (6)$$

Misic and Thodos developed a two-part correlation for hydrocarbons [5,6]. For $T_r$ values less than 1.0:

$$\lambda = (4.45 \times 10^{-6})T_r(C_p/\Gamma) \qquad (7)$$

For $T_r$ values greater than 1, they proposed:

$$\lambda = (10^{-6})(14.52T_r - 5.14)^{2/3}(C_p/\Gamma) \qquad (8)$$

The program selects between these automatically, depending on the temperature given.

The program calculates thermal conductivity at 1 atm and the given temperature by all four of these correlations. Table III lists the program's results for methane at various temperatures, and compares them with experimental ones published by Carmichael, Reamer and Sage [4]. The required data are also listed.

## Roy-Thodos correlation

The last correlation that was programmed for calculating $\lambda$ values at low pressure is that of Roy and Thodos [7,8]:

$$\lambda\Gamma = (\lambda\Gamma)_{tr} + (\lambda\Gamma)_{int} \qquad (9)$$

when

$$(\lambda\Gamma)_{tr} = (99.6 \times 10^6)(e^{0.0464T_r} - e^{-0.2412T_r}) \qquad (10)$$

and

$$(\lambda\Gamma)_{int} = C''f(T_r) \qquad (11)$$

Both terms on the right hand side of Eq. (11) are functions of the types of compounds involved. Data on $f(T_r)$ are given in Table IV. The structural constant, $C''$, is determined via a group contribution method. However, in many classes of compounds, $C'' \times 10^5$ may be expressed very well as a function of molecular weight, in the form:

$$C'' = AM^B \qquad (12)$$

Values for $A$ and $B$ are given in Table V. Roy and Thodos reported $C''$ values for other types of compounds—nitriles, halides, and cyclic compounds—as well as the types noted in Table V.

To use the Roy-Thodos equation, enter "R" when prompted, then the constants requested: $a$, $b$ and $c$, the coefficients of $T_r$, $T_r^2$, and $T_r^3$ from Table IV; and $A$ and $B$, the constants of Eq. (12) from Table V. Then enter the rest of the data, as for the other correlations, and get the thermal conductivity at 1 atm and the specified temperature.

Table III compares thermal conductivities calculated by the Roy-Thodos correlation with experimental ones from Carmichael et al. [4].

Adapted from an article by James H. Weber, University of Nebraska, originally published January 12, 1981.

## Thermal conductivity vs. pressure

The effect of pressure on thermal conductivity may be taken into account by relationships proposed by Stiel and Thodos [9].

For $\rho_r < 0.5$:

$$(\lambda - \lambda^o)\Gamma Z_c^5 = (14.0 \times 10^{-8})(e^{0.535\rho_r} - 1) \quad (13)$$

For $0.5 < \rho_r < 2.0$:

$$(\lambda - \lambda^o)\Gamma Z_c^5 = (13.1 \times 10^{-8})(e^{0.67\rho_r} - 1.069) \quad (14)$$

For $2.0 < \rho_r < 2.8$:

$$(\lambda - \lambda^o)\Gamma Z_c^5 = (2.976 \times 10^{-8})(e^{1.155\rho_r} + 2.016) \quad (15)$$

Since the selection of the proper relationship depends upon the value of $\rho_r$, the Redlich-Kwong equation of state [10] has been programmed, in addition to Eq. (13) through (15). The user has the option of using a density value obtained from another source.

The program chooses the correct relationship automatically. Required data are: critical temperature, pressure, and compressibility factor; molecular weight; thermal conductivity at 1 atm and the temperature of interest; pressure; and density at the temperature and pressure of interest. If density is not entered, the program will calculate it by the Redlich-Kwong equation; then temperature is also required.

Table III shows calculated values of thermal conductivity at various pressures, and compares them with experimental ones from Carmichael et al. [4].

## Nomenclature

- $A$, $B$ — Constants of Redlich-Kwong equation of state, or of Eq. (12)
- $a$, $b$, $c$, $d$ — Constants of Eq. (5)
- $C_p$ — Heat capacity at constant pressure, cal/(g-mol)(K)
- $C_v$ — Heat capacity at constant volume, cal/(g-mol)(K)
- $C''$ — Structural constant of Roy-Thodos correlation
- $h$ — $BP/Z$
- $M$ — Molecular weight
- $P$ — Pressure, atm
- $R$ — Gas law constant, 1.987 cal/(g-mol)(K), 0.08207 (L)(atm)/(g-mol)(K)
- $T$ — Absolute temperature, K
- $t$ — Temperature, °C
- $V$ — Volume, L/g-mol
- $Z$ — Compressibility factor
- $\Gamma$ — $(T_c^{1/6}M^{1/2})/P_c^{2/3}$
- $\eta$ — Viscosity, $\mu P$
- $\lambda$ — Thermal conductivity, cal/(cm)(s)(K)
- $\xi$ — $T_c^{1/6}/(M^{1/2}P_c^{2/3})$
- $\rho$ — Density, g-mol/L

**Subscripts**

- $c$ — Critical
- $r$ — Reduced

**Superscript**

- $o$ — Refers to ideal gas state

## References

1. Liley, P. E., Symposium on Thermal Properties, Purdue University, Lafayette, Ind., 1959, p. 40.
2. Stiel, L. I., and Thodos, G., *AIChE J.*, Vol. 10, 1964, p. 26.
3. Yoon, P., and Thodos, G., *AIChE J.*, Vol. 16, 1970, p. 300.
4. Carmichael, L. T., Reamer, H. H., and Sage, B. H., *J. of Chem. Eng. Data*, Vol. 11, 1966, p. 52.
5. Misic, D., and Thodos, G., *AIChE J.*, Vol. 7, 1961, p. 264.
6. Misic, D., and Thodos, G., *J. of Chem. Eng. Data*, Vol. 9, 1963, p. 540.
7. Roy, D., and Thodos, G., *Ind. Eng. Chem. Fundamentals*, Vol. 7, 1968, p. 529.
8. Roy, D., and Thodos, G., *Ibid.*, Vol. 9, 1968, p. 71.
9. Stiel, L. T., and Thodos, G., *AIChE J.*, Vol. 10, 1964, p. 26.
10. Redlich, O., and Kwong, J. N. S., *Chem. Rev.*, Vol. 44, 1949, p. 233.

**Program for predicting thermal conductivities of pure gases** **Table I**

```
10 REM   THERMAL CONDUCTIVITIES OF PURE GASES
20 REM   FROM CHEMICAL ENGINEERING, JANUARY 12, 1981
30 REM   BY JAMES H. WEBER
40 REM  TRANSLATED BY WILLIAM VOLK
50 REM   COPYRIGHT (C) 1984
60 REM  BY CHEMICAL ENGINEERING
70 HOME : VTAB 4
80 REM  SET DISPLAY AT FOUR DECIMALS.
90 DEF  FN P(X) =  INT (1E4 * (X + .00004)) / 1E4
100 PRINT "PROGRAM CALCULATES THE THERMAL CONDUC-"
110 PRINT "TIVITY OF GASES AT LOW PRESSURES BY"
120 PRINT "SEVERAL CORRELATIONS:"
130 A$(1) = "EUKEN EQUATION"
140 A$(2) = "MODIFIED EUKEN EQUATION"
150 A$(3) = "STIEL-THODOS CORRELATION"
160 A$(4) = "MISIC-THODOS CORRELATION"
170 A$(5) = "ROY-THODOS CORRELATION"
180 FOR I = 1 TO 5
190 PRINT
200 PRINT I;". ";A$(I)
210 NEXT I
220 PRINT "   REQUIRING SPECIAL CONSTANTS."
230 PRINT
240 PRINT "IT ALSO PROVIDES THE STIEL-THODOS"
250 PRINT "CORRELATION FOR THE EFFECT OF PRESSURE"
260 PRINT "ON THERMAL CONDUCTIVITY."
270 PRINT
280 INPUT "PRESS RETURN TO CONTINUE     ";Q$
290 HOME : VTAB 4
300 PRINT "FOR A THERMAL CONDUCTIVITY CALCULATION"
310 PRINT "THE FOLLOWING DATA ARE REQUIRED:"
320 PRINT
330 PRINT "CRITICAL TEMPERATURE AND PRESSURE,"
340 PRINT "MOLECULAR WEIGHT, EITHER THE"
350 PRINT "HEAT CAPACITY AT CONSTANT PRESSURE, OR"
360 PRINT "CONSTANTS FOR THE FOLLOWING EQUATION:"
370 PRINT
380 PRINT "CP = A + B*T + C*T^2 + D*T^3"
```

**Program for predicting thermal conductivities of pure gases (continued)** **Table I**

```
390  PRINT
400  PRINT "AND THE GAS TEMPERATURE."
410  PRINT
420  PRINT
430  PRINT "FOR THE EFFECT OF PRESSURE:"
440  PRINT
450  PRINT "THE CRITICAL TEMPERATURE, PRESSURE,"
460  PRINT "AND COMPRESSIBILITY FACTOR ARE RE-"
470  PRINT "QUIRED; AND ALSO THE THERMAL CONDUC-"
480  PRINT "TIVITY AT 1 ATM, THE GAS DENSITY, AND"
490  PRINT "THE GAS PRESSURE."
500  PRINT : PRINT
510  INPUT "PRESS RETURN TO CONTINUE.  ";Q$
520  HOME : VTAB 4
530  PRINT "ENTER: 1 OR 2 "
540  PRINT
550  PRINT "1. TO CALCULATE THERMAL CONDUCTIVITY"
560  PRINT "   AT ATMOSPHERIC PRESSURE."
570  PRINT
580  PRINT "2. TO FIND THE EFFECT OF PRESSURE."
590  PRINT : PRINT
600  INPUT "ENTER THE OPTION NOW:           ";OP
610  ON OP GOTO 620,1340
620  HOME : VTAB 4
630  PRINT "IF YOU WISH TO USE THE ROY-THODOS"
640  PRINT "CORRELATION REQUIRING SPECIAL"
650  PRINT "CONSTANTS FOR SPECIFIC COMPOUNDS,"
660  PRINT "ENTER 'R'.  IF NOT, SIMPLY PRESS THE"
670  PRINT "RETURN KEY TO ENTER DATA FOR THE"
680  INPUT "OTHER CORRELATIONS.         ";Q$
690  IF Q$ = "R" GOTO 1810
700  HOME : VTAB 4
710  PRINT "TO CALCULATE THERMAL CONDUCTIVITY"
720  PRINT "AT ONE ATMOSPHERE PRESSURE"
730  PRINT "ENTER THE FOLLOWING DATA:"
740  PRINT
750  PRINT
760  INPUT "CRITICAL TEMPERATURE, DEG. K     ";TC
770  INPUT "CRITICAL PRESSURE, ATM           ";PC
780  INPUT "MOLECULAR WEIGHT                 ";MW
790  PRINT
800  IF Q$ = "R" GOTO 960
810  PRINT "IF YOU KNOW THE HEAT CAPACITY AT CON-"
820  PRINT "STANT PRESSURE, CAL/(G-MOL)(K)"
830  PRINT "ENTER THAT VALUE.  IF NOT ENTER"
840  PRINT "ZERO AND THEN THE CONSTANTS FOR THE"
850  PRINT "EQUATION:"
860  PRINT
870  PRINT " CP = A + B*T + C*T^2 + D*T^3"
880  PRINT
890  PRINT "ENTER CP IF YOU HAVE IT, "
900  INPUT "OR ELSE ENTER ZERO      ";CP
910  IF CP > 0 GOTO 960
920  INPUT "CONSTANT A              ";A
930  INPUT "CONSTANT B              ";B
940  INPUT "CONSTANT C              ";C
950  INPUT "CONSTANT D              ";D
960  PRINT
970  INPUT "TEST TEMPERATURE, DEG. C         ";TI
980 T = TI + 273.16
990 TR = T / TC
1000  IF Q$ = "R" GOTO 1040
1010  IF CP > 0 GOTO 1040
1020 CP = A + B * T + C * T ^ 2 + D * T ^ 3
1030 F1 = 1
1040 XI = TC ^ (1 / 6) / ( SQR (MW) * PC ^ (2 / 3))
1050 GM = XI * MW
1060  IF Q$ = "R" GOTO 1970
1070 CV = CP - 1.987
1080 TP = 4.610 * TR ^ (.618) - 2.04 *  EXP ( - .449 * TR)
     + 1.94 *  EXP ( - 4.058 * TR) + .1
1090 ET = TP / XI
1100  REM  EUCKEN EQUATION
1110 LM(1) = (CV + 4.47) * ET / MW
1120  REM  MODIFIED EUCKEN EQUATION
1130 LM(2) = (1.32 * CV + 3.52) * ET / MW
1140  REM   STIEL-THODOS CORRELATION
1150 LM(3) = (1.15 * CV + 4.04) * ET / MW
1160  REM   MISIC-THODOS CORRELATION
1170  IF TR < 1 GOTO 1200
1180 LM(4) = (14.52 * TR - 5.14) ^ (2 / 3) * (CP / GM)
1190  GOTO 1210
1200 LM(4) = 4.45 * TR * CP / GM
1210  HOME : VTAB 4
1220  PRINT "THE THERMAL CONDUCTIVITIES IN"
1230  PRINT "CAL/(CM)(S)(DEG. C) * 10^6 ARE AS"
1240  PRINT "FOLLOWS:"
1250  PRINT
1260  FOR I = 1 TO 4
1270  PRINT "BY ";A$(I); TAB( 31); FN P(LM(I))
1280  NEXT I
1290  IF F1 = 0 GOTO 2110
1300  PRINT
1310  PRINT "THE CALCULATED HEAT CAPACITY,"
1320  PRINT "CAL/(G-MOLE)(DEG. C) IS      "; FN P(CP)
1330  GOTO 2110
1340  HOME : VTAB 4
1350  PRINT "TO FIND THE EFFECT OF PRESSURE,"
1360  PRINT "ENTER THE FOLLOWING DATA:"
1370  PRINT
1380  INPUT "CRITICAL TEMPERATURE, DEG. K      ";TC
1390  INPUT "CRITICAL PRESSURE, ATM            ";PC
1400  INPUT "CRITICAL COMPRESSIBILITY FACTOR   ";ZC
1410 VC = .08207 * ZC * TC / PC
1420  INPUT "MOLECULAR WEIGHT                  ";MW
1430  PRINT "THERMAL CONDUCTIVITY AT 1 ATM"
1440  INPUT "       CAL/(CM)(S)(DEG. K) * 10^6 ";LM
1450  INPUT "GAS PRESSURE, ATM                 ";P
1460  PRINT
1470  PRINT "IF YOU HAVE THE GAS DENSITY, G-MOL/L,
1480  PRINT "AT TEST PRESSURE, ENTER THAT VALUE."
1490  INPUT "IF NOT, ENTER ZERO.               ";RO
1500  IF RO > 0 GOTO 1660
1510  PRINT
```

**Program for predicting thermal conductivities of pure gases (continued)** **Table I**

```
1520 PRINT "THE PROGRAM WILL ESTIMATE THE GAS"
1530 PRINT "DENSITY USING THE REDLICH-KWONG EQUA-"
1540 PRINT "TION, FROM THE TEMPERATURE AND PRESSURE."
1550 INPUT "ENTER THE TEMPERATURE, DEG. C      ";TI
1560 T = TI + 273.16
1570 D = .4278 * TC ^ 2.5 / (PC * T ^ 2.5)
1580 NI = 1
1590 E = .0867 * TC / PC / T
1600 I = E * P / NI
1610 NO = 1 / (1 - I) - (I * D / E) / (1 + I)
1620  IF  ABS (NO - NI) < .0001 GOTO 1650
1630 NI = NO
1640  GOTO 1600
1650 RO = P / (NO * .08207 * T)
1660 GM = TC ^ (1 / 6) *  SQR (MW) / PC ^ (2 / 3)
1670 RC = RO * VC
1680  IF RC > 2.8 GOTO 1790
1690  IF RC > 2 GOTO 1770
1700  IF RC > .5 GOTO 1750
1710 TP = .14 * ( EXP (.535 * RC) - 1)
1720 TP = TP / GM / ZC ^ 5
1730 LM = TP + LM
1740  GOTO 2080
1750 TP = .131 * ( EXP (.67 * RC) - 1.069)
1760  GOTO 1720
1770 TP = .02976 * ( EXP (1.155 * RC) + 2.016)
1780  GOTO 1720
1790  PRINT "GAS DENSITY IS OUTSIDE RANGE OF"
1800  PRINT "THE CORRELATION."
1810  HOME : VTAB 4
1820  PRINT "ENTER THE CONSTANTS FOR THE ROY-THODOS"
1830  PRINT "CORRELATION:"
1840  PRINT
1850  PRINT "  F(TR) = A1*TR + B1*TR^2 + C1*TR^3"
1860  PRINT
1870  INPUT "ENTER A1     ";A1
1880  INPUT "ENTER B1     ";B1
1890  INPUT "ENTER C1     ";C1
1900  PRINT
1910  PRINT "ENTER TWO STRUCTURAL CONSTANTS"
1920  PRINT "FOR THE RELATION C' = A2*MW^B2"
1930  PRINT
1940  INPUT "ENTER A2     ";A2
1950  INPUT "ENTER B2     ";B2
1960  GOTO 700
1970 TP = 99.6 * ( EXP (.0464 * TR) -  EXP ( - .2412 *
     TR))
1980 FT = A1 * TR + B1 * TR ^ 2 + C1 * TR ^ 3
1990 TP = TP + 10 * A2 * MW ^ B2 * FT
2000 LM = TP / GM
2010  HOME : VTAB 4
2020  PRINT "THE THERMAL CONDUCTIVITY"
2030  PRINT "CAL/(CM)(S)(DEG. C) * 10^6"
2040  PRINT "BY ROY-THODOS CORRELATION IS"
2050  PRINT
2060  PRINT  TAB( 13); FN P(LM)
2070  GOTO 2110
2080  PRINT
2090  PRINT "AT ";P;" ATM. THE THERMAL CONDUCTIVITY"
2100  PRINT "IS "; FN P(LM);" CAL/(CM)(S)(DEG. C) * 10^6"
2110  PRINT
2120  PRINT
2130  PRINT  TAB( 13);"END OF PROGRAM"
2140  END
```

**Example for: Thermal conductivities of pure gases** **Table II**

**(Start of first display)**

```
PROGRAM CALCULATES THE THERMAL CONDUC-
TIVITY OF GASES AT LOW PRESSURES BY
SEVERAL CORRELATIONS:

1. EUKEN EQUATION

2. MODIFIED EUKEN EQUATION

3. STIEL-THODOS CORRELATION

4. MISIC-THODOS CORRELATION

5. ROY-THODOS CORRELATION
   REQUIRING SPECIAL CONSTANTS.

IT ALSO PROVIDES THE STIEL-THODOS
CORRELATION FOR THE EFFECT OF PRESSURE
ON THERMAL CONDUCTIVITY.

PRESS RETURN TO CONTINUE
```

**(Start of next display)**

```
FOR A THERMAL CONDUCTIVITY CALCULATION
THE FOLLOWING DATA ARE REQUIRED:

CRITICAL TEMPERATURE AND PRESSURE,
MOLECULAR WEIGHT, EITHER THE
HEAT CAPACITY AT CONSTANT PRESSURE, OR
CONSTANTS FOR THE FOLLOWING EQUATION:

CP = A + B*T + C*T^2 + D*T^3

AND THE GAS TEMPERATURE.


FOR THE EFFECT OF PRESSURE:

THE CRITICAL TEMPERATURE, PRESSURE,
AND COMPRESSIBILITY FACTOR ARE RE-
QUIRED; AND ALSO THE THERMAL CONDUC-
TIVITY AT 1 ATM, THE GAS DENSITY, AND
THE GAS PRESSURE.
```

**Example for: Thermal conductivities of pure gases (continued)** **Table II**

```
PRESS RETURN TO CONTINUE.
```

**(Start of next display)**

```
ENTER: 1 OR 2

1. TO CALCULATE THERMAL CONDUCTIVITY
   AT ATMOSPHERIC PRESSURE.

2. TO FIND THE EFFECT OF PRESSURE.

ENTER THE OPTION NOW:          1
```

**(Start of next display)**

```
IF YOU WISH TO USE THE ROY-THODOS
CORRELATION REQUIRING SPECIAL
CONSTANTS FOR SPECIFIC COMPOUNDS,
ENTER 'R'.  IF NOT, SIMPLY PRESS THE
RETURN KEY TO ENTER DATA FOR THE
OTHER CORRELATIONS.
```

**(Start of next display)**

```
TO CALCULATE THERMAL CONDUCTIVITY
AT ONE ATMOSPHERE PRESSURE
ENTER THE FOLLOWING DATA:


CRITICAL TEMPERATURE, DEG. K    190.6
CRITICAL PRESSURE, ATM          45.4
MOLECULAR WEIGHT                16.043

IF YOU KNOW THE HEAT CAPACITY AT CON-
STANT PRESSURE, CAL/(G-MOL)(K)
ENTER THAT VALUE.  IF NOT ENTER
ZERO AND THEN THE CONSTANTS FOR THE
EQUATION:

 CP = A + B*T + C*T^2 + D*T^3

ENTER CP IF YOU HAVE IT,
OR ELSE ENTER ZERO      0
CONSTANT A              4.598
CONSTANT B              1.245E-2
CONSTANT C              2.86E-6
CONSTANT D              -2.7E-9

TEST TEMPERATURE, DEG. C          71.1
```

**(Start of next display)**

```
THE THERMAL CONDUCTIVITIES IN
CAL/(CM)(S)(DEG. C) * 10^6 ARE AS
FOLLOWS:

BY EUKEN EQUATION               89.6601
BY MODIFIED EUKEN EQUATION      99.9459
BY STIEL-THODOS CORRELATION     94.6
BY MISIC-THODOS CORRELATION     92.1203

THE CALCULATED HEAT CAPACITY,
CAL/(G-MOLE)(DEG. C) IS         9.1128


                END OF PROGRAM
```

**Comparison of calculated and experimental values for the thermal conductivity of methane** **Table III**

| $t$,°C (P=1 atm) | Calculated Eq. (1) | Eq. (2) | Eq. (3) | Eq. (7), (8) | Eq. (9) | | Experimental Carmichael, et al. |
|---|---|---|---|---|---|---|---|
| 37.73 | 79.12* | 87.55 | 83.18 | 80.39 | 83.53 | 86.09† | 84.15 |
| 71.1 | 89.66 | 99.95 | 94.60 | 92.12 | 95.29 | 98.43 | 94.73 |
| 104.4 | 100.43 | 112.72 | 106.32 | 104.26 | 107.54 | 111.33 | 106.55 |
| 137.8 | 111.48 | 125.91 | 118.38 | 116.86 | 120.34 | 124.82 | 119.61 |
| 171.1 | 122.73 | 139.42 | 130.42 | 130.70 | 133.60 | 138.84 | 133.83 |
| 204.4 | 134.20 | 153.28 | 143.30 | 143.22 | 147.36 | 153.39 | 149.03 |

**Effect of pressure on the thermal conductivity of methane ($t$ = 37.73°C)**

| $P$, atm | Calculated Eq. (13), (14), (15) $(\lambda - \lambda^o)$* | $\lambda$* | Experimental Carmichael, et al. $(\lambda - \lambda^o)$* | $\lambda$* |
|---|---|---|---|---|
| 68.05 | 16 | 100 | 12 | 97 |
| 136.09 | 38 | 123 | 33 | 118 |
| 204.14 | 68 | 152 | 56 | 140 |
| 272.18 | 95 | 180 | 75 | 159 |
| 340.2 | 120 | 204 | 91 | 175 |

*$\lambda$, [cal/(cm)(s)°C)] $(10^6)$

†$C''(0.90 \times 10^5)$, calculated from Eq. (12)
The Roy-Thodos reported value is $0.83 \times 10^5$.

**Physical properties of methane used in calculations**

$M$ = 16.043
$T_c$ = 190.6 K
$P_c$ = 45.4 atm
$V_c$ = 99.0 cm$^3$/g-mol
$Z_c$ = 0.288
$a$ = 4.598
$b$ = $1.245 \times 10^{-2}$
$c$ = $2.86 \times 10^{-6}$
$d$ = $-2.7 \times 10^{-9}$
($a$, $b$, $c$, $d$: Constants in Eq. (5))

Source: Reid, R.C., Prausnitz, J.M., and Sherwood, T.K., "The Properties of Gases and Liquids," 3rd ed., McGraw-Hill, New York, 1977.

**$f(T_r)$ relationships for Eq. (11)** **Table IV**

| Type of compound | Equation |
|---|---|
| Saturated hydrocarbons* | $-0.152T_r + 1.191T_r^2 - 0.039T_r^3$ |
| Olefins | $-0.255T_r + 1.065T_r^2 + 0.190T_r^3$ |
| Acetylenes | $-0.068T_r + 1.251T_r^2 - 0.183T_r^3$ |
| Naphthenes and aromatics | $-0.354T_r + 1.501T_r^2 - 0.147T_r^3$ |
| Alcohols | $1.000T_r^2$ |
| Aldehydes, ketones, ethers, esters | $-0.082T_r + 1.045T_r^2 + 0.037T_r^3$ |
| Amines and nitriles | $0.633T_r^2 + 0.367T_r^3$ |
| Halides | $-0.107T_r + 1.330T_r^2 - 0.223T_r^3$ |
| Cyclic compounds† | $-0.354T_r + 1.501T_r^2 - 0.147T_r^3$ |

*Not recommended for methane
†E.g., pyridine, thiophene, ethylene oxide, dioxane, piperidine

**Constants of Eq. (12)** **Table V**

| Type of compound | A | B |
|---|---|---|
| Hydrocarbons | 0.0049197 | 1.87708 |
| Oxygen-containing compounds (except diethyl ether) | 0.019067 | 1.50338 |
| Amines | 0.011291 | 1.70436 |
| Normal alcohols (including methanol and ethanol) | 0.42499 | 1.38200 |
| Iso alcohols (plus methanol and ethanol) | 0.047533 | 1.34656 |

# Section IV
# Fluid Flow

Pressure-drop calculations
Friction factor
Piping design
Centrifugal-pump hydraulics
Orifice sizes for gas flows
Hole-area distribution for liquid spargers

# Pressure-drop calculations

## Here is a program that will find the pressure drop of a fluid moving through a pipe as a vapor, liquid, or combination of these two phases.

☐ The program presented in this section is intended to provide quick, accurate estimates of pressure losses for the most common types of flow. Written for the Apple II, the program has these main features:

- Vapor-phase calculations take into account the compressibility of the fluid. The program also adjusts for differential changes in velocity and density as the pressure decreases.
- Two-phase calculations utilize Lockhart-Martinelli empirical relations for pressure-loss estimation.
- Liquid-phase pressure loss is determined by the Bernoulli equation.
- The program determines whether the flow is laminar, transitional or turbulent, and calculates an appropriate friction factor for slightly corroded steel pipe.
- An optional friction factor may be inputted to override the program's calculated value.

## Methodology

The program listing appearing in Table I is actually comprised of three programs. Each program can be run independently. A summary of correlations used by three programs appears in Table II.

In calculating the liquid-phase pressure loss, the program determines the Reynolds number to see whether the flow is laminar, turbulent or transitional. The friction factor for slightly corroded commercial steel pipe is then estimated as a function of the Reynolds number, using an equation appropriate for the flow regime [*5,6*]. An alternative friction factor may be inputted by the user and will override the program's calculated value. Pressure loss is calculated using the Bernoulli equation.

Vapor-phase calculations of pressure loss are more complex, since the change in pressure causes the vapor to accelerate and the density to decrease along the pipe length. Consequently, pressure loss must be described by a differential equation in which density and velocity appear as variables. These can in turn be expressed as functions of pressure. Solution of the differential equation yields a transcendental function (Table II).

The outlet pressure is obtained from the transcendental function by utilizing a golden-section search routine [*8*]. The friction factor for vapor-phase calculations is evaluated at inlet conditions and is constant, since it is a function of the Reynolds number only (i.e., the product of velocity and density is constant).

The vapor-flow calculations are rigorous for isothermal ideal gases, and are in error for nonideal gases only by the inaccuracy of the compressibility factor. When the flow is under high pressure and the pressure loss is substantial, it is recommended that the user compare the inlet and outlet compressibility factors. The two should be averaged if the difference is large. When pressure losses are small ($\Delta P/P < 0.1$), the liquid-phase loss correlation can generally be used as a good approximation of the vapor-phase loss.

The two-phase pressure loss is determined using the empirical correlations of Lockhart and Martinelli [*7*]. The correlations used in the program assume that turbulent flow exists and that pressure loss is less than 10% of the inlet pressure.

## Using the program

The first display informs the user of the three different pressure-drop calculation-program options. These are:

1. liquid-phase flow,
2. vapor-phase flow, or
3. two-phase flow.

The second display (reached by pressing the "RETURN" key) lists the data requirements for each flow option (see Table III). The user is then requested to select, by number, the type of flow calculation to be done.

For the first of the two examples (Table III and IV, respectively), the pressure drop for a liquid-phase flow problem is calculated. For this case, the user must input values for pipe diameter, equivalent pipe length, flowrate, liquid density, and liquid viscosity. (Note: For data values greater than 999—for example, for flowrate—do not use commas in entering these numbers.)

The user is then offered the option to enter a value for the Fanning friction factor. If the user does not wish to enter an optional value, the program will estimate a value, assuming flow conditions for a slightly corroded pipe. To do this, the user should enter a zero ("0").

Once the friction factor is entered, the program calculates pressure drop, bulk velocity, Reynolds number, and Fanning friction factor.

Similarly, in the second example (see Table IV), for liquid-phase flow, the same four pressure-drop variables are calculated. In addition, for this case, two new pieces of data are determined: (1) pressure drop, if the vapor is incompressible, and (2) maximum velocity.

**Adapted from an article by James M. Meyer, Koch Engineering Co., originally published March 10, 1980.**

## Nomenclature

| | | | |
|---|---|---|---|
| $D$ | Inside pipe dia., ft | $MUL$ | Liquid viscosity, cP |
| $DP2$ | Two-phase-flow pressure drop, psi | $MUV$ | Vapor viscosity, cP |
| $DPL$ | Differential pressure, liquid, psi | $MW$ | Molecular weight |
| $DPV$ | Differential pressure, vapor, psi | $P$ | Pressure, atm |
| $F$ | Fanning friction factor | $RHOL$ | Liquid density, lb/ft$^3$ |
| $Flow$ | Flowrate, lb/h | $RHOV$ | Vapor density, lb/ft$^3$ |
| $g_c$ | Acceleration of gravity, 32.17 $(\text{lb}_m)(\text{ft})/(\text{lb}_f)(\text{s}^2)$ | $T$ | Temperature, °F |
| $L$ | Equivalent pipe length, ft | $V$ | Linear flowrate, ft/s |
| | | $X$ | Martinelli parameter |
| | | $Y$ | Weight fraction vapor |
| | | $Z$ | Compressibility factor |

## References


1. Kern, R., Piping Design for Two-phase Flow, *Chem. Eng.*, June 23, 1975.
2. McCabe, W. L., and Smith J. C., "Unit Operations of Chemical Engineering," McGraw-Hill, New York, 1976.
3. Greenkorn, R. A., and Kessler, D. P., "Transfer Operations," McGraw-Hill, New York, 1972.
4. Perry, R. H., "Chemical Engineers' Handbook," 5th edition, McGraw-Hill, New York, 1973.
5. Nikuradse, H., Laws of Flow in Rough Pipes, Translation in Natl. Advisory Committee for Aeronautics (NACA) Tech. Mem. 1292, 1950.
6. Sams, E., "Friction Coefficients in Circular Tubes Having Square-Thread Roughness," NACA, 1952.
7. Lockhart, R. W., and Martinelli, R. C., Proposed Correlation of Data for Isothermal Two-Phase, Two-Component Flow in Pipe, *Chem. Eng. Prog.*, Jan. 1949.
8. Rudd, D. F., and Watson, C. C., "Strategy of Process Engineering," John Wiley, New York, 1968, p. 162.

## The author

**James M. Meyer** is a process engineer specializing in distillation with Koch Engineering Co., P.O. Box 8127, Wichita, KS 67208. Previously, he served as a process engineer for Celanese Chemical Co. and did conceptual design work for the Phillips Petroleum Research Center. He received his B.S. degree in chemical engineering from Iowa State University, and is a member of Tau Beta Pi.

**Program listing for computing pressure-drop** — **Table I**

```
10  REM   PRESSURE-DROP CALCULATIONS
20  REM   FROM CHEMICAL ENGINEERING, MARCH 10, 1980
30  REM   BY JAMES M. MEYER
40  REM  TRANSLATED BY WILLIAM VOLK
50  REM   COPYRIGHT (C) 1984
60  REM  BY CHEMICAL ENGINEERING
70 PI = 3.14159265
80  REM  SET DISPLAY AT FOUR DECIMALS
90  DEF  FN P(X) =  INT (1E4 * (X + .00005)) / 1E4
100  HOME : VTAB 8
110  PRINT "PROGRAM TO CALCULATE LINEAR PRESSURE"
120  PRINT "DROP FOR:"
130  PRINT
140  PRINT "1. LIQUID PHASE FLOW,"
150  PRINT "2. TWO-PHASE FLOW, OR"
160  PRINT "3. VAPOR PHASE FLOW."
170  PRINT
180  INPUT "PRESS RETURN KEY TO SEE DATA REQUIRED.  ";Q$
190  HOME
200  PRINT "YOU WILL NEED THE FOLLOWING DATA:"
210  PRINT
220  PRINT  TAB( 30);"LIQ"; TAB( 34);"VAP"; TAB( 38);"MIX"
230  PRINT "CALCULATION"; TAB( 31);"1"; TAB( 35);"2"; TAB( 39);"3"
240  PRINT
250  PRINT "INSIDE PIPE DIAMETER, IN."; TAB( 31);"*"; TAB( 35);"*"; TAB( 39);"*"
260  PRINT "EQUIVALENT PIPE LENGTH, FT"; TAB( 31);"*"; TAB( 35);"*"; TAB( 39);"*
"
270  PRINT "FLOW RATE, LB/HR"; TAB( 31);"*"; TAB( 35);"*"; TAB( 39)"*"
280  PRINT "LIQUID DENSITY, LB/CU.FT"; TAB( 31);"*"; TAB( 39);"*"
290  PRINT "INLET VAP. DENSITY,LB/CU.FT"; TAB( 31);"*"; TAB( 35);"*"
300  PRINT "LIQUID VISCOSITY, CP"; TAB( 31);"*"; TAB( 39);"*"
```

**Program listing for computing pressure-drop (continued)** **Table I**

```
310 PRINT "VAPOR VISCOSITY, CP"; TAB( 35);"*";
    TAB( 39);"*"
320 PRINT "INLET PRESSURE, ATM"; TAB( 35);"*";
    TAB( 39);"*"
330 PRINT "VAPOR DENSITY, LB/CU.FT"; TAB( 39);"*"
340 PRINT "WEIGHT FRACTION VAPOR"; TAB( 39);"*"
350 PRINT "MOLECULAR WEIGHT"; TAB( 35);"*"
360 PRINT "TEMPERATURE, DEG. F"; TAB( 35);"*"
370 PRINT "COMPRESSIBILITY FACTOR"; TAB( 35);"*"
380 PRINT
390 PRINT "SELECT ONE OF THE CALCULATIONS BY NUMBER."
400 INPUT "ENTER 1, 2, OR 3.  ";OP
410 IF OP > 3 OR OP < 1 GOTO 430
420 GOTO 480
430 HOME : VTAB 4
440 PRINT "YOUR OPTIONS WERE 1, 2, OR 3.  YOU"
450 PRINT "ENTERED ";OP;".  TRY AGAIN."
460 PRINT
470 GOTO 380
480 HOME
490 PRINT "ENTER:"
500 PRINT
510 INPUT "PIPE DIAMETER, IN.                 ";D
520 D = D / 12
530 INPUT "EQUIVALENT PIPE LENGTH, FT         ";L
540 INPUT "FLOW RATE, LB/H                    ";FL
550 ON OP GOTO 560,590,560
560 INPUT "LIQUID DENSITY, LB/CU.FT           ";LD
570 INPUT "LIQUID VISCOSITY, CP               ";LU
580 ON OP GOTO 760,590,610
590 INPUT "INLET VAPOR DENSITY, LB/CU.FT      ";IV
600 LD = IV
610 INPUT "VAPOR VISCOSITY, CP                ";VU
620 INPUT "INLET PRESSURE, ATM                ";P
630 IF OP = 3 GOTO 680
640 INPUT "MOLECULAR WEIGHT                   ";MW
650 INPUT "TEMPERATURE, DEG. F                ";T
660 INPUT "COMPRESSIBILITY FACTOR             ";Z
670 GOTO 760
680 INPUT "VAPOR DENSITY, LB/CU.FT            ";VD
690 INPUT "WEIGHT FRACTION VAPOR              ";Y
700 PRINT
710 PRINT "ENTER THE CRITICAL PRESSURE, ATM
720 PRINT "IF YOU DO NOT HAVE THE CRITICAL"
730 PRINT "PRESSURE, ENTER ZERO.  THE PROGRAM WILL"
740 INPUT "ESTIMATE A VALUE.                  ";PC
750 IF PC = 0 THEN PC = 1E35
760 PRINT
770 PRINT "THE PROGRAM WILL ESTIMATE A FRICTION"
780 PRINT "FACTOR ASSUMING SLIGHTLY CORRODED PIPE."
790 PRINT "IF YOU HAVE A FRICTION FACTOR, YOU MAY
800 PRINT "ENTER IT NOW.  IF NOT, ENTER ZERO AND"
810 INPUT "THE PROGRAM WILL CALCULATE A VALUE. ";F
820 IF OP = 2 THEN LU = VU
830 V = FL / (LD * PI * D ^ 2 / 4 * 3600)
840 IF F > 0 GOTO 930
850 RE = D * V * LD / (LU * 6.72E - 4)
860 IF RE > 2100 GOTO 890
870 F = 16 / RE
880 GOTO 930
890 IF RE > 3500 GOTO 920
900 F = .0001 * RE ^ .575
910 GOTO 930
920 F = .0035 + .264 * RE ^ ( - .42)
930 DP = 2 * F * L * V ^ 2 * LD / (D * 32.17 * 144)
940 ON OP GOTO 1440,990,950
950 X = ((1 - Y) / Y) ^ .9 * (VD / LD) ^ .5 * (LU / VU)
    ^ .1
960 CF = (1 + 3 * (1 - P / PC) * X ^ ( - .555) + X ^
    ( - 1.11)) ^ 1.75
970 DP = DP * CF
980 GOTO 1440
990 SX = DP
1000 PRINT
1010 PRINT "CALCULATION MAY TAKE A FEW SECONDS"
1020 SV = 14.4 * SX
1030 ZP = P * 2116.8 - SV
1040 GOTO 1100
1050 ZP = ZP - SV
1060 IF ZP > 0 GOTO 1100
1070 ZP = ZP + SV
1080 SV = SV / 10
1090 GOTO 1050
1100 I = ZP
1110 GOSUB 1290
1120 FP = TP
1130 IF - FP > 0 GOTO 1050
1140 P1 = 2116.8 * P
1150 TW = (P1 - ZP) * .382 + ZP
1160 TH = (P1 - ZP) * .618 + ZP
1170 I = TW
1180 GOSUB 1290
1190 FP = ABS (TP)
1200 I = TH
1210 GOSUB 1290
1220 FV = ABS (TP)
1230 IF FV > FP GOTO 1260
1240 ZP = TW
1250 GOTO 1270
1260 P1 = TH
1270 IF (P1 - ZP) / (144 * SX) > .01 GOTO 1150
1280 GOTO 1400
1290 TP = 83184 * D ^ 5 * MW
1300 TP = TP * ((2116.8 * P) ^ 2 - I ^ 2)
1310 TP = TP / (Z * (T + 460) * FL ^ 2)
1320 TP = TP - D * LOG (2116.8 * P / I) - 2 * F * L
1330 CO = CO + 1
1340 IF CO > 100 GOTO 1360
1350 RETURN
1360 HOME : VTAB 4
1370 PRINT "CALCULATION IS NOT CONVERGING."
1380 PRINT "CHECK YOUR DATA AND START AGAIN."
1390 GOTO 1560
1400 PD = 14.7 * P - (P1 + ZP) / 288
```

**Program listing for computing pressure-drop (continued)** **Table I**

```
1410 XD = DP
1420 DP = PD
1430 MX =  SQR (P * 14.7 * 32.17 * 144 / IV)
1440 F = ( INT (1000 * F)) / 1000
1450  HOME : VTAB 8
1460  PRINT "PRESSURE DROP, PSI           "; FN P(DP)
1470  PRINT "BULK VELOCITY, FT/S          "; FN P(V)
1480  PRINT "REYNOLDS NUMBER, RE          "; FN P(RE)
1490  PRINT "FRICTION FACTOR, F           ";F
1500  ON OP GOTO 1560,1510,1560
1510  PRINT
1520  PRINT "PRESSURE DROP IF VAPOR"
1530  PRINT "WAS INCOMPRESSIBLE, PSI     "; FN P(XD)
1540  PRINT
1550  PRINT "MAXIMUM VELOCITY, FT/S      "; FN P(MX)
1560  PRINT
1570  PRINT  TAB( 13);"END OF PROGRAM"
1580  END
```

**Equations used in the calculation routines** **Table II**

**Liquid-phase flow**

$$V = \frac{Flow}{(RHOL)\,(\frac{1}{4}\pi D^2)\,(3{,}600)} \quad (1)$$

$$Re = \frac{D\,V\,RHOL}{(MUL)\,(6.72 \times 10^{-4})} \quad (2)$$

$$F = 0.0001\,Re^{0.575} \qquad 2{,}100 < Re < 3{,}500 \qquad \text{Ref. [5,6]} \quad (3)$$

$$F = 0.0035 + 0.264\,Re^{-0.420} \qquad Re > 3{,}500 \qquad \text{Ref. [5,6]} \quad (4)$$

$$F = \frac{16}{Re} \qquad Re < 2{,}100 \quad (5)$$

$$DPL = \frac{2\,F\,L\,V^2\,RHOL}{(D)\,(32.17)\,(144)} \quad (6)$$

**Two-phase flow**

$$X = \left(\frac{1-Y}{Y}\right)^{0.9} \left(\frac{RHOV}{RHOL}\right)^{0.5} \left(\frac{MUL}{MUV}\right)^{0.1} = \left[\frac{\Delta P\ (\text{liquid only})}{\Delta P\ (\text{vapor only})}\right]^{1/2} \qquad \text{Ref. [7]} \quad (7)$$

$$COEFF = \left[1 + 3\left(1 - \frac{P}{P_c}\right) X^{-0.55} + X^{-1.11}\right]^{1.75} \qquad \text{Ref. [7]} \quad (8)$$

$$DP2 = (COEFF)\,(DPL) \qquad \text{Ref. [7]} \quad (9)$$

**Vapor-phase flow**

$$\text{Max. velocity, ft/s} = \left[\frac{P\,(14.7)\,(32.17)\,(144)}{RHOV}\right]^{0.5} \qquad \text{Ref. [2]} \quad (10)$$

$$DPV = \frac{2\,F\,L\,V^2\,RHOV}{D\,(32.17)\,(144)} \quad (11)$$

Vapor-phase pressure drop:

$$\frac{dP}{RHOV} + \left(\frac{d(V^2)}{2g_c}\right) - \frac{2\,F\,V^2 dL}{D g_c} = 0 \quad (12)$$

$$RHOV = \frac{P\,(MW)}{(Z)\,(0.7302)\,(T+460)} \quad (13)$$

$$V = \frac{(3.54 \times 10^{-4})\,(Flow)\,(0.7302)\,(Z)\,(T+460)}{(D^2)\,(P)\,(MW)} \quad (14)$$

The solution of the above differential equation after substitution is:

$$\frac{(83{,}184)\,(D^5)\,(MW)\,[(P(14.7)\,(144))^2 - P_{out}^2]}{(Z)\,(T+460)\,(Flow)^2} \quad (15)$$

$$- D \ln\left(\frac{P(14.7)\,(144)}{P_{out}}\right) - 2\,F\,L = 0 \quad (16)$$

In Eq. (18) only, $P$ is assigned units of lb/ft$^2$. Elsewhere, it is in psi.

**Example 1 for: Pressure-drop calculations** **Table III**

**(Start of first display)**

```
PROGRAM TO CALCULATE LINEAR PRESSURE
DROP FOR:

1. LIQUID PHASE FLOW,
2. TWO-PHASE FLOW, OR
3. VAPOR PHASE FLOW.

PRESS RETURN KEY TO SEE DATA REQUIRED.
```

**(Start of next display)**

```
YOU WILL NEED THE FOLLOWING DATA:

                               LIQ VAP MIX
CALCULATION                     1   2   3

INSIDE PIPE DIAMETER, IN.       *   *   *
EQUIVALENT PIPE LENGTH, FT      *   *   *
FLOW RATE, LB/HR                *   *   *
LIQUID DENSITY, LB/CU.FT        *       *
INLET VAP. DENSITY,LB/CU.FT     *   *
LIQUID VISCOSITY, CP            *       *
VAPOR VISCOSITY, CP                 *   *
INLET PRESSURE, ATM                 *   *
VAPOR DENSITY, LB/CU.FT                 *
WEIGHT FRACTION VAPOR                   *
MOLECULAR WEIGHT                    *
TEMPERATURE, DEG. F                 *
COMPRESSIBILITY FACTOR              *

SELECT ONE OF THE CALCULATIONS BY NUMBER.
ENTER 1, 2, OR 3.  1
```

**(Start of next display)**

```
ENTER:

PIPE DIAMETER, IN.                4
EQUIVALENT PIPE LENGTH, FT        100
FLOW RATE, LB/H                   150000
LIQUID DENSITY, LB/CU.FT          62.37
LIQUID VISCOSITY, CP              1.1

THE PROGRAM WILL ESTIMATE A FRICTION
FACTOR ASSUMING SLIGHTLY CORRODED PIPE.
IF YOU HAVE A FRICTION FACTOR, YOU MAY
ENTER IT NOW.  IF NOT, ENTER ZERO AND
THE PROGRAM WILL CALCULATE A VALUE. 0
PRESSURE DROP, PSI          2.3764
BULK VELOCITY, FT/S         7.6554
REYNOLDS NUMBER, RE         215307.012
FRICTION FACTOR, F          5E-03

          END OF PROGRAM
```

**Example 2 for: Pressure-drop calculations** **Table IV**

**(Start of first display)**

```
PROGRAM TO CALCULATE LINEAR PRESSURE
DROP FOR:

1. LIQUID PHASE FLOW,
2. TWO-PHASE FLOW, OR
3. VAPOR PHASE FLOW.

PRESS RETURN KEY TO SEE DATA REQUIRED.
```

**(Start of next display)**

```
YOU WILL NEED THE FOLLOWING DATA:

                               LIQ VAP MIX
CALCULATION                     1   2   3

INSIDE PIPE DIAMETER, IN.       *   *   *
EQUIVALENT PIPE LENGTH, FT      *   *   *
FLOW RATE, LB/HR                *   *   *
```

**Example 2 for: Pressure-drop calculations (continued)** **Table III**

```
LIQUID DENSITY, LB/CU.FT          *       *
INLET VAP. DENSITY,LB/CU.FT       *   *
LIQUID VISCOSITY, CP              *       *
VAPOR VISCOSITY, CP                   *   *
INLET PRESSURE, ATM                   *   *
VAPOR DENSITY, LB/CU.FT                   *
WEIGHT FRACTION VAPOR                     *
MOLECULAR WEIGHT                      *
TEMPERATURE, DEG. F                   *
COMPRESSIBILITY FACTOR                *

SELECT ONE OF THE CALCULATIONS BY NUMBER.
ENTER 1, 2, OR 3.  2
```

**(Start of next display)**

```
ENTER:

PIPE DIAMETER, IN.                4
EQUIVALENT PIPE LENGTH, FT        100
FLOW RATE, LB/H                   9896
INLET VAPOR DENSITY, LB/CU.FT     .14
VAPOR VISCOSITY, CP               .01
INLET PRESSURE, ATM               2
MOLECULAR WEIGHT                  28.59
TEMPERATURE, DEG. F               100
COMPRESSIBILITY FACTOR            1

THE PROGRAM WILL ESTIMATE A FRICTION
FACTOR ASSUMING SLIGHTLY CORRODED PIPE.
IF YOU HAVE A FRICTION FACTOR, YOU MAY
ENTER IT NOW.  IF NOT, ENTER ZERO AND
THE PROGRAM WILL CALCULATE A VALUE. 0

CALCULATION MAY TAKE A FEW SECONDS
PRESSURE DROP, PSI          4.4072
BULK VELOCITY, FT/S         224.9996
REYNOLDS NUMBER, RE         1562497.34
FRICTION FACTOR, F          4E-03
PRESSURE DROP IF VAPOR
WAS INCOMPRESSIBLE, PSI     3.8197

MAXIMUM VELOCITY, FT/S      986.3168

           END OF PROGRAM
```

# Friction factor

## This program quickly determines $f$ for either gas or liquid flow, eliminating the need for a friction-factor chart.

☐ This program uses a binary-search method to determine the friction-factor value, $f$.

The friction factor, $f$, for laminar flow ($N_{Re} \leq 2{,}000$) may be determined by the equation [1,2]:

$$f = \frac{64}{N_{Re}} = \frac{d\mu}{49.3\,QS} \tag{1}$$

The Reynolds number may be calculated from [2]:

$$N_{Re} = 50.6\frac{Q\rho}{d\mu} = 3{,}157.44\frac{QS}{d\mu} \tag{2}$$

For Reynolds numbers between about 2,000 to 4,000, the flow may be laminar or turbulent. Turbulent flow exists when the Reynolds number is around 4,000 and higher. In turbulent flow, the friction factor is dependent not only upon the Reynolds number, but also upon the relative roughness of the pipe internal wall, which is $\epsilon/D$.

A Moody chart is used to correlate $f$ and $\epsilon/D$ to determine the friction factor. The Moody chart is a presentation of the Colebrook equation [1,3]:

$$\frac{1}{\sqrt{f}} = -2\log\left[\frac{2.51}{N_{Re}\sqrt{f}} + \frac{\epsilon}{3.7D}\right] \tag{3}$$

The program we will develop uses $\epsilon$ for steel and wrought-iron pipes. The absolute roughness for such pipes is $\epsilon = 0.00015$ ft ($\epsilon' = 0.0018$ in.).

Substituting $\epsilon' = 0.0018$ in., and $N_{Re}$ from Eq. (2) into Eq. (3), and rearranging:

$$x = \frac{1}{\sqrt{f}} + 2\log\left[\frac{d\mu}{1{,}257\,QS\sqrt{f}} + \frac{1}{2{,}055.6\,d}\right] \tag{4}$$

where:

$$f = \frac{1}{2}(f_1 + f_2) \tag{5}$$

Choosing the proper value of $f$ will reduce $x$ to zero. By using the binary-search method, starting with $f = 0.1$ (designated as $f_1$), which yields a negative $x$ and $f = 0.01$ (designated as $f_2$), which gives a positive $x$, the calculator will perform successive trials of $f$ until $x$ approaches zero—or a set tolerance. A tolerance of $\pm 0.00001$ is chosen here. Once $f$ is found, it can be used to calculate the pressure drop [2]:

$$\Delta P = 0.000216\frac{fL\rho Q^2}{d^5} = 0.0135\frac{fLSQ^2}{d^5} \tag{6}$$

The logic diagram for the program is given in the figure. The program listing is presented in Table I.

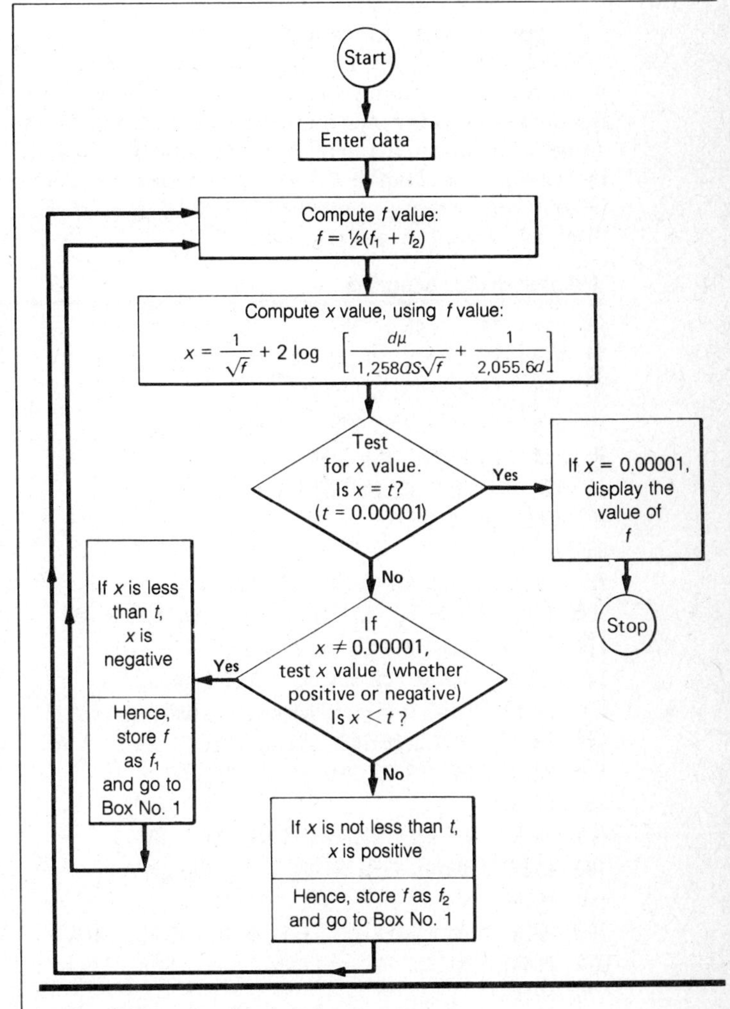


**Logic diagram for friction-factor program**

### Nomenclature

| | | | |
|---|---|---|---|
| $d$ | Internal pipe dia., in. | $N_{Re}$ | Reynolds number |
| $D$ | Internal pipe dia., ft | $\Delta P$ | Pressure drop, psi |
| $f$ | Friction factor | $Q$ | Liquid flowrate, gpm |
| $f_1, f_2$ | Upper and lower limits of friction factor—see Eq. (5) | $S$ | Specific gravity |
| | | $x$ | As defined by Eq. (4) |
| $L$ | Equivalent pipe length, ft | | |

**Greek letters**

| | | | |
|---|---|---|---|
| $\epsilon$ | Roughness parameter, ft | $\mu$ | Viscosity, cP |
| $\epsilon'$ | Roughness parameter, in. | $\rho$ | Density, lb/ft$^3$ |

**Adapted from an article by Mamerto A. Irasga, MMIC—Surigao Nickel Project, originally published April 21, 1980.**

## Using the program

As seen in the first display of the example (Table II), the user can use this program to calculate the friction factor, *f*, for either gas or liquid flow. A pipe roughness factor of 0.0018 in. units, which normally applies to commercial pipe, has been assumed. However, the user is free to change this value, and given the opportunity when inputting data.

Flow data can be entered as any of three different sets of units:

1. lb/min,
2. $ft^3$/min and lb/$ft^3$, or
3. gpm and specific gravity.

Upon selecting (by number) the set of units to be used, the user is then prompted to input the values of the data. This is for pipe diameter (which should be the value of the inside diameter, although the nominal diameter will be acceptable), flowrate (whose value should be entered without commas for values greater than 999), and liquid or gas viscosity.

At this point, the user can input a roughness factor that is different than the assigned value of 0.0018 in. However, if the assigned value is acceptable, enter "0".

The program then calculates the friction factor, using the binary-search method described earlier.

## References


1. Noyer, John J., Calculating Pressure Drops by Calculator or Computer, *Chem. Eng.*, Feb. 18, 1974, p. 154.
2. Crane Co., Technical paper No. 410-C (1969), Flow of Fluids, pp. 3-2 and A-24.
3. Baumeister, Theodore, et al., ed., "Marks' Standard Handbook for Mechanical Engineers," Eighth Ed., 1978, McGraw-Hill Book Co., New York City, p. 3-54.


## The author


**Mamerto A. Irasga** is an engineer in the process engineering department of the MMIC—Surigao Nickel Project, Nonoc Island, Surigao City, Philippines. He graduated from the University of Mindanao (Davao City, Philippines) with a B.S. degree in chemical engineering in 1974, and then joined MMIC as a plant field-operator. He currently does process evaluation, accounting, research and development, and process design. He is a registered engineer in his country, and a member of the Philippine Institute of Chemical Engineers.


**Friction-factor program** **Table I**

```
10 REM   FRICTION FACTOR
20 REM   FROM CHEMICAL ENGINEERING, APRIL 21, 1980
30 REM   BY MAMERTO A. IRASGA
40 REM  TRANSLATED BY WILLIAM VOLK
50 REM   COPYRIGHT (C) 1984
60 REM  BY CHEMICAL ENGINEERING
70 HOME : VTAB 4
80 PI = 3.14159265
90 REM  SET DISPLAY TO FOUR DECIMALS
100 DEF  FN P(X) =  INT (1E4 * (X + .00005)) / 1E4
110 PRINT "PROGRAM WILL CALCULATE THE FRICTION"
120 PRINT "FACTOR FOR GAS OR LIQUID FLOW"
130 PRINT "USING A PIPE ROUGHNESS FACTOR"
140 PRINT "OF 0.0018 INCH UNITS, WHICH"
150 PRINT "APPLIES TO COMMERCIAL PIPE."
160 PRINT
170 PRINT "YOU WILL HAVE AN OPPORTUNITY TO"
180 PRINT "CHANGE THIS VALUE IF YOU WISH."
190 PRINT
200 PRINT "YOU CAN ENTER THE FLOW IN THE FOLLOWING"
210 PRINT "UNITS:"
220 PRINT
230 PRINT "1. LB/MIN"
240 PRINT "2. CU. FT/MIN AND LB/CU. FT"
250 PRINT "3. GAL/MIN AND SPECIFIC GRAVITY."
260 PRINT
270 INPUT "SELECT YOUR METHOD BY NUMBER.   ";OP
280 HOME : VTAB 8
290 PRINT "ENTER THE FOLLOWING DATA:"
300 PRINT : PRINT
310 INPUT "PIPE DIAMETER, IN.              ";D
320 ON OP GOTO 330,350,390
330 INPUT "FLOW RATE, LB/MIN               ";G
340 GOTO 420
350 INPUT "FLOW RATE, CU.FT/MIN            ";CF
360 INPUT "DENSITY, LB/CU.FT               ";DN
370 G = CF * DN
380 GOTO 420
390 INPUT "FLOW RATE, GAL/MIN              ";GA
400 INPUT "SPECIFIC GRAVITY                ";SG
410 G = GA * SG * 8.33
420 INPUT "VISCOSITY, CP                   ";MU
430 PRINT
440 PRINT "IF YOU WISH A DIFFERENT ROUGHNESS"
450 PRINT "FACTOR THAN 0.0018 IN., ENTER THAT"
460 INPUT "VALUE.  IF NOT, ENTER ZERO.     ";R
470 IF R = 0 THEN R = .0018
480 PRINT
490 REM  REYNOLDS NUMBER
500 RE = G / D / MU * .8E4 / 6.72 / PI
510 IF RE > 2000 GOTO 560
520 REM  FOR REYNOLDS NUMBERS 2000 OR LESS
530 F = 64 / RE
540 GOTO 690
550 REM  FOR REYNOLDS NUMBERS GREATER THAN 2000
560 F2 = 1
570 SQ = 2.51 / RE /  SQR (F2) + R / 3.7 / D
580 RT =  - 2 *  LOG (SQ) /  LOG (10)
590 F1 = (1 / RT) ^ 2
600 IF  ABS (F1 - F2) < .0001 GOTO 680
610 F2 = (F1 + F2) / 2
620 CO = CO + 1
630 IF CO < 100 GOTO 570
640 HOME : VTAB 4
650 PRINT "CALCULATION DOES NOT CONVERGE.
660 PRINT "CHECK YOUR DATA."
670 GOTO 710
680 F = (F1 + F2) / 2
690 F = ( INT (F * 1E5)) / 1E5
700 PRINT "FRICTION FACTOR IS "; FN P(F)
710 PRINT
720 PRINT  TAB( 13);"END OF PROGRAM"
730 END
```

**Example for: Friction-factor program** **Table II**

**(Start of first display)**

```
PROGRAM WILL CALCULATE THE FRICTION
FACTOR FOR GAS OR LIQUID FLOW
USING A PIPE ROUGHNESS FACTOR
OF 0.0018 INCH UNITS, WHICH
APPLIES TO COMMERCIAL PIPE.

YOU WILL HAVE AN OPPORTUNITY TO
CHANGE THIS VALUE IF YOU WISH.

YOU CAN ENTER THE FLOW IN THE FOLLOWING
UNITS:

1. LB/MIN
2. CU. FT/MIN AND LB/CU. FT
3. GAL/MIN AND SPECIFIC GRAVITY.

SELECT YOUR METHOD BY NUMBER.   1
```

**(Start of next display)**

```
ENTER THE FOLLOWING DATA:


PIPE DIAMETER, IN.          4.03
FLOW RATE, LB/MIN           3200
VISCOSITY, CP               .3

IF YOU WISH A DIFFERENT ROUGHNESS
FACTOR THAN 0.0018 IN., ENTER THAT
VALUE.  IF NOT, ENTER ZERO.     0

FRICTION FACTOR IS .0168

            END OF PROGRAM
```

**(Start of first display)**

```
PROGRAM WILL CALCULATE THE FRICTION
FACTOR FOR GAS OR LIQUID FLOW
USING A PIPE ROUGHNESS FACTOR
OF 0.0018 INCH UNITS, WHICH
APPLIES TO COMMERCIAL PIPE.

YOU WILL HAVE AN OPPORTUNITY TO
CHANGE THIS VALUE IF YOU WISH.

YOU CAN ENTER THE FLOW IN THE FOLLOWING
UNITS:

1. LB/MIN
2. CU. FT/MIN AND LB/CU. FT
3. GAL/MIN AND SPECIFIC GRAVITY.

SELECT YOUR METHOD BY NUMBER.   3
```

**(Start of next display)**

```
ENTER THE FOLLOWING DATA:


PIPE DIAMETER, IN.          4.03
FLOW RATE, GAL/MIN          415
SPECIFIC GRAVITY            .936362
VISCOSITY, CP               .3

IF YOU WISH A DIFFERENT ROUGHNESS
FACTOR THAN 0.0018 IN., ENTER THAT
VALUE.  IF NOT, ENTER ZERO.     .0036

FRICTION FACTOR IS .0195

            END OF PROGRAM
```

# Piping design

Good for both laminar and turbulent flow, this rapid-converging program will predict any one of the following variables if the other two are known: pressure drop, flowrate, and piping diameter.

☐ The laborious calculations and tedious charts and nomographs once associated with pipe sizing have given way in recent years to streamlined computations via card-programmable calculators. With their increasing sophistication, these calculators have proved more convenient for line sizing than any other tool available, including large main-frame computers. Now, this methodology is offered for the microcomputer.

A number of piping-hydraulics programs are available in the literature [2,4]. Crowley's compilation [3] of over 850 programs that perform hydrologic and hydraulic computations is particularly worthy of note. Most of the available pressure-drop programs use the implicit Colebrook/White friction factor equation [5] and are designed for specific sets of units or consistent units. Many are not suitable for the critical range of Reynolds numbers (i.e., the transition region between laminar and turbulent flow), and are not written for laminar flow or do not converge in that flow regime.

The accompanying program, originally designed for the HP 67/97 calculators, was designed to solve a broad range of fluid flow problems with the utmost convenience. It is based on sizing techniques discussed by the author in the Apr. 3, 1978, article [1], "Design of Plant Piping." The program's main features are:

- Any convenient and consistent set of units can be used. (See Table III for a summary of input and output variables and their appropriate units.)
- Given any two of the following three input parameters, the program will solve for the third: pressure drop or fluid head, mass or volumetric flowrate, and piping diameter.
- Computations for both laminar and turbulent flow are built into the program. The general Churchill equation [7], which spans the two flow regimes, is incorporated into the program as a Weisbach friction factor [1]. The program has been devised so as to circumvent an overflow problem that develops with the equation at low Reynolds numbers.
- The program uses an Aitken $\delta^2$ iterative technique [6], which converges more rapidly and uses less storage than the usual Newton scheme.
- The user can specify the number of velocity heads, and piping length or equivalent length. Program outputs include velocity, Reynolds number, and friction factor. Again, any convenient units can be used for velocity.
- The program can accommodate different sets of units, as well as default values for density, viscosity, pipe roughness, length, and number of velocity heads.

## Using the program

This piping design program calculates either pressure drop, flowrate or pipe diameter, depending on the input of the other two variables. See Table I for the program listing.

As seen in the first display (Table II), the program uses default values for pipe fitting losses (i.e., number of velocity heads), pipe length, fluid density and velocity, and pipe roughness factor. Or, it allows the user to input values in place of any of the default values.

The program also permits the use of a variety of units—5 different "conventional unit" systems, or 4 consistent metric-unit sets—corresponding to the nomenclature of Table III. The second display offers the user to select a set of units: "C" for conventional, or "M" for metric. For the example shown, "C" is chosen, which results in a display of the five (numbered) system options available for conventional units.

At this point, and for this case, the user selection is "2," which is consistent with the units in the example.

The next display prompts the user to enter the number for each type of pipe fitting, so that these losses can be calculated. Values of velocity head are preassigned for the following pipe-fitting losses: pipe elbows, check valves, entrance losses, exit losses, and butterfly valves.

The next display allows the user the option to use, or change, the following default values, which are for this example:

| | |
|---|---|
| Length | 100 ft |
| Density | 62.32 lb/ft$^3$ |
| Viscosity | 1.005 cP |
| Roughness factor | 0.0018 in. |

Adapted from an article by Larry L. Simpson, Union Carbide Corp., originally published January 29, 1979.

If you wish to change any of the above default values, enter your value. If not, press the "RETURN" key.

At this point, the new display requests the user to input values for two of the following three variables:

Pressure drop
Flowrate
Pipe diameter.

The third variable (for which a "0" value should be inputted) will be calculated.

The final display summarizes the data entered, and offers the calculated results. In this case, pipe diameter, flow velocity, Reynolds number, and friction factor are calculated values.

## Example

Size a steel pipe for 500 gpm water at ambient conditions. Pipe is to be 250 ft long with 8 elbows, one swing check, one entrance, one exit, and one butterfly valve (open). Physical properties of water are $\rho = 62.32$ lb/ft$^3$ and $\mu = 1.005$ cp. A pressure drop of 5 psi is available. A roughness of $\epsilon = 0.0018$ is used as the default value, but will be changed to $\epsilon = 0.0036$ for this problem.

## References


1. Simpson, L., and Weirick, M., *Chem. Eng.*, Apr. 3, 1978, p. 35.
2. Benenati, R. F., *Chem. Eng.*, Feb. 28, 1977, pp. 201–206.
3. Crowley, T. E., "Hydrologic and Hydraulic Computations on Small Programmable Calculators," Iowa Institute of Hydraulic Research, Iowa City, Iowa, 1977.
4. Users' Library, "Thermal and Transport Sciences," Hewlett-Packard, 1977.
5. Moody, L. F., *Trans. ASME*, Vol. 66, 1944, pp. 671–684.
6. Abramowitz, M., and Stegun, I. A., "Handbook of Mathematical Functions," National Bureau of Standards, Washington, D.C., 1964.
7. Churchill, S. W., *Chem. Eng.*, Nov. 7, 1977, pp. 91–92.

## The author


**Larry L. Simpson** is manager of heat transfer and fluid dynamics in the Engineering Dept. at Union Carbide's Technical Center, Box 8364, South Charleston, WV 25303. His past work includes plant design and startup, and for the last eleven years, fluid dynamics specialization. He has written several articles on different aspects of this subject, including a chapter in the book "Turbulence in Mixing Operations." He has a B.S.Ch.E. from Purdue University and an M.S.Ch.E. from West Virginia University. He is a registered professional engineer in West Virginia and is active in AIChE.


**Program for piping design** **Table I**

```
10  REM   PIPING DESIGN
20  REM   FROM CHEMICAL ENGINEERING, JANUARY 19, 1979
30  REM   BY LARRY L. SIMPSON
40  REM  TRANSLATED BY WILLIAM VOLK
50  REM   COPYRIGHT (C) 1984
60  REM  BY CHEMICAL ENGINEERING
70  REM  PROGRAM DEFAULT VALUES
80 PI = 3.14159265
90 L = 100
100 L$ = "FT"
110 RO = 62.32
120 RO$ = "LB/CU.FT"
130 L$ = "FT"
140 MU = 1.005
150 MU$ = "CP"
160 R = .0018
170 R$ = "IN."
180 DI$ = "IN."
190  HOME : VTAB 4
200  REM  SET DISPLAY AT FOUR DECIMALS
210  DEF  FN P(X) =  INT (1E4 * (X + .00005)) / 1E4
220  PRINT "THIS PROGRAM CALCULATES PRESSURE DROP,"
230  PRINT "FLOW RATE OR DIAMETER FROM INPUT OF"
240  PRINT "THE OTHER TWO PARAMETERS."
250  PRINT
260  PRINT "IT USES DEFAULT VALUES FOR PIPE FITTING"
270  PRINT "LOSSES, PIPE ROUGHNESS FACTOR,
280  PRINT "FLUID DENSITY, AND VISCOSITY."
300  PRINT
310  PRINT "IT ALLOWS INPUT OF YOUR VALUES IN PLACE"
320  PRINT "OF ANY OF THE DEFAULT VALUES."
330  PRINT
340  PRINT "IT PERMITS A VARIETY OF UNITS FOR THE"
350  PRINT "SYSTEM PARAMETERS."
360  PRINT
370  INPUT "PRESS RETURN TO CONTINUE.   ";Q$
380  HOME : VTAB 4
390  PRINT "DATA MAY BE ENTERED IN A NUMBER OF"
410  PRINT "COMBINATIONS OF CONVENTIONAL UNITS:"
420  PRINT "LB/H, GPH, FT, IN., PSI, ETC.
430  PRINT
440  PRINT "OR"
450  PRINT
460  PRINT "COMBINATIONS OF METRIC UNITS:"
470  PRINT "KG/S, L/S, M, MM, BAR, PA, ETC.
480  PRINT
490  PRINT "MAKE YOUR CHOICE:"
500  INPUT "C FOR CONVENTIONAL OR M FOR METRIC. ";CH$
510  IF CH$ = "C" GOTO 570
520  IF CH$ = "M" GOTO 1120
530  PRINT
540  PRINT "YOUR CHOICES WERE C OR M."
550  PRINT "YOU ENTERED "CH$".  TRY AGAIN."
560  GOTO 480
570  HOME
580 U$ = "CONVENTIONAL"
590  PRINT "CONSISTENT CONVENTIONAL UNITS ARE:"
600  PRINT
610  PRINT "SYSTEM"
620  PRINT "NUMBER"; TAB( 12);"1"; TAB( 18);"2"; TAB( 24)
     ;"3"; TAB( 29);"4"; TAB( 36);"5"
630  PRINT  TAB( 35);"GAS"
```

**Program for piping design (continued)** **Table I**

```
640  PRINT
650  PRINT "VEL.HEAD"; TAB( 11);"PSI"; TAB( 17);"PSI"; TAB( 23);"FT"; TAB( 28);"
FT"; TAB( 34);"IN H2O"
660  PRINT "FLOW RATE"; TAB( 11);"LB/H"; TAB( 17);"GPM"; TAB( 22);"LB/H"; TAB( 2
8);"GPM"; TAB( 34);"ACFM"
670  PRINT "DIAMETER"; TAB( 11);"IN"; TAB( 17);"IN"; TAB( 23);"IN"; TAB( 28);"IN
"; TAB( 34);"IN
680  PRINT "ROUGHNESS"; TAB( 11);"IN"; TAB( 17);"IN"; TAB( 23);"IN"; TAB( 28);"I
N"; TAB( 34);"IN"
690  PRINT "LENGTH"; TAB( 11);"FT"; TAB( 17);"FT"; TAB( 23);"FT"; TAB( 28);"FT";
 TAB( 34);"FT"
700  PRINT "DENSITY"; TAB( 10);"LB/CF"; TAB( 16);"LB/CF"; TAB( 22);"LB/CF"; TAB(
 28);"LB/CF"; TAB( 34);"LB/CF"
710  PRINT "VISCOSITY"; TAB( 11);"CP"; TAB( 17);"CP"; TAB( 23);"CP"; TAB( 28);"C
P"; TAB( 34);"CP"
720  PRINT "VELOCITY"; TAB( 11);"F/S"; TAB( 17);"F/S"; TAB( 23);"F/S"; TAB( 28);
"F/S"; TAB( 34);"F/M"
730  PRINT
740  INPUT "MAKE YOUR CHOICE BY NUMBER.     ";OP
750 B = 12
760  ON OP GOTO 770,840,910,980,1050
770 A = 2.799E - 7
780 C = 6.316
790 D = .05093
800 DP$ = "PSI"
810 W$ = "LB/H"
820 V$ = "FT/S"
830  GOTO 1530
840 A = 1.801E - 5
850 C = 50.66
860 D =  - .4085
870 DP$ = "PSI"
880 W$ = "GPM"
890 V$ = "FT/S"
900  GOTO 1530
910 A = (4.031E - 5) / 12
920 C =  - 6.316
930 D = .05093
940 DP$ = "FT"
950 W$ = "LB/H"
960 V$ = "FT/S"
970  GOTO 1530
980 A = (2.593E - 3) / 12
990 C =  - 50.66
1000 D =  - .4085
1010 DP$ = "FT"
1020 W$ = "GPM"
1030 V$ = "FT/S"
1040  GOTO 1530
1050 A = .02792
1060 C = 379
1070 D =  - 183.3
1080 DP$ = "IN H2O"
1090 W$ = "ACFM"
1100 V$ = "FT/MIN"
1110  GOTO 1530
1120  HOME
```

**Program for piping design (continued)** **Table I**

```
1130 U$ = "METRIC"
1140 L$ = "M"
1150 RO = 998.4
1160 RO$ = "KG/CU.M"
1170 MU = .001005
1180 MU$ = "PA.S"
1190 R = 4.57E - 5
1200 R$ = "M"
1210  PRINT "CONSISTENT METRIC UNITS ARE:"
1220  PRINT
1230  PRINT "SYSTEM NUMBER"; TAB( 17);"1"; TAB( 23);"2"; TAB( 29);"3"; TAB( 35);
"4"
1240  PRINT
1250  PRINT "VEL.HEAD"; TAB( 16);"BAR"; TAB( 22);"BAR"; TAB( 29);"PA"; TAB( 35);
"M"
1260  PRINT "FLOW RATE"; TAB( 16);"KG/S"; TAB( 22);"L/S"; TAB( 28);"KG/S"; TAB(
33);"CU.M/S"
1270  PRINT "DIAMETER"; TAB( 16);"MM"; TAB( 22);"MM"; TAB( 29);"M"; TAB( 35);"M"

1280  PRINT "ROUGHNESS"; TAB( 16);"MM"; TAB( 22);"MM"; TAB( 29);"M"; TAB( 35);"M
"
1290  PRINT "LENGTH"; TAB( 16);"- - - - - M - - - - -"
1300  PRINT "DENSITY"; TAB( 16);"- - - - KG/CU.M - - -"
1310  PRINT "VISCOSITY"; TAB( 16);"- M.PA.S -"; TAB( 29);"- PA.S -"
1320  PRINT "VELOCITY"; TAB( 16);"- - - - - M/S - - - -"
1330  PRINT
1340  INPUT "MAKE YOUR CHOICE BY NUMBER.     ";OP
1350 C = 1.273:V$ = "M/S"
1360  ON OP GOTO 1370,1370,1460,1500
1370 A = 8.265E6
1380 B = 1000
1390 C = C * 1E6
1400 D = C
1410 MU = 1.005:MU$ = "MPA.S"
1420 R = .04573:R$ = "MM":DI$ = "MM"
1430 DP$ = "BAR":W$ = "KG/S"
1440  IF OP = 2 THEN W$ = "L/S"
1450  GOTO 1530
1460 A = .8106:B = 1
1470 DP$ = "PA":W$ = "KG/S"
1480 DI$ = "M"
1490  GOTO 1520
1500 A = .08265:B = 1
1510 DP$ = "M":W$ = "CU.M/S"
1520 D = C:DI$ = "M"
1530  HOME : VTAB 4
1540  PRINT "THE CALCULATION WILL ALLOW FOR PIPE"
1550  PRINT "FITTING LOSSES.
1560  PRINT
1570  PRINT "IF YOU HAVE ANY OF THE FOLLOWING,
1580  PRINT "ENTER THE NUMBER.  IF NOT, ENTER ZERO."
1590  PRINT
1600  INPUT "   PIPE ELBOWS          ";EL
1610  INPUT "   CHECK VALVES         ";CK
1620  INPUT "   ENTRANCE LOSS        ";EN
1630  INPUT "   EXIT LOSS            ";EX
1640  INPUT "   BUTTERFLY VALVES     ";BU
```

**Program for piping design (continued)** **Table I**

```
1650 K = K + .3 * EL + 2 * CK + .5 * EN + EX + .2 * BU
1660  HOME : VTAB (4)
1670  PRINT "THE PROGRAM USES THE FOLLOWING DEFAULT"
1680  PRINT "VALUES.  IF YOU WISH TO CHANGE ANY,"
1690  PRINT "ENTER YOUR VALUES.  IF NOT, JUST PRESS"
1700  PRINT "THE 'RETURN' KEY."
1710  PRINT
1720  PRINT "THE DEFAULT DENSITY AND VISCOSITY ARE"
1730  PRINT "BASED ON WATER AT 20 DEG. F IN THE UNITS"
1740  PRINT "YOU SELECTED; AND THE ROUGHNESS FACTOR"
1750  PRINT "IS FOR CLEAN IRON PIPE."
1760  PRINT
1770  PRINT "                DEFAULT VALUE"
1780  PRINT
1790  PRINT "LENGTH"; TAB( 20);"100 ";L$;: INPUT "      ";A$
1800  IF  VAL (A$) <  > 0 THEN L =  VAL (A$)
1810  PRINT "DENSITY"; TAB( 20); STR$ (RO) + " " + RO$;: INPUT "   ";B$
1820  IF  VAL (B$) <  > 0 THEN RO =  VAL (B$)
1830  PRINT "VISCOSITY"; TAB( 20); STR$ (MU) + " " + MU$;: INPUT "    ";C$
1840  IF  VAL (C$) <  > 0 THEN MU =  VAL (C$)
1850  PRINT "ROUGHNESS FACTOR"; TAB( 20); STR$ (R) + " " + R$;: INPUT "   ";D$
1860  IF  VAL (D$) <  > 0 THEN R =  VAL (D$)
1870  HOME : VTAB (4)
1880  PRINT "WITH INPUT OF TWO OF THE FOLLOWING, THE PROGRAM WILL CALCULATE THE
THIRD."
1890  PRINT
1900  PRINT "ENTER THE VALUES YOU HAVE AND ZERO"
1910  PRINT "FOR THE VALUE YOU WANT CALCULATED."
1920  PRINT
1930  PRINT "USE THE ";U$;" UNITS"
1940  PRINT "SELECTED EARLIER."
1950  PRINT
1960  PRINT "PRESSURE DROP, ";DP$; TAB( 30);" ";: INPUT " ";DP
1970  PRINT "FLOW RATE, ";W$; TAB( 30);: INPUT " ";W
1980  PRINT "PIPE DIAMETER, ";DI$; TAB( 30);: INPUT " ";DI
1990 D(1) = DP
2000 D(2) = W
2010 D(3) = DI
2020  IF DP = 0 GOTO 2060
2030  IF C > 0 GOTO 2050
2040 D(1) = D(1) * RO
2050  IF W = 0 GOTO 2160
2060  IF D > 0 GOTO 2080
2070 D(2) = D(2) * RO
2080  IF DI = 0 GOTO 2210
2090  REM  CALCULATE DP
2100 I = 1
2110  GOSUB 2390
2120 DP = SB
2130  IF C > 0 GOTO 2500
2140 DP = DP / RO
2150  GOTO 2500
2160  REM  CALCULATE W
2170 I = 2
2180 D(I) = DI ^ 2 *  SQR (RO * DP / A)
2190  IF D < 0 THEN FL = 1
2200  GOTO 2240
```

**Program for piping design (continued)** **Table I**

```
2210  REM  CALCULATE DI
2220 I = 3
2230 D(I) =  SQR (D(2) *  SQR (A / RO / DP))
2240 NI = D(I)
2250  IF I = 3 GOTO 2300
2260  GOSUB 2470
2270 D(I) = XH
2280  GOSUB 2470
2290  GOTO 2330
2300  GOSUB 2360
2310 D(I) = XH
2320  GOSUB 2360
2330  IF  ABS (D(I) - NI) < 1E - 4 GOTO 2500
2340 D(I) = NI - (D(I) - NI) ^ 2 / (XH - D(I) - D(I)
     + NI)
2350  GOTO 2240
2360  GOSUB 2390
2370 XH = D(I) * (SB / DP) ^ (1 / 4)
2380  RETURN
2390 RE =  ABS (C) * D(2) / MU / D(3)
2400  IF RE =  < 1 GOTO 2440
2410 TM = 2 * ( LOG ((7 / RE) ^ .9 + (R / D(3)) * .27)
     /  LOG (10))
2420 SM = ((TM) ^ 16 + ((115) ^ 2 / RE) ^ 16) ^ 1.5
2430 SM = 1 / SM
2440 F = (SM + (64 / RE) ^ 12) ^ (1 / 12)
2450 SB = A / RO * (F * B * L / D(3) + K) * (D(2) /
     D(3) ^ 2) ^ 2
2460  RETURN
2470  GOSUB 2390
2480 XH = D(2) /  SQR (SB / DP)
2490  RETURN
2500  HOME : VTAB (4)
2510  IF I = 2 GOTO 2530
2520  GOTO 2560
2530 W = D(I)
2540  IF D > 0 GOTO 2560
2550 W = W / RO
2560 D(1) = DP
2570 DI = D(3)
2580 V =  ABS (D(2) * D / RO) / DI ^ 2
2590  PRINT "WITH YOUR DATA:"
2600  PRINT
2610  PRINT "PIPE LENGTH,";L$; TAB( 30);L
2620  PRINT "FLUID DENSITY,";RO$; TAB( 30);RO
2630  PRINT "FLUID VISCOSITY,";MU$; TAB( 30);MU
2640  PRINT "ROUGHNESS FACTOR,";R$; TAB( 30);R
2650  PRINT
2660  PRINT "A PIPE FITTING FACTOR, K:" TAB( 30);K
2670  PRINT
2680  PRINT "THE CALCULATED RESULTS ARE:"
2690  PRINT
2700  PRINT "FLOW RATE,";W$; TAB( 30); FN P(W)
2710  PRINT "DIAMETER, ";DI$; TAB( 30); FN P(DI)
2720  PRINT "VELOCITY, ";V$; TAB( 30); FN P(V)
2730  PRINT "PRESSURE DROP,";DP$; TAB( 30); FN P(DP)
2740  PRINT "REYNOLDS NUMBER"; TAB( 30); INT (RE)
2750  PRINT "FRICTION FACTOR"; TAB( 30); FN P(F)
2760  PRINT
2770  PRINT  TAB( 13);"END OF PROGRAM"
2780  END
```

**Example for: Piping design** **Table II**

**(Start of first display)**

```
THIS PROGRAM CALCULATES PRESSURE DROP,
FLOW RATE OR DIAMETER FROM INPUT OF
THE OTHER TWO PARAMETERS.

IT USES DEFAULT VALUES FOR PIPE FITTING
LOSSES, PIPE ROUGHNESS FACTOR,
FLUID DENSITY, AND VISCOSITY.

IT ALLOWS INPUT OF YOUR VALUES IN PLACE
OF ANY OF THE DEFAULT VALUES.

IT PERMITS A VARIETY OF UNITS FOR THE
SYSTEM PARAMETERS.

PRESS RETURN TO CONTINUE.
```

**(Start of next display)**

```
DATA MAY BE ENTERED IN A NUMBER OF
COMBINATIONS OF CONVENTIONAL UNITS:
LB/H, GPH, FT, IN., PSI, ETC.

OR

COMBINATIONS OF METRIC UNITS:
KG/S, L/S, M, MM, BAR, PA, ETC.

MAKE YOUR CHOICE:
C FOR CONVENTIONAL OR M FOR METRIC. C
```

**Example for: Piping design (continued)** **Table II**

**(Start of next display)**

```
CONSISTENT CONVENTIONAL UNITS ARE:

SYSTEM
NUMBER     1     2     3     4      5
                                   GAS

VEL.HEAD  PSI   PSI   FT    FT     IN H20
FLOW RATE LB/H  GPM   LB/H  GPM    ACFM
DIAMETER  IN    IN    IN    IN     IN
ROUGHNESS IN    IN    IN    IN     IN
LENGTH    FT    FT    FT    FT     FT
DENSITY  LB/CF LB/CF LB/CF LB/CF LB/CF
VISCOSITY CP    CP    CP    CP     CP
VELOCITY  F/S   F/S   F/S   F/S    F/M

MAKE YOUR CHOICE BY NUMBER.     2
```

**(Start of next display)**

```
THE CALCULATION WILL ALLOW FOR PIPE
FITTING LOSSES.

IF YOU HAVE ANY OF THE FOLLOWING,
ENTER THE NUMBER.  IF NOT, ENTER ZERO.

   PIPE ELBOWS          8
   CHECK VALVES         1
   ENTRANCE LOSS        1
   EXIT LOSS            1
   BUTTERFLY VALVES     1
```

**(Start of next display)**

```
THE PROGRAM USES THE FOLLOWING DEFAULT
VALUES.  IF YOU WISH TO CHANGE ANY,
ENTER YOUR VALUES.  IF NOT, JUST PRESS
THE 'RETURN' KEY.

THE DEFAULT DENSITY AND VISCOSITY ARE
BASED ON WATER AT 20 DEG. F IN THE UNITS
YOU SELECTED; AND THE ROUGHNESS FACTOR
IS FOR CLEAN IRON PIPE.

               DEFAULT VALUE

LENGTH              100 FT       250
DENSITY             62.32 LB/CU.FT
VISCOSITY           1.005 CP
ROUGHNESS FACTOR    1.8E-03 IN.   .0036
```

**(Start of next display)**

```
WITH INPUT OF TWO OF THE FOLLOWING, THE PROGRAM WILL
CALCULATE THE THIRD.

ENTER THE VALUES YOU HAVE AND ZERO
FOR THE VALUE YOU WANT CALCULATED.

USE THE CONVENTIONAL UNITS
SELECTED EARLIER.

PRESSURE DROP, PSI             5
FLOW RATE, GPM               500
PIPE DIAMETER, IN.             0
```

**(Start of next display)**

```
WITH YOUR DATA:

PIPE LENGTH,FT               250
FLUID DENSITY,LB/CU.FT       62.32
FLUID VISCOSITY,CP           1.005
ROUGHNESS FACTOR,IN.         3.6E-03

A PIPE FITTING FACTOR, K:    6.1

THE CALCULATED RESULTS ARE:

FLOW RATE,GPM                500
DIAMETER, IN.                5.5199
VELOCITY, FT/S               6.7034
PRESSURE DROP,PSI            5
REYNOLDS NUMBER              284553
FRICTION FACTOR              .0192

            END OF PROGRAM
```

## Nomenclature and dimensions of input and output variables

**Table III**

**Pressure drop, Reynolds number, and velocity**

| Flowrate | $-\Delta P_f$ or $\rho H_f$ | $N_{Re}$ | $V$ |
|---|---|---|---|
| $W$ | $\left(\frac{bfL}{D}+\Sigma K_i\right)\frac{aW^2}{\rho D^4}$ | $\frac{cW}{\mu D}$ | $\frac{dW}{\rho D^2}$ |
| $Q$ | $\left(\frac{bfL}{D}+\Sigma K_i\right)\frac{a\rho Q^2}{D^4}$ | $\frac{cQ\rho}{\mu D}$ | $\frac{dQ}{D^2}$ |

**Units and constants**

| | Conventional units | | | | | Metric units | | | |
|---|---|---|---|---|---|---|---|---|---|
| | | | | | [Gas] | | | | |
| $-\Delta P_f$ or $H_f$ | psi | psi | ft | ft | in. $H_2O$ (60°F) | bar | bar | Pa | m |
| $W$ or $Q$ | lb/h | gpm | lb/h | gpm | acfm | kg/s | L/s | kg/s | $m^3/s$ |
| $D$ | in. | in. | in. | in. | in. | mm | mm | m | m |
| $\epsilon$ | in. | in. | in. | in. | in. | mm | mm | m | m |
| $L$ | ft | ft | ft | ft | ft | m | m | m | m |
| $\rho$ | $lb/ft^3$ | $lb/ft^3$ | $lb/ft^3$ | $lb/ft^3$ | $lb/ft^3$ | $kg/m^3$ | $kg/m^3$ | $kg/m^3$ | $kg/m^3$ |
| $\mu$ | cP | cP | cP | cP | cP | mPa·s (cP) | mPa·s (cP) | Pa·s | Pa·s |
| $V$ | ft/s | ft/s | ft/s | ft/s | ft/min | m/s | m/s | m/s | m/s |
| $a$ | $2.799 \times 10^{-7}$ | $1.801 \times 10^{-5}$ | $4.031 \times 10^{-5}$ | $2.593 \times 10^{-3}$ | 0.02792 | $8.106 \times 10^6$ | 8.106 | 0.8106 | 0.08265 |
| $b$ | 12 | 12 | 12 | 12 | 12 | 1,000 | 1,000 | 1 | 1 |
| $c$ | 6.316 | 50.66 | 6.316 | 50.66 | 379.0 | $1.273 \times 10^6$ | 1,273 | 1.273 | 1.273 |
| $d$ | 0.05093 | 0.4085 | 0.05093 | 0.4085 | 183.3 | $1.273 \times 10^6$ | 1,273 | 1.273 | 1.273 |

$a, b, c, d$ Constants for consistent units
$D$ Pipe diameter
$f$ Friction factor [7]
$H_f$ Frictional head loss
$K$ Number of velocity heads
$L$ Pipe length
$-\Delta P_f$ Frictional pressure drop
$Q$ Volumetric flowrate
$N_{Re}$ Reynolds number
$V$ Velocity
$W$ Mass flowrate
$\epsilon$ Pipe roughness
$\mu$ Fluid viscosity
$\rho$ Fluid density

# Centrifugal-pump hydraulics

This computer program simplifies the writing of a pump specification by asking all the right questions. The questions prompt the user to furnish the data needed to carry out head calculations in an efficient, step-by-step manner.

☐ The program described here performs all head calculations needed to write a pump specification suitable for a vendor inquiry or purchase order. The program is written for use with the Apple II computer.

A unique feature of the program (Table I) is that it is self-prompting. It asks a series of questions that elicit the necessary data from the user. The program proceeds in this manner until all the parameters required to specify a pump have been calculated.

## Methodology

Fig. 1 shows a typical calculation worksheet for a process pump.

The calculations proceed along the flow circuit shown in the figure, using the vessel pressure as the starting point (or "origin"). To this is added the static head, which is entered in feet, and converted to psi. (All liquid static heads are automatically converted to psi.) Line pressure-drop in the pump suction is then subtracted, and the calculated suction-pressure printed.

A similar procedure is used to compute NPSH (net-positive suction head) and discharge pressure. Once the discharge pressure has been determined, the total pump pressure-rise, $\Delta P$ in psi or feet of liquid, and the hydraulic horsepower are calculated automatically. Actual brake horsepower is easily obtained by dividing by the pump efficiency.

The program can be applied to any centrifugal pump, provided the proper input data are available. The amount of data required is the same as for other methods used to compute a pump's hydraulics, but the step-by-step layout of the program reduces the possibilities for omitting information, overlooking a step, or introducing mathematical errors.

## Using the program

This program (Table I) will request that the following data be entered:

1 Flow, gpm
2 Specific gravity
3 Vapor pressure, psia
4 Origin pressure, psia
5 Suction line losses, psi
6 Suction static head, ft
7 Delivery pressure, psi
8 Discharge static head, ft
9 Control valve ΔP, psi
10 Exchanger ΔP, psi
11 Orifice ΔP, psi
12 Discharge line losses, psi
13 Other losses, psi.

Please note that an index number is assigned each item, thus allowing the user to correct data by entering the index number and the new data value.

At this point, the user is asked if there are any corrections to be made. Corrections should be entered as index number, followed by a comma, and the correct value of that item.

Once all the correct values are inputted, the next display supplies the calculated results for suction static head, static pressure, available NPSH, discharge static head, discharge pressure, total pump pressure drop (in both psi and ft), and hydraulic brake horsepower (HBHP).

Once again, the user can change data and make another calculation, as described above (i.e., via index number and new data value inputs).

## An example

The program was used to find the hydraulics of a pump circulating water to a quench column (Table II).

The flowsheet in Fig. 1 summarizes the operating conditions in the system, as well as the pressure drops across the various components. (Input data required to run the program have been marked with an asterisk.)

### The author

**W. Wayne Blackwell** is a senior process engineer with Ford, Bacon & Davis Texas, Inc., P.O. Box 30209, Dallas TX 75238. His present responsibilities include process design of tail-gas cleanup systems, and the hydraulic and utility designs for several refineries. He has twenty years' experience in gas-process design, including four years as senior process engineer with Ralph M. Parsons Co. in Pasadena, Calif., and sixteen years as process design engineer with the U.S. Bureau of Mines Helium Operations in Amarillo, Tex. He holds a B.S. in chemical engineering from Texas Technological University, and is a registered P.E. in that state.

Adapted from an article by W. Wayne Blackwell, Ford, Bacon & Davis Texas, Inc., originally published January 28, 1980.

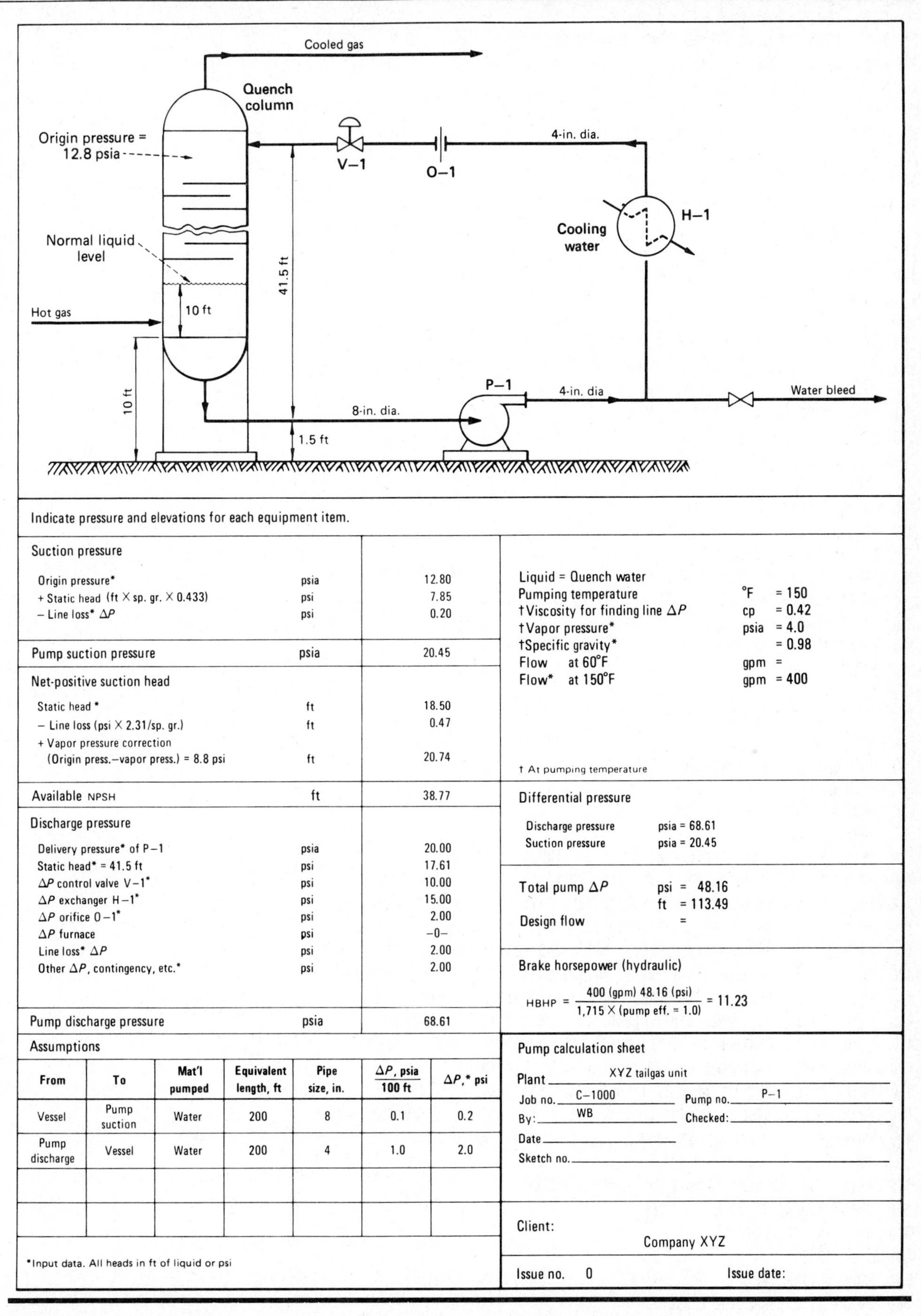


Indicate pressure and elevations for each equipment item.

| Suction pressure | | |
|---|---|---|
| Origin pressure* | psia | 12.80 |
| + Static head (ft × sp. gr. × 0.433) | psi | 7.85 |
| − Line loss* $\Delta P$ | psi | 0.20 |
| **Pump suction pressure** | psia | 20.45 |
| **Net-positive suction head** | | |
| Static head * | ft | 18.50 |
| − Line loss (psi × 2.31/sp. gr.) | ft | 0.47 |
| + Vapor pressure correction (Origin press.−vapor press.) = 8.8 psi | ft | 20.74 |
| **Available NPSH** | ft | 38.77 |
| **Discharge pressure** | | |
| Delivery pressure* of P−1 | psia | 20.00 |
| Static head* = 41.5 ft | psi | 17.61 |
| $\Delta P$ control valve V−1* | psi | 10.00 |
| $\Delta P$ exchanger H−1* | psi | 15.00 |
| $\Delta P$ orifice O−1* | psi | 2.00 |
| $\Delta P$ furnace | psi | −0− |
| Line loss* $\Delta P$ | psi | 2.00 |
| Other $\Delta P$, contingency, etc.* | psi | 2.00 |
| **Pump discharge pressure** | psia | 68.61 |

Liquid = Quench water
Pumping temperature °F = 150
†Viscosity for finding line $\Delta P$ cp = 0.42
†Vapor pressure* psia = 4.0
†Specific gravity* = 0.98
Flow at 60°F gpm =
Flow* at 150°F gpm = 400

† At pumping temperature

**Differential pressure**

Discharge pressure psia = 68.61
Suction pressure psia = 20.45

Total pump $\Delta P$ psi = 48.16
ft = 113.49
Design flow =

**Brake horsepower (hydraulic)**

$$\text{HBHP} = \frac{400\ (\text{gpm})\ 48.16\ (\text{psi})}{1{,}715 \times (\text{pump eff.} = 1.0)} = 11.23$$

**Assumptions**

| From | To | Mat'l pumped | Equivalent length, ft | Pipe size, in. | $\Delta P$, psia 100 ft | $\Delta P$,* psi |
|---|---|---|---|---|---|---|
| Vessel | Pump suction | Water | 200 | 8 | 0.1 | 0.2 |
| Pump discharge | Vessel | Water | 200 | 4 | 1.0 | 2.0 |
| | | | | | | |
| | | | | | | |

*Input data. All heads in ft of liquid or psi

**Pump calculation sheet**

Plant XYZ tailgas unit
Job no. C−1000 Pump no. P−1
By: WB Checked:
Date
Sketch no.

Client: Company XYZ

Issue no. 0 Issue date:

**Typical pump-calculation worksheet. Pump circulates water to quench column (see text) Fig. 1**

**Program for centrifugal-pump hydraulics** **Table I**

```
10  REM   CENTRIFUGAL-PUMP HYDRAULICS
20  REM   FROM CHEMICAL ENGINEERING, JANUARY 28, 1980
30  REM   BY W. WAYNE BLACKWELL
40  REM  TRANSLATED BY WILLIAM VOLK
50  REM   COPYRIGHT (C) 1984
60  REM  BY CHEMICAL ENGINEERING
70  REM  DIMENSION OF INPUT VARIABLES AND SET DISPLAY TO 4 DECIMALS.
80  DIM R(15)
90  DEF  FN P(X) =  INT (1E4 * (X + .00005)) / 1E4
100  HOME : PRINT
110  PRINT "PROGRAM CALCULATES DATA REQUIRED FOR"
120  PRINT "CENTRIFUGAL PUMP SPECIFICATION."
130  PRINT
140  PRINT "PROGRAM WILL REQUEST THE FOLLOWING"
150  PRINT "DATA, AND WILL ASSIGN AN INDEX NUMBER"
160  PRINT "TO EACH ITEM SO THAT CHANGES MAY"
170  PRINT "BE MADE BY ENTERING THE INDEX NUMBER"
180  PRINT "AND THE CORRECTED VALUE."
190  PRINT
200  PRINT "FLOW, GPM"
210  PRINT "SPECIFIC GRAVITY"
220  PRINT "VAPOR PRESSURE, PSIA"
230  PRINT "ORIGIN PRESSURE, PSIA"
240  PRINT "SUCTION LINE LOSSES, PSIA"
250  PRINT "SUCTION STATIC HEAD, FT"
260  PRINT "DELIVERY PRESSURE, PSI"
270  PRINT "DISCHARGE STATIC HEAD, FT"
280  PRINT "LINE LOSSES: CONTROL VALVE, EXCHANGER,"
290  PRINT "ORIFICE, ETC."
300  PRINT
310  INPUT "PRESS RETURN TO ENTER DATA   ";Q$
320  HOME : VTAB 4
330  PRINT "ENTER THE FOLLOWING DATA:"
340  PRINT
350  PRINT "INDEX             DATA"
360  PRINT
370  PRINT "  1   FLOW, GPM"; TAB( 33);" ";: INPUT " ";R(1)
380  PRINT "  2   SPECIFIC GRAVITY"; TAB( 33);" ";: INPUT " ";R(2)
390  PRINT "  3   VAPOR PRESSURE, PSIA"; TAB( 33);" ";: INPUT " ";R(3)
400  PRINT "  4   ORIGIN PRESSURE, PSIA"; TAB( 33);" ";: INPUT " ";R(4)
410  PRINT "  5   SUCTION LINE LOSSES, PSI"; TAB( 33);" ";: INPUT " ";R(5)
420  PRINT "  6   SUCTION STATIC HEAD, FT"; TAB( 33);" ";: INPUT " ";R(6)
430  PRINT "  7   DELIVERY PRESSURE, PSI"; TAB( 33);" ";: INPUT " ";R(7)
440  PRINT "  8   DISCHARGE STATIC HEAD, FT"; TAB( 33);" ";: INPUT " ";R(8)
450  PRINT "  9   CONTROL VALVE DP, PSI"; TAB( 33);" ";: INPUT " ";R(9)
460  PRINT " 10   EXCHANGER DP, PSI"; TAB( 33);" ";: INPUT " ";R(10)
470  PRINT " 11   ORIFICE DP, PSI"; TAB( 33);" ";: INPUT " ";R(11)
480  PRINT " 12   DISCHARGE LINE LOSSES, PSI"; TAB( 33);" ";: INPUT " ";R(12)
490  PRINT " 13   OTHER LOSSES, PSI"; TAB( 33);" ";: INPUT " ";R(13)
500  PRINT
510  PRINT "DO YOU WANT TO MAKE ANY CORRECTIONS?
520  INPUT "ANSWER Y OR N.    ";X$
530  IF X$ = "Y" GOTO 900
540 R1 = .433
550  HOME : VTAB (4)
560  PRINT "THESE ARE THE RESULTS:"
570  PRINT
```

**Program for centrifugal-pump hydraulics (continued)** **Table I**

```
580 R9 = R(6) * R(2) * R1
590 R9 =  FN P(R9)
600  PRINT "SUCTION STATIC HEAD, PSI"; TAB( 33);R9
610 R0 = R9 + R(4) - R(5)
620 R0 =  FN P(R0)
630  PRINT "SUCTION PRESSURE, PSI"; TAB( 33);R0
640 R2 = R(6) + (R(4) - R(3)) / R1 / R(2) - (R(5) / R1
    / R(2))
650 R2 =  FN P(R2)
660  PRINT "AVAILABLE NPSH, FT"; TAB( 33);R2
670 DF = R(8) * R1 * R(2)
680 DF =  FN P(DF)
690  PRINT "DISCHARGE STATIC HEAD, FT"; TAB( 33);DF
700 R(7) = R(7) + DF
710  FOR I = 9 TO 13
720 R(7) = R(7) + R(I)
730  NEXT I
740 R(7) =  FN P(R(7))
750  PRINT "DISCHARGE PRESSURE, PSIA"; TAB( 33);R(7)
760 RD = R(7) - R0
770  PRINT "TOTAL PUMP DP, PSI"; TAB( 33);RD
780 DF = RD / R1 / R(2)
790 DF =  FN P(DF)
800  PRINT "                FEET"; TAB( 33);DF
810 HB = RD * R(1) / (1715)
820 HB =  FN P(HB)
830  PRINT "                HBHP"; TAB( 33);HB
840  PRINT
850  PRINT "IF YOU WISH ANOTHER CALCULATION, YOU"
860  PRINT "MAY CHANGE THE DATA WITH THE CORRECTION"
870  PRINT "ROUTINE.  DO YOU WISH TO DO SO?"
880  INPUT "ANSWER Y OR N.    ";X$
890  IF X$ = "N" GOTO 1070
900  REM  CORRECTION ROUTINE
910  HOME : VTAB (8)
920  PRINT "ENTER THE DATA INDEX AND THE CORRECT"
930  PRINT "VALUE.    FOR EXAMPLE:
940  PRINT
950  PRINT
960  PRINT "         2, 0.975"
970  PRINT
980  PRINT "WOULD CHANGE THE SECOND DATA ITEM,"
990  PRINT "THE SPCIFIC GRAVITY, TO 0.975."
1000  PRINT
1010  INPUT "ENTER YOUR CORRECTIONS. ";I,X
1020 R(I) = X
1030  PRINT
1040  INPUT "ANY OTHERS?  ANSWER Y OR N.    ";X$
1050  IF X$ = "Y" GOTO 1000
1060  GOTO 540
1070  PRINT
1080  PRINT  TAB( 13);"END OF PROGRAM"
1090  END
```

**Example for: Centrifugal-pump hydraulics** **Table II**

**(Start of first display)**

```
PROGRAM CALCULATES DATA REQUIRED FOR
CENTRIFUGAL PUMP SPECIFICATION.

PROGRAM WILL REQUEST THE FOLLOWING
DATA, AND WILL ASSIGN AN INDEX NUMBER
TO EACH ITEM SO THAT CHANGES MAY
BE MADE BY ENTERING THE INDEX NUMBER
AND THE CORRECTED VALUE.

FLOW, GPM
SPECIFIC GRAVITY
VAPOR PRESSURE, PSIA
ORIGIN PRESSURE, PSIA
SUCTION LINE LOSSES, PSIA
SUCTION STATIC HEAD, FT
DELIVERY PRESSURE, PSI
DISCHARGE STATIC HEAD, FT
LINE LOSSES: CONTROL VALVE, EXCHANGER,
ORIFICE, ETC.

PRESS RETURN TO ENTER DATA
```

**(Start of next display)**

```
ENTER THE FOLLOWING DATA:

INDEX          DATA

  1   FLOW, GPM                   400
  2   SPECIFIC GRAVITY            .98
  3   VAPOR PRESSURE, PSIA        4
  4   ORIGIN PRESSURE, PSIA       2112.8
  5   SUCTION LINE LOSSES, PSI    .2
  6   SUCTION STATIC HEAD, FT     18.5
  7   DELIVERY PRESSURE, PSI      20
  8   DISCHARGE STATIC HEAD, FT   15
  9   CONTROL VALVE DP, PSI       10
 10   EXCHANGER DP, PSI           15
 11   ORIFICE DP, PSI             2
 12   DISCHARGE LINE LOSSES, PSI  2
 13   OTHER LOSSES, PSI           2

DO YOU WANT TO MAKE ANY CORRECTIONS?
ANSWER Y OR N.    Y
```

**Example for: Centrifugal-pump hydraulics (continued)** **Table II**

**(Start of next display)**

```
ENTER THE DATA INDEX AND THE CORRECT
VALUE.    FOR EXAMPLE:


        2, 0.975

WOULD CHANGE THE SECOND DATA ITEM,
THE SPCIFIC GRAVITY, TO 0.975.

ENTER YOUR CORRECTIONS. 4,12.8

ANY OTHERS?  ANSWER Y OR N.    Y

ENTER YOUR CORRECTIONS. 8,41.5

ANY OTHERS?  ANSWER Y OR N.    N
```

**(Start of next display)**

```
THESE ARE THE RESULTS:

SUCTION STATIC HEAD, PSI          7.8503
SUCTION PRESSURE, PSI            20.4503
AVAILABLE NPSH, FT               38.7668
DISCHARGE STATIC HEAD, FT        17.6101
DISCHARGE PRESSURE, PSIA         68.6101
TOTAL PUMP DP, PSI               48.1598
               FEET             113.4934
               HBHP              11.2326

IF YOU WISH ANOTHER CALCULATION, YOU
MAY CHANGE THE DATA WITH THE CORRECTION
ROUTINE.  DO YOU WISH TO DO SO?
ANSWER Y OR N.    N

              END OF PROGRAM
```

# Orifice sizes for gas flows

This program yields the orifice diameter directly. Gases are accounted for by a determination of the expansion factor.

☐ In process design, an orifice must often be sized for a given flowrate and pressure drop. Frequently, the orifice is designed to develop a full-scale pressure-differential at maximum flow. A convenient formula that yields the flowrate for the orifice pressure-drop is (see Bean, Howard S., Fluid meters—their theory and application, "Report of ASME Research Committee on Fluid Meters," 6th ed., Amer. Soc. of Mech. Engrs., 1971, p. 208):

$$m = 0.52502\left(\frac{CYd^2F_a}{\sqrt{1-\beta^4}}\right)\sqrt{\rho\Delta P} \qquad (1)$$

For liquid flow, the expansion factor, $Y$, is 1.0, and Eq. (1) is solved easily for the orifice size at maximum flow for the full-scale $\Delta P$.

With gas flow, however, $Y$ is a function of $\beta$. For ideal gases:

$$Y = 1 - (0.41 + 0.35\beta^4)\frac{\Delta P}{Pk} \qquad (2)$$

Although Eq. (2) is for ideal gases, it will be assumed accurate enough for real gases in sizing orifices.

Determining the orifice size involves either a trial-and-error calculation or solution of a cubic equation. The solution of the cubic equation is demonstrated below.

## Cubic equation

Eq. (1) and (2) may be simplified by assuming that $C = 0.62$ and $F_a = 1.0$. This value of $C$ is a commonly used one for high Reynolds numbers. (Rarely is the flow through an orifice at low Reynolds numbers.) The program may be modified for use with other values of $C$ if desired by replacing the constant 0.3255 by 0.525 $C$ in program steps 168 through 172. The following substitutions are made:

Let:

$$y = \beta^4 \qquad (3)$$

$$l = \frac{m}{0.3255D^2\sqrt{\rho\Delta P}} \qquad (4)$$

$$n = 1 - 0.41\frac{\Delta P}{Pk} \qquad (5)$$

$$t = 0.35\frac{\Delta P}{Pk} \qquad (6)$$

Eq. (1) and (2) can be modified and combined:

$$l = (ny^{1/2} - ty^{3/2})/(1 - y)^{1/2} \qquad (7)$$

Squaring both sides and dividing by $t^2$ puts Eq. (7) in cubic form:

$$y^3 - 2\frac{n}{t}y^2 + \frac{n^2 + l^2}{t^2}y - \frac{l^2}{t^2} = 0 \qquad (8)$$

Eq. (8) may be solved for $y$ using the conventional method for finding roots to cubic equations.

Let:

$$p = 2n/t \qquad (9)$$

$$q = (n^2 + l^2)/t^2 \qquad (10)$$

$$r = -l^2/t^2 \qquad (11)$$

and:

$$a = (3q - p^2)/3 = (3l^2 - n^2)/3t^2 \qquad (12)$$

$$b = (2p^3 - 9pq + 27r)/27 = -(2n^3 + 18nl^2 + 27tl^2)/27t^3 \qquad (13)$$

The solution to Eq. (8) yielding a real number is:

$$y = [-b/2 + (b^2/4 + a^3/27)^{1/2}]^{1/3} + [-b/2 - (b^2/4 + a^3/27)^{1/2}]^{1/3} - 2n/3t \qquad (14)$$

The orifice diameter is given by:

$$d = Dy^{1/4} \qquad (15)$$

As can be seen, calculation of the orifice diameter by the above equations can be quite tedious. However, the use of a programmable calculator greatly minimizes the work involved.

## The Apple II program

A program has been developed for the Apple II to perform this calculation. The program (Table I) permits determination of the orifice size, using full-scale flowrates in either lb/s or actual $ft^3$/min and full-scale pressure drops in either psi or in. $H_2O$, using the relationships:

Adapted from an article by William H. Mink, Battelle Columbus Laboratories, originally published August 25, 1980.

$$\Delta P = 0.036063 h_w \tag{16}$$

$$m = \frac{(ACFM)\rho}{60} \tag{17}$$

The program applies to corner, flange, vena-contracta, and diameter and ½-diameter taps.

## Instructions and example

The result of a sample calculation is shown in Table II. In this example, the orifice size to give a full-scale reading of 20 in. $H_2O$ for 0.25 lb/s flow of gas through a 4-in.-I.D. pipe is determined. The upstream gas density is 0.07 lb/ft$^3$, the specific heat ratio is 1.4, and the upstream pressure is 14.7 psi. As a check, substituting the calculated orifice size, 1.8405, in Eq. (1) and (2) yields a value of 0.250 for *m*. The program (see Table I for listing) prompts the user to enter the data required, which is shown in the first display (Table II). After hitting the "RETURN" key, the next display asks the user which units are being used for orifice pressure drop. The choices are:

"1" for psi, or
"2" for in. $H_2O$.

Then, the user should enter the value of the orifice pressure drop in these units.

Similarly, for flowrate, the user can enter units as:

"1" for lb/s, or
"2" for acfm.

Once again, after inputting the correct number, the flowrate value for these units should be entered.

At this point, the other data required are:

pipe diameter, in.
gas density, lb/ft$^3$
upstream pressure, psi
specific heat ratio, k.

From this data, the orifice diameter is calculated, and the user can enter new data by inputting "Y" to the query, "DO YOU WANT TO MAKE ANOTHER CALCULATION?".

Be careful when using values of *k* from a table of data. Specific heat ratios are a function of both temperature and pressure. Temperature has a greater effect. For example, *k* for carbon dioxide decreases from 1.37 at $-100°F$ to about 1.30 at 60°F. The pressure effect is not great except near saturation.

## Nomenclature

| | |
|---|---|
| $a$ | As defined in Eq. (12) |
| $ACFM$ | Full-scale flowrate at upstream conditions, ft$^3$/min |
| $b$ | As defined in Eq. (13) |
| $C$ | Coefficient of discharge |
| $D$ | Pipe I.D., in. |
| $d$ | Orifice dia. for full-scale $\Delta P$, in. |
| $F_a$ | Area thermal-expansion factor. Takes into account the expansion of the orifice with temperature changes |
| $h_w$ | Full-scale orifice pressure-differential, in. $H_2O$ at 68°F |
| $k$ | Specific heat ratio (constant pressure to constant volume) |
| $l$ | As defined in Eq. (4) |
| $m$ | Full-scale flowrate, lb/s |
| $n$ | As defined in Eq. (5) |
| $P$ | Upstream pressure, psia |
| $\Delta P$ | Full-scale orifice pressure-differential, psi |
| $p$ | As defined in Eq. (9) |
| $q$ | As defined in Eq. (10) |
| $r$ | As defined in Eq. (11) |
| $t$ | As defined in Eq. (6) |
| $Y$ | Expansion factor |
| $y$ | As defined in Eq. (3) |
| **Greek letters** | |
| $\beta$ | $d/D$, ratio of orifice dia. to pipe I.D. |
| $\rho$ | Upstream fluid density, lb/ft$^3$ |

## The author

**William H. Mink** is Senior Research Engineer at Battelle Columbus Laboratories, 505 King Ave., Columbus, OH 43201. He is active in the process development of hazardous-waste treatment, flue-gas desulfurization, and coal gasification. He holds a B.S. and an M.S. in chemical engineering from the University of Wisconsin. He is a member of AIChE and a registered professional engineer in the state of Ohio.

**Program for orifice sizes for gas flows** **Table I**

```
10 REM  ORIFICE SIZES FOR GAS FLOWS
20 REM  FROM CHEMICAL ENGINEERING, AUGUST 25, 1980
30 REM  BY WILLIAM H. MINK
40 REM  TRANSLATED BY WILLIAM VOLK
50 REM  COPYRIGHT (C) 1984
60 REM  BY CHEMICAL ENGINEERING
70 HOME : VTAB 4
80 ONERR  GOTO 850
90 PRINT "CALCULATION OF ORIFICE DIAMETER"
100 PRINT "FOR GAS FLOW."
110 PRINT
120 PRINT "DATA REQUIRED ARE:"
130 PRINT
140 PRINT "ORIFICE PRESSURE DROP, IN. H2O OR PSI"
150 PRINT "FLOW RATE, ACFM OR LB/S"
160 PRINT "PIPE ID, IN."
170 PRINT "GAS DENSITY, LB/CU. FT"
180 PRINT "UPSTREAM PRESSURE, PSI"
190 PRINT "SPECIFIC HEAT RATIO, K"
200 PRINT
210 INPUT "PRESS RETURN KEY TO ENTER THE DATA.";X$
220 HOME : VTAB (4)
230 PRINT "IF ORIFICE DP IS PSI, ENTER 1."
240 INPUT "IF IT IS IN. H2O,      ENTER 2.    ";DP
250 PRINT
260 IF DP = 2 GOTO 360
```

**Program for orifice sizes for gas flows (continued)** **Table I**

```
270  IF DP = 1 GOTO 330
280  PRINT "YOUR CHOICES WERE 1 OR 2.  YOU ENTERED"
290  PRINT DP;".  TRY AGAIN."
300  PRINT
310  GOTO 230
320  IF DP = 2 GOTO 360
330  INPUT "ENTER ORIFICE DP, PSI              ";R1
340 A$(1) = "PSI"
350  GOTO 380
360  INPUT "ENTER ORIFICE DP, IN. H20          ";R5
370 A$(2) = "IN. H20"
380  PRINT
390  PRINT "IF FLOW IS LB/S       ENTER 1."
400  INPUT "IF IT IS ACFM,        ENTER 2.    ";FL
410  PRINT
420  IF FL = 2 GOTO 510
430  IF FL = 1 GOTO 480
440  PRINT "YOUR CHOICES WERE 1 OR 2.  YOU ENTERED"
450  PRINT FL;".  TRY AGAIN."
460  GOTO 380
470  IF FL = 2 GOTO 510
480  INPUT "ENTER FLOW, LB/S                   ";R2
490 B$(1) = "LB/S"
500  GOTO 530
510  INPUT "ENTER FLOW, ACFM                   ";R6
520 B$(2) = "ACFM"
530  PRINT
540  INPUT "ENTER PIPE DIAMETER, IN.           ";R3
550  INPUT "ENTER GAS DENSITY, LB/CU. FT       ";R4
560  INPUT "ENTER UPSTREAM PRESSURE, PSI       ";R7
570  INPUT "ENTER SPECIFIC HEAT RATIO, K       ";R9
580  IF R5 = 0 GOTO 600
590 R1 = .036063 * R5
600  IF R6 = 0 GOTO 620
610 R2 = R6 * R4 / 60
620 N = 1 - .41 * R1 / R7 / R9
630 T = .35 * R1 / R7 / R9
640 L = R2 / .3255 / R3 ^ 2 /  SQR (R4 * R1)
650 P = 2 * N / T
660 Q = (N ^ 2 + L ^ 2) / T ^ 2
670 R =  - (L ^ 2 / T ^ 2)
680 A = (3 * Q - P ^ 2) / 3
690 B = (2 * P ^ 3 - 9 * P * Q + 27 * R) / 27
700 Y = ( - B / 2 + (B ^ 2 / 4 + A ^ 3 / 27) ^ .5) ^
   (1 / 3)
710 Y = Y + ( - B / 2 - (B ^ 2 / 4 + A ^ 3 / 27) ^ .5)
   ^ (1 / 3)
720 Y = Y - 2 * N / 3 / T
730 D = R3 * Y ^ .25
740  PRINT
750 D =  INT (1E4 * (D + .00005)) / 1E4
760  PRINT "ORIFICE DIAMETER, INCHES IS       ";D
770  PRINT
780  PRINT "DO YOU WANT TO MAKE ANOTHER CALCULATION?"
790  INPUT "ANSWER Y OR N.                 ";AN$
800  IF AN$ = "Y" GOTO 220
810  GOTO 1050
820  PRINT
830  PRINT  TAB( 13);"END OF PROGRAM"
840  GOTO 1000
850  HOME : PRINT
860  PRINT "THERE APPEARS TO BE SOMETHING WRONG"
870  PRINT "WITH YOUR DATA."
880  PRINT
890  PRINT "THE VALUES ARE:"
900  PRINT
910  IF DP = 2 GOTO 940
920  PRINT "ORIFICE DP ";R1;" ";A$(DP)
930  GOTO 950
940  PRINT "ORIFICE DP ";R5;" ";A$(DP)
950  IF FL = 2 GOTO 980
960  PRINT "FLOW      ";R2;" ";B$(FL)
970  GOTO 990
980  PRINT "FLOW      ";R6;" ";B$(FL)
990  PRINT "PIPE DIA.  ";R3;" IN."
1000  PRINT "DENSITY    ";R4;" LB/CU. FT"
1010  PRINT "PRESSURE   ";R7;" PSI"
1020  PRINT "K VALUE    ";R9
1030  PRINT
1040  GOTO 780
1050  PRINT
1060  PRINT  TAB( 13);"END OF PROGRAM"
1070  END
```

**Example for: Orifice sizes for gas flows** **Table II**

**(Start of first display)**

```
CALCULATION OF ORIFICE DIAMETER
FOR GAS FLOW.

DATA REQUIRED ARE:

ORIFICE PRESSURE DROP, IN. H20 OR PSI
FLOW RATE, ACFM OR LB/S
PIPE ID, IN.
GAS DENSITY, LB/CU. FT
UPSTREAM PRESSURE, PSI
SPECIFIC HEAT RATIO, K

PRESS RETURN KEY TO ENTER THE DATA.
```

**(Start of next display)**

```
IF ORIFICE DP IS PSI, ENTER 1.
IF IT IS IN. H20,     ENTER 2.    2

ENTER ORIFICE DP, IN. H20         20

IF FLOW IS LB/S       ENTER 1.
IF IT IS ACFM,        ENTER 2.    1

ENTER FLOW, LB/S                  .25

ENTER PIPE DIAMETER, IN.          4
ENTER GAS DENSITY, LB/CU. FT      .07
ENTER UPSTREAM PRESSURE, PSI      14.7
ENTER SPECIFIC HEAT RATIO, K      1.4

ORIFICE DIAMETER, INCHES IS       1.8405

DO YOU WANT TO MAKE ANOTHER CALCULATION?
ANSWER Y OR N.              N

           END OF PROGRAM
```

# Hole-area distribution for liquid spargers

This program does the tedious calculations involved in designing spargers, and prints (or displays) the results.

☐ In many operations, it is necessary to sparge a liquid or gas into another fluid; frequently, a uniform distribution is desired. Since, owing to frictional losses, the pressure decreases as the fluid flows down the sparger pipe, the hole area must increase along the length of the sparger to maintain uniform flow from it.

H.W. Cooper† has developed a series of relationships that are useful in calculating sparger hole area for uniform distribution. It has been pointed out that these equations are in error because the originator neglected to include the effect of velocity head. This velocity-head effect can be significant. Only when the pipe $L/D$ ratio is high or when flowrates are low do the Cooper equations provide an approximation. The author has developed a new program, based on a stepwise calculation, which yields correct hole sizes regardless of $L/D$ ratio or flowrates.

## Calculational approach

The pressure causing flow through any orifice is the difference between the static pressure at the point of the orifice and the ambient pressure. The static pressure is the total pressure less the velocity head and any frictional loss:

$$P_s = P_t - P_{vh} - P_f \tag{1}$$

For the sparger, the total inlet pressure is:

$$P_{t,in} = P_{s,in} + P_{vh,in} \tag{2}$$

In Eq. (1) and (2),

$$P_{vh} = \frac{V^2\rho}{2g(144)} = \frac{V^2\rho}{9{,}274} \tag{3}$$

$$V = \frac{0.4085Q}{d^2} \tag{4}$$

and

$$P_f = \frac{FV^2\rho L/10}{d(193)} \tag{5}$$

†Cooper, Herbert W., Area Allocation for Distributor Pipes, *Chem. Eng.*, Oct. 28, 1963, p. 148.

The friction factor, $F$, depends on the Reynolds number, $N_{Re}$:

$$F = 16/N_{Re}, \text{ if } N_{Re} < 2{,}100 \tag{6}$$

$$F = 0.0035 + 0.264(N_{Re})^{-0.42}, \text{ if } N_{Re} \geq 2{,}100 \tag{7}$$

where

$$N_{Re} = \frac{124\ Vd\rho}{\mu} \tag{8}$$

The orifice area can be calculated by the orifice equation, which can be simplified to:

$$a = Q/1830[(P_s - P_e)/\rho]^{1/2} \tag{9}$$

A program has been developed for the Apple II, using the above procedure. The program calculates the hole area for each 10% of sparger length (using Eq. (9)). The program is shown in Table I.

## Procedure

For the calculation of the hole area in the first section (first 10% of length), the flowrate down the pipe is set at the entered value. For each subsequent hole-area calculation, the flowrate down the pipe is reduced by 10% of the entered value.

The velocity is calculated by Eq. (4), the velocity head by Eq. (3), and the Reynolds number by Eq. (8).* The appropriate equation for friction factor is selected [Eq. (6) or (7)]. The frictional loss is calculated by Eq. (5).

The pressure difference, $P_s - P_e$, is calculated using Eq. (1) and the entered ambient pressure. This difference is then used to calculate the area, using Eq. (9).

It should be noted that, since the entered $P_{in}$ is a static pressure, $P_{t,in}$ must be calculated using Eq. (2). (Since holes are assumed to be located at the center of each section, the frictional loss for the first section takes

*The Reynolds number is increased by 1 in the program so that Eq. (5) does not give an infinite result (causing the computer to indicate an error condition at the end of the calculation) when $V = 0$. The addition of 1 has negligible effect on the calculation results.

Adapted from an article by William H. Mink, Battelle Columbus Laboratories, originally published April 6, 1981.

place over only 5% of the pipe length rather than 10% as for the subsequent sections. To adjust for the program, which subtracts the frictional loss for 10% of the pipe length for each hole calculation, the initial total pressure used in the calculation is increased by one-half of the initial frictional loss.)

## Sample problem

Cooper demonstrated the calculation of hole area with the following example:

Consider 2,000 gal/min of water at 25 psia and 95°F, which is to be sprayed onto a distributor plate via holes in an internal pipe that is 10 ft long. Determine the area allocation for a 6-in. pipe. Tower pressure is 24.5 psia. In this case, $\rho = 62.4$ lb/cu ft, $\mu = 0.76$ cP.

Table II shows the printout for this example. The first display shows the data required, which are:

flow to sparger, gpm
pressure to sparger, psia
ambient pressure, psia
sparger I.D., in.
sparger fluid viscosity, cP
sparger fluid density, lb/ft³
sparger length, ft.

After pressing "RETURN", the user is prompted to enter the values for the above variables.

The calculated results are for the sparger area for each 10% of the length, and for the total area. In some cases, the sparger diameter might be too small for the specified flow. The program then indicates this fact, and also repeats your data, as well as asking whether you wish to enter new data.

The areas calculated differ substantially from those calculated by Cooper; in this example, velocity head is an important factor.

Fig. 1 shows, diagramatically, the application of the calculated results. (The hole size and spacing shown are for illustrative purposes and do not relate to the sample problem.) In this example, the hole area could be obtained by several combinations of hole sizes and numbers of holes. One possible arrangement would be to have 16 1-in. holes over the first 10% of length, 11 1-in. holes over the next 10% of length, and finally 6 1-in. holes over the last 10% of length.

### Nomenclature

| | |
|---|---|
| $a$ | Orifice area, in.$^2$ |
| $d$ | Sparger pipe I.D., in. |
| $D$ | Sparger pipe I.D., ft |
| $L$ | Pipe length, ft |
| $P_e$ | Ambient pressure (pressure sparger is discharging into), psia |
| $P_{in}$ | Initial (entered) static pressure, psia |
| $P_f$ | Frictional loss, psi |
| $P_s$ | Static pressure, psia |
| $P_t$ | Total pressure, psia |
| $P_{vh}$ | Velocity head, psi |
| $Q$ | Flowrate entering sparger pipe, gpm |
| $V$ | Velocity in sparger pipe, ft/s |
| $\rho$ | Fluid density, lb/ft$^3$ |
| $\mu$ | Fluid viscosity, cP |

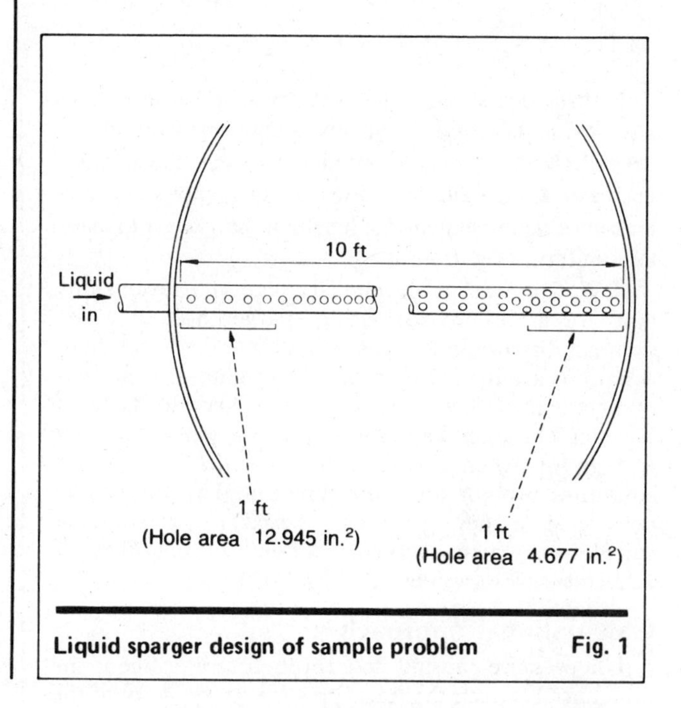


Liquid sparger design of sample problem **Fig. 1**

**Program for hole-area distribution for liquid spargers** **Table I**

```
10  REM  HOLE-AREA DISTRIBUTION FOR LIQUID SPARGERS
20  REM   FROM CHEMICAL ENGINEERING, APRIL 6, 1981
30  REM   BY WILLIAM H. MINK
40  REM  TRANSLATED BY WILLIAM VOLK
50  REM   COPYRIGHT (C) 1984
60  REM  BY CHEMICAL ENGINEERING
70  HOME : VTAB 4
80  PRINT "CALCULATION OF HOLE-AREA DISTRIBUTION"
90  PRINT "FOR LIQUID SPARGERS."
100  PRINT
110  PRINT "DATA REQUIRED ARE:"
120  PRINT
130  PRINT "FLOW TO SPARGER, GPM"
140  PRINT "PRESSURE TO SPARGER, PSIA"
150  PRINT "AMBIENT PRESSURE, PSIA"
160  PRINT "SPARGER I.D., IN."
170  PRINT "SPARGER FLUID VISCOSITY, CP"
180  PRINT "SPARGER FLUID DENSITY, LB/CU. FT"
190  PRINT "SPARGER LENGTH, FT"
200  PRINT
210  INPUT "PRESS RETURN KEY TO ENTER THE DATA.  ";X$
220  HOME : VTAB 8
230  PRINT "ENTER THE FOLLOWING DATA."
240  PRINT
250  INPUT "FLOW TO SPARGER, GPM                    ";R1
260  INPUT "PRESSURE TO SPARGER, PSIA               ";R2
270 R8 = R2
280  INPUT "AMBIENT PRESSURE, PSIA                  ";R3
290  INPUT "SPARGER I.D., IN.                       ";R4
300  INPUT "SPARGER FLUID VISCOSITY, CP             ";R5
310 F1 = 1
320  INPUT "SPARGER FLUID DENSITY, LB/CU. FT  ";R6
```

**Program for hole-area distribution for liquid spargers (continued)** **Table I**

```
330  INPUT "SPARGER LENGTH, FT                    ";R7
340 P1 = R1
350 R8 = 0
360 P3 = .4085 * P1 / R4 ^ 2
370 P4 = P3 ^ 2 * R6 / 9274
380 P5 = 1 + 124 * P3 * R4 * R6 / R5
390  IF P5 < 2100 GOTO 420
400 P6 = .0035 + (.264 * P5 ^ ( - .42))
410  GOTO 430
420 P6 = 16 / P5
430 P7 = R7 * R6 * P6 * P3 ^ 2 / R4 / 1930
440  IF F1 = 1 GOTO 560
450 P8 = P8 - P7
460 TP = P8 - P4 - R3
470  IF TP < 0 GOTO 700
480 P0 = R1 / 1830 /  SQR (TP / R6)
490 SM = P1 - R1 / 10
500  IF SM < 0 GOTO 590
510 P1 = SM
520 H(C) = P0
530 C = C + 1
540 R8 = R8 + P0
550  GOTO 360
560 P8 = P7 / 2 + P4 + R2
570 F1 = 0
580  GOTO 450
590  HOME : VTAB 4
600  PRINT "SPARGER AREA FOR EACH 10% OF LENGTH."
610  PRINT
620  PRINT  TAB( 12);"SQ. IN."
630  PRINT
640  FOR C = 0 TO 9
650  PRINT  TAB( 12); INT (1E4 * (H(C) + .00005)) / 1E4
660  NEXT C
670  PRINT
680  PRINT "TOTAL AREA "; INT (1E4 * (R8 + .000005)) /
     1E4;" SQ. IN."
690  GOTO 910
700  HOME : VTAB 8
710  PRINT "SPARGER DIAMETER TOO SMALL FOR THE"
720  PRINT "SPECIFIED FLOW.
730  PRINT
740  PRINT "YOUR DATA WERE:"
750  PRINT
760  PRINT "LIQUID RATE,        "R1;" GPM"
770  PRINT "INLET PRESSURE,     "R2;" PSI"
780  PRINT "AMBIENT PRESSURE,   "R3;" PSI"
790  PRINT "SPARGER DIAMETER,   "R4;" IN."
800  PRINT "FLUID VISCOSITY,    "R5;" CP"
810  PRINT "FLUID DENSITY,      "R6;" LB/CU. FT"
820  PRINT "SPARGER LENGTH,     "R7;" IN."
830  PRINT
840  PRINT "DO YOU WISH TO ENTER NEW DATA?"
850  INPUT "ANSWER Y OR N. ";X$
860  IF X$ = "N" GOTO 910
870  IF X$ = "Y" GOTO 220
880  PRINT "YOUR CHOICES WERE (Y)ES OR (N)O."
890  PRINT "YOU ANSWERED "X$". TRY AGAIN."
900  GOTO 850
910  PRINT
920  PRINT  TAB( 13);"END OF PROGRAM"
930  END
```

**Example for: Hole-area distribution for liquid spargers** **Table II**

**(Start of first display)**

```
CALCULATION OF HOLE-AREA DISTRIBUTION
FOR LIQUID SPARGERS.

DATA REQUIRED ARE:

FLOW TO SPARGER, GPM
PRESSURE TO SPARGER, PSIA
AMBIENT PRESSURE, PSIA
SPARGER I.D., IN.
SPARGER FLUID VISCOSITY, CP
SPARGER FLUID DENSITY, LB/CU. FT
SPARGER LENGTH, FT

PRESS RETURN KEY TO ENTER THE DATA.
```

**(Start of next display)**

```
ENTER THE FOLLOWING DATA.

FLOW TO SPARGER, GPM                2000
PRESSURE TO SPARGER, PSIA           25
AMBIENT PRESSURE, PSIA              24.5
SPARGER I.D., IN.                   6.0648
SPARGER FLUID VISCOSITY, CP         .76
SPARGER FLUID DENSITY, LB/CU. FT    62.4
SPARGER LENGTH, FT                  10
```

**(Start of next display)**

```
SPARGER AREA FOR EACH 10% OF LENGTH.

           SQ. IN.

           12.9448
           8.6968
           7.1014
           6.2299
           5.6795
           5.3068
           5.0463
           4.8661
           4.7463
           4.6774

TOTAL AREA 65.2958 SQ. IN.

             END OF PROGRAM
```

# Section V
# Heat Transfer

Heat-exchanger performance by successive summation
Heat transfer through composite walls
Radiant heat flux in direct-fired heaters
Quench-tower design

# Heat-exchanger performance by successive summation

This programmable method lets the engineer calculate heat-transfer coefficients and operating temperatures without resorting to the correction factors associated with the classical, log-mean temperature-difference approach.

☐ Many plant heat-exchange operations are not accurately analyzed by the log-mean temperature-difference, $\Delta\bar{T}_{LM}$, method for calculating heat transfer. To improve this situation, the method presented here was developed. Today, it is being used to calculate heat-transfer coefficients on a time-shared terminal system available in one of Du Pont's plants. It has replaced the classical method in many applications because it is compatible with programmable calculators and requires less data input. This is significant in plant situations where complete sets of data are rare.

## $\Delta\bar{T}_{LM}$ method

The classical method of calculating the overall heat-transfer coefficient for an exchanger is to integrate the basic differential equation describing heat transfer as shown in Eq. (1) and (2).

$$dq = UdA(T - t) \qquad (1)$$

$$q = UA(\Delta T - \Delta t)/\ln(\Delta T/\Delta t) = UA\Delta\bar{T}_{LM} \qquad (2)$$

$\Delta\bar{T}_{LM}$ results from this integration.

For single-pass, cocurrent and countercurrent flow, calculation of $\Delta\bar{T}_{LM}$ is straightforward. Complications arise when multipass tube-bundles, condenser or reboiler liquid-levels, or cooling medium crossflows (such as on air coolers) must be considered. In such cases, correction factors must be used to modify $\Delta\bar{T}_{LM}$ to fit the particular design configuration. Correction factors are empirical and lead the engineer away from Eq. (1), the mathematical base of heat transfer. In effect, the engineer is reacting to his equipment instead of predicting its performance.

## Successive summation method

Correction factors are avoided by directly using the differential equation. By programmed calculation, Eq. (1) can be manipulated in differential elements, and then, through successive summation, the numerical value of the integral is obtained without using a $\Delta\bar{T}_{LM}$.

As an example of this method, a program will be constructed to determine (a) the overall heat-transfer coefficient, $U$, or (b) the exit gas temperature, $T_o$, for a two-pass heat exchanger.

The two-pass tube bundle shown in Fig. 1 permits a hot gas to pass in a U pattern through the exchanger. A baffled head controls the gas flow. The shell side contains one-pass cooling water, which enters at the right and leaves at the left. (For simplicity, water crossflow due to shell-side baffles has been neglected but could be incorporated into the program.) Water enters the shell at temperature $t_i$, and increases in temperature as it proceeds through each element by the amount:

$$\Delta t = \Delta q/m_w Cp_w \qquad (3)$$

Written in accordance with Fig. 1, the temperature of the water leaving the differential element, j, is equal to the entering temperature plus $\Delta t$:

$$t_{(j+1)} = t_j + \Delta t \qquad (4)$$

The rate of heat transfer in the differential element, $dq_{(j)}$, is equal to the sum of the rates in the upper and lower sections:

$$dq_{(j)} = dq_{(j)}\,(\text{upper}) + dq_{(j)}\,(\text{lower}) \qquad (5)$$

The sections must be analyzed separately because they contain gas at two different temperatures. Referring to Fig. 1, $T_i$ is the inlet gas temperature, and $T_{i(j)}$ is the first-pass gas temperature through the $j$ element. $T_o$ is the exit gas temperature and $T_{o(j)}$ is the second-pass gas temperature through the $j$ element. In the upper section of the element, the rate of heat transfer is:

$$dq_{(j)}\,(\text{upper}) = U(A/2)(T_{i(j)} - t_{(j)})/N \qquad (6)$$

and in the lower section it is:

$$dq_{(j)}\,(\text{lower}) = U(A/2)(T_{o(j)} - t_{(j)})/N \qquad (7)$$

The $\Delta t$ of the cooling water through the differential element is calculated from Eq. (3), (4) and (5) to be:

$$\Delta t = t_{(j+1)} - t_{(j)} = \frac{dq_{(j)}\,(\text{upper}) + dq_{(j)}\,(\text{lower})}{m_w Cp_w} \qquad (8)$$

Adapted from an article by Robert A. Spencer, Jr., E. I. du Pont de Nemours & Co., originally published December 4, 1978.

Similarly, for the upper section of the element:

$$\Delta T_i = T_{i(j+1)} - T_{ij} = \frac{dq_{(j)}(\text{upper})}{m_g Cp_g} = \frac{U(A/2)(T_{i(j)} - t_{(j)})}{N m_g Cp_g} \quad (9)$$

For the lower section:

$$\Delta T_o = T_{o(j)} - T_{o(j+1)} = \frac{dq_{(j)}\text{ lower}}{m_g Cp_g} = \frac{U(A/2)(T_{o(j)} - t_{(j)})}{N m_g Cp_g} \quad (10)$$

Eq. (4 to 10) are the components of the calculation, which will now be programmed.

## Program structure

The structure of the program is shown in Fig. 2. After calling for general input, it sets up the information needed for the type of calculation to be done, for either $T_o$ or $U$. An initial guess of the intermediate temperature, $T_{int}$ (see Fig. 1), is used to start the successive summation at the right (water inlet) and progress to the left (water exit). For the first element, $j = 1$, $T_{int}$ is equal to both $T_{o(1)}$ and $T_{i(1)}$. The calculation then proceeds through $N$ elements (15 to 20 elements are practical, based on computing time and accuracy considerations) to produce values for the gas inlet temperature, $T_{i(N)}$, and gas exit temperature, $T_{o(N)}$.

Next, the program compares the calculated $T_{i(N)}$ with the given input $T_i$. If they are not equal, a new $T_{int}$ is calculated and the program is restarted. Convergence of $T_{int}$ with the correct value will allow $T_{i(N)}$ to be equal to $T_i$. If they are equal (or within convergence criteria limits), then $T_{o(N)}$ is the predicted gas exit temperature and the calculation is finished.

If $U$ is to be calculated, $T_{o(N)}$ is compared to the given (input) $T_o$. If $U$ is correct, $T_{o(N)}$ will be equal to $T_o$. If they are not equal, a new value for $U$ is calculated and the program is restarted. Once $U$ converges with the correct value, $T_{o(N)}$ will be equal to $T_o$, $T_{i(N)}$ will be equal to $T_i$, and the calculation is finished.

See worked-out sample problem below.

The described calculation has been specifically structured for a two-pass heat exchanger, but its concept is applicable to many other heat-transfer problems. In general, if three of the four inlet and exit temperatures are known, the successive summation method will calculate the fourth. If two of the four are known, the third can be determined by trial and error.

Computing time depends on the number of differential elements and trial-and-error loops, and on the success of the convergence routines.

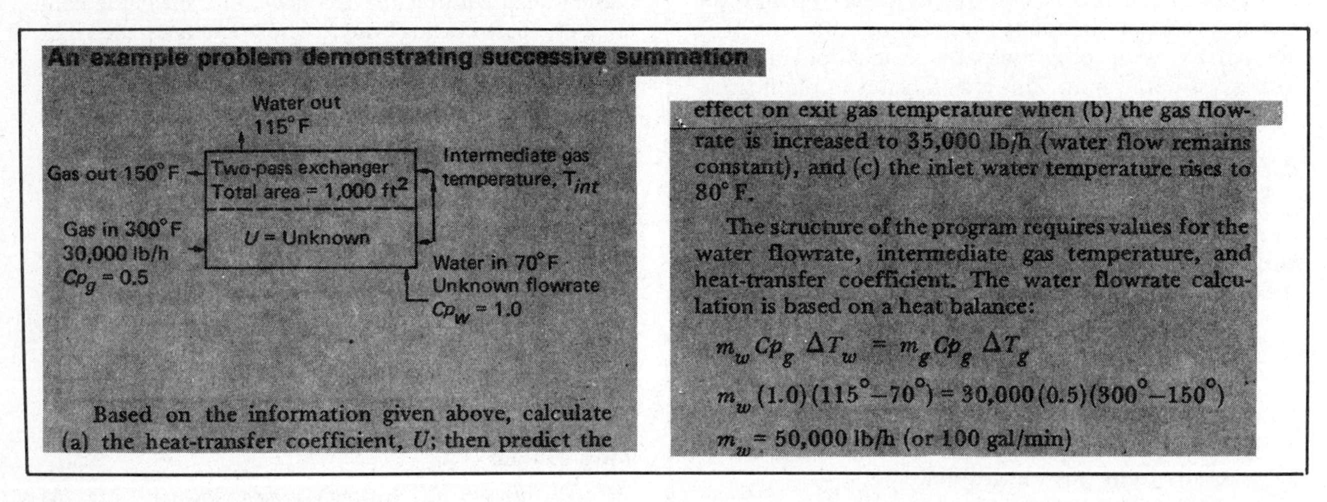
**An example problem demonstrating successive summation**



Based on the information given above, calculate (a) the heat-transfer coefficient, $U$; then predict the effect on exit gas temperature when (b) the gas flowrate is increased to 35,000 lb/h (water flow remains constant), and (c) the inlet water temperature rises to 80° F.

The structure of the program requires values for the water flowrate, intermediate gas temperature, and heat-transfer coefficient. The water flowrate calculation is based on a heat balance:

$$m_w Cp_g \Delta T_w = m_g Cp_g \Delta T_g$$

$$m_w (1.0)(115° - 70°) = 30,000(0.5)(300° - 150°)$$

$$m_w = 50,000 \text{ lb/h (or 100 gal/min)}$$

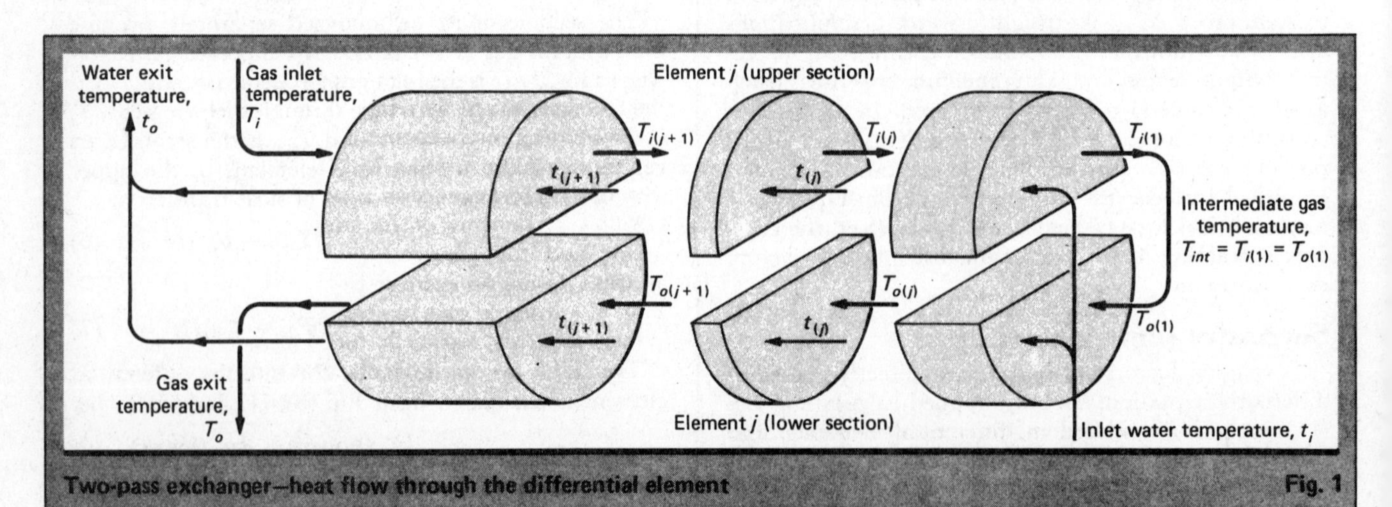


Two-pass exchanger—heat flow through the differential element — Fig. 1

Accuracy depends on the number of differential elements and the temperature used to represent each element. Using an average element temperature rather than the inlet or exit temperature of the element increases accuracy, especially for cocurrent flow. Relative errors of 1% are not difficult to achieve.

## Predicting performance

The predictive nature of the method is an important tool. Consider the cooling of a recycle gas stream by 9 exchangers arranged in 3 sets of 3 parallel exchangers connected in series by manifold for gas stream intermixing between successive sets. Gas recycle rate is the limiting factor in production. The exchangers must be periodically cleaned. Which ones and when?

The successive summation method can be used to make the decision. Heat-transfer coefficients are calculated for each of the heat exchangers, based on plant data. These coefficients are then used to predict the consequences of removing a particular exchanger from service. The optimum arrangement is the one that provides the maximum heat-transfer capacity.

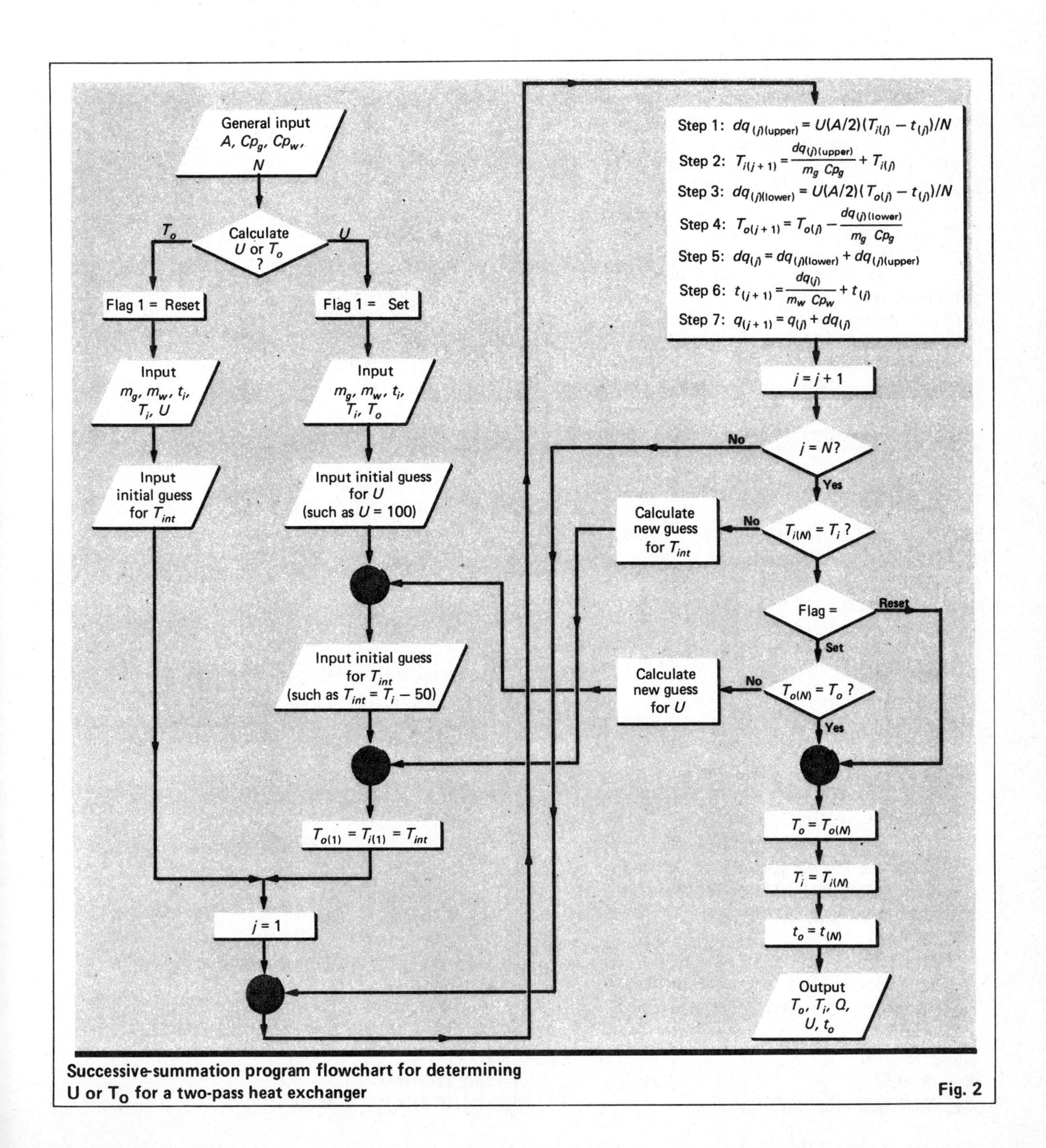


**Successive-summation program flowchart for determining U or $T_o$ for a two-pass heat exchanger**

**Fig. 2**

## Using the program

This program calculates the performance of a two-pass gas-flow, single-pass water-flow heat exchanger. A listing of the program is presented in Table I.

The first display (Table II) shows which variables can be calculated:

1) either the overall heat-transfer coefficient, U, or
2) the surface area required, A, or
3) the gas outlet temperature, $T_0$.

Data required are:
for U—surface area and gas temperatures,
for A—overall heat-transfer coefficient and gas temperatures,
for $T_0$—surface area and overall heat-transfer coefficient, and for all cases—gas flowrate, gas heat capacity, gas inlet temperature, and water inlet temperature and outlet temperature.

The next display asks the user which calculation is to be made. The user should then enter either "U", "A", or "T".

Data is entered upon prompting from the program. Once all the data is put in, the requested variable is calculated. As well, the exchanger duty and water requirement are determined.

In some cases, the calculations cannot be made. The user is informed of this, and then requested to check the data and start again.

## Nomenclature

| | |
|---|---|
| $A$ | area of heat transfer, ft$^2$ |
| $Cp$ | heat capacity, Btu/(h)(°F) |
| $D$ | dia. of tube, ft |
| $g$ | subscript denoting gas |
| $i$ | inlet condition, or first tube-pass condition |
| $j$ | denotes differential element |
| $l$ | length of tube, ft |
| $m$ | mass flowrate, lb/h |
| $N$ | number of differential elements |
| $o$ | exit condition, or second tube-pass condition |
| $q$ | heat-transfer rate, Btu/h |
| $t$ | temperature, shellside (water), °F |
| $T$ | temperature, tubeside (gas), °F |
| $U$ | heat-transfer coefficient, Btu/(h)(ft$^2$)(°F) |
| $w$ | subscript denoting water |

## The author

**Robert A. Spencer, Jr.**, is a developmental engineer for Du Pont's Explosive Products Div., Repauno Plant, Gibbstown, NJ 08027. He is currently developing composite diamond coatings for various end-uses. He has been a process engineer and a manufacturing supervisor. His experience includes production of intermediate chemicals for nylon and polyethylene manufacture.
Mr. Spencer holds a B.S. in chemical engineering from Virginia Polytechnic Institute and State University.

**Program for predicting heat-exchanger performance by successive summation** **Table I**

```
10 REM   HEAT-EXCHANGER PERFORMANCE BY SUCCESSIVE
   SUMMATION
20 REM   FROM CHEMICAL ENGINEERING, DECEMBER 4, 1978
30 REM   BY ROBERT A. SPENCER, JR.
40 REM  TRANSLATED BY WILLIAM VOLK
50 REM   COPYRIGHT (C) 1984
60 REM  BY CHEMICAL ENGINEERING
70 DEF  FN P(X) =  INT (1E4 * (X + .00005)) / 1E4
80 HOME : PRINT
90 ONERR  GOTO 1140
100  PRINT "THIS PROGRAM CALCULATES THE PERFORMANCE"
110  PRINT "OF A TWO-PASS GAS-FLOW, SINGLE-PASS"
120  PRINT "WATER-FLOW HEAT EXCHANGER.
130  PRINT
140  PRINT "IT CALCULATES EITHER THE"
150  PRINT "OVERALL TRANSFER COEFFICIENT, U; OR"
160  PRINT "THE SURFACE REQUIRED, A; OR"
170  PRINT "GAS OUTLET TEMPERATURE, T(0).
180  PRINT "DATA REQUIRED ARE:"
190  PRINT
200  PRINT "FOR U: SURFACE AND GAS TEMPERATURES."
210  PRINT "FOR A: COEFFICIENT AND GAS TEMPS."
220  PRINT "FOR T(0): SURFACE AND COEFFICIENT."
230  PRINT
240  PRINT "AND"
250  PRINT
260  PRINT "GAS FLOW RATE, LB/HR"
270  PRINT "GAS HEAT CAPACITY, BTU/(LB)(DEG. F)
280  PRINT "GAS INLET TEMPERATURE, DEG. F"
290  PRINT "WATER INLET AND OUTLET TEMPERATURES."
300  PRINT
310  INPUT "PRESS RETURN KEY TO SELECT CALCULATION.";X$
320  HOME : VTAB 8
330  PRINT "ENTER U, A, OR T TO CALCULATE EITHER:"
340  PRINT
350  PRINT "U, FOR TRANSFER COEFFICIENT,"
360  PRINT "A, FOR SURFACE REQUIRED, OR"
370  PRINT "T, FOR GAS OUTLET TEMPERATURE."
380  PRINT
390  INPUT "      ";P$
400  IF P$ = "U" OR P$ = "A" OR P$ = "T" GOTO 460
410  PRINT
420  PRINT "YOUR CHOICES WERE U, A, OR T"
430  PRINT "YOU ENTERED "P$".  TRY AGAIN."
440  PRINT
450  GOTO 330
460  HOME : VTAB 4
470  PRINT "ENTER THE FOLLOWING DATA:"
480  PRINT
490  IF P$ = "U" OR P$ = "A" GOTO 530
500  INPUT "COEFFICIENT, BTU/(HR)(SQ.FT)(DEG.F)";R9
510 F1 = 1
520  GOTO 560
530  IF P$ = "U" GOTO 560
```

**Program for predicting heat-exchanger performance by successive summation (continued)** **Table I**

```
540  INPUT "COEFFICIENT, BTU/(HR)(SQ.FT)(DEG.F)";R1
550  GOTO 570
560  INPUT "EXCHANGER SURFACE, SQ.FT              ";R1
570  INPUT "GAS RATE, LB/HR                       ";R4
580  INPUT "GAS HEAT CAPACITY, BTU/LB/DEG.F       ";R5
590  INPUT "GAS INLET TEMPERATURE, DEG.F          ";R2
600  IF P$ = "T" GOTO 620
610  INPUT "GAS OUTLET TEMPERATURE, DEG.F         ";R3
620  INPUT "WATER INLET TEMPERATURE, DEG.F        ";R7
630  INPUT "WATER OUTLET TEMPERATURE, DEG.F       ";R8
640  IF R8 =  < R7 GOTO 1140
650  IF R3 =  > R2 GOTO 1140
660  IF R7 =  > R2 GOTO 1140
670  IF R8 =  > R3 GOTO 1140
680 R6 = 1: REM  WATER HEAT CAPACITY
690 R0 = 20: REM  PARTITIONS FOR CALCULATION.
700 D = (R2 + R3) / 2
710 A = R4 * R5
720 E = A * (R2 - R3)
730  IF F1 = 1 GOTO 780
740 I = R2 - R8
750 B = R3 - R7
760 R9 = E / ((I - B) /  LOG (I / B)) / R1
770  IF R9 < 0 GOTO 1140
780 B = E / (R8 - R7) / R6
790 C = B * R6
800 F2 = 0
810 E = (R9 * R1) / 2 / R0
820 P3 = D
830 P2 = D
840 P9 = R7
850  FOR J = 1 TO R0
860 P7 = E * (P2 - P9)
870 P2 = P2 + P7 / A
880 P8 = E * (P3 - P9)
890 P3 = P3 - P8 / A
900 P0 = P7 + P8
910 P9 = P9 + (P7 + P8) / C
920  NEXT J
930 I = P3
940  IF  ABS (P2 - R2) < 1 GOTO 1100
950  IF F1 = 0 GOTO 970
960 R3 = I
970  IF F2 = 1 GOTO 1040
980 P4 = D
990 P5 = P2
1000 D = R2 - (P2 - D) / (P2 - R3) * (R2 - R3)
1010 F2 = 1
1020  IF F1 = 1 GOTO 710
1030  GOTO 820
1040 D = D + (R2 - P2) * (D - P4) / (P2 - P5)
1050 P5 = P2
1060 P4 = D
1070 F2 = 1
1080  IF F1 = 1 GOTO 710
1090  GOTO 820
1100  IF F1 = 1 GOTO 1150
1110  IF  ABS (R3 - I) < 1 GOTO 1150
1120 R9 = (I / R3) ^ 2 * R9
1130  GOTO 800
1140 F1 = 1
1150  PRINT
1160  PRINT "WITH THE ABOVE DATA:"
1170  IF F1 = 0 GOTO 1230
1180  PRINT
1190  PRINT "THE CALCULATION CAN NOT BE MADE."
1200  PRINT
1210  PRINT "CHECK THE DATA AND START AGAIN.
1220  GOTO 1350
1230  PRINT "
1240  IF P$ = "A" OR P$ = "T" GOTO 1270
1250  PRINT "COEFFICIENT, BTU/(HR)(SQ.FT)(DEG.F)
          "; FN P(R9)
1260  GOTO 1310
1270  IF P$ = "T" GOTO 1300
1280  PRINT "SURFACE REQUIRED, SQ.FT
          "; INT (R9)
1290  GOTO 1310
1300  PRINT "GAS OUTLET TEMPERATURE, DEG.F
          "; FN P(R3)
1310 Q = A * (R2 - R3)
1320  PRINT "EXCHANGER DUTY, BTU/HR
          "; INT (Q)
1330 W = Q / (R8 - R7)
1340  PRINT "WATER REQUIREMENT, LB/HR
          "; INT (W)
1350  PRINT
1360  PRINT  TAB( 13);"END OF PROGRAM"
1370  END
```

**Example for: Heat-exchanger performance by successive summation** **Table II**

**(Start of first display)**

```
THIS PROGRAM CALCULATES THE PERFORMANCE
OF A TWO-PASS GAS-FLOW, SINGLE-PASS
WATER-FLOW HEAT EXCHANGER.

IT CALCULATES EITHER THE
OVERALL TRANSFER COEFFICIENT, U; OR
THE SURFACE REQUIRED, A; OR
GAS OUTLET TEMPERATURE, T(O).
DATA REQUIRED ARE:

FOR U: SURFACE AND GAS TEMPERATURES.
FOR A: COEFFICIENT AND GAS TEMPS.
```

**Example for: Heat-exchanger performance by successive summation (continued)** **Table II**

```
FOR T(O): SURFACE AND COEFFICIENT.

AND

GAS FLOW RATE, LB/HR
GAS HEAT CAPACITY, BTU/(LB)(DEG. F)
GAS INLET TEMPERATURE, DEG. F
WATER INLET AND OUTLET TEMPERATURES.

PRESS RETURN KEY TO SELECT CALCULATION.
```

**(Start of next display)**

```
ENTER U, A, OR T TO CALCULATE EITHER:

U, FOR TRANSFER COEFFICIENT,
A, FOR SURFACE REQUIRED, OR
T, FOR GAS OUTLET TEMPERATURE.

     U
```

**(Start of next display)**

```
ENTER THE FOLLOWING DATA:

EXCHANGER SURFACE, SQ.FT               1000
GAS RATE, LB/HR                        30000
GAS HEAT CAPACITY, BTU/LB/DEG.F        .5
GAS INLET TEMPERATURE, DEG.F           300
GAS OUTLET TEMPERATURE, DEG.F          150
WATER INLET TEMPERATURE, DEG.F         70
WATER OUTLET TEMPERATURE, DEG.F        115

WITH THE ABOVE DATA:

COEFFICIENT, BTU/(HR)(SQ.FT)(DEG.F)        19.1367
EXCHANGER DUTY, BTU/HR                     2250000
WATER REQUIREMENT, LB/HR                   50000

              END OF PROGRAM
```

# Heat transfer through composite walls

This program computes the heat-transfer rate through walls composes of any two materials for which coefficients are known. It also determines the average and surface temperatures of the walls.

☐ The heat transferring through each wall of a composite wall via conduction, $q' = (K/X)\Delta t$, is identical to the heat being lost at the outside surface via convection and radiation, $q' = h\Delta t$.

From the figure:

$$q' = (K_1/X_1)(t_i - t_1) = (K_2/X_2)(t_1 - t_2) = h(t_2 - t_a) \quad (1)$$

Solving for $t_1$, $t_2$ and $q'$:

$$t_1 = t_2 + (X_2/K_2)\, h(t_2 - t_a) \quad (2)$$

$$t_2 = t_a + (q'/h) \quad (3)$$

$$q' = (t_i - t_a)/[(X_1/K_1) + (X_2/K_2) + (1/h)] \quad (4)$$

Via regression analysis of the film coefficient data in Table II, an equation for $h$ is derived:

$$h = 1.535 + 0.00582\,(\,|t_2 - t_a|\,) \quad (5)$$

The correlation coefficient for Eq. (5) is 0.9998.

From the thermal conductivity data in Table II, a value for $K_1$ is similarly determined:

$$K_1 = 4680/[(t_1 + t_i)/2] + 0.00425 \times [(t_i + t_1)/2] \quad (6)$$

The correlation coefficient for Eq. (6) is 0.9992.

An equation for $K_2$ is also derived from data in Table II:

$$K_2 = 0.988\, e^{0.00033\,[(t_1 + t_2)/2]} \quad (7)$$

The correlation coefficient for Eq. (7) is 0.9885.

## How the program converges

A value for $t_2$ is assumed, and with $t_i$ and $t_a$ known, $h$ is calculated via Eq. (5), and $K_2$ via Eq. (7). Initially, $t_1$ is taken as zero; however, a value for $t_1$ is inserted during the first calculation sequence.

With the values for $h_1$, $K_2$ and $t_2$, $t_1'$ is calculated by means of Eq. (2). This value of $t_1'$ is compared to the initial $t_1$. If the absolute value of $t_1 - t_1' \geqq 0.5$, the program repeats the calculation for $t_1'$, setting $t_1 = t_1'$. The program converges when the absolute value of $t_1 - t_1' < 0.5$. Then the program continues, to test $t_2$.

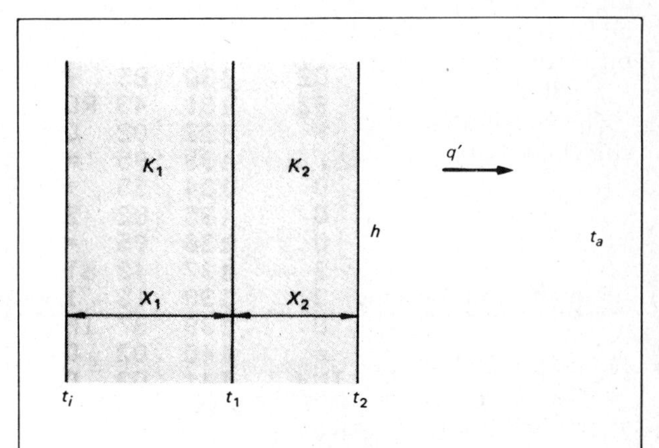


**Variables involved in heat transfer through a composite wall**

With the assumed value for $t_2$, $K_1$ is calculated by means of Eq. (6), and $q'$ by Eq. (4).

With $t_a$ known, and $q'$ and $h$ now having been determined, $t_2'$ is calculated. If the absolute value of $t_2' - t_2 < 0.5$, the program concludes with $t_2 = t_2'$ and $t_1 = t_1'$.

However, if the absolute value of $t_2' - t_2 \geqq 0.5$, the program returns to its beginning, with $t_2 = t_2'$, and $t_1$ is calculated again.

Eventually, the program will converge, with the differences between $t_1$ and $t_1'$ and $t_2$ and $t_2'$ reduced to less than the 0.5 difference allowed by the program (Table 1).

Adapted from an article by Calvin R. Brunner, Malcolm Pirnie, Inc., originally published June 16, 1980.

## Using the program

This program calculates the heat loss through composite walls. The user has the option to enter data for thermal conductivity ($K_1$ and $K_2$) and/or film coefficient ($h$), or use values calculated by the program. In the case of thermal conductivity, the calculated values are based on fireclay brick for the inner wall ($K_1$), and lightweight castable for the outer wall ($K_2$). The program will also determine a film coefficient for the outer wall ($h$), if no value is entered by the user.

To enter user data for $K_1$, $K_2$, or $h$, the user should enter a "Y" to the query ending the first display (see Table III), which asks "DO YOU WISH TO ENTER VALUES OF THERMAL CONDUCTIVITY OR FILM COEFFICIENT?".

At this point, the user can enter values for any (or all) of these three variables. However, if the calculated values for $K_1$, $K_2$, or h are to be used, enter a "0" when prompted by the program.

In the example (see Table III), the user opted to have the calculated values of $K_1$, $K_2$, and h used, by entering a "N" to the first display's query.

The next display permits the entering of the remaining data. For example, inner wall thickness ($X_1$) is 2 in., and outer wall length ($X_2$) is 4.5 in. The inside temperature ($t_i$) is 1,300°F, and the outside ambient temperature ($t_a$) is 80°F.

Once all the data is put in, the results are then displayed. The variables calculated are as follows:

Insulation interface temperature, $t_1$
Outside wall temperature, $t_2$
Average interior insulation temperature (the average temperature along length $X_1$)
Average exterior insulation temperature (the average temperature along length $X_2$)
Film-radiation coefficient, h
Inside insulation transfer coefficient, $K_1$ (the coefficient of thermal conductivity for the inner layer)
Outside insulation transfer coefficient, $K_2$ (the coefficient of thermal conductivity for the outer layer)
Heat transfer rate, $q'$

### The author

**Calvin R. Brunner** is a principal engineer with Malcolm Pirnie, Inc. (2 Corporate Park Dr., White Plains, NY 10602). His experience includes the design of thermal reduction equipment for solid waste and sludges, and also operator assistance with incineration. He has presented seminars on incineration and on programming the TI-59, and has developed programs for heat transfer, steam flow, and the design of waste-disposal equipment. Holder of a B.M.E. from City College of New York and an M. Eng. from Pennsylvania State University, he is a licensed engineer in four states and a member of ASME.

**Program for calculating heat transfer through composite walls** **Table I**

```
10 REM   HEAT TRANSFER THROUGH COMPOSITE WALLS
20 REM   FROM CHEMICAL ENGINEERING, JUNE 16, 1980
30 REM   BY CALVIN R. BRUNNER
40 REM  TRANSLATED BY WILLIAM VOLK
50 REM   COPYRIGHT (C) 1984
60 REM  BY CHEMICAL ENGINEERING
70 DEF  FN P(X) =  INT (1E4 * (X + .00005)) / 1E4
80 HOME : VTAB 3
90 PRINT "PROGRAM CALCULATES HEAT LOSS THROUGH"
100 PRINT "A WALL OF TWO LAYERS OF INSULATION."
110 PRINT
120 PRINT "IF NO VALUES FOR THERMAL CONDUCTIVITY"
130 PRINT "ARE ENTERED, THE PROGRAM CALCULATES"
140 PRINT "VALUES BASED ON FIRECLAY BRICK FOR THE"
150 PRINT "INNER LAYER AND LIGHTWEIGHT CASTABLE"
160 PRINT "FOR THE OUTER LAYER."
170 PRINT
180 PRINT "THE PROGRAM WILL ALSO CALCULATE A FILM"
190 PRINT "COEFFICIENT FOR THE OUTER WALL IF NONE"
200 PRINT "IS ENTERED."
210 PRINT
220 PRINT "DO YOU WISH TO ENTER VALUES OF THERMAL"
230 PRINT "CONDUCTIVITY OR FILM COEFFICIENT?"
240 INPUT "ANSWER Y OR N.    ";X$
250 PRINT
260 IF X$ = "N" GOTO 460
270 IF X$ = "Y" GOTO 320
280 PRINT "YOUR CHOICES WERE (Y)ES OR (N).  YOU"
290 PRINT "ENTERED "X$".  TRY AGAIN."
300 PRINT
310 GOTO 240
320 HOME : VTAB (8)
330 PRINT "YOU CAN ENTERED VALUES FOR THE THERMAL"
340 PRINT "CONDUCTIVITY OF THE INNER AND OUTER"
350 PRINT "INSULATION LAYERS, AND FOR THE OUTER"
360 PRINT "FILM COEFFICIENT.  IF YOU DO NOT"
370 PRINT "WISH TO ENTER ANY OF THESE, ENTER ZERO."
380 PRINT
390 INPUT "YOUR VALUE OF K FOR INNER LAYER.   ";R9
400 INPUT "YOUR VALUE OF K FOR OUTER LAYER.   ";R7
410 INPUT "YOUR VALUE OF FILM COEFFICIENT.    ";R5
420 IF R9 <  > 0 THEN F1 = 1
430 IF R7 <  > 0 THEN F2 = 1
440 IF R5 <  > 0 THEN F3 = 1
450 PRINT
460 PRINT "ENTER THE FOLLOWING DATA:"
470 PRINT
480 INPUT "INNER THICKNESS, INCHES              ";R0
490 INPUT "OUTER THICKNESS, INCHES              ";R1
500 INPUT "INSIDE TEMPERATURE, DEG. F           ";R2
```

**Program for calculating heat transfer through composite walls (continued)** **Table I**

```
510 INPUT "OUTSIDE TEMPERATURE, DEG. F      ";R3
520 R6 = (R3 + R2) / 2
530 R4 = (R6 + R3) / 2
540  IF F1 = 1 GOTO 550
550  IF F3 = 1 GOTO 570
560  GOSUB 700
570  IF F2 = 1 GOTO 590
580  GOSUB 730
590 I = R4 + (R1 * R5 * (R4 - R3)) / R7
600  IF  ABS (I - R6) <  ABS (.001 * R6) GOTO 630
610 R6 = I
620  GOTO 570
630  IF F1 = 1 GOTO 650
640  GOSUB 760
650 A = (R2 - R3) / (1 / R5 + R0 / R9 + R1 / R7)
660 I = R3 + A / R5
670  IF  ABS (I - R4) <  ABS (.001 * R4) GOTO 790
680 R4 = I
690  GOTO 550
700  REM  H:FILM COEFFICIENT
710 R5 =  ABS (R4 - R3) * .00582 + 1.535
720  RETURN
730  REM  K2:OUTER LAYER
740 R7 = 0.988 *  EXP (.00033 * (R6 + R4) / 2)
750  RETURN
760  REM  K1:INNER LAYER
770 R9 = 4680 / (R6 + R2) * 2 + (R6 + R2) / 2 * .00425
780  RETURN
790  HOME : PRINT
800  PRINT "RESULTS ARE:"
810  PRINT
820  PRINT "TEMERATURES, DEG. F:"
830  PRINT
840  PRINT "INSULATION INTERFACE,          "; FN P(R6)
850  PRINT "OUTSIDE WALL,                  "; FN P(R4)
860 TA =  FN P((R2 + R6) / 2)
870  PRINT "AVG. INTERIOR INSULATION,      "; FN P(TA)
880 TB =  FN P((R4 + R6) / 2)
890  PRINT "AVG. EXTERIOR INSULATION,      "; FN P(TB)
900  PRINT
910  PRINT "FILM COEFFICIENT"
920  PRINT "     BTU/(HR)(SQ.FT)(DEG.F)      "; FN P(R5)
930  PRINT
940  PRINT "TRANSFER COEFFICIENTS,"
950  PRINT "     BTU/(HR)(SQ.FT)(DEG.F)      "
960  PRINT
970  PRINT "INSIDE INSULATION                "; FN P(R9)
980  PRINT "OUTSIDE INSULATION,              "; FN P(R7)
990  PRINT
1000  PRINT "HEAT TRANSFER RATE,"
1010  PRINT "     BTU/(HR)(SQ.FT)            "; FN P(A)
1020  PRINT
1030  PRINT  TAB( 13);"END OF PROGRAM"
1040  END
```

**Data for derivation of Eq. (5), (6) and (7)** **Table II**

Eq. (5), film coefficient ($h$) for vertical flat plate:

| $\lvert t_2 - t_a \rvert$, °F | 50 | 100 | 150 | 200 | 250 |
|---|---|---|---|---|---|
| **Handbook** $h^*$ | 1.82 | 2.13 | 2.40 | 2.70 | 2.99 |
| **Derived** $h$ | 1.83 | 2.12 | 2.41 | 2.70 | 2.99 |

Eq. (6), thermal conductivity ($K_1$) for fireclay brick:

| ½ $(t_i + t_1)$, °F | 1,000 | 1,400 | 1,800 | 2,200 | 2,600 |
|---|---|---|---|---|---|
| **Catalog** $K_1{}^\dagger$ | 9.2 | 9.4 | 10.0 | 11.3 | 13.0 |
| **Derived** $K_1$ | 8.9 | 9.3 | 10.3 | 11.5 | 12.9 |

Eq. (7), thermal conductivity ($K_2$) for lightweight castable:

| ½ $(t_i + t_1)$, °F | 200 | 600 | 1,000 | 1,400 |
|---|---|---|---|---|
| **Catalog** $K_2{}^{**}$ | 1.1 | 1.2 | 1.3 | 1.5 |
| **Derived** $K_2$ | 1.1 | 1.2 | 1.4 | 1.6 |

* From Baumeister, T. and Marks, L., "Standard Handbook for Mechanical Engineers," 7th ed. McGraw-Hill, Inc., 1967, p. 4-106.

† From "Harbison-Walker Superduty Fireclay Brick," Harbison-Walker Refractories Div. of Dresser Industries, Inc.

** From "Harbison-Walker H-W Lightweight Castable 22," Harbison-Walker Refractories Div. of Dresser Industries, Inc.

**Example for: Heat transfer through composite walls** **Table III**

**(Start of first display)**

```
PROGRAM CALCULATES HEAT LOSS THROUGH
A WALL OF TWO LAYERS OF INSULATION.

IF NO VALUES FOR THERMAL CONDUCTIVITY
ARE ENTERED, THE PROGRAM CALCULATES
VALUES BASED ON FIRECLAY BRICK FOR THE
INNER LAYER AND LIGHTWEIGHT CASTABLE
FOR THE OUTER LAYER.

THE PROGRAM WILL ALSO CALCULATE A FILM
COEFFICIENT FOR THE OUTER WALL IF NONE
IS ENTERED.

DO YOU WISH TO ENTER VALUES OF THERMAL
CONDUCTIVITY OR FILM COEFFICIENT?
ANSWER Y OR N.    N
```

**Example for: Heat transfer through composite walls (continued)** **Table III**

**(Start of next display)**

```
ENTER THE FOLLOWING DATA:

INNER THICKNESS, INCHES            2
OUTER THICKNESS, INCHES            4.5
INSIDE TEMPERATURE, DEG. F         1300
OUTSIDE TEMPERATURE, DEG. F        80
```

**(Start of next display)**

```
RESULTS ARE:

TEMERATURES, DEG. F:

INSULATION INTERFACE,         1236.4606
OUTSIDE WALL,                 206.3592
AVG. INTERIOR INSULATION,     1268.2303
AVG. EXTERIOR INSULATION,     721.4099

FILM COEFFICIENT
     BTU/(HR)(SQ.FT)(DEG.F)   2.2704

TRANSFER COEFFICIENTS,
     BTU/(HR)(SQ.FT)(DEG.F)

INSIDE INSULATION             9.0802
OUTSIDE INSULATION,           1.2536

HEAT TRANSFER RATE,
     BTU/(HR)(SQ.FT)         287.0273

            END OF PROGRAM
```

# Radiant heat flux in direct-fired heaters

In these heaters, tubes can rupture if they get too hot. Here is a program to determine the heat flux and the tube wall temperature so that rupture can be avoided.

☐ One of the main causes for tube rupture in direct-fired heaters is the high tube-metal temperature ($T_{TM}$)—or tube-skin temperature—experienced by some of the tubes in the furnace's radiant section.

Although it is customary for vendors of heaters to assume an average radiant heat flux in the unit's radiant section, such an assumption is incorrect. The radiant heat flux is a function of the tube metal temperature, and varies through the radiant coil according to the following equation, which is a modification of the Stefan-Boltzmann law:

$$q = K(T_{BW}^4 - T_{TM}^4) \qquad (1)$$

Here, $K = 654.4 \times 10^{-12}$ Btu/(h)(ft²)(R⁴). This value is based on assuming a radiant heat flux of 10,000 Btu/(h)(ft²) for a $T_{BW}$ of 1,620°F (2,080 R) and a $T_{TM}$ of 880°F (1,340 R). This value of $K$ is also for one row of tubes, and assumes that tubes are spaced on center lines equal to two times their nominal diameters, and are 1½ nominal dia. from the refractory surface.

## Trial-and-error calculations

To obtain the radiant heat flux for the tubes, a two-way trial-and-error method is usually used:

1. A radiant heat flux, $q$, is assumed at the coil outlet.
2. The tube metal temperature is calculated:

$$T_{TM} = T_{FO} + \Delta T_{TOTAL} \qquad (2)$$

where:

$$\Delta T_{TOTAL} = \Delta T_{FILM} + \Delta T_{COKE} + \Delta T_{METAL} \qquad (3)$$

$$\Delta T_{TOTAL} = \frac{q}{h_i}\left(\frac{\text{O.D.}}{\text{I.D.} - 2t_c}\right) + \frac{q \times t_c}{k_c}\left(\frac{\text{O.D.}}{\text{I.D.} - t_c}\right) + \frac{q \times t_a}{k_w}\left(\frac{\text{O.D.}}{\text{O.D.} - t_a}\right)$$

$$= q \times \text{O.D.}\left[\frac{1}{h_i(\text{I.D.} - 2t_c)} + \frac{t_c}{k_c(\text{I.D.} - t_c)} + \frac{t_a}{k_w(\text{O.D.} - t_a)}\right] \qquad (4)$$

Film coefficients are calculated by a method given by API [1]. In Eq. (4), values of $q_{ASSD}$ are first put in for $q$, then a value is calculated and is used.

3. The radiant heat flux at each tube outlet is then calculated from Eq. (1).
4. If the calculated heat flux is close enough to the assumed value, we move back one tube in the coil and repeat the same steps. If it is not, we go back to step 1 and assume a new heat flux.
5. The heat absorbed by each tube is calculated, based on the average of the heat fluxes at the inlet and outlet of the tube.
6. The total duty calculated for the whole radiant coil should be close to the heat absorbed in the radiant section. If not, start again at the coil outlet with a different guess for the radiant flux.
7. Repeat this two-way iterative process until the solution converges to a stable value.

The program is divided into two sections: mainline and subroutine "del-tee." The logic diagram for the program appears in the figure. The program is shown in Table I, and an example in Table II.

### Nomenclature

| | |
|---|---|
| $A$ | Radiant-coil total heat-transfer area, ft² |
| $A_T$ | Surface area per tube, ft² |
| $h_{io}$ | Inside film coeff. at coil outlet, Btu/(h)(ft²)(°F) |
| $h_{ii}$ | Inside film coeff. at coil inlet, Btu/(h)(ft²)(°F) |
| $I$ | Number of tubes per pass |
| I.D. | Tube inside dia., in. |
| $K$ | Constant in Eq. (1), $6.454 \times 10^{-10}$ Btu/(h)(ft²)(R⁴) |
| $k_c$ | Coke thermal conductivity, Btu/(h)(ft²)(°F/ft) |
| $k_w$ | Tube-metal conductivity, Btu/(h)(ft²)(°F/in.) |
| $L$ | Effective tube length, ft |
| O.D. | Tube outside dia., in. |
| $Q$ | Total heat absorbed in radiant coil, Btu/h |
| $q_{ASSD}$ | Assumed heat flux, Btu/(h)(ft²) |
| $q_{CALC}$ | Calculated heat flux, Btu/(h)(ft²) |
| $T_{BW}$ | Bridge wall temperature, °F |
| $T_{FI}$ | Fluid temperature at coil inlet, °F |
| $T_{FO}$ | Fluid temperature at coil outlet, °F |
| $T_{TM}$ | Tube-metal temperature, °F |
| $t_a$ | Average tube-wall thickness, in. |
| $t_c$ | Thickness of coke deposit, in. |
| $\Delta T_{TOTAL}$ | As defined by Eq. (3) |
| $\Delta T_{FILM}$, $\Delta T_{COKE}$, $\Delta T_{METAL}$ | Temperature differences across layers, °F |

Adapted from an article by Tayseer A. Abdel-Halim, KTI Corp., originally published December 17, 1979.

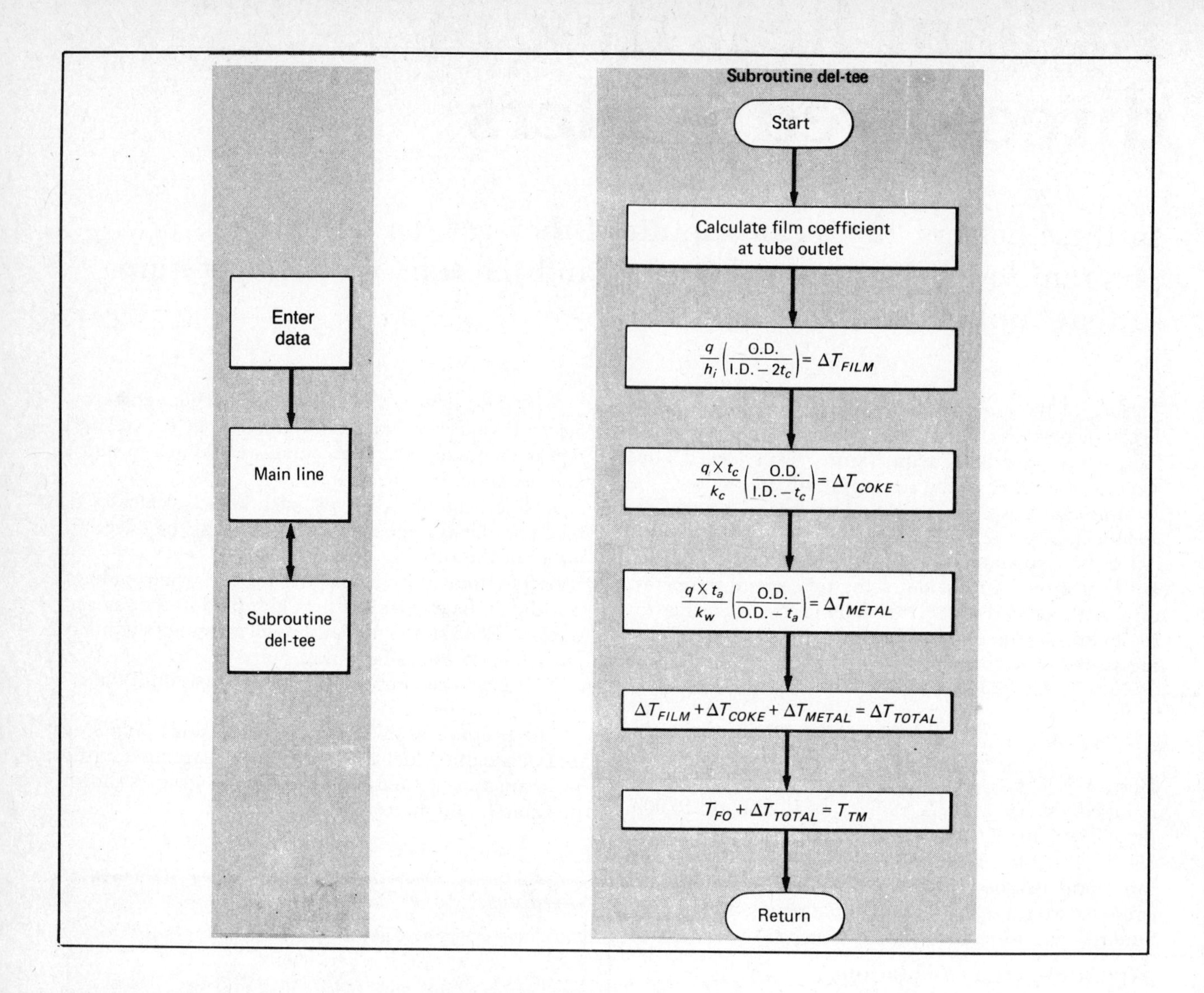


## Notes

The author has found out that a good first guess that forces the program to converge readily is to use a value somewhat less than that of the average heat flux for the whole coil.

A Newton-Raphson or interval-halving [2] subroutine could have been used to force the program to the next guess. Since this requires a substantial increase in both program size and execution time, and since by the above method of guessing we always know the correctness of the second assumption made, the addition of a convergence subroutine cannot be justified.

To obtain the maximum tube-metal temperatures (defined as the temperature of the front 60 deg of the tube), multiply the above TMTs by 1.8. This number, given by API [1], is for tubes arranged in single rows against a wall, and on center-to-center spacing equal to twice the nominal tube diameter.

## Example

Data for an example are given in Table III. The heater has 64 tubes arranged in an 8-pass flow. The coil tubes are Sch. 40A pipe, with a thickness of 0.280 in. The thermal conductivity of coke—3.0 Btu/(h)(ft$^2$)(°F/ft)—is built into the program. Note $T_{BW}$ is 1,728°F.

The heat flux was guessed to be 8,000 Btu/(h)(ft$^2$). Data for the example are shown in Table III. ABS stands for heat absorbed per tube in Btu/h. RATE is heat flux, Btu/(h)(ft$^2$). Note that $T_{MT}$ is symbolized by TMT, and $T_{FO}$ is symbolized by FOT. Similar substitutions are made elsewhere.

## References


1. Amer. Petroleum Inst., publication RP-530, "Recommended Practice for the Calculation of Heater Tube Thickness," 2nd ed., Dec. 1976.
2. Carnahan, Brice, et al., "Applied Numerical Methods," John Wiley & Sons, Inc., New York, 1969.

## The author


**Tayseer A. Abdel-Halim** is a process engineer with the Process Plant Div. of KTI Corp., 221 East Walnut St., Pasadena, CA 91101. Tel: (213) 577-1600, x284. He is involved with the design of ethylene plants. He has also worked for Born, Inc., where he was responsible for the design and cost estimation of direct-fired heaters, and for Heat Research Corp., as a senior process engineer. Abdel-Halim obtained a B.Sc. degree in refinery engineering from the Egyptian High Inst. for Petroleum Engineering, and an M.Sc. degree in chemical engineering from the University of Tulsa. He is a member of AIChE.

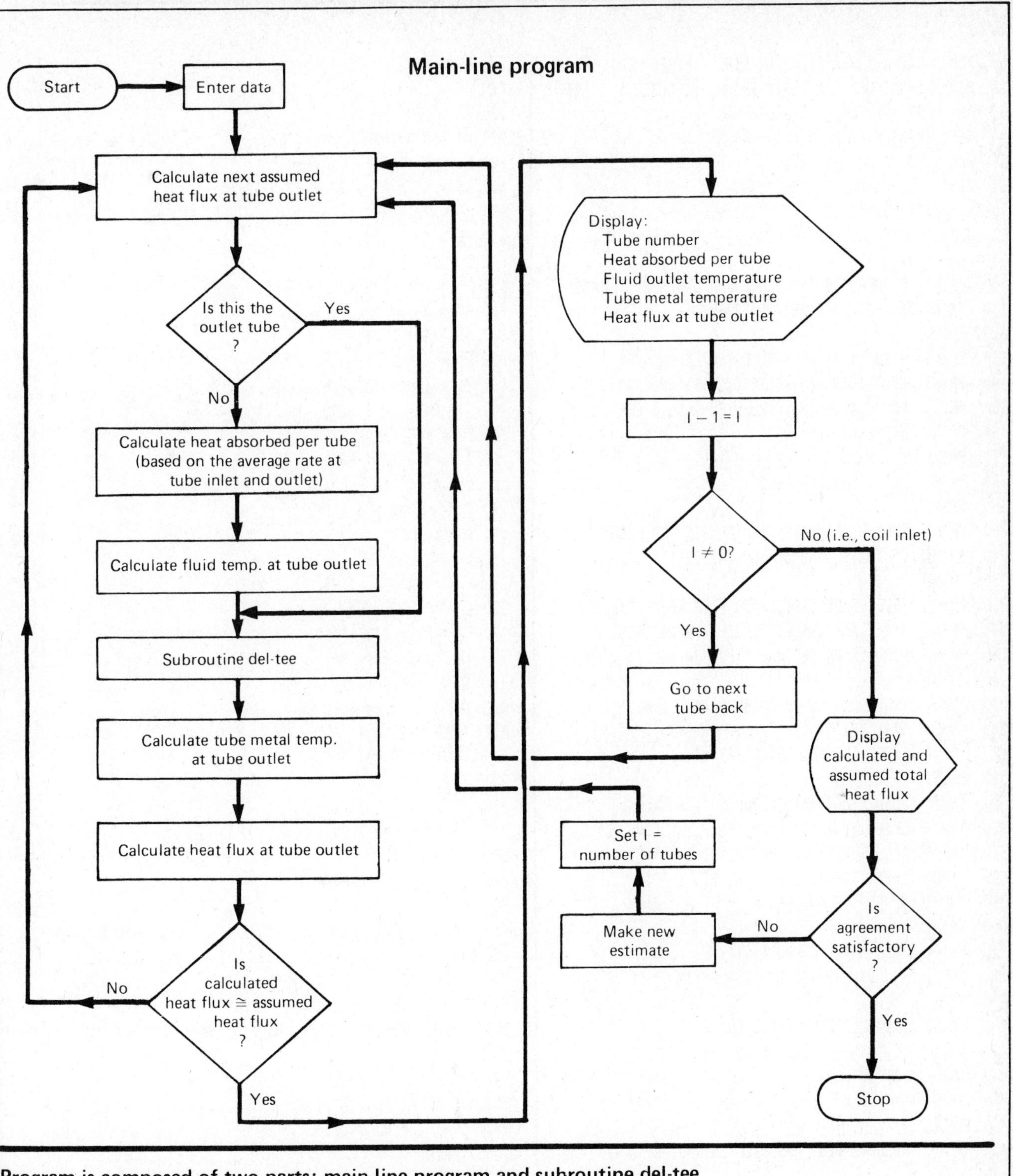


**Program is composed of two parts: main-line program and subroutine del-tee**

**Program predicts radiant heat flux in direct-fired heaters** **Table I**

```
10  REM   RADIANT HEAT FLUX IN DIRECT-FIRED HEATERS
20  REM   FROM CHEMICAL ENGINEERING, DECEMBER 17, 1979
30  REM   BY TAYSEER A. ABDEL-HALIM
40  REM  TRANSLATED BY WILLIAM VOLK
50  REM   COPYRIGHT (C) 1984
60  REM  BY CHEMICAL ENGINEERING
70 PI = 3.14159265
80  DEF  FN P(X) =  INT (1E4 * (X + .00005)) / 1E4
90  HOME : VTAB 4
100  PRINT "PROGRAM TO CALCULATE RADIANT HEAT FLUX"
110  PRINT "IN DIRECT FIRED HEATERS."
120  PRINT
130  PRINT "WITH INPUT OF TUBE LAYOUT DATA AND"
140  PRINT "HEAT TRANSFER COEFFICIENTS, PROGRAM"
150  PRINT "CALCULATES INDIVIDUAL TUBE DUTY AND"
160  PRINT "TEMPERATURE."
170  PRINT
180  PRINT "DATA REQUIRED ARE:"
190  PRINT
200  PRINT "NUMBER, DIAMETER, THICHKESS AND LENGTH"
210  PRINT "OF TUBES.  NUMBER OF PASSES."
220  PRINT
230  PRINT "TUBE CONDUCTIVITY AND FILM COEFFICIENTS."
240  PRINT "INLET AND OUTLET FLUID TEMPERATURES."
250  PRINT "FURNACE BRIDGE WALL TEMPERATURE."
260  PRINT "
270  INPUT "PRESS RETURN TO ENTER DATA  ";Q$
280  HOME : VTAB 3
290  PRINT "ENTER THE FOLLOWING DATA:"
300  PRINT
310  INPUT "NUMBER OF RADIANT TUBES       ";R3
320  INPUT "OUTER TUBE DIAMETER, IN.      ";R5
330  INPUT "AVG. TUBE WALL THICKNESS, IN. ";F3
340  INPUT "EFFECTIVE TUBE LENGTH, FT     ";R7
350  INPUT "NUMBER OF PASSES              ";R9
360  PRINT
370  PRINT "TUBE-METAL CONDUCTIVITY,"
380  INPUT "     BTU/(H)(SQ.FT)(DEG.F)/IN. ";P0
390  PRINT
400  PRINT "INSIDE FILM COEFFICIENT"
410  PRINT "     BTU/(H)(SQ.FT)(DEG.F)"
420  INPUT " AT OUTLET                     ";P1
430  INPUT " AT INLET                      ";P2
440  PRINT
450  PRINT "ASSUMED HEAT ABSORBED IN RADIANT COIL"
460  INPUT "     BTU/H                     ";P3
470  PRINT
480  PRINT "FLUID TEMPERATURE, DEG.F"
490  INPUT " AT COIL OUTLET                ";P4
500  INPUT " AT COIL INLET                 ";P5
510  PRINT
520  INPUT "COKE DEPOSIT, IN.              ";F9
530  INPUT "BRIDGE WALL TEMP., DEG.F       ";D
540 CO = P4
550 P6 = 6.454E - 10
560 E = 0
570 R0 = 1000
580 R2 = 1000
590 R4 = R3 / R9
600 A = R5 / 12 - F3 / 6
610 R8 = R3 * R5 * R7 * PI / 12
620 P7 = R8 / R4
630 P8 = P3 / (P4 - P5)
640 B = P4
650 FP = (P1 - P2) / (P4 - P5)
660 P9 = P4 + FP
670 I = R4 + 1
680 R1 = I
690 R0 = 100 + (R2 + R0) / 2
700  IF R1 =  < I GOTO 730
710 R6 = (R2 + R0) * P7 / 2
720 B = P4 - R6 / P8
730 X = 6 / (((B - P5) * FP) + P2)
740 X = X / (A * 6 - F9)
750 X = X + F9 / ((A * 12 - F9) * 3)
760 X = X / 12
770 X = X + (F3 / ((R5 - F3) * P0))
780 C = X * R0 * R5 + B
790 R2 = P6 * ((D + 460) ^ 4 - (C + 460) ^ 4)
800  IF  ABS (R2 - R0) > 100 GOTO 690
810  IF I = 1 GOTO 860
820  IF I =  < R4 GOTO 880
830  HOME : VTAB 4
840  PRINT "CONDITIONS AT COIL OUTLET:"
850  GOTO 880
860  PRINT
870  PRINT "CONDITIONS AT COIL INLET:"
880  PRINT
890  PRINT "TUBE NUMBER                   ";I - 1
900  PRINT
910  PRINT "HEAT ABSORBED, BTU/H          "; FN P(R6)
920 E = E + R6
930  PRINT "FLUID TEMPERATURE, DEG.F      "; FN P(B)
940  PRINT "TUBE WALL TEMP., DEG.F        "; FN P(C)
950  PRINT "ABSORPT. RATE, BTU/(H)(SQ.FT) "; FN P(R2)
960  PRINT
970 P4 = B
980 I = I - 1
990  IF I = 0 GOTO 1030
1000  PRINT
1010  INPUT "PRESS RETURN KEY TO CONTINUE ";X$
1020  GOTO 690
1030  PRINT "TOTAL ABSORPTION, BTU/H      "; FN P(E)
1040  PRINT "ESTIMATED VALUE WAS:         ";P3
1050  PRINT
1060  PRINT "DO YOU WANT TO MAKE ANOTHER HEAT FLUX"
1070  PRINT "ESTIMATE?  IF SO, ENTER THE VALUE.  IF"
1080  INPUT "NOT, ENTER ZERO.    ";A$
1090  IF  VAL (A$) = 0 GOTO 1130
1100 P3 =  VAL (A$)
```

**Program predicts radiant heat flux in direct-fired heaters (continued)** **Table I**

```
1110 P4 = CO:R6 = 0
1120  GOTO 550
1130  PRINT
1140  PRINT  TAB( 13);"END OF PROGRAM"
1150  END
```

**Example for: Radiant heat flux in direct-fired heaters** **Table II**

**(Start of first display)**

```
PROGRAM TO CALCULATE RADIANT HEAT FLUX
IN DIRECT FIRED HEATERS.

WITH INPUT OF TUBE LAYOUT DATA AND
HEAT TRANSFER COEFFICIENTS, PROGRAM
CALCULATES INDIVIDUAL TUBE DUTY AND
TEMPERATURE.

DATA REQUIRED ARE:

NUMBER, DIAMETER, THICHKESS AND LENGTH
OF TUBES.  NUMBER OF PASSES.

TUBE CONDUCTIVITY AND FILM COEFFICIENTS.
INLET AND OUTLET FLUID TEMPERATURES.
FURNACE BRIDGE WALL TEMPERATURE.

PRESS RETURN TO ENTER DATA
```

**(Start of next display)**

```
ENTER THE FOLLOWING DATA:

NUMBER OF RADIANT TUBES         64
OUTER TUBE DIAMETER, IN.        6.625
AVG. TUBE WALL THICKNESS, IN. .28
EFFECTIVE TUBE LENGTH, FT       45.5
NUMBER OF PASSES                8

TUBE-METAL CONDUCTIVITY,
    BTU/(H)(SQ.FT)(DEG.F)/IN. 200

INSIDE FILM COEFFICIENT
    BTU/(H)(SQ.FT)(DEG.F)
 AT OUTLET                      152
 AT INLET                       145

ASSUMED HEAT ABSORBED IN RADIANT COIL
    BTU/H                       58.522E6

FLUID TEMPERATURE, DEG.F
 AT COIL OUTLET                 1094
 AT COIL INLET                  730

COKE DEPOSIT, IN.               .0625
BRIDGE WALL TEMP., DEG.F        1728
```

**(Start of next display)**

```
CONDITIONS AT COIL OUTLET:

TUBE NUMBER                   8

HEAT ABSORBED, BTU/H          0
FLUID TEMPERATURE, DEG.F      1094
TUBE WALL TEMP., DEG.F        1200.3346
ABSORPT. RATE, BTU/(H)(SQ.FT) 9887.016


PRESS RETURN KEY TO CONTINUE
```

**(Start of next display)**

```
TUBE NUMBER                   7

HEAT ABSORBED, BTU/H          6496633.82
FLUID TEMPERATURE, DEG.F      1053.5917
TUBE WALL TEMP., DEG.F        1163.9418
ABSORPT. RATE, BTU/(H)(SQ.FT) 10303.1056


PRESS RETURN KEY TO CONTINUE
```

**(Start of next display)**

```
TUBE NUMBER                   6

HEAT ABSORBED, BTU/H          6745297.35
FLUID TEMPERATURE, DEG.F      1011.6367
TUBE WALL TEMP., DEG.F        1126.5844
ABSORPT. RATE, BTU/(H)(SQ.FT) 10702.0965


PRESS RETURN KEY TO CONTINUE
```

**(Start of next display)**

```
TUBE NUMBER                   5

HEAT ABSORBED, BTU/H          6989544.32
FLUID TEMPERATURE, DEG.F      968.1626
TUBE WALL TEMP., DEG.F        1087.7678
ABSORPT. RATE, BTU/(H)(SQ.FT) 11087.8653


PRESS RETURN KEY TO CONTINUE
```

**Example for: Radiant heat flux in direct-fired heaters (continued)** **Table II**

**(Start of next display)**

```
TUBE NUMBER                    4

HEAT ABSORBED, BTU/H           7225847.44
FLUID TEMPERATURE, DEG.F       923.2187
TUBE WALL TEMP., DEG.F         1047.4214
ABSORPT. RATE, BTU/(H)(SQ.FT)  11459.2239


PRESS RETURN KEY TO CONTINUE
```

**(Start of next display)**

```
TUBE NUMBER                    3

HEAT ABSORBED, BTU/H           7452425.91
FLUID TEMPERATURE, DEG.F       876.8654
TUBE WALL TEMP., DEG.F         1005.5634
ABSORPT. RATE, BTU/(H)(SQ.FT)  11814.2348


PRESS RETURN KEY TO CONTINUE
```

**(Start of next display)**

```
TUBE NUMBER                    2

HEAT ABSORBED, BTU/H           7667959.16
FLUID TEMPERATURE, DEG.F       829.1716
TUBE WALL TEMP., DEG.F         962.2364
ABSORPT. RATE, BTU/(H)(SQ.FT)  12151.0227


PRESS RETURN KEY TO CONTINUE
```

**(Start of next display)**

```
TUBE NUMBER                    1

HEAT ABSORBED, BTU/H           7871355.68
FLUID TEMPERATURE, DEG.F       780.2127
TUBE WALL TEMP., DEG.F         917.4948
ABSORPT. RATE, BTU/(H)(SQ.FT)  12467.9578


PRESS RETURN KEY TO CONTINUE
```

**(Start of next display)**

```
CONDITIONS AT COIL INLET:

TUBE NUMBER                    0

HEAT ABSORBED, BTU/H           8061724.94
FLUID TEMPERATURE, DEG.F       730.0697
TUBE WALL TEMP., DEG.F         871.4016
ABSORPT. RATE, BTU/(H)(SQ.FT)  12763.7175

TOTAL ABSORPTION, BTU/H        58510788.6
ESTIMATED VALUE WAS:           58522000

DO YOU WANT TO MAKE ANOTHER HEAT FLUX
ESTIMATE? IF SO, ENTER THE VALUE.  IF
NOT, ENTER ZERO.    0

           END OF PROGRAM
```

**Storage information for calculating radiant flux. Values are given for a typical problem** **Table III**

| Data | Example |
|---|---|
| $q_{ASSD}$ | |
| Index | |
| $q_{CALC}$ | |
| No. of radiant tubes | 64 |
| $l$ | |
| O.D. | 6.625 in. |
| $t_a$ | 0.280 in. |
| $L$ | 45.5 ft |
| $A$ | |
| No. of passes | 8 |
| $k_w$ | 200 Btu/(h)(ft$^2$)(°F/in) |
| $h_{io}$ | 152 Btu/(h)(ft$^2$)(°F) |
| $h_{ii}$ | 145 Btu/(h)(ft$^2$)(°F) |
| $Q$ | $58.522 \times 10^6$ Btu/h |
| $T_{FO}$ | 1,094°F |
| $T_{FI}$ | 730°F |
| $K$ | $6.454 \times 10^{-10}$ Btu/(h)(ft$^2$)(R$^4$) |
| $A_T$ | |
| Temp. gradient | |
| $t_c$ | 0.0625 in. |
| Sum of heat absorbed | |
| $T_{FO}$ at each tube | |
| $T_{MT}$ | |
| $T_{BW}$ | 1,728°F |
| Film coeff. gradient | |
| Film coeff. at each tube outlet | |

# Quench-tower design

An essential step in designing evaporative cooling equipment is finding the adiabatic saturation temperature of the gas. This program streamlines the computations.

☐ Adiabatic humidification is commonly used in the chemical process industries as an economical means of cooling hot gases. In the design of such evaporative cooling systems as quench towers, the engineer must frequently determine the adiabatic saturation temperature of the gas from its temperature, pressure and composition.

If the gas is air at atmospheric pressure, psychrometric charts or tables make this task straightforward and relatively simple. Charts are even available to accommodate slight changes in composition and pressure.

However, if the gas composition varies significantly from that of air (as in the case of flue gas), or if the pressure varies significantly from atmospheric, a rather tedious iterative calculation is required. The calculation is derived as follows:

The saturation humidity, $W_s$, at the adiabatic-saturation temperature, $t_s$, is given by [1]:

$$W_s = W_t + \frac{h_g + h_w W_t}{r_s}(t - t_s) \quad (1)$$

where: $W_t$ = humidity at temperature $t$, lb water vapor/lb dry air; $t$ = air temperature, °F; $t_s$ = saturation temperature, °F; $r_s$ = latent heat of vaporization of water at the adiabatic-saturation temperature, Btu/lb; $h_w$ = specific heat of water vapor, Btu/lb/°F; and $h_g$ = specific heat of the dry gas, Btu/lb/°F.

The latent heat of vaporization can be expressed by [2]:

$$r_s = 91.86(705.56 - t_s)^{0.38} \quad (2)$$

The saturation humidity is also given by [1]:

$$W_s = \frac{18.016\, p_s}{M\,(P - p_s)} \quad (3)$$

which may be expressed as:

$$W_s = \frac{p_s}{(P - p_s)[1.6096(1 - f_{CO_2}) + 2.4428\, f_{CO_2}]} \quad (4)$$

where: $p_s$ = vapor pressure of water at $t_s$, psia; $P$ = total pressure, psia; $M$ = molecular weight of dry gas; and $f$ = mole fraction.

The vapor pressure of water can be calculated from [3]:

$$\log p_s = 15.092 - \frac{5079.6}{T} - 1.6908 \log T - 3.193 \times 10^{-3} T + 1.234 \times 10^{-6} T^2 \quad (5)$$

where $T$ = temperature, °R.

The specific heat of the gas can be determined using equations of the form [4]:

$$h = a + bt + ct^2 \quad (6)$$

For an $N_2/O_2$ mixture similar to air, $a = 0.239$, $b = 1.288 \times 10^{-5}$, and $c = 1.4 \times 10^{-9}$. For $CO_2$, $a = 0.2045$, $b = 6.135 \times 10^{-5}$, and $c = 5.8 \times 10^{-9}$. For $H_2O$, $a = 0.4633$, $b = 2.581 \times 10^{-5}$, and $c = 2.3 \times 10^{-8}$.

The mean specific heat is given by:

$$h\ (mean) = \frac{h_2 t_2 - h_1 t_1}{t_2 - t_1}. \quad (7)$$

The mean specific heat of the gas mixture can be determined using mole fractions:

$$h_{mix}\ (mean) = (1 - f_{CO_2} - f_{H_2O}\, h_{N_2,O_2} + f_{CO_2} h_{CO_2} + f_{H_2O} h_{H_2O}) \quad (8)$$

The adiabatic saturation temperature can now be calculated by the following iterative process:

1. Assume $t_s$.
2. Calculate $r_s$, $p_s$, and $h$, using Eq. (2), (4), (6), (7) and (8).
3. Calculate $W_s$, using Eq. (1).
4. Calculate $W_s$, using Eq. (3).
5. If $W_s$ by Eq. (1) is not equal to $W_s$ by Eq. (3), assume a new value of $t_s$ and repeat the above.

## Using the program

The program is shown in Table I. This program calculates the gas temperature in a cooling tower, in which the gas—a mixture of nitrogen, oxygen and carbon dioxide—is quenched with water.

The calculation balances the cooling effect of water evaporation and the temperature drop of the gas. The program also calculates the water concentration of the cooled gas.

Data required are:

- system pressure, psia
- entering gas temperature, °F
- entering humidity, lb water/lb dry gas
- carbon dioxide content, mole % (based on moisture-free gas).

The user is prompted to enter the above variables. Once all the data is put in, a "CALCULATION TAKES A FEW SECONDS" message is displayed. The results calculated are adiabatic-saturation temperature ($t_s$) and saturation humidity ($W_s$).

Adapted from an article by William H. Mink, Battelle Columbus Laboratories, originally published December 3, 1979.

Results of a sample calculation are shown in Table II. In this example, the adiabatic saturation temperature and humidity are calculated for air (0% $CO_2$) at atmospheric pressure (0 psig). Psychrometric tables [1] give values of 82.0 and 0.02387, respectively, for saturation temperature and saturated humidity at the above conditions.

Small variations between the results calculated by this program and those available in tables are likely, since specific heats, latent heats, and partial pressures are calculated from polynomial formulas rather than from tabular data. However, the error in the saturation temperature is less than 0.5°F over a wide range of input data and is, for the most part, less than 0.2°F. The error in saturated humidity is less than 1½%.

## References


1. Zimmerman, O. T., and Lavine, I., "Psychrometric Tables and Charts," Mack Printing Co., Easton, Pa., 1964.
2. Thakore, S. B., Miller, J. W., Jr., and Yaws, C. L., Heats of Vaporization, *Chem. Eng.*, Aug. 16, 1976, p. 85.
3. Patel, P. M., Schorr, G. R., Shah, P. N., and Yaws, C. L., Vapor Pressure, *Chem. Eng.*, Nov. 22, 1976, p. 159.
4. Hougen, O. A., and Watson, K. M., "Industrial Chemical Calculations," John Wiley and Sons, 1936.


**Program for quench-tower design** **Table I**

```
10 REM  QUENCH-TOWER DESIGN
20 REM  FROM CHEMICAL ENGINEERING, DECEMBER 3, 1979
30 REM  BY WILLIAM H. MINK
40 REM  TRANSLATED BY WILLIAM VOLK
50 REM  COPYRIGHT (C) 1984
60 REM  BY CHEMICAL ENGINEERING
70 HOME : VTAB 4
80 PRINT "PROGRAM CALCULATES THE GAS TEMPERATURE"
90 PRINT "IN A COOLING TOWER IN WHICH THE GAS,"
100 PRINT "A MIXTURE OF N2, O2, AND CO2, IS"
110 PRINT "QUENCHED WITH WATER."
120 PRINT
130 PRINT "THE CALCULATION BALANCES THE COOLING"
140 PRINT "EFFECT OF WATER EVAPORATION AND"
150 PRINT "THE TEMPERATURE DROP OF THE GAS."
160 PRINT
170 PRINT "THE PROGRAM ALSO CALCULATES THE WATER"
180 PRINT "CONCENTRATION OF THE COOLED GAS."
190 PRINT
200 PRINT "DATA REQUIRED ARE:"
210 PRINT
220 PRINT "PRESSURE,"
230 PRINT "TEMPERATURE OF THE GAS,"
240 PRINT "ENTERING HUMIDITY,"
250 PRINT "CO2 CONCENTRATION."
260 PRINT "PROGRAM ASSUMES BALANCE OF DRY GAS"
270 PRINT "IS AIR AT NORMAL O2/N2 RATIO."
280 PRINT
290 INPUT "PRESS THE RETURN KEY TO ENTER DATA.  ";X$
300 HOME : VTAB (4)
310 PRINT "ENTER THE FOLLOWING DATA:"
320 PRINT
330 INPUT "PRESSURE, PSIG                      ";R1
340 R1 = R1 + 14.7
350 INPUT "GAS TEMPERATURE, DEG. F             ";R2
360 INPUT "GAS HUMIDITY, LB H2O/LB DRY GAS   ";R3
370 INPUT "MOL% CO2 BASED ON DRY GAS          ";R4
380 PRINT
390 PRINT "CALCULATION TAKES A FEW SECONDS."
400 R4 = R4 / 100
410 R0 = 800
420 R0 = R0 - 100
430 GOSUB 710
440 IF X > 0 GOTO 420
450 GOSUB 800
460 IF X > 0 GOTO 420
470 R0 = R0 + 100
480 R0 = R0 - 10
490 GOSUB 710
500 IF X > 0 GOTO 480
510 GOSUB 800
520 IF X > 0 GOTO 480
530 R0 = R0 + 10
540 R0 = R0 - 1
550 GOSUB 710
560 IF X > 0 GOTO 540
570 GOSUB 800
580 IF X > 0 GOTO 540
590 R0 = R0 + 1
600 R0 = R0 - .1
610 GOSUB 710
620 IF X > 0 GOTO 600
630 GOSUB 800
640 IF X > 0 GOTO 600
650 IF R0 - R2 < 0 GOTO 690
660 R0 = R2
670 GOSUB 710
680 GOSUB 800
690 A =  INT (1E4 * A) / 1E4
700 GOTO 920
710 A = R2 - R0
720 B = R2 ^ 2 - R0 ^ 2
730 C = R2 ^ 3 - R0 ^ 3
740 D = R0 + 460
750 E = 15.092 - 5079.6 / D - 1.6908 *  LOG (D) /  LOG
    (10) + D * ((1.234E - 6 * D) - 3.193E - 3)
760 E = 10 ^ E
770 X = E - R1
780 RETURN
790 REM  THE FOLLOWING IS FOR CO2
800 Y = A * .204499 + B * 6.135E - 5 + C * 5.8E - 9
810 Y = Y * R4
820 REM  THE FOLLOWING IS FOR N2/O2
830 Z = A * .239 + B * 1.2875E - 5 + C * 1.4E - 9
840 Y = Y + Z * (1 - R4)
850 REM  THE FOLLOWING IS FOR H2O
```

**Program for quench-tower design (continued)** **Table I**

```
860 Z = A * .463255 + B * 2.581E - 5 + C * 2.3E - 9
870 Z = Z * R3 + Y
880 A1 = R3 + Z / (91.85747 * (705.56 - R0) ^ .38)
890 A = E / (R1 - E) / (1.60785 * (1 - R4)) + (2.44283
    * R4)
900 X = A - A1
910  RETURN
920  PRINT
930  PRINT "SATURATION TEMP., DEG. F      ";R0
940  PRINT
950  PRINT "SATURATION HUMIDITY"
960  PRINT "LB H20/LB DRY GAS             ";A
970  PRINT
980  PRINT
990  PRINT  TAB( 13);"END OF PROGRAM"
1000  END
```

**Example for: Quench-tower design** **Table II**

**(Start of first display)**

```
PROGRAM CALCULATES THE GAS TEMPERATURE
IN A COOLING TOWER IN WHICH THE GAS,
A MIXTURE OF N2, O2, AND CO2, IS
QUENCHED WITH WATER.

THE CALCULATION BALANCES THE COOLING
EFFECT OF WATER EVAPORATION AND
THE TEMPERATURE DROP OF THE GAS.

THE PROGRAM ALSO CALCULATES THE WATER
CONCENTRATION OF THE COOLED GAS.

DATA REQUIRED ARE:

PRESSURE,
TEMPERATURE OF THE GAS,
ENTERING HUMIDITY,
CO2 CONCENTRATION.
PROGRAM ASSUMES BALANCE OF DRY GAS
IS AIR AT NORMAL O2/N2 RATIO.

PRESS THE RETURN KEY TO ENTER DATA.
```

**(Start of next display)**

```
ENTER THE FOLLOWING DATA:

PRESSURE, PSIG                    0
GAS TEMPERATURE, DEG. F           160
GAS HUMIDITY, LB H20/LB DRY GAS  .005747
MOL% CO2 BASED ON DRY GAS         0

CALCULATION TAKES A FEW SECONDS.

SATURATION TEMP., DEG. F     82.2

SATURATION HUMIDITY
LB H20/LB DRY GAS            .0237


             END OF PROGRAM
```

# Section VI
# Mass Transfer

Flash computations
Packed-tower design
Sour-water-stripper design
Kinetics of fixed-bed sorption processes
Design of spouted beds
Cyclone-efficiency equations

# Flash computations

## Flash vaporization of a continuous feed is the starting point for countless distillation and fractionation operations. This convenient program performs flash calculations on mixtures containing up to 100 components.

☐ The simplest continuous-distillation technique is the single-stage flashing of a feed liquid. Once the temperature and pressure within the flash tank have been set, the equilibrium phase concentrations in the liquid and gaseous product streams can then be determined.

**This program, written for the Apple II computer, performs flash calculations for multicomponent mixtures containing up to 100 components. In order to define the equilibrium properties of the mixture, the user must know the *K*-values of the components at the flash conditions.**

### Calculation method

Assume that $F$ moles/h of an $n$-component stream are introduced as feed into the flash vaporization tank shown in the figure. The resulting vapor and liquid streams are withdrawn at the rates of $V$ and $L$ moles/h. The mole fractions of the components in the feed, vapor and liquid streams are designated as $z_i$, $y_i$ and $x_i$, respectively (where $i = 1, 2, \ldots n$).

Assuming a vapor-liquid equilibrium and steady-state operation, we have the following series of algebraic relationships:

Overall balance: $F = V + L$

Component balance: $z_i F = x_i L + y_i V$

Equilibrium: $K_i = y_i / x_i$

$K_i$ is the equilibrium constant for the $i$th component, at the temperature and pressure in the tank. (The $K$-values can be found in Ref. [2].) From the above equations, and the fact that

$$\sum_{i=1}^{n} x_i = \sum_{i=1}^{n} y_i = 1$$

it can be shown that:

$$\sum_{i=1}^{n} \frac{z_i(K_i - 1)}{V(K_i - 1) + F} = 0 \qquad (1)$$

Newton's method will be used to solve this equation for $V$. The algorithm for this iterative method can be written as:

$$V_{k+1} = V_k - \frac{f(V_k)}{f'(V_k)} \qquad (2)$$

where
$$f(V_k) = \sum_{i=1}^{n} \frac{z_i(K_i - 1)}{V_k(K_i - 1) + F} \qquad (3)$$

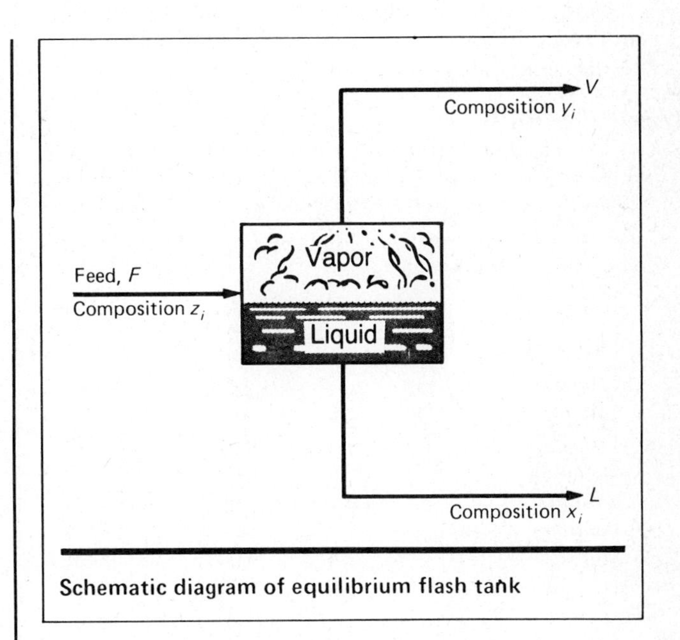


**Schematic diagram of equilibrium flash tank**

$$f'(V_k) = \sum_{i=1}^{n} \frac{z_i(K_i - 1)^2}{[V_k(K_i - 1) + F]^2} \qquad (4)$$

**A reasonable first estimate for $V$ is needed to start the iteration. (Note: The program makes the first vapor estimate. No input from the user is required.)**

We will use the following test criterion for terminating the iteration procedure:

$$\left| \frac{V_{k+1} - V_k}{V_{k+1}} \right| \leqslant \epsilon \qquad (5)$$

Here $\epsilon$ is a small positive number. For $\epsilon = 10^{-N}$, the final value of $V$ should be accurate to approximately $N$ significant figures.

Note that Eq. 1 has $n$ roots ($n$ is number of components). Only one root lying within the interval $0 < V < F$ is of interest to us.

After computing the vapor flowrate $V$, we can find the values of $x_i$ and $y_i$, the mole fractions of a component in the equilibrium phases, from the following equations:

$$x_i = \frac{z_i F}{V(K_i - 1) + F}$$

$$y_i = x_i K_i$$

**Adapted from an article by Sohrab Mansouri, University of Michigan, originally published August 27, 1979.**

## Using the program

This program (see listing in Table I) calculates the equilibrium-flash compositions for mixtures. There is no limit on the number of components in the mixture. If you need to use more than 100 components, change the dimension statement on program line 230.

Data required are:
- feed rate, $F$, mol/h
- feed composition, mol fraction ($z_i$) for individual components
- equilibrium constants, $K_i$, also for individual components.

Output will be the vapor and liquid rates (in mol/h), and compositions (in mol fractions, $x_i$ and $y_i$) for each component.

## An example

The composition of the feed to a natural-gas liquefaction plant [3] is given below. The feed will be flashed at 600 psia and 20°F. If the flowrate is 1,000 moles/h, determine the flowrates of the liquid and vapor streams and their compositions.

| Component | $i$ | $z_i$ | $K_i$ |
|---|---|---|---|
| Carbon dioxide | 1 | 0.0112 | 0.90 |
| Methane | 2 | 0.8957 | 2.70 |
| Ethane | 3 | 0.0526 | 0.38 |
| Propane | 4 | 0.0197 | 0.098 |
| Isobutane | 5 | 0.0068 | 0.038 |
| $n$-Butane | 6 | 0.0047 | 0.024 |
| Pentanes | 7 | 0.0038 | 0.0075 |
| Hexanes | 8 | 0.0031 | 0.0019 |
| Heptanes and heavier | 9 | 0.0024 | 0.0007 |
| | | 1.0000 | |

As shown in Table II, the user is prompted by the program to enter a value for feed rate. Then, the composition data for each component should be put in. The values for component name, $z_i$, and $K_i$, are entered, each separated with a comma. When all the data are put in, enter "0,0,0".

If the sum of the mol fractions does not add up to 1.000, the user is shown a display that indicates the incorrect total, and the complete data set for the individual components. At this point, the user can correct the existing data, enter new data, or exit the program. The options are displayed as follows;

"TO CORRECT THE DATA ENTER C."
"TO START WITH NEW DATA ENTER N."
"TO EXIT PROGRAM ENTER E."

To correct the existing data, the user should enter "C." Then, changes are made by entering item number, column number, and correct value. For instance, "3,2,.45" will change the mol fraction (the second column) for the third item (component) to 0.45. The user is now asked "ANY OTHER CHANGES Y OR N." If there are additional corrections, input changes as shown above.

The compositions of the liquid and vapor streams are as follows:

| Component $i$ | $x_i$ | $y_i$ |
|---|---|---|
| 1 | 0.0124 | 0.0111 |
| 2 | 0.3405 | 0.9193 |
| 3 | 0.1298 | 0.0493 |
| 4 | 0.1461 | 0.0143 |
| 5 | 0.0880 | 0.0033 |
| 6 | 0.0736 | 0.0018 |
| 7 | 0.0791 | 0.0006 |
| 8 | 0.0727 | 0.0001 |
| 9 | 0.0578 | $4.05 \times 10^{-5}$ |

## References


1. Carnahan, B., Luther, H. A., and Wilkes, J. O., "Applied Numerical Methods," John Wiley and Sons, 1969.
2. Katz, D. L., et al., "Handbook of Natural Gas Engineering," McGraw-Hill, New York, 1959.
3. Woicik, J. F., Equilibrium-flash calculation, *Chem. Eng.*, Aug. 16, 1976, p. 89.


## The author

**Sohrab Mansouri**, 1230 Hubbard St., Stanley 2203, Ann Arbor, MI 48109, is completing work toward a Ph.D. in bioengineering at the University of Michigan. He took his M.S. in chemical engineering at the same university and a B.S. in chemical engineering at Tehran University in Iran. He is a member of AIChE.

**Program for flash computations** **Table I**

```
10 REM   FLASH COMPUTATIONS
20 REM   FROM CHEMICAL ENGINEERING, AUGUST 17, 1979
30 REM   BY SOHRAB MANSOURI
40 REM  TRANSLATED BY WILLIAM VOLK
50 REM   COPYRIGHT (C) 1984
60 REM  BY CHEMICAL ENGINEERING
70 HOME : VTAB 4
80 DEF  FN P(X) =  INT (1E4 * (X + .00005)) / 1E4
90 PRINT "PROGRAM CALCULATES THE EQUILIBRIUM"
100 PRINT "FLASH COMPOSITIONS FOR ANY NUMBER"
110 PRINT "OF COMPONENTS."
120 PRINT
130 PRINT "DATA REQUIRED ARE:"
140 PRINT
150 PRINT "FEED RATE, MOL/HR"
160 PRINT "FEED COMPOSITION, MOL FRACTION"
170 PRINT "INDIVIDUAL COMPONENT K VALUES"
180 PRINT
190 PRINT "OUTPUT WILL BE VAPOR AND LIQUID RATES,"
200 PRINT "AND COMPOSITIONS."
210 PRINT
220 INPUT "TO ENTER DATA, PRESS RETURN KEY.   ";Q$
230 DIM N$(100),F(100),L(100),V(100),K(100)
240 HOME : VTAB 8
250 INPUT "ENTER THE FEED RATE, MOL/HR      ";F1
260 PRINT
270 PRINT "FOR EACH COMPONENT, ENTER NAME,"
280 PRINT "MOL FRACTION IN FEED AND K VALUE."
```

**Program for flash computations (continued)** **Table I**

```
290  PRINT
300  PRINT "SEPARATE THE ITEMS WITH COMMAS."
310  PRINT "WHEN ALL THE DATA ARE IN, ENTER THREE"
320  PRINT "ZEROS.: 0.0,0"
330  PRINT
340  PRINT "ENTER FIRST DATA: NAME. F(I), K(I)"
350  PRINT
360 I = 1
370  INPUT "                    ";N$(I),F(I),K(I)
380 T = T + F(I)
390 I = I + 1
400  INPUT "ENTER NEXT DATA:  ";N$(I),F(I),K(I)
410 T = T + F(I)
420  IF N$(I) = "0" GOTO 440
430  GOTO 390
440 N = I - 1
450  IF  ABS (T - 1) < .001 GOTO 880
460  HOME : VTAB 3
470  PRINT "TOTAL OF FEED MOL FRACTIONS IS ";T
480  PRINT "IT SHOULD BE 1.000"
490  PRINT
500  PRINT "YOUR INPUT WAS:
510  PRINT "COLUMN"; TAB( 12);"1"; TAB( 24);"2";
     TAB( 32);"3"
520  PRINT
530  PRINT "ITEM"; TAB( 10);"NAME"; TAB( 20);"MOL FRAC"
     ; TAB( 32);"K"
540  PRINT
550  FOR I = 1 TO N
560  PRINT I; TAB( 5);N$(I); TAB( 20);F(I); TAB( 30);
     K(I)
570  NEXT I
580  PRINT
590  PRINT "YOUR MAY START OVER WITH NEW DATA, OR"
600  PRINT "YOU MAY CORRECT THE ABOVE."
610  PRINT "TO CORRECT THE DATA      ENTER C."
620  PRINT "TO START WITH NEW DATA   ENTER N."
630  INPUT "TO EXIT PROGRAM          ENTER E.    ";X$
640  IF X$ = "N" GOTO 240
650  IF X$ = "C" GOTO 720
660  IF X$ = "E" GOTO 1270
670  PRINT
680  PRINT "YOUR CHOICES WERE C, E, OR N.  YOU"
690  PRINT "ENTERED;"X$".  TRY AGAIN."
700  PRINT
710  GOTO 610
720  PRINT
730  PRINT "ENTER ITEM NUMBER, COLUMN NUMBER,"
740  PRINT "AND CORRECT VALUE: 3,2,.45 WOULD"
750  PRINT "CHANGE THE MOL FRACTION OF THE THIRD"
760  PRINT "ITEM TO .45."
770  PRINT
780  INPUT "ENTER YOUR CORRECTION         ";A,B,C
790  IF B = 2 THEN T = T - F(A)
800 F(A) = C
810 T = T + F(A)
820  GOTO 840
830 K(A) = C
840  PRINT
850  INPUT "ANY OTHER CHANGES, Y OR N     ";X$
860  IF X$ = "Y" GOTO 780
870  GOTO 450
880 V = 0
890 F = 1000
900 A = 100
910 B = A
920  FOR I1 = A TO A + 9 * B + P1 STEP B
930 V = I1
940 V0 = 0
950  FOR I = 1 TO N
960 V0 = V0 + F(I) * K(I) * V * F / (F + V * (K(I) - 1))
970  NEXT I
980  IF V = F GOTO 1040
990  IF  ABS (V0 - V) < .001 GOTO 1090
1000  IF V0 < V GOTO 1040
1010  NEXT I1
1020 A = A * 10
1030  GOTO 910
1040 A = I1 - B
1050  IF A = 0 THEN A = B / 10
1060 B = B / 10
1070 P1 = B
1080  GOTO 920
1090  HOME : PRINT
1100  PRINT  TAB( 13);"THE RESULTS ARE:"
1110  PRINT
1120 V = F1 / 1000 * V
1130 F = F1
1140 L = F - V
1150  PRINT  TAB( 20);"VAPOR"; TAB( 30);"LIQUID"
1160  PRINT
1170  PRINT "MOL/HR"; TAB( 20); FN P(V); TAB( 30); FN P(L)
1180  PRINT
1190  PRINT  TAB( 20);"MOL FRACTION"
1200  PRINT "  COMPONENT"; TAB( 18);"IN VAPOR"; TAB( 28);
      "IN LIQUID"
1210  PRINT
1220  FOR I = 1 TO N
1230 L(I) = F(I) * F / (V * (K(I) - 1) + F)
1240 V(I) = L(I) * K(I)
1250  PRINT I; TAB( 3);N$(I); TAB( 20); FN P(V(I)); TAB
      ( 30); FN P(L(I))
1260  NEXT I
1270  PRINT
1280  PRINT  TAB( 13);"END OF PROGRAM"
1290  END
```

**Example for: Flash computations** **Table II**

**(Start of first display)**

```
PROGRAM CALCULATES THE EQUILIBRIUM
FLASH COMPOSITIONS FOR ANY NUMBER
OF COMPONENTS.

DATA REQUIRED ARE:

FEED RATE, MOL/HR
FEED COMPOSITION, MOL FRACTION
INDIVIDUAL COMPONENT K VALUES

OUTPUT WILL BE VAPOR AND LIQUID RATES,
AND COMPOSITIONS.

TO ENTER DATA, PRESS RETURN KEY.
```

**(Start of next display)**

```
ENTER THE FEED RATE, MOL/HR      1000

FOR EACH COMPONENT, ENTER NAME,
MOL FRACTION IN FEED AND K VALUE.

SEPARATE THE ITEMS WITH COMMAS.
WHEN ALL THE DATA ARE IN, ENTER THREE
ZEROS.: 0,0,0

ENTER FIRST DATA: NAME, F(I), K(I)

                  CO2,.0112,.9
ENTER NEXT DATA:  CH4,.8957,2.7
ENTER NEXT DATA:  C2H6,.0526,.38
ENTER NEXT DATA:  C3H8,.0197,.098
ENTER NEXT DATA:  ISO-C4,.0068,.038
ENTER NEXT DATA:  N-C4,.0047,.024
ENTER NEXT DATA:  C5'S,.0038,.0075
ENTER NEXT DATA:  C6'S,.0031,.0019
ENTER NEXT DATA:  C7+,.0024,.0007
ENTER NEXT DATA:  0,0,0
```

**(Start of next display)**

```
              THE RESULTS ARE:

                      VAPOR     LIQUID

MOL/HR                959.17    40.83

                      MOL FRACTION
  COMPONENT         IN VAPOR  IN LIQUID

1 CO2                 .0111     .0124
2 CH4                 .9193     .3405
3 C2H6                .0493     .1298
4 C3H8                .0143     .1461
5 ISO-C4              3.3E-03   .088
6 N-C4                1.8E-03   .0736
7 C5'S                6E-04     .0791
8 C6'S                1E-04     .0727
9 C7+                 0         .0578

              END OF PROGRAM
```

# Packed-tower design

Computers are useful for solving the trial-and-error calculations involved in designing packed towers. This program permits quick, accurate determinations of tower diameter or flooding rates.

☐ A correlation of flooding velocity for gas and liquid flowing through packed towers was first given by Sherwood, Shipley and Holloway [1] as a plot of:

$$\left(\frac{G^2 a_v \mu_L{}^{0.2}}{g_c e^3 \rho_G \rho_L}\right) \text{ vs. } \left[\frac{L}{G}(\rho_G/\rho_L)^{1/2}\right]$$

As modified by Lobo and coworkers on the basis of experimental data [2], the correlation curve has been used widely for predicting flooding rates in gas-liquid packed towers.

Because $G$, the superficial mass velocity of the gas, appears in both coordinates, it must be calculated by trial and error (Fig. 3). To eliminate this, Zenz plotted the terms on log-log paper (Fig. 1) [3]:

$$\left(\frac{L a_v \mu_L{}^{0.2}}{3{,}600^2 g_c e^3 \rho_L{}^2}\right) \text{ vs. } \left[\frac{L}{G}(\rho_G/\rho_L)^{1/2}\right]$$

Later, Zenz and Eckert plotted the terms [4]:

$$\left(\frac{V}{(\rho_L/\rho_G)^{1/2}}\right)\left(\frac{a_v \mu_L{}^{0.2}}{e^3}\right) \text{ vs. } Q\left(\frac{a_v \mu_L{}^{0.2}}{e^3}\right)^{1/2}$$

However, using these correlation curves is still time consuming, and some error can result from reading from the curves.

The computer is, of course, ideally suited to solving trial-and-error calculations. For this problem, therefore, a correlation and a computer program are developed.

## The design procedure

Based on the data taken from the curve of Lobo and coworkers [4], the following equation was developed by the least-squares method of Chen [5], for calculating the flooding rate in gas-liquid packed towers:

$$\ln(AG^2) = 0.9729 \ln(G/B) - 0.084472[\ln(B/G)]^2 - 3.995851 \quad (1)$$

In Eq. (1), $A = a_v \mu_L{}^{0.2}/g_c e^3 \rho_G \rho_L$, and $B = L(\rho_G/\rho_L)^{1/2}$. Eq. (1) agrees with values from the Lobo curve for the range of $L/G(\rho_G/\rho_L)^{1/2}$ from 0.02 to 7.00. The mean deviation is ±0.06%, and the standard deviation is ±3.20%.

For solving $G$, it is assumed that specific surface area, superficial liquid rate, and fraction of free void space in the packing are specified, and that the density of the gas and the density and viscosity of the liquid are known.

The design procedure for the computer program follows the Newton-Raphson method, which searches for the root of a function as:

$$Y(G) = 0$$

This root is determined by trials and is given as:

$$G_{n+1} = G_n - Y(G_n)/Y'(G_n)$$

Here, $Y'$ is the derivative with respect to $G$.

For this particular design, Eq. (1) is rearranged to give:

$$Y = \ln(AG^2) + 0.083472[\ln(B/G)]^2 + 0.9729 \ln(B/G) + 3.995851 \quad (2)$$

$$Y' = 1.0271(1/G) - 0.166944(1/G)\ln(B/G)$$

The computer program is written in BASIC. The flow-

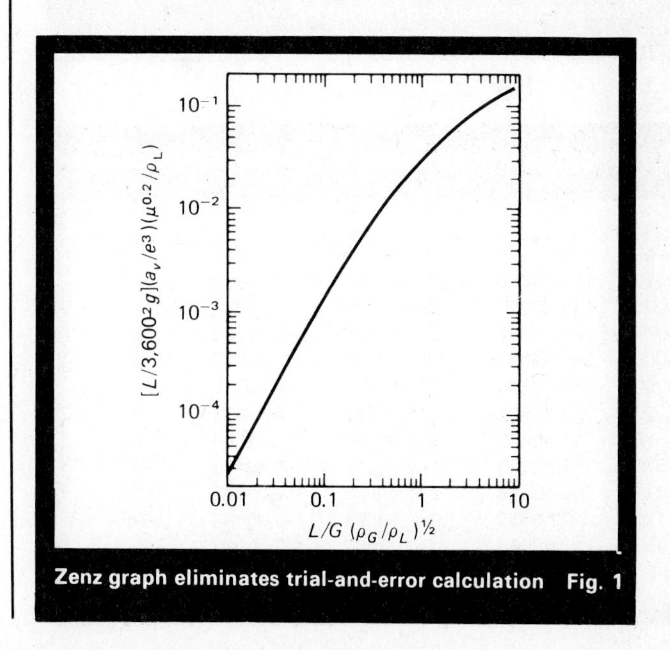


**Zenz graph eliminates trial-and-error calculation Fig. 1**

Adapted from an article by Hung Xuan Nguyen, Allied Chemical Corp., originally published November 20, 1978.

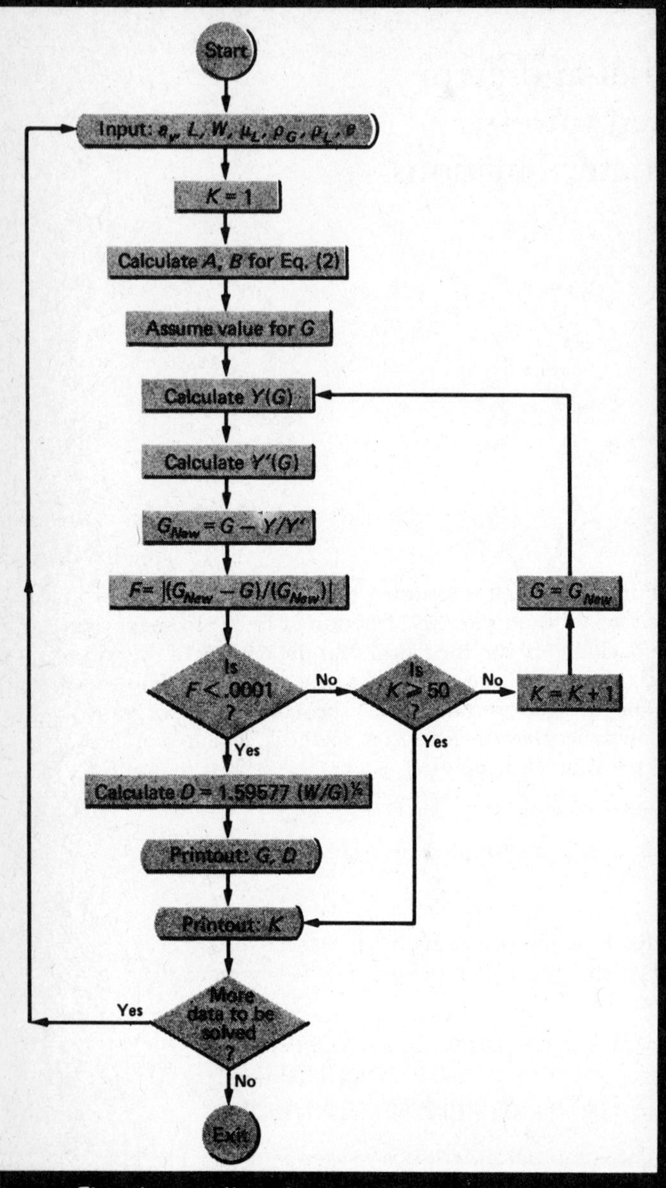


Flowchart outlines determination of flooding rate

Fig. 2

Trial-and-error procedure determines superficial gas velocity

| f+ | f− | $\bar{G}$=½[(f+) + (f−)] | Y | \|Y\|< 0.0001? |
|---|---|---|---|---|
| 600.00000 | 150.00000 | 375.00000 | +0.38625 | No |
| 375.00000 | 150.00000 | 262.50000 | +0.05187 | No |
| 262.50000 | 150.00000 | 206.25000 | −0.16218 | No |
| 262.50000 | 206.25000 | 234.37500 | −0.04992 | No |
| 262.50000 | 234.37500 | 248.43750 | +0.00215 | No |
| 248.43750 | 234.37500 | 241.40625 | −0.02358 | No |
| 248.43750 | 241.40625 | 244.92188 | −0.01064 | No |
| 248.43750 | 244.92188 | 246.68315 | −0.00422 | No |
| 248.43750 | 246.68315 | 247.56033 | −0.00103 | No |
| 248.43750 | 247.56033 | 247.99891 | +0.00056 | No |
| 247.99891 | 247.56033 | 247.77962 | −0.00023 | No |
| 247.99891 | 247.77962 | 247.88927 | +0.00016 | No |
| 247.88927 | 237.77962 | 247.83444 | −0.00004 | Yes |

## Nomenclature

| | |
|---|---|
| $A$ | As defined, Eq. (1) |
| $a_v$ | Specific surface area of dry packing, ft²/ft³ |
| $B$ | As defined, Eq. (1) |
| $c_1$, $c_2$ | As defined, Eq. (4) |
| $D$ | Packed-tower dia., ft |
| $e$ | Fraction of free void space in packing |
| $G$ | Superficial mass velocity of gas, lb/(h)(ft²) |
| $g_c$ | Newton's Law conversion factor, (lbm)(ft)/(lbf)(h²) |
| $L$ | Superficial liquid flowrate, lb/(h)(ft²) |
| $Q$ | Liquid flowrate, gal/(min)(ft² of superficial cross-sectional area of tower) |
| $V$ | Gas flowrate, ft³/(min)(ft² of superficial cross-sectional area of tower) |
| $W$ | Weight flowrate of gas, lb/h |
| $X$ | As defined, Eq. (4), ln $G$ |
| $\alpha$, $\beta$, $\gamma$ | As defined, Eq. (4) |
| $\mu_L$ | Viscosity of liquid, cP |
| $\rho_G$ | Density of gas, lb/ft³ |
| $\rho_L$ | Density of liquid, lb/ft³ |

chart and the program are presented in Fig. 2 and Table I, respectively.

Although an upper limit on the number of iterations in the Newton-Raphson procedure is imposed in the program, 5 to 8 iterations are usually required to arrive at answers.

Because a tower is usually designed for a gas velocity of 50% of the superficial flooding rate, the required tower diameter, after $G$ has been determined, can be computed from:

$$D = 1.59577(W/G)^{1/2} \tag{3}$$

The computer program provides accurate results for many gas-liquid packed-tower calculations quickly.

## Example illustrates procedure

A tower randomly packed with 1-in. Raschig rings is to recover an organic substance from an air mixture by countercurrent scrubbing with a hydrocarbon oil. Determine the tower diameter when $W = 1{,}800$ lb/h and $L = 1{,}516$ lb/(h)(ft²). The actual gas velocity is assumed to be 50% of the superficial gas velocity. The fluid properties are: $\rho_G = 0.075$ lbm/ft³, $\rho_L = 55.66$ lbm/ft³, $\mu_L = 33.9$ cP, $a_v = 51$ ft²/ft³, and $e = 0.683$ ft³/ft³.

Putting these values into the computer program (see Table II), yields a tower diameter of 2.3 ft, which closely approaches the 2.4 ft dia. obtained with the same example by Badger and Banchero [8].

## Using the program

This program calculates the gas rate for flooding of a packed tower, and then determines the tower diameter for 50% of flooding.

The following data are required:

specific surface area of dry packing, ft²/ft³
superficial liquid flowrate, lb/(h) (ft²)

liquid density, lb/ft$^3$
liquid viscosity, cP
weight flowrate of gas, lb/h
gas density, lb/ft$^3$
free-void-space fraction in packing, to be entered as a decimal fraction.

The program prompts the user to input the values for the above variables. Once all the data are entered, the next display repeats the packed-tower conditions, and provides the calculated results:

flooding gas rate, lb/(h) (ft$^2$)
packed-tower diameter at 50% of flooding, ft.

## References


1. Sherwood, T. K., Shipley, G. H. and Holloway, F. A. L., *Ind. Eng. Chem.*, Vol. 30, 1938.
2. Lobo, W. E., Friend, L., Hashmall, F. and Zenz, F. A., Limiting Capacity of Dumped Tower Packings, *Trans. AIChE,* Vol. 41, No. 693, 1945.
3. Zenz, F. A., What Every Engineer Should Know About Packed Tower Operations, *Chem. Eng.,* Aug. 1953, pp. 176–184.
4. Zenz, F. A., Eckert, R. A., New Chart for Packed Tower Flooding, *Pet. Ref.,* Feb. 1961, p. 130.
5. Chen, N. H., Equations for Flooding Rate in Packed Towers, *Ind. Eng. Chem.,* Vol. 53, No. 6, 1961.
6. Perry, J. H., "Chemical Engineers' Handbook," 3rd ed., McGraw-Hill, New York, p. 685.
7. Nguyen, H. X., Simplify Calculation of Economic Pipe Size, *Hydrocarbon Proc.,* Feb. 1978, pp. 143–144.
8. Badger, W. L., and Banchero, J. T., "Introduction to Chemical Engineering," McGraw-Hill, New York, 1955, p. 427.

## The author


**Hung Xuan Nguyen** has been a process engineer for Allied Chemical Corp., responsible for solving technical problems and making process improvements involving heat transfer, fluid dynamics, mass transfer, energy conservation and computer control. On Nov. 1, he joined Philip Morris, Inc. (Research Center, P.O. Box 26583, Richmond, VA 23261). He holds two degrees in chemical engineering, a B.S. from Lowell University and an M.S. from University of Massachusetts, and is an associate member of AIChE.


**Program for packed-tower design** **Table I**

```
10 REM  PACKED-TOWER DESIGN
20 REM  FROM CHEMICAL ENGINEERING, NOVEMBER 20, 1978
30 REM  BY HUNG XUAN NGUYEN
40 REM TRANSLATED BY WILLIAM VOLK
50 REM  COPYRIGHT (C) 1984
60 REM BY CHEMICAL ENGINEERING.
70 DEF FN P(X) = INT (1E4 * (X + .00005)) / 1E4
80 HOME : VTAB 4
90 PRINT "PROGRAM CALCULATES THE GAS RATE FOR"
100 PRINT "FLOODING OF A PACKED TOWER, AND THEN"
110 PRINT "CALCULATES THE TOWER DIAMETER FOR 50%"
120 PRINT "OF FLOODING
130 PRINT
140 PRINT "THE FOLLOWING DATA ARE REQUIRED:
150 PRINT
160 PRINT "PACKING SPECIFIC AREA, SQ. FT/CU. FT"
170 PRINT "LIQUID RATE, LB/(HR)(SQ. FT)"
180 PRINT "LIQUID DENSITY, LB/CU. FT"
190 PRINT "LIQUID VISCOSITY, CP"
200 PRINT "WEIGHT RATE OF GAS, LB/HR"
210 PRINT "GAS DENSITY, LB/CU. FT"
220 PRINT "FRACTION VOIDS IN PACKING"
230 PRINT
240 INPUT "PRESS RETURN KEY TO ENTER DATA. ";X$
250 HOME : PRINT
260 PRINT "ENTER THE FOLLOWING DATA:"
270 PRINT
280 PRINT "PACKING SPECIFIC SURFACE,"
290 INPUT "        SQ. FT/CU. FT         ";A
300 PRINT
310 PRINT "SUPERFICIAL LIQUID RATE,"
320 INPUT "        LB/(HR)(SQ. FT)       ";L
330 PRINT
340 PRINT "LIQUID DENSITY,"
350 INPUT "        LB/CU. FT             ";D2
360 PRINT
370 INPUT "LIQUID VISCOSITY, CP          ";V
380 PRINT
390 INPUT "WEIGHT FLOW OF GAS, LB/HR     ";W
400 PRINT
410 INPUT "GAS DENSITY, LB/CU. FT        ";D1
420 PRINT
430 INPUT "FRACTION PACKING VOID SPACE. ";E
440 IF E = 0 GOTO 460
450 IF E < 1 GOTO 530
460 HOME : VTAB 8
470 PRINT "PACKING VOIDS SHOULD BE ENTERED AS"
480 PRINT "A DECIMAL FRACTION, GREATER THAN"
490 PRINT "ZERO AND LESS THAN ONE.  YOU ENTERED"
500 PRINT E;".  TRY AGAIN."
510 PRINT
520 GOTO 420
530 K = 1
540 A1 = A * (V ^ .2) / (4.17E8 * D1 * D2 * (E ^ 3))
550 B = L * ((D1 / D2) ^ .5)
560 G = 10
570 Y = LOG (A1 * G * G) + .083472 * (( LOG (B / G)) ^ 2)
    + .9729 * LOG (B / G) + 3.99585
580 Y1 = 1.0271 * (1 / G) - .166944 * (1 / G) * LOG
    (B / G)
590 G1 = G - Y / Y1
600 F = ABS ((G1 - G) / G1)
610 IF F < .0001 GOTO 660
620 IF K > 50 GOTO 820
630 K = K + 1
640 G = G1
650 GOTO 570
660 D = 1.59577 * ((W / G) ^ .5)
670 HOME : PRINT
680 PRINT "FOR THE FOLLOWING CONDITIONS:"
690 PRINT
700 PRINT "PACKING"; TAB( 18);A;" SQ. FT/CU. FT"
```

**Program for packed-tower design (continued)** **Table I**

```
710  PRINT "LIQUID"; TAB( 18);L;" LB/(HR)(SQ. FT)"
720  PRINT "LIQUID DENSITY"; TAB( 18);D2;" LB/CU. FT"
730  PRINT "LIQUID VISCOSITY"; TAB( 18);V;" CP"
740  PRINT "GAS"; TAB( 18);W;" LB/HR"
750  PRINT "GAS DENSITY"; TAB( 18);D1;" LB/CU. FT"
760  PRINT "FRACTION VOIDS"; TAB( 18);E
770  PRINT
780  PRINT "THE RESULTS ARE:"
790  PRINT
800  PRINT "FLOODING GAS RATE,"
810  PRINT "     LB/(HR)(SQ. FT)           "; FN P(G)
820  PRINT
830  PRINT "PACKED TOWER DIAMETER,"
840  PRINT "    AT 50% OF FLOODING       "; FN P(D);" FT"
850 K = (L / W) * (D1 / D2) ^ .5 * (3.14159 * D ^ 2 / 4)
860 K = K / 2
870  IF K =  > .02 AND K =  < 7 GOTO 930
880  PRINT
890  PRINT "CALCULATION BEYOND RANGE OF RELIABILITY."
900  PRINT "(LIQUID/GAS) RATIO TIMES SQUARE ROOT OF "
910  PRINT "GAS DENSITY SHOULD BE IN RANGE OF 0.02  "
915  PRINT "TO 7.  THESE DATA GIVE"
920  PRINT "          "; FN P(K)
930  PRINT
940  PRINT  TAB( 13);"END OF PROGRAM"
```

**Example for: Packed-tower design** **Table II**

**(Start of first display)**

```
PROGRAM CALCULATES THE GAS RATE FOR
FLOODING OF A PACKED TOWER, AND THEN
CALCULATES THE TOWER DIAMETER FOR 50%
OF FLOODING

THE FOLLOWING DATA ARE REQUIRED:

PACKING SPECIFIC AREA, SQ. FT/CU. FT
LIQUID RATE, LB/(HR)(SQ. FT)
LIQUID DENSITY, LB/CU. FT
LIQUID VISCOSITY, CP
WEIGHT RATE OF GAS, LB/HR
GAS DENSITY, LB/CU. FT
FRACTION VOIDS IN PACKING

PRESS RETURN KEY TO ENTER DATA.
```

**(Start of next display)**

```
ENTER THE FOLLOWING DATA:

PACKING SPECIFIC SURFACE,
      SQ. FT/CU. FT          51

SUPERFICIAL LIQUID RATE,
      LB/(HR)(SQ. FT)        1516

LIQUID DENSITY,
      LB/CU. FT              55.66

LIQUID VISCOSITY, CP         33.9

WEIGHT FLOW OF GAS, LB/HR    1800

GAS DENSITY, LB/CU. FT       .075

FRACTION PACKING VOID SPACE. .683
```

**(Start of next display)**

```
FOR THE FOLLOWING CONDITIONS:

PACKING          51 SQ. FT/CU. FT
LIQUID           1516 LB/(HR)(SQ. FT)
LIQUID DENSITY   55.66 LB/CU. FT
LIQUID VISCOSITY 33.9 CP
GAS              1800 LB/HR
GAS DENSITY      .075 LB/CU. FT
FRACTION VOIDS   .683

THE RESULTS ARE:

FLOODING GAS RATE,
     LB/(HR)(SQ. FT)         874.9231

PACKED TOWER DIAMETER,
AT 50% OF FLOODING       2.2889 FT

          END OF PROGRAM
```

# Sour-water-stripper design

Computers can take the drudgery out of lengthy design mathematics. This program determines the number of trays required to reduce the ammonia and hydrogen sulfide content of a sour-water stream within given specifications.

☐ Designing a sour-water stripper can be a laborious procedure, involving lengthy tray-to-tray calculations and trial-and-error methods. This section outlines the functions of a sour-water stripper and the theory of the design calculations, and presents the Apple II program that has been devised. With this program, it is possible to punch in the basic design data in a few seconds and then have the computer display the number of trays required. A worked example is given and explained.

## Sour-water strippers

Sour-water strippers are widely used in oil refineries and petrochemical plants to clean up fouled water streams prior to sending these to rivers or reusing them in the plant. Such waters commonly arise from the washing of reactor products that have been hydrogen treated in hydrodesulfurization or hydrocracking operations. The waters usually contain ammonia ($NH_3$) and hydrogen sulfide ($H_2S$). Other components present may include phenols and cyanides, but removal of these is beyond the scope of this article.

A stripper may consist merely of a packed or trayed column down which the sour water is passed countercurrently to "open" steam. The excess steam and foul vapors from the top of such a column may then be burned. Legislation restricting atmospheric pollution, and especially the high cost of energy, has rendered such a crude arrangement unsuitable for new designs.

To conserve the steam condensate in pure form, it is preferable to energize the stripper, using a reboiler instead of "open" steam. Also, passage of the hot overhead vapors directly to a furnace is likely to have an adverse effect on combustion there, due to the diluent steam present. Thus an overhead condenser will usually be provided to remove the excess steam, even though the resultant reflux is rich in $NH_3$ and $H_2S$ and must be returned to the stripper, thereby increasing the stripping load.

The reflux may be returned to a separate rectifying section above the main stripping column. This is theoretically more efficient, but it complicates the design somewhat since the column diameter for the rectifying zone should be reduced in accordance with the reduced liquid loading. Also, severe corrosion and foaming problems have sometimes been reported in rectifying sections where $NH_3$ and $H_2S$ concentrations are necessarily high.

In general, it is preferable to return the reflux directly on to the feed tray, which is then designated Tray 1, the top tray.

The feed is conveniently preheated by external heat exchange with the bottoms before entering the stripper.

A stripper designed along these lines is shown in Fig. 1, with sample data written in. This forms the basis for the calculation procedure of this article. Naturally, the procedure may be modified to agree with other basic types of stripper design.

## Background to calculations

The Beychok/Van Krevlin correlations [*1* (p. 163)], covering ammonia / hydrogen-sulfide / steam / water equilibria appear to provide the most practical basis for designing sour-water strippers. Researchers for the American Petroleum Institute [*2*] have more recently produced other correlations but these are difficult to use for an initial design. However they could possibly be applied at a later stage for checking equilibria at critical points.

Commercially-available computer programs use the Beychok method. However, those examined by the author did not calculate the number of trays required for stripping to a specification (which is the designer's problem) but merely tested the effect of a given number of trays.

By using Beychok's suggestions (p. 173) for simplifying the equations to avoid using trial-and-error charts, a computer program was devised for determining the number of theoretical stripping trays required to meet given specifications under given conditions.

The program calculates the number of trays required directly. It also gives the reboiler duty. It is possible to optimize the design by repeated computer runs.

## Design correlations

The method of tray-to-tray design calculation to determine the number of stripping trays required is a development of Beychok's manual procedure (p. 182). Since the known conditions are in the reflux drum and at the top of the stripper, the partial pressures for $NH_3$ and $H_2S$ above liquid at a known temperature are easily found. The Beychok/Van Krevlin equations then give the $NH_3$ and $H_2S$ concentrations in the equilibrium liquid phase as parts per million. By a mass balance, the partial pressures for $NH_3$ and $H_2S$ in the

Adapted from an article by Norman H. Wild, Davy International Ltd., originally published February 12, 1979.

## Nomenclature

*Data Constants* (given and directly-derived data)

- $a$ — $NH_3$ in feed stream, lb/h
- $b$ — $H_2S$ in feed stream, lb/h
- $c$ — $H_2O$ in feed stream, lb/h
- $d$ — $NH_3$ in bottoms product, ppm (specified)
- $e$ — $H_2S$ in bottoms product, ppm (specified)
- $f$ — $H_2O$ in bottoms product, ppm (derived)
- $h$ — $NH_3$ in tail gas, lb/h (derived)
- $j$ — $H_2S$ in tail gas, lb/h (derived)
- $k$ — $H_2O$ in tail gas, lb/h (derived)
- $p$ — $NH_3$ in bottoms product, lb/h
- $q$ — $H_2S$ in bottoms product, lb/h
- $r$ — $H_2O$ in bottoms product, lb/h
- $stm$ — $H_2O$ in stripper overhead vapor, lb/h (i.e., stripping-stream rate)
- $P_n$ — Pressure above Tray "n", psia
- $P_{tg}$ — Pressure of tail gas above reflux, psia
- $T_{tg}$ — Temperature of tail gas/reflux in reflux drum, °F
- $T_{fd}$ — Temperature of feed entering stripper, °F (after external preheat)
- $T_{bm}$ — Temperature of bottoms leaving stripper reboiler, °F
- $\Delta P_{tt}$ — Pressure drop across one theoretical tray, psi

*Data Variables* (indirectly derived data)

- $L_r$ — $H_2O$ liquid content of reflux, lb/h
- $L_n$ — $H_2O$ content of liquid leaving Tray "n", lb/h
- $V_r$ — $H_2O$ content of vapor leaving reflux drum, lb/h (note $V_r = k$)
- $V_n$ — $H_2O$ content of vapor leaving Tray "n", lb/h (note $V_1 = stm$)
- $x_r$ — $NH_3$ content of liquid leaving reflux drum, lb/h
- $x_n$ — $NH_3$ content of vapor leaving Tray "n", lb/h
- $y_r$ — $H_2S$ content of liquid leaving reflux drum, lb/h
- $y_n$ — $H_2S$ content of vapor leaving Tray "n", lb/h
- $A_r$ — $NH_3$ content of liquid leaving reflux drum, ppm
- $A_n$ — $NH_3$ content of liquid leaving Tray "n", ppm
- $S_r$ — $H_2S$ content of liquid leaving reflux drum, ppm
- $S_n$ — $H_2S$ content of liquid leaving Tray "n", ppm
- $AP_r$ — $NH_3$ partial pressure above reflux, psia
- $AP_n$ — $NH_3$ partial pressure above Tray "n", psia
- $SP_r$ — $H_2S$ partial pressure above reflux, psia
- $SP_n$ — $H_2S$ partial pressure above Tray "n", psia
- $WP_r$ — $H_2O$ partial pressure above reflux, psia
- $WP_n$ — $H_2O$ partial pressure above Tray "n", psia
- $T_n$ — Temperature of liquid leaving Tray "n", °F

*Miscellaneous*

- $K$, $C$, $H_o$ — Temperature-dependent variables used in Beychok/Van Krevlin equations

vapor from the next tray are found as, in turn, are the concentrations in the liquid. Thus it is possible to work down the stripper until sufficient tray equilibria have been calculated to achieve the specified concentrations of $NH_3$ and $H_2S$ in the bottoms liquid.

Two points concerning the correlations should be noted. First, Beychok proposes (p. 173) that his equations may be simplified, provided that the $H_2S$ concentration is less than about 5,000 ppm. Second, the Van Krevlin work is only held to be valid for molar ratios of $NH_3$ to $H_2S$ exceeding 1.5.

For practical use, both of these restrictions are relatively unimportant, since by the time the feed and reflux have passed through about two theoretical trays such limitations will have been removed anyway. The $H_2S$ is relatively so volatile that the molar ratio of 1.5 for $NH_3$ to $H_2S$ will be rapidly exceeded; also, the concentration of $H_2S$ will be quickly brought below 5,000 ppm.

In cases of doubt—where few trays are being employed in a design; where the feed is very rich; or where an inadequate safety factor is being used to cover for variations in feed composition—the equilibria occurring at the top of the stripper should be checked out by independent correlations, such as those derived by Miles and Wilson [*2*]. These correlations, although derived more recently than Beychok's, are, unfortunately, very difficult to use when attempting to obtain the liquid composition from that of the vapor—instead of the reverse.

For convenience, the correlations given in Beychok's book have been modified to use temperature in °F throughout, instead of a mixture of °F and °C. Thus, from Fig. 25 and 26, page 163 and 164, and also from

Eq. (14) and (15) may be derived:

$$K = 121{,}000/(1.0163)^T \quad (1)$$

$$C = \left(\frac{2.84 \times 10^{-7}}{K}\right)(1.27)^{[(T-32)^{0.648}]} \quad (2)$$

where $T$ is the temperature of the liquid/vapor equilibrium mixture in °F, and $K$, $C$ are temperature-dependent variables. Also, Beychok's Eq. 16 applies:

$$S = \sqrt{\frac{AP \times SP}{C}} \quad (3)$$

where $S$ = ppm $H_2S$ in the liquid, and $AP$ and $SP$ are the partial pressures (psia) of $NH_3$ and $H_2S$ above the liquid. By substituting, $K = (8.8 \times 10^5)H_o$ into Beychok's equations, one obtains:

$$A = K \cdot AP + S/2 \quad (4)$$

where $A$ = ppm $NH_3$ in the liquid and $H_o$ is a temperature-dependent variable. In order to correlate water vapor pressure (which frequently occurs under the description, $H_2O$ partial pressure or $WP$ psia) with temperature ($T$°F), the equation quoted by Miles and Wilson, page 24, relating mm Hg to °C, has been modified and also adjusted to conform better with steam-table data, to give:

$$T = 7{,}007/(14.465 - \log_e WP) - 383 \quad (5)$$

Other equations used include, for tray-to-tray mass balance:

$$x = A\left(\frac{L}{10^6 - A - S}\right) - p \quad (6)$$

$$y = S\left(\frac{L}{10^6 - A - S}\right) - q \quad (7)$$

where $x$ = $NH_3$ content of vapor (lb/h)
$y$ = $H_2S$ content of vapor (lb/h)
$p$ = $NH_3$ in bottoms product (lb/h)
$q$ = $H_2S$ in bottoms product (lb/h)

From the $x$, $y$ and $V$ values, partial pressures are calculated by, for example:

$$AP = \frac{x}{17}\left(\frac{P}{x/17 + y/34 + V/18}\right)$$

where $P$ = pressure, psia and $V$ = $H_2O$ content of vapor, lb/h. Molecular weight of $NH_3$ = 17, $H_2S$ = 34 and $H_2O$ = 18.

This reduces to:

$$AP = 2x\left(\frac{P}{2x + y + 1.89V}\right) \quad (8)$$

Similarly:

$$SP = y\left(\frac{P}{2x + y + 1.89V}\right) \quad (9)$$

$$WP = 1.89V\left(\frac{P}{2x + y + 1.89V}\right) \quad (10)$$

and:

$$AP_1 = 2x_1\left(\frac{P_1 - WP_1}{2x_1 + y_1}\right) \quad (11)$$

$$SP_1 = y_1\left(\frac{P_1 - WP_1}{2x_1 + y_1}\right) \quad (12)$$

Equations used for the overall mass balance include:

$$k = \frac{(2a + b - Mc)N}{1 - MN} \quad (13)$$

where $M = \frac{2d + e}{f}$ and $N = \frac{18}{34}\left(\frac{WP_r}{P_{tg} - WP_r}\right)$

$$p = d\left(\frac{c - k}{f}\right) \quad (14)$$

$$q = e\left(\frac{c - k}{f}\right) \quad (15)$$

$$r = c - k \quad (16)$$

## Using the program

**This program (Table I) calculates the number of trays required for the stripping of hydrogen sulfide and ammonia from sour water. It makes individual-tray material and heat balances, and provides the results for each tray.**

The data required are:

- sour water flowrate, lb/h
- hydrogen sulfide flowrate, lb/h
- ammonia flowrate, lb/h
- feed temperature, °F
- stripped-liquid concentration, hydrogen sulfide in clear water, ppm
- stripped-liquid concentration, ammonia in clear water, ppm
- tower top pressure, psia
- tower bottom pressure, psia
- reflux drum pressure, psia
- reflux drum temperature, °F
- steam rate, lb/h
- pressure drop per tray, psia

## Example

The proposed arrangement for the stripper is shown in Fig. 1, in which the figures shown are given data for the worked example.

The example illustrates the running of a program for the tray-to-tray calculation of a sour-water stripper, the most tedious part of the design being the calculation of the number of theoretical trays required.

## Problem

A sour-water stripper is to be designed to remove $H_2S$ and $NH_3$ down to levels of 5 ppm and 20 ppm, respectively, from a feed stream of 100,000 lb/h containing 0.73 wt % $H_2S$ and 0.556 wt % $NH_3$.

Stripped overheads are to be condensed, with the liquid portion to be returned as reflux to the top tray (feed tray) of the tower, while the reflux and tail gas will separate in the reflux drum at a pressure of 27.2 psia and temperature of 184°F. The tail gas will be sent to a sulfur recovery plant.

It is estimated that a tower-top operating pressure of 30 psia will be suitable, allowing for reboiler operation at 35 psia and a bottoms temperature (ex-reboiler) of 259°F. The pressure drop per theoret-

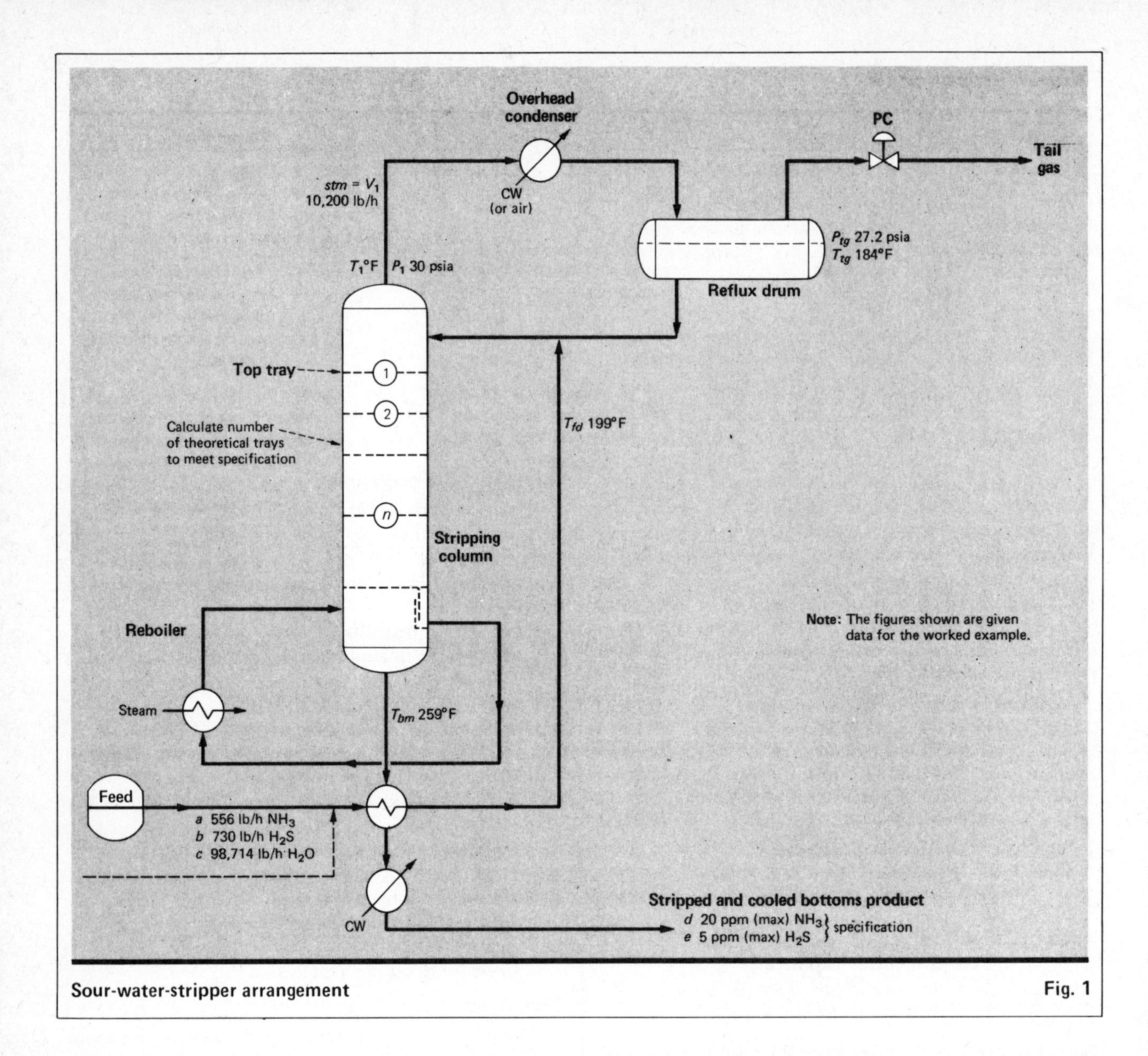


Sour-water-stripper arrangement Fig. 1

ical tray is assumed to be 0.45 psi.

The stripping steam rate is to be taken as 10,200 lb/h (i.e., 10.2% of feed).

In consideration of heat exchange with the bottoms stream, the feed is assumed to be preheated, to 199°F, with no vaporization, before entering the feed tray (Tray 1).

## Output

As seen in the example (Table II), once all the data are entered, the initial output display offers the first calculated results. Data are printed out in three groups of three items, making nine lines altogether. In each group, ammonia is first, hydrogen sulfide second and water third (when there is a third item). In the first group are the feed quantities (lb/h) as entered, the bottoms quantities (lb/h), and tail-gas quantities (lb/h). In the second group are the reflux-drum partial pressures (psia) and compositions (ppm). The third group contains the ammonia, hydrogen sulfide, and water (lb/h) in the reflux, and the overhead. Also displayed are the reflux conditions (pressure and temperature). The user is then prompted to press "RETURN" to continue.

The next display offers the thermal loading data. This includes the heat absorbed (Btu/h) on and below Tray 1, and equivalent steam condensed there (lb/h).

The next group of data offers the following:

$AP_1$, the partial pressure of ammonia (psia) above Tray 1

$SP_1$, the partial pressure of hydrogen sulfide (psia) above Tray 1

$A_1$, the ammonia in the liquid (ppm) on Tray 1

$S_1$, the hydrogen sulfide in the liquid (ppm) on Tray 1.

Once again, the user is prompted to press "RETURN" to continue.

The next display, and subsequent displays, provide the tray-by-tray calculations. The calculations continue until values for ammonia and hydrogen sulfide in the liquid are found to be negative (indicating that the spe-

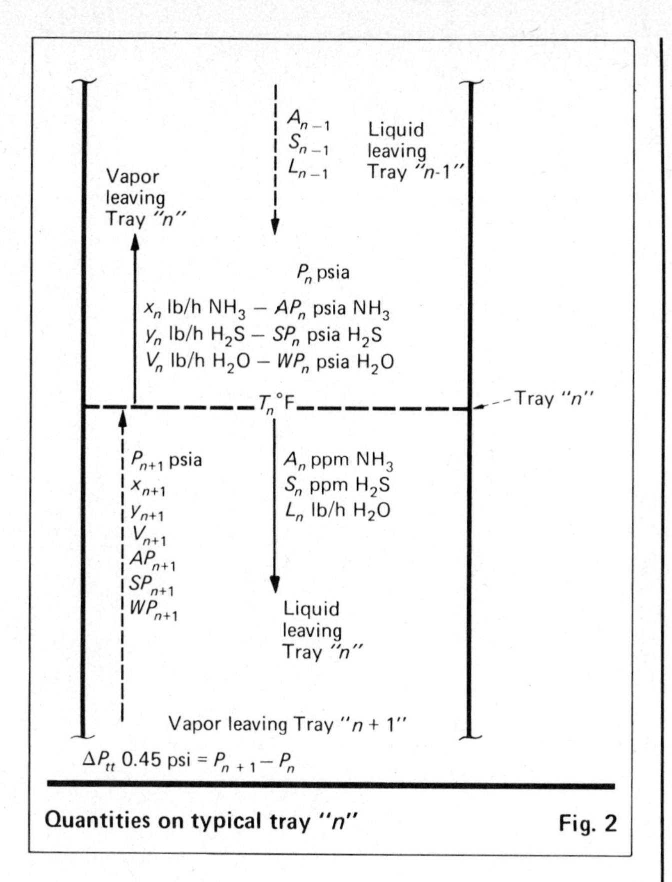


Quantities on typical tray "n" Fig. 2

cifications have been met). In this example, this occurs after twelve trays have been calculated, and means that no more than twelve are necessary.

Each tray-by-tray calculation provides the following information:

tray number
water in liquid, lb/h
ammonia in vapor, lb/h
hydrogen sulfide in vapor, lb/h
water in vapor, lb/h
pressure above tray, psia
ammonia pressure above tray, psia
hydrogen sulfide pressure above tray, psia
water pressure above tray, psia
temperature on tray, °F
ammonia in liquid, ppm
hydrogen sulfide in liquid, ppm.

## References

1. Beychok, M. R., "Aqueous Wastes from Petroleum and Petrochemical Plants," John Wiley & Sons, New York. 1967.
2. Miles, D. H., and Wilson, G. M., Vapor-Liquid Equilibrium Data for Design of Sour Water Strippers, Annual Report for 1974 (sponsored by the American Petroleum Institute), Center for Thermochemical Studies and Dept. of Chemical Engineering, Brigham Young University, October 1975.

## The author

**Norman H. Wild** is Process Group Head—Offsites, in the Technical Operations department of Davy International Ltd., Baker Street, London W.1, England. Tel: 01-486-6677. He has covered a wide range of refinery and petrochemical plant operations and design and also has had experience in designing and operating pilot plants. He obtained a London University external B.Sc. in Chemistry and later chose to take up chemical engineering as a process engineer in the south of England with Esso Petroleum Co. at the Fawley Oil Refinery. He is a member of the Royal Institute of Chemistry and the Institution of Chemical Engineers in England.

**Program for sour-water-stripper design** **Table I**

```
10 REM   SOUR-WATER-STRIPPER DESIGN
20 REM   FROM CHEMICAL ENGINEERING, FEBRUARY 12, 1979
30 REM   BY NORMAN H. WILD
40 REM  TRANSLATED BY WILLIAM VOLK
50 REM   COPYRIGHT (C) 1984
60 REM  BY CHEMICAL ENGINEERING
70 HOME : VTAB 4
80 DEF  FN P(X) =  INT (100 * (X + .005)) / 100
90 REM  THE ABOVE 'DEF' SETS THE DISPLAY TO TWO
   DECIMALS.
100 PRINT "PROGRAM CALCULATES THE NUMBER OF TRAYS"
110 PRINT "REQUIRED FOR THE STRIPPING OF H2S AND"
120 PRINT "NH3 FROM SOUR WATER."
130 PRINT
140 PRINT "IT MAKES INDIVIDUAL TRAY HEAT AND"
150 PRINT "MATERIAL BALANCES, AND GIVES THE"
160 PRINT "RESULTS FOR EACH TRAY.
170 PRINT
180 PRINT "THE DATA REQUIRED ARE:
190 PRINT
200 PRINT "SOUR WATER, H2S AND NH3 RATES, LB/HR"
210 PRINT
220 PRINT "REFLUX DRUM PRESSURE AND TEMPERATURE."
230 PRINT "TOWER TOP PRESSURE, BOTTOM TEMPERATURE."
240 PRINT "STRIPPED LIQUID H2S AND NH3, PPM."
250 PRINT
260 PRINT "STEAM RATE, LB/HR"
270 PRINT "PRESSURE DROP PER TRAY, PSIA.
280 PRINT
290 INPUT "PRESS RETURN TO ENTER DATA.  ";Q$
300 HOME : PRINT
310 PRINT "ENTER THE FOLLOWING DATA:"
320 PRINT
330 INPUT "WATER FLOW RATE, LB/HR        ";R2
340 INPUT "H2S FLOW RATE, LB/HR          ";R1
350 INPUT "NH3 FLOW RATE, LB/HR          ";R0
360 INPUT "FEED TEMPERATURE, DEG. F      ";P2
370 INPUT "H2S IN CLEAR WATER, PPM       ";R4
380 INPUT "NH3 IN CLEAR WATER, PPM       ";R3
390 INPUT "TOWER TOP PRESSURE, PSIA      ";P1
400 INPUT "TOWER BOTTOM TEMP., DEG. F    ";P3
410 INPUT "REFLUX DRUM PRESSURE, PSIA    ";R8
420 INPUT "REFLUX DRUM TEMP., DEG. F     ";D
430 INPUT "STEAM RATE, LB/HR             ";P0
440 INPUT "PRESSURE DROP PER TRAY, PSIA ";P7
450 R5 = 1E6 - R3 - R4
```

**Program for sour-water-stripper design (continued)** **Table I**

```
460 C =  EXP (14.465 - (7007 / (D + 383)))
470 A = R8 - C
480 R9 = (2 * R3 + R4) / R5
490 TP = ((2 * R0 + R1) - (R2 * R9)) * (18 * C / 34 / A)
500 R9 = TP / (1 - R9 * (18 * C / 34 / A))
510 R6 = (R2 - R9) / R5
520 I = R6 * R3
530 R4 = R6 * R4
540 R5 = R2 - R9
550 R6 = R0 - (R3 * R6)
560 R7 = R1 - R4
570 B = A / (1 + R6 / R7 * 2)
580 A = A - B
590  HOME : PRINT
600  PRINT "LB/HR"; TAB( 10);"FEED"; TAB( 20);"BOTTOMS"; TAB( 30);"TAIL GAS"
610  PRINT
620  PRINT " NH3"; TAB( 10); FN P(R0); TAB( 20); FN P(I); TAB( 30); FN P(R6)
630  PRINT " H2S"; TAB( 10); FN P(R1); TAB( 20); FN P(R4); TAB( 30); FN P(R7)
640  PRINT " H20"; TAB( 10); FN P(R2); TAB( 20); FN P(R5); TAB( 30); FN P(R9)
650  PRINT
660 R3 = I
670 K1 = 1.21E5 / (1.0163) ^ D
680 C2 = (2.84E - 7 / K1) * ((1.27) ^ ((D - 32) ^ .648))
690 E =  SQR (A * B / C2)
700 I = K1 * A + E / 2
710  PRINT  TAB( 18);"REFLUX DRUM"
720  PRINT  TAB( 8);"PARTIAL PRESSURE"; TAB( 27);"COMPOSITION"
730  PRINT  TAB( 14);"PSIA"; TAB( 31);"PPM"
740  PRINT
750  PRINT " NH3"; TAB( 14); FN P(A); TAB( 30); FN P(I)
760  PRINT " H2S"; TAB( 14); FN P(B); TAB( 30); FN P(E)
770  PRINT " H20"; TAB( 14); FN P(C)
780 P4 = P0 - R9
790 R8 = P4 / (1E6 - I - E)
800 A = I * R8
810  PRINT
820 B = R8 * E
830  PRINT "   REFLUX"; TAB( 19);"OVERHEAD"
840  PRINT "           LB/HR"; TAB( 20);"LB/HR"; TAB( 30);"CONDITIONS"
850 P5 = R6 + A
860  PRINT
870 P6 = R7 + B
880 P8 = P1
890 C = P1 * (1.89 * P0) / (2 * P5 + P6 + 1.89 * P0)
900 I = 7007 / (14.465 -  LOG (C)) - 383
910  PRINT " NH3"; TAB( 10); FN P(A); TAB( 20); FN P(P5); TAB( 30); FN P(C);" PS
IA"
920  PRINT " H2S"; TAB( 10); FN P(B); TAB( 20); FN P(P6); TAB( 30); FN P(I);" F"

930  PRINT " H20"; TAB( 10); FN P(P4); TAB( 20); FN P(P0)
940  PRINT
950  INPUT "PRESS RETURN TO CONTINUE.    ";X$
960 TP = ((R2 + .2 * R1 + .5 * R0) * (I - P2) + (P4 + .5 * A + .2 * B) * (I - D)
)
970 E = P5 * 350
980 TQ = .7 * E + 200 * P6
990 TW = TP + TQ
```

**Program for sour-water-stripper design (continued)** **Table I**

```
1000  HOME : PRINT
1010 P9 = TW / 950
1020  PRINT  TAB( 15);"THERMAL LOADING"
1030 P4 = P4 + R2 + P9
1040 TP = P4 * (P3 - I) + .3 * E
1050  PRINT
1060 E = TP / 950
1070  PRINT  TAB( 15);"ON TRAY 1"; TAB( 28);"BELOW TRAY 1"
1080  PRINT "HEAT ABSORBED"
1090  PRINT "    BTU/HR"; TAB( 15); FN P(TW); TAB( 30); FN P(TP)
1100  PRINT "CONDENSED STEAM"
1110  PRINT "     LB/HR"; TAB( 15); FN P(P9); TAB( 30); FN P(E)
1120 TP = P7
1130 P7 = P9 + P0
1140 P9 = TP
1150 P1 = R3
1160 P2 = R4
1170 P3 = E / (P3 - I)
1180 P5 = 2 * P5
1190 B = (P8 - C) / (P5 + P6)
1200 A = B * P5
1210 B = B * P6
1220  PRINT
1230 D = I
1240 TP = (2.84E - 7) * (1.27 ^ ((D - 32) ^ .648)) / (121000) * 1.0163 ^ D
1250 E =  SQR (A * B / TP)
1260 P0 = E / 2 + A * (121000 / (1.0163 ^ D))
1270  IF F1 = 1 GOTO 1370
1280  PRINT  TAB( 15);"PARTIAL"; TAB( 27);"LIQUID"
1290  PRINT  TAB( 15);"PRESSURE"; TAB( 25);"COMPOSITION"
1300  PRINT  TAB( 16);"PSIA"; TAB( 28);"PPM"
1310  PRINT
1320  PRINT " NH3"; TAB( 15); FN P(A); TAB( 25); FN P(P0)
1330  PRINT " H2S"; TAB( 15); FN P(B); TAB( 25); FN P(E)"
1340  PRINT
1350  INPUT "PRESS RETURN TO CONTINUE.   ";X$
1360  IF F1 = 0 GOTO 1400
1370  PRINT "NH3 IN LIQUID, PPM"; TAB( 30); FN P(P0)
1380  PRINT "H2S IN LIQUID, PPM"; TAB( 30); FN P(E)
1390  RETURN
1400 I = 1
1410 I = I + 1
1420  HOME
1430  PRINT "TRAY BY TRAY CALCULATION"
1440  PRINT
1450  PRINT "TRAY NUMBER:"; TAB( 30);I
1460  PRINT
1470  PRINT "H2O IN LIQUID, LB/HR"; TAB( 30); FN P(P4)
1480 P6 = P4 / (1E6 - P0 - E)
1490 P5 = 2 * (P6 * P0 - P1)
1500 Q5 = P5 / 2
1510  PRINT "NH3 IN VAPOR, LB/HR"; TAB( 30); FN P(Q5)
1520 P6 = P6 * E - P2
1530  PRINT "H2S IN VAPOR, LB/HR"; TAB( 30); FN P(P6)
1540 P8 = P8 + P9
1550  PRINT "H2O IN VAPOR, LB/HR"; TAB( 30); FN P(P7)
1560  PRINT "PRESSURE ABOVE TRAY, PSIA"; TAB( 30); FN P(P8)
```

**Program for sour-water-stripper design (continued)** **Table I**

```
1570 E = 1.89 * P7
1580 P0 = P8 / (P6 + P5 + E)
1590 A = P0 * P5
1600  PRINT "NH3 PRESS. ABOVE TRAY, PSIA"; TAB( 30); FN P(A)
1610 B = P0 * P6
1620  PRINT "H2S PRESS. ABOVE TRAY, PSIA"; TAB( 30); FN P(B)
1630 C = P0 * E
1640  PRINT "H20 PRESS. ABOVE TRAY. PSIA"; TAB( 30); FN P(C)
1650 TP = 7007 / (14.465 -  LOG (C)) - 383
1660  PRINT "TEMPERATURE ON TRAY, F"; TAB( 30); FN P(TP)
1670 P4 = P4 + (TP - D) * P3
1680 P7 = P7 + (TP - D) * P3
1690 D = TP
1700 F1 = 1
1710 F1 = 1
1720  GOSUB 1240
1730  PRINT
1740  INPUT "PRESS RETURN TO CONTINUE.   ";X$
1750  IF P5 < 0 AND P6 < 0 GOTO 1770
1760  GOTO 1410
1770  PRINT
1780  PRINT I;" TRAYS REQUIRED."
1790  PRINT
1800  PRINT "CALCULATION COMPLETED."
1810  PRINT
1820  PRINT  TAB( 13);"END OF PROGRAM"
1830  END
```

**Example for: Sour-water-stripper design** **Table II**

**(Start of first display)**

```
PROGRAM CALCULATES THE NUMBER OF TRAYS
REQUIRED FOR THE STRIPPING OF H2S AND
NH3 FROM SOUR WATER.

IT MAKES INDIVIDUAL TRAY HEAT AND
MATERIAL BALANCES, AND GIVES THE
RESULTS FOR EACH TRAY.

THE DATA REQUIRED ARE:

SOUR WATER, H2S AND NH3 RATES, LB/HR

REFLUX DRUM PRESSURE AND TEMPERATURE.
TOWER TOP PRESSURE, BOTTOM TEMPERATURE.
STRIPPED LIQUID H2S AND NH3, PPM.

STEAM RATE, LB/HR
PRESSURE DROP PER TRAY, PSIA.

PRESS RETURN TO ENTER DATA.
```

**(Start of next display)**

```
ENTER THE FOLLOWING DATA:

WATER FLOW RATE, LB/HR         98714
H2S FLOW RATE, LB/HR           730
NH3 FLOW RATE, LB/HR           556
FEED TEMPERATURE, DEG. F       199
H2S IN CLEAR WATER, PPM        5
NH3 IN CLEAR WATER, PPM        20
TOWER TOP PRESSURE, PSIA       30
TOWER BOTTOM TEMP., DEG. F     259
REFLUX DRUM PRESSURE, PSIA     27.2
REFLUX DRUM TEMP., DEG. F      184
STEAM RATE, LB/HR              10200
PRESSURE DROP PER TRAY, PSIA .45
```

**(Start of next display)**

```
LB/HR    FEED     BOTTOMS    TAIL GAS

 NH3     556      1.97       554.03
```

**Example for: Sour-water-stripper design (continued)** **Table II**

```
H2S     730       .49        729.51
H2O     98714     98292.43   421.57

                 REFLUX DRUM
        PARTIAL PRESSURE   COMPOSITION
              PSIA             PPM

NH3           11.44            101549.83
H2S           7.53             61741.12
H2O           8.22

  REFLUX         OVERHEAD
        LB/HR      LB/HR      CONDITIONS

NH3     1186.79    1740.82    23.89 PSIA
H2S     721.55     1451.06    237.55 F
H2O     9778.43    10200

PRESS RETURN TO CONTINUE.
```

**(Start of next display)**

```
              THERMAL LOADING

              ON TRAY 1     BELOW TRAY 1
HEAT ABSORBED
    BTU/HR    5101400.29    2625365.55
CONDENSED STEAM
    LB/HR     5369.9        2763.54

              PARTIAL     LIQUID
              PRESSURE  COMPOSITION
               PSIA         PPM

NH3           4.31      14286.28
H2S           1.8       6150.78

PRESS RETURN TO CONTINUE.
```

**(Start of next display)**

```
TRAY BY TRAY CALCULATION

TRAY NUMBER:                 2

H2O IN LIQUID, LB/HR         113862.33
NH3 IN VAPOR, LB/HR          1658.64
H2S IN VAPOR, LB/HR          714.46
H2O IN VAPOR, LB/HR          15569.9
PRESSURE ABOVE TRAY, PSIA    30.45
NH3 PRESS. ABOVE TRAY, PSIA  3.02
H2S PRESS. ABOVE TRAY, PSIA  .65
H2O PRESS. ABOVE TRAY. PSIA  26.78
TEMPERATURE ON TRAY, F       243.89
NH3 IN LIQUID, PPM           8443.49
H2S IN LIQUID, PPM           2726.99

PRESS RETURN TO CONTINUE.
```

**(Start of next display)**

```
TRAY BY TRAY CALCULATION

TRAY NUMBER:                 3

H2O IN LIQUID, LB/HR         114679.95
NH3 IN VAPOR, LB/HR          977.27
H2S IN VAPOR, LB/HR          315.77
H2O IN VAPOR, LB/HR          16387.52
PRESSURE ABOVE TRAY, PSIA    30.9
NH3 PRESS. ABOVE TRAY, PSIA  1.82
H2S PRESS. ABOVE TRAY, PSIA  .29
H2O PRESS. ABOVE TRAY. PSIA  28.79
TEMPERATURE ON TRAY, F       247.98
NH3 IN LIQUID, PPM           4643.98
H2S IN LIQUID, PPM           1310.96

PRESS RETURN TO CONTINUE.
```

**(Start of next display)**

```
TRAY BY TRAY CALCULATION

TRAY NUMBER:                 4

H2O IN LIQUID, LB/HR         115205.97
NH3 IN VAPOR, LB/HR          536.25
H2S IN VAPOR, LB/HR          151.44
H2O IN VAPOR, LB/HR          16913.54
PRESSURE ABOVE TRAY, PSIA    31.35
NH3 PRESS. ABOVE TRAY, PSIA  1.01
H2S PRESS. ABOVE TRAY, PSIA  .14
H2O PRESS. ABOVE TRAY. PSIA  30.19
TEMPERATURE ON TRAY, F       250.7
NH3 IN LIQUID, PPM           2452.2
H2S IN LIQUID, PPM           647.7

PRESS RETURN TO CONTINUE.
```

**(Start of next display)**

```
TRAY BY TRAY CALCULATION

TRAY NUMBER:                 5

H2O IN LIQUID, LB/HR         115556.06
NH3 IN VAPOR, LB/HR          282.28
H2S IN VAPOR, LB/HR          74.59
H2O IN VAPOR, LB/HR          17263.63
PRESSURE ABOVE TRAY, PSIA    31.8
NH3 PRESS. ABOVE TRAY, PSIA  .54
H2S PRESS. ABOVE TRAY, PSIA  .07
H2O PRESS. ABOVE TRAY. PSIA  31.19
TEMPERATURE ON TRAY, F       252.56
NH3 IN LIQUID, PPM           1261.01
H2S IN LIQUID, PPM           321.72

PRESS RETURN TO CONTINUE.
```

**Example for: Sour-water-stripper design (continued)** **Table II**

**(Start of next display)**

TRAY BY TRAY CALCULATION

| TRAY NUMBER: | 6 |
|---|---|
| H2O IN LIQUID, LB/HR | 115796.17 |
| NH3 IN VAPOR, LB/HR | 144.29 |
| H2S IN VAPOR, LB/HR | 36.82 |
| H2O IN VAPOR, LB/HR | 17503.73 |
| PRESSURE ABOVE TRAY, PSIA | 32.25 |
| NH3 PRESS. ABOVE TRAY, PSIA | .28 |
| H2S PRESS. ABOVE TRAY, PSIA | .04 |
| H2O PRESS. ABOVE TRAY. PSIA | 31.94 |
| TEMPERATURE ON TRAY, F | 253.93 |
| NH3 IN LIQUID, PPM | 634.93 |
| H2S IN LIQUID, PPM | 158.88 |

PRESS RETURN TO CONTINUE.

**(Start of next display)**

TRAY BY TRAY CALCULATION

| TRAY NUMBER: | 7 |
|---|---|
| H2O IN LIQUID, LB/HR | 115972.28 |
| NH3 IN VAPOR, LB/HR | 71.73 |
| H2S IN VAPOR, LB/HR | 17.95 |
| H2O IN VAPOR, LB/HR | 17679.84 |
| PRESSURE ABOVE TRAY, PSIA | 32.7 |
| NH3 PRESS. ABOVE TRAY, PSIA | .14 |
| H2S PRESS. ABOVE TRAY, PSIA | .02 |
| H2O PRESS. ABOVE TRAY. PSIA | 32.54 |
| TEMPERATURE ON TRAY, F | 255.02 |
| NH3 IN LIQUID, PPM | 312.33 |
| H2S IN LIQUID, PPM | 77.23 |

PRESS RETURN TO CONTINUE.

**(Start of next display)**

TRAY BY TRAY CALCULATION

| TRAY NUMBER: | 8 |
|---|---|
| H2O IN LIQUID, LB/HR | 116112.93 |
| NH3 IN VAPOR, LB/HR | 34.31 |
| H2S IN VAPOR, LB/HR | 8.48 |
| H2O IN VAPOR, LB/HR | 17820.5 |
| PRESSURE ABOVE TRAY, PSIA | 33.15 |
| NH3 PRESS. ABOVE TRAY, PSIA | .07 |
| H2S PRESS. ABOVE TRAY, PSIA | .01 |
| H2O PRESS. ABOVE TRAY. PSIA | 33.07 |
| TEMPERATURE ON TRAY, F | 255.96 |
| NH3 IN LIQUID, PPM | 148.2 |
| H2S IN LIQUID, PPM | 36.34 |

PRESS RETURN TO CONTINUE.

**(Start of next display)**

TRAY BY TRAY CALCULATION

| TRAY NUMBER: | 9 |
|---|---|
| H2O IN LIQUID, LB/HR | 116234.35 |
| NH3 IN VAPOR, LB/HR | 15.26 |
| H2S IN VAPOR, LB/HR | 3.73 |
| H2O IN VAPOR, LB/HR | 17941.91 |
| PRESSURE ABOVE TRAY, PSIA | 33.6 |
| NH3 PRESS. ABOVE TRAY, PSIA | .03 |
| H2S PRESS. ABOVE TRAY, PSIA | 0 |
| H2O PRESS. ABOVE TRAY. PSIA | 33.57 |
| TEMPERATURE ON TRAY, F | 256.82 |
| NH3 IN LIQUID, PPM | 65.47 |
| H2S IN LIQUID, PPM | 15.94 |

PRESS RETURN TO CONTINUE.

**(Start of next display)**

TRAY BY TRAY CALCULATION

| TRAY NUMBER: | 10 |
|---|---|
| H2O IN LIQUID, LB/HR | 116345.29 |
| NH3 IN VAPOR, LB/HR | 5.65 |
| H2S IN VAPOR, LB/HR | 1.36 |
| H2O IN VAPOR, LB/HR | 18052.86 |
| PRESSURE ABOVE TRAY, PSIA | 34.05 |
| NH3 PRESS. ABOVE TRAY, PSIA | .01 |
| H2S PRESS. ABOVE TRAY, PSIA | 0 |
| H2O PRESS. ABOVE TRAY. PSIA | 34.04 |
| TEMPERATURE ON TRAY, F | 257.64 |
| NH3 IN LIQUID, PPM | 24.08 |
| H2S IN LIQUID, PPM | 5.81 |

PRESS RETURN TO CONTINUE.

**(Start of next display)**

TRAY BY TRAY CALCULATION

| TRAY NUMBER: | 11 |
|---|---|
| H2O IN LIQUID, LB/HR | 116450.36 |
| NH3 IN VAPOR, LB/HR | .84 |
| H2S IN VAPOR, LB/HR | .19 |
| H2O IN VAPOR, LB/HR | 18157.93 |
| PRESSURE ABOVE TRAY, PSIA | 34.5 |
| NH3 PRESS. ABOVE TRAY, PSIA | 0 |
| H2S PRESS. ABOVE TRAY, PSIA | 0 |
| H2O PRESS. ABOVE TRAY. PSIA | 34.5 |
| TEMPERATURE ON TRAY, F | 258.43 |
| NH3 IN LIQUID, PPM | 3.54 |
| H2S IN LIQUID, PPM | .82 |

PRESS RETURN TO CONTINUE.

**Example for: Sour-water-stripper design (continued)** **Table II**

**(Start of next display)**

```
TRAY BY TRAY CALCULATION

TRAY NUMBER:                12

H2O IN LIQUID, LB/HR        116551.94
NH3 IN VAPOR, LB/HR         -1.55
H2S IN VAPOR, LB/HR         -.4
H2O IN VAPOR, LB/HR         18259.51
PRESSURE ABOVE TRAY, PSIA   34.95
NH3 PRESS. ABOVE TRAY, PSIA 0
H2S PRESS. ABOVE TRAY, PSIA 0
H2O PRESS. ABOVE TRAY. PSIA 34.95
TEMPERATURE ON TRAY, F      259.2
NH3 IN LIQUID, PPM          -4.95
H2S IN LIQUID, PPM          1.62

PRESS RETURN TO CONTINUE.
```

**(Start of next display)**

```
12 TRAYS REQUIRED.

CALCULATION COMPLETED.

            END OF PROGRAM
```

# Kinetics of fixed-bed sorption processes

## Here is a computer program for solving the second-order kinetics used to calculate breakthrough curves for fixed-bed sorption processes.

☐ In the design of fixed-bed sorption processes, the solution to the second-order kinetics first derived by Thomas [1] and modified by Hiester and Vermeulen [2] is often used. The solution is given by the following dimensionless equation:

$$X = \frac{J(RN,NT)}{J(RN,NT) + [1 - J(N,RNT)]\exp[(R-1)N(T-1)]} \quad (1)$$

where $X$ = normalized fluid-phase composition.
$R$ = separation factor.
$N$ = number of transfer units.
$T$ = throughput parameter.

$$J(u,v) = 1 - \int_0^u \exp(-v-\lambda) I_0(2\sqrt{v\lambda})\,d\lambda.$$

$I_0$ = modified Bessel function of zero order.

In a previous article (*Chem. Eng.*, Oct. 24, 1977, p. 158), an HP-25 program for generating $J$ functions was presented. The same algorithm can be used for solving Eq. (1) by a pocket calculator. However, when the values of $R$, $N$ and $T$ are large, the convergence of the exact series solution for the $J$ function is very slow, and asymptotic expansion in terms of error function is usually applied. For $\sqrt{uv} > 7$, the following approximation will result in no more than a 1% error [3]:

$$J(u,v) = \frac{1}{2}\left\{1 - \text{erf}(\sqrt{u} - \sqrt{v}) + \frac{\exp[-(\sqrt{u}-\sqrt{v})^2]}{\sqrt{\pi}[\sqrt{v} - (uv)^{1/4}]}\right\}$$

$$\text{where } \text{erf}(z) = \frac{2}{\sqrt{\pi}}\int_0^z \exp(-\lambda^2)\,d\lambda$$

Thus, for $\sqrt{(RN)(NT)} > 7$, the values of $X$ can be calculated from Eq. 2 (at bottom of page).

The Apple II program given here calculates values of $X$, using both the exact series expansion of the $J$-function and the approximate solution of Eq. (2). Generation of error function is incorporated in the subroutines of the program. The program is applicable to a wide range of parameter values of $R$, $N$ and $T$, as illustrated by the sample calculated results that are shown in Table III.

There are limitations in adapting Eq. (1) to the program. However, the program indicates when these limits are exceeded. It makes the calculation, but prints a warning that the results exceed the limits of reliability. No overflow or interruption occurs.

### Using the program

This program (Table I) calculates the normalized fluid-phase composition for second-order reaction kinetics from calculated $J$ factors.

Data required are:
- separation factor, $R$
- number of transfer units, $N$
- throughput parameter, $T$

The user is prompted to enter data by the program (see Table II for examples). Once all the data are put in, the calculated normalized fluid-phase composition, $X$, is displayed.

If the data entered is beyond the limitations of Eq. (1), a value for $X$ is calculated, but the following warning is given:

"THE CONDITIONS ARE OUTSIDE THE RANGE OF PRECISE CALCULATION, AND THE RESULT IS AN APPROXIMATION."

$$X = \frac{g(RN,NT) + [(RN^2T)^{1/4} + \sqrt{NT}]^{-1}}{g(RN,NT) + [(RN^2T)^{1/4} + \sqrt{NT}]^{-1} + g(RNT,N) - [(RN^2T)^{1/4} + \sqrt{RNT}]^{-1}} \quad (2)$$

$$\text{where } g(u,v) = \sqrt{\pi}[1 - \text{erf}(\sqrt{u} - \sqrt{v})]\exp[(\sqrt{u} - \sqrt{v})^2]$$

Adapted from an article by Henry K. S. Tan, University of Toronto, Toronto, Canada, originally published March 24, 1980.

## References


1. Thomas, H. C., *J. Am. Chem. Soc.*, Vol. 66, p. 1,664 (1944).
2. Hiester, N. K., and Vermeulen, T., in Perry's "Chemical Engineers' Handbook," Sect. 16, 5th ed., McGraw-Hill, New York, 1973.
3. Hiester, N. K., and Vermeulen, T., *Chem. Eng. Prog.*, Vol. 48, p. 505 (1952).

## The Author

Henry K. S. Tan is a research associate in the Dept. of Chemical Engineering and Applied Chemistry, University of Toronto, Toronto, Ontario, M5S 1A4 Canada. He has published several papers on process dynamics and fixed-bed ion exchange. He holds both M.A.Sc. and Ph.D. degrees in chemical engineering from the University of Toronto.

**Program for kinetics of fixed-bed sorption processes** **Table I**

```
10  REM  KINETICS OF FIXED-BED SORPTION PROCESES
20  REM   FROM CHEMICAL ENGINEERING, MARCH 24, 1980
30  REM   BY HENRY K. S. TAN
40  REM  TRANSLATED BY WILLIAM VOLK
50  REM   COPYRIGHT (C) 1984
60  REM  BY CHEMICAL ENGINEERING
70  HOME : VTAB 4
80 PI = 3.14159265
90  DEF  FN P(X) =  INT (1E4 * (X + .00005)) / 1E4
100  REM  ABOVE 'DEF' SETS DISPLAY TO FOUR DECIMALS.
110  PRINT "THIS PROGRAM CALCULATES THE"
120  PRINT "NORMALIZED FLUID-PHASE COMPOSITION"
130  PRINT "FOR SECOND-ORDER REACTION KINETICS"
140  PRINT "FROM CALCULATED 'J' FACTORS."
150  PRINT
160  PRINT "THE DATA REQUIRED ARE:"
170  PRINT
180  PRINT "SEPARATION FACTOR,"
190  PRINT "NUMBER OF TRANSFER UNITS, AND"
200  PRINT "THROUGHPUT PARAMETER."
210  PRINT
220  INPUT "PRESS RETURN TO ENTER DATA.  ";Q$
230  HOME : VTAB 8
240  PRINT "ENTER THE FOLLOWING DATA:"
250  PRINT
260  INPUT "SEPARATION FACTOR, R          ";R
270  INPUT "NUMBER OF TRANSFER UNITS, N  ";N
280  INPUT "THROUGHPUT PARAMETER, T      ";T
290 P0 = N
300 P1 = N * R
310 P0 = P0 - P1
320 P3 =  SQR (P1)
330 P2 = T * N
340 P0 = P0 - P2
350 P4 =  SQR (P2)
360  IF P4 * P3 > 7 GOTO 740
370  GOSUB 560
380 P8 = X
390 P9 = X
400 P1 = P1 / R
410 P2 = P2 * R
420 F2 = 1
430  GOSUB 560
440 P9 = P9 + (1 - X) *  EXP (P2 + P0)
450  PRINT : PRINT
460  PRINT "WITH THE ABOVE DATA,"
470 AN = P8 / P9
480  PRINT "THE NORMALIZED FLUID-PHASE"
490  PRINT "COMPOSITION, X IS              "; FN P(AN)
500  IF F3 = 0 GOTO 550
510  PRINT
520  PRINT "THE CONDITIONS ARE OUTSIDE THE RANGE"
530  PRINT "OF PRECISE CALCULATION, AND THE"
540  PRINT "RESULT IS AN APPROXIMATION."
550  GOTO 1220
560 P6 = 1
570 P3 = 1 /  EXP (P1)
580 P4 = P6 - P3
590 P5 = P4 /  EXP (P2)
600 P4 = P4 * P6
610 P4 = P4 - P3 * (P1 ^ P6)
620 F = P6
630  GOSUB 1170
640 TP = (P4 / XF) * (P2 ^ P6) / XF /  EXP (P2)
650 P5 = P5 + TP
660  IF TP = (1 - P5) GOTO 700
670  IF (1 - P5) * (1E - 5) > (TP / (1 - P5)) GOTO 720
680 P6 = P6 + 1
690  GOTO 600
700 X = 1 - R5
710  RETURN
720 X = 1 - P5
730  RETURN
740 P7 =  SQR (P4 * P3)
750 P8 = 1 / (P7 + P4)
760  GOSUB 830
770 P8 = P8 + DX
780 P3 = P3 *  SQR (T)
790 P4 = P4 /  SQR (T)
800  GOSUB 830
810 P9 = P8 + (DX - 1 / (P7 + P3))
820  GOTO 450
830 P1 = 3
840 P0 = P3 - P4
850 P5 = P0
860 P6 = P0
870  IF P6 = 0 GOTO 960
880  IF P6 < 0 GOTO 1090
```

**Program for kinetics of fixed-bed sorption processes (continued)** **Table I**

```
890  IF P6 > 3 GOTO 980
900 P5 = P5 * 2 * P0 ^ 2
910 P5 = P5 / P1
920 P6 = P6 + P5
930  IF 1E - 9 > (P5 / P6) GOTO 960
940 P1 = P1 + 2
950  GOTO 900
960 DX =  SQR (PI) *  EXP (P0 ^ 2) - 2 * P6
970  RETURN
980 P1 = 1
990 P5 = P1 / P0
1000 P6 = P5
1010 P5 = P5 / ( - 2 * P0 ^ 2)
1020 P5 = P5 * P1
1030 P6 = P6 + P5
1040  IF 1E - 3 >  ABS (P5 / P6) GOTO 1070
1050 P1 = P1 + 2
1060  GOTO 1010
1070 DX = P6
1080  RETURN
1090  IF  ABS (P0) =  < 3 GOTO 900
1100  IF 9 >  ABS (P0) GOTO 1140
1110 TP = 9
1120 F3 = 1
1130  GOTO 1150
1140 TP =  ABS (P0)
1150 DX = 2 *  SQR (PI) *  EXP (TP ^ 2)
1160  RETURN
1170 XF = 1
1180  FOR I = 1 TO F
1190 XF = XF * I
1200  NEXT I
1210  RETURN
1220  PRINT
1230  PRINT  TAB( 13);"END OF PROGRAM"
1240  END
```

**Example for: Kinetics of fixed-bed sorption processes** **Table II**

**(Start of first display)**

```
THIS PROGRAM CALCULATES THE
NORMALIZED FLUID-PHASE COMPOSITION
FOR SECOND-ORDER REACTION KINETICS
FROM CALCULATED 'J' FACTORS.

THE DATA REQUIRED ARE:

SEPARATION FACTOR,
NUMBER OF TRANSFER UNITS, AND
THROUGHPUT PARAMETER.

PRESS RETURN TO ENTER DATA.
```

**(Start of next display)**

```
ENTER THE FOLLOWING DATA:

SEPARATION FACTOR, R           10
NUMBER OF TRANSFER UNITS, N   .1
THROUGHPUT PARAMETER, T         2


WITH THE ABOVE DATA,
THE NORMALIZED FLUID-PHASE
COMPOSITION, X IS            .9262

            END OF PROGRAM
```

**(Start of first display)**

```
THIS PROGRAM CALCULATES THE
NORMALIZED FLUID-PHASE COMPOSITION
FOR SECOND-ORDER REACTION KINETICS
FROM CALCULATED 'J' FACTORS.

THE DATA REQUIRED ARE:

SEPARATION FACTOR,
NUMBER OF TRANSFER UNITS, AND
THROUGHPUT PARAMETER.

PRESS RETURN TO ENTER DATA.
```

**(Start of next display)**

```
ENTER THE FOLLOWING DATA:

SEPARATION FACTOR, R          2
NUMBER OF TRANSFER UNITS, N  5
THROUGHPUT PARAMETER, T      .1


WITH THE ABOVE DATA,
THE NORMALIZED FLUID-PHASE
COMPOSITION, X IS            .0528

            END OF PROGRAM
```

**Sample calculated results** **Table III**

| *N* | *T* | *R* | *X* |
|---|---|---|---|
| 0.1 | 2 | 10 | 0.9262 |
| 0.5 | 0.2 | 0.3 | 0.6318 |
| 1 | 0.001 | 0.01 | 0.3681 |
| 5 | 0.1 | 2 | 0.0528 |
| 10 | 0.2 | 0.5 | 0.0031 |
| 50 | 0.5 | 1.5 | 0.0428 |
| 100 | 1 | 10 | 0.7600 |
| 200 | 0.8 | 1.1 | 0.0634 |
| 500 | 2 | 2.5 | 0.9191 |
| 1,000 | 1 | 0.25 | 0.5000 |
| 5,000 | 10 | 200 | 0.9826 |

# Design of spouted beds

Spouted beds have found numerous applications in drying, coating and granulating. With this program one can estimate the maximum spoutable bed depth, minimum fluid velocity for spouting and mean spout diameter.

☐ A spouted bed is a fluid-solids contacting and mixing device that was developed primarily for solids too coarse for fluidized bed techniques. As shown in Fig. 1, solids are entrained by the fluid and conveyed to the top of the bed. They then flow downward in the surrounding annulus, countercurrent to fluid flow.

**Design is based on determining the maximum spoutable bed depth, the minimum fluid velocity for spouting, and the mean spout diameter. The Apple II computer program presented here estimates these parameters using the same basic design equations as employed by Zanker [1].**

To estimate the spoutable bed depth, Malek and Lu [2] proposed that:

$$\frac{H_m}{D_c} = 0.105 \left(\frac{D_c}{d_p}\right)^{0.75} \left(\frac{D_c}{D_i}\right)^{0.4} \left(\frac{\lambda^2}{\rho_s^{1.2}}\right) \quad (1)$$

where λ, the shape factor, is:

$$\lambda = 0.205\,(A/V^{2/3}) \quad (2)$$

Mathur and Gishler [3] have defined the minimum fluid velocity for spouting by:

$$U_{ms} = (d_p/D_c)\,(D_i/D_c)^{1/3}\,[2gH\,(\rho_s - \rho_f)/\rho_f]^{1/2} \quad (3)$$

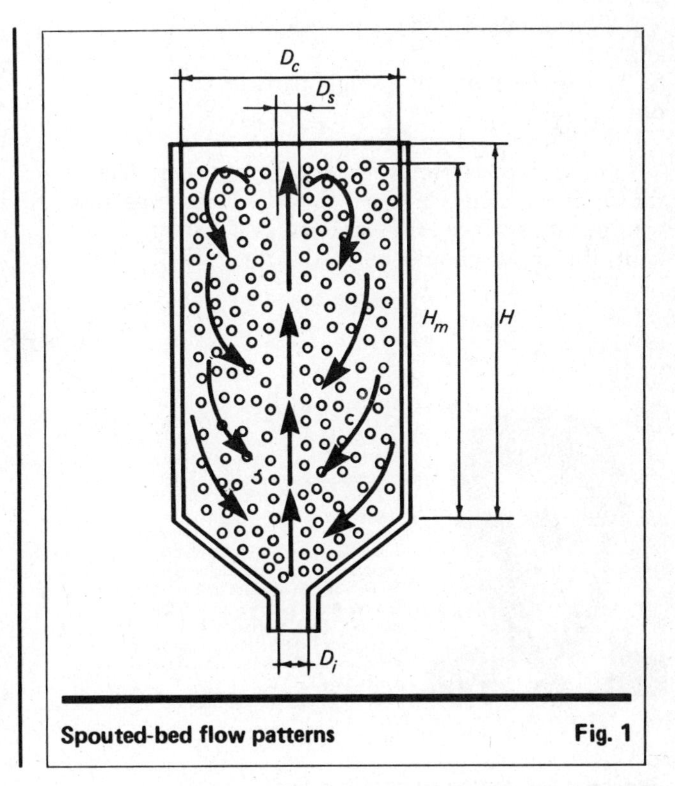


**Spouted-bed flow patterns** **Fig. 1**

## Nomenclature

| | | | |
|---|---|---|---|
| $A$ | Surface area of a single particle, cm$^2$ | $H_m$ | Maximum spoutable bed depth, cm |
| $D_c$ | Column inside dia., cm | $U_{ms}$ | Minimum superficial fluid velocity for spouting, cm/s |
| $D_i$ | Diameter of fluid inlet orifice, cm | $V$ | Volume of a single particle, cm$^3$ |
| $D_s$ | Mean spout dia., cm | $\lambda$ | Particle shape factor |
| $d_p$ | Average particle dia., cm | $\rho_f$ | Fluid density, g/cm$^3$ |
| $G$ | Fluid mass flowrate, kg/h-m$^2$ | $\rho_s$ | Particle density, g/cm$^3$ |
| g | Acceleration of gravity, 981 cm/s$^2$ | | |
| $H$ | Depth of the bed, cm | | |

**Adapted from an article by Domingo Melé and Julián Martínez, Universidad Politécnica de Valencia, Spain, originally published October 20, 1980.**

The mean spout diameter is determined by using Mikhailik's [4] equation:

$$D_s = 1.45\,[0.115 \log D_c - 0.077]\,(G/\rho_s)^{1/2} \quad (4)$$

The above formulas are accurate to within ±12% for the following ranges:

$$5\text{cm} < D_c < 50\text{cm},\ H < 300\text{cm},\ d_p < 0.1\text{cm}$$

## Using the program

This program (Table I) calculates:

$H_m$, maximum spoutable bed depth, cm
$U_{ms}$, minimum superficial velocity, cm/s
$D_s$, mean spout dia., cm

Data required are:

$D_i$, dia. of fluid inlet orifice, cm
$D_c$, column inside dia., cm
$\rho_s$, particle density, g/cm$^3$
$d_p$, average particle dia., cm
$\rho_f$, fluid density, g/cm$^3$
$G$, fluid mass flowrate, kg/(h)(m$^2$)

and either

$V$, volume of a single particle, cm$^3$

and

$A$, surface area of a single particle, cm$^2$

or,

$\lambda$, particle shape factor.

The user is prompted by the program to enter data, as shown in the example (see Table II). For the particle volume and surface area variables, or the shape factor item, the input should be in the form of "$V$, $A$, $\lambda$" separated by commas. If particle volume and surface area are known, enter those two values and a zero for the shape factor. Conversely, enter zeros for the volume and surface area items, and the known value of the shape factor, if the latter is the only variable whose value is known.

After entering all the data, the next display offers the data supplied and the calculated results. Please note that in the case where volume and surface area items are entered, only the shape factor (which is calculated) is shown as data.

In cases where data are beyond the limits indicated earlier for $D_c$, $H$, or $d_p$, the calculated results will be displayed, but the added message that "DATA ARE OUTSIDE RANGE OF RELIABLE CALCULATION. RESULTS MAY BE IN ERROR BY MORE THAN 12%.".

## References


1. Zanker, A., Designing Spouted Beds, *Chem. Eng.*, Nov. 21, 1977, pp. 207-209.
2. Malek, M. A., and Lu, Benjamin C. Y., Pressure Drop and Spoutable Bed Heights in Spouted Beds, *Ind. Eng. Chem. Process Des. Dev.*, Vol. 4, No. 1, pp.123-128 (1965).
3. Mathur, K. B., and Gishler, P. E., A Technique for Contacting Gases With Coarse Solid Particles, *AIChE J.*, Vol. 1, p. 157 (1955).
4. Mikhailik, V. D., "Research on Heat and Mass Transfer in Technological Processes", p. 37, Nauka Tekhnika BSSR, Minsk, 1966.


## The authors

**Domingo Melé** is chairman of the Dept. of Chemical Engineering at the Universidad Politécnica de Valencia, Mendizábal 14, Valencia, Spain. Previously, he was professor of unit operations and chemical reactors at the Politécnica University of Barcelona for eight years.
Mr. Melé has a degree in chemistry from the University of Barcelona and a doctorate in industrial engineering from the Universidad Politécnica de Barcelona. He is a registered engineer in Valencia, a member of the Asociación Nacional de Ingenieros Industriales de España and of the Spanish Water Pollution Control Federation.

**Julián Martínez** is an assistant professor of Chemical Engineering at the Universidad Politécnica de Valencia, from which he was graduated as a chemical industrial engineer. He has worked at Compañía Española del Gas in Valencia. He is preparing for a Ph.D.

**Program for design of spouted beds** **Table I**

```
10 REM   DESIGN OF SPOUTED BEDS
20 REM   FROM CHEMICAL ENGINEERING, OCTOBER 20, 1980
30 REM   BY DOMINGO MELE AND JULIAN MARTINEZ
40 REM  TRANSLATED BY WILLIAM VOLK
50 REM   COPYRIGHT (C) 1984
60 REM  BY CHEMICAL ENGINEERING
70 HOME : VTAB 8
80 REM  SET DISPLAY TO TWO DECIMALS.
90 DEF  FN P(X) =  INT (100 * (X + .005)) / 100
100 PRINT "PROGRAM CALCULATES:
110 PRINT
120 PRINT "MAXIMUM SPOUTABLE BED DEPTH, CM"
130 PRINT "MINIMUM SUPERFICIAL VELOCITY, CM/S
140 PRINT "MEAN SPOUT DIAMETER, CM"
150 PRINT
160 PRINT "DATA REQUIRED ARE:"
170 PRINT
180 PRINT "DIAMETER OF FLUID INLET, CM"
190 PRINT "COLUMN DIAMETER, CM"
200 PRINT "PARTICLE DENSITY, G/CU. CM"
210 PRINT "PARTICLE DIAMETER, CM"
220 PRINT "FLUID DENSITY, G/CU. CM"
230 PRINT "FLUID FLOW, KG/(HR)(SQ. M)
240 PRINT
250 PRINT "AND:"
260 PRINT
```

**Program for design of spouted beds (continued)** **Table I**

```
270  PRINT "EITHER THE VOLUME AND SURFACE AREA"
280  PRINT "OF A SINGLE PARTICLE, OR THE PARTICLE"
290  PRINT "SHAPE FACTOR."
300  PRINT
310  INPUT "PRESS RETURN TO ENTER DATA   ";X$
320  HOME : VTAB 4
330  PRINT "ENTER THE FOLLOWING DATA:"
340  PRINT
350  INPUT "DIAMETER OF FLUID INLET, CM       ";P1
360  INPUT "COLUMN INSIDE DIAMETER, CM        ";P0
370  INPUT "PARTICLE DENSITY, G/CU. CM        ";P2
380  INPUT "SINGLE PARTICLE DIAMETER, CM      ";P3
390  IF P3 < .1 THEN F3 = 1
400  INPUT "FLUID DENSITY, G/CU. CM           ";P7
410  INPUT "FLUID MASS FLOWRATE, KG/(HR)(SQ.M)";P8
420  PRINT
430  PRINT "ENTER THE PARTICLE VOLUME AND SURFACE,"
440  PRINT "AND A ZERO FOR THE SHAPE FACTOR."
450  PRINT
460  PRINT "OR ENTER ZEROS FOR THE VOLUME AND"
470  PRINT "SURFACE, AND ENTER THE SHAPE FACTOR."
480  PRINT
490  PRINT "VOLUME AND SURFACE IN CM UNITS."
500  PRINT "SEPARATE YOUR VALUES WITH COMMAS."
510  PRINT
520  INPUT "ENTER THE DATA NOW: ";P6,P5,P4
530  IF P0 =  > 5 AND P0 =  < 50 GOTO 680
540  HOME : VTAB (4)
550  PRINT "CORRELATION IS ONLY GOOD FOR COLUMN"
560  PRINT "DIAMETER IN THE RANGE 5 TO 50 CM."
570  PRINT
580  PRINT "YOUR VALUE IS ";P0;"."
590  PRINT
600  PRINT "YOU MAY CHANGE YOUR VALUE OR CONTINUE"
610  PRINT "WITH THE CALCULATION USING YOUR VALUE."
620  PRINT
630  PRINT "IF YOU WISH TO CHANGE THE VALUE,"
640  PRINT "ENTER THE NEW VALUE.  IF NOT, ENTER"
650  INPUT "ZERO.   ";CH
660  IF CH = 0 GOTO 680
670 P0 = CH
680  IF P0 < 5 OR P0 > 50 THEN F3 = 1
690  IF P4 <  > 0 GOTO 710
700 P4 = .205 * (P5 / (P6 ^ (2 / 3)))
710 HM = P0 * (.105) * (P0 / P3) ^ .75 * (P0 / P1) ^ .4
    * (P4 ^ 2 / P2 ^ 1.2)
720 UM = (P3 / P0) * (P1 / P0) ^ (1 / 3) * ( SQR (1962 *
    HM * (P2 - P7) / P7))
730 DS = 1.45 * (.115 *  LOG (P0) /  LOG (10) - .077) *
    SQR (P8 / P2)
740  HOME : VTAB (4)
750  PRINT "WITH THE FOLLOWING DATA:"
760  PRINT
770  PRINT "INLET DIAMETER          ";P1;" CM"
780  PRINT "COLUMN DIAMETER         ";P0;" CM"
790  PRINT "PARTICLE DENSITY        ";P2;" G/CU. CM"
800  PRINT "PARTICLE DIAMETER       ";P3;" CM"
810  PRINT "FLUID DENSITY           ";P7;" G/CU. CM"
820  PRINT "FLUID RATE              ";P8;" KG/(HR)(SQ. M)"
830  IF R5 = 0 GOTO 880
840  PRINT "PARTICLE AREA           ";P5;" SQ. CM"
850  PRINT "PARTICLE VOLUME         ";P6;" CU. CM"
860  PRINT "CALCULATED SHAPE FACTOR ";P4
870  GOTO 890
880  PRINT "SHAPE FACTOR            ";P4
890  PRINT
900  PRINT "THE CALCULATED RESULTS ARE:"
910  PRINT
920  PRINT "MAXIMUM SPOUTABLE BED DEPTH, CM  "; FN P(HM)
930  IF HM > 300 THEN F3 = 1
940  PRINT "MINIMUM SUPERFICIAL VEL., CM/S   "; FN P(UM)
950  PRINT "MEAN SPOUT DIAMETER, CM          "; FN P(DS)
960  IF F3 = 0 GOTO 1010
970  PRINT
980  PRINT "DATA ARE OUTSIDE RANGE OF RELIABLE"
990  PRINT "CALCULATION.  RESULTS MAY BE IN ERROR"
1000  PRINT "BY MORE THAN 12%."
1010  PRINT
1020  PRINT  TAB( 13);"END OF PROGRAM"
1030  END
```

**Example for: Design of spouted beds** **Table II**

**(Start of first display)**

```
PROGRAM CALCULATES:

MAXIMUM SPOUTABLE BED DEPTH, CM
MINIMUM SUPERFICIAL VELOCITY, CM/S
MEAN SPOUT DIAMETER, CM

DATA REQUIRED ARE:

DIAMETER OF FLUID INLET, CM
COLUMN DIAMETER, CM
PARTICLE DENSITY, G/CU. CM
PARTICLE DIAMETER, CM
FLUID DENSITY, G/CU. CM
FLUID FLOW, KG/(HR)(SQ. M)

AND:

EITHER THE VOLUME AND SURFACE AREA
OF A SINGLE PARTICLE, OR THE PARTICLE
SHAPE FACTOR.

PRESS RETURN TO ENTER DATA
```

**Example for: Design of spouted beds (continued)** **Table II**

**(Start of next display)**

```
ENTER THE FOLLOWING DATA:

DIAMETER OF FLUID INLET, CM        3
COLUMN INSIDE DIAMETER, CM         20
PARTICLE DENSITY, G/CU. CM         2.2
SINGLE PARTICLE DIAMETER, CM       .6
FLUID DENSITY, G/CU. CM            1.227E-3
FLUID MASS FLOWRATE, KG/(HR)(SQ.M)3000

ENTER THE PARTICLE VOLUME AND SURFACE,
AND A ZERO FOR THE SHAPE FACTOR.

OR ENTER ZEROS FOR THE VOLUME AND
SURFACE, AND ENTER THE SHAPE FACTOR.

VOLUME AND SURFACE IN CM UNITS.
SEPARATE YOUR VALUES WITH COMMAS.

ENTER THE DATA NOW: 0,0,1
```

**(Start of next display)**

```
WITH THE FOLLOWING DATA:

INLET DIAMETER          3 CM
COLUMN DIAMETER         20 CM
PARTICLE DENSITY        2.2 G/CU. CM
PARTICLE DIAMETER       .6 CM
FLUID DENSITY           1.227E-03 G/CU. CM
FLUID RATE              3000 KG/(HR)(SQ. M)
SHAPE FACTOR            1

THE CALCULATED RESULTS ARE:

MAXIMUM SPOUTABLE BED DEPTH, CM  24.16
MINIMUM SUPERFICIAL VEL., CM/S   146.9
MEAN SPOUT DIAMETER, CM          3.89

              END OF PROGRAM
```

# Cyclone-efficiency equations

This program calculates cyclone efficiency, based on cyclone dimensions and on fluid and particle characteristics.

☐ The procedure* for determining cyclone fractional (grade) efficiency is based on the cyclone dimensions and the flow characteristics of particle-laden gases. It involves progressively solving several equations and using the results to calculate fractional efficiency. The equations are complex, containing exponential and logarithmic functions.

The figure gives the input variables for the efficiency equations. These are either used directly in the final grade efficiency equation, or their derivatives are required for solving it.

Table I lists the program, and Table II shows example displays. The required data are the cyclone dimensions, flows, and fluid and particle characteristics shown in the figure. The program prompts for all these data, and then calculates efficiency.

## How the program works

The equations for calculating cyclone fractional efficiencies are:

For $l$—Natural length (the distance below the gas outlet where the vortex turns), ft:

$$l = 2.3D_e(D_c^2/ab)^{1/3} \tag{1}$$

For $V_{nl}$—Volume at natural length (excluding the core), ft$^3$:

$$V_{nl} = \frac{\pi D_c^2}{4}(h - S) + \frac{(\pi D_c^2)}{4}\frac{(l + S - h)}{3}\left(1 + \frac{d}{D_c} + \frac{d^2}{D_c^2}\right) - \frac{\pi D_e^2 l}{4} \tag{2}$$

For $V_H$—Volume below exit duct (excluding the core), ft$^3$:

$$V_H = \frac{\pi D_c^2}{4}(h - S) + \left(\frac{\pi D_c^2}{4}\right)\left(\frac{H - h}{3}\right) \times \left(1 + \frac{B}{D_c} + \frac{B^2}{D_c^2}\right) - \frac{\pi D_e^2}{4}(H - S) \tag{3}$$

For $d$—Dia. of central core at point where vortex turns, ft:

$$d = D_c - (D_c - B)\left(\frac{S + l - h}{H - h}\right) \tag{4}$$

For $n$—Vortex exponent:

$$n = 1 - \left[1 - \frac{(12D_c)^{0.14}}{2.5}\right]\left[\frac{T + 460}{530}\right]^{0.3} \tag{5}$$

For $\tau_i$—Relaxation time, $s$:

$$\tau_i = \rho_p(d_{pi})^2/(18\mu) \tag{6}$$

For $G$—Cyclone configuration factor (specified by the geometric ratios that describe the cyclone's shape):

$$G = 8K_c/K_a^2K_b^2$$

Substituting values of $K_a$, $K_b$, and $K_c$ from the Koch and Licht article gives:

$$G = \{2[\pi(S - a/2)(D_c^2 - D_e^2)] + 4V_{H,nl}\}\frac{D_c}{a^2b^2} \tag{7}$$

In equation (7), use $V_H$ when $S + l > H$, and $V_{nl}$ when $S + l \leq H$.

For $\eta_i$—Fractional efficiency:

$$\eta_i = 1 - exp\left\{-2\left[\frac{G\tau_i Q}{D_c^3}(n + 1)\right]^{0.5/(n+1)}\right\} \tag{8}$$

To use the program, simply enter the data, and get the single result: percent efficiency for the given conditions and particles. Table II shows some example data, and the calculated efficiency for those data—63.31%.

Input variables for efficiency equations

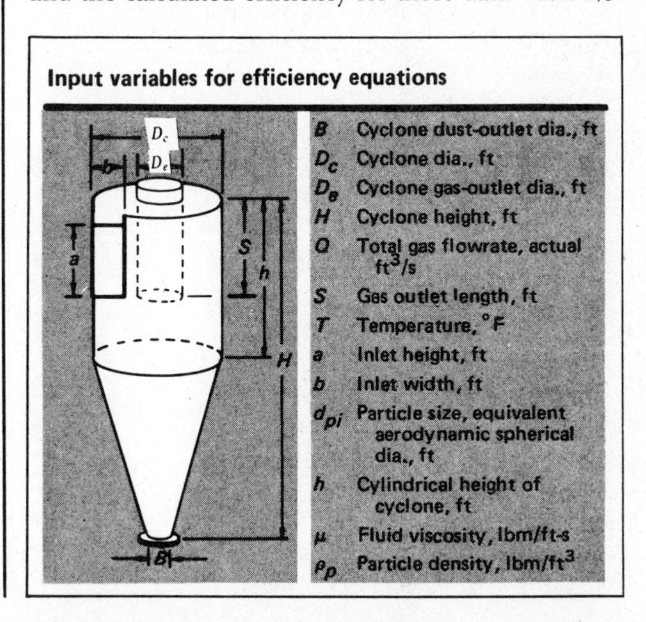


*Koch, W. H. and Licht, W., New Design Approach Boosts Cyclone Efficiency, *Chem. Eng.*, Nov. 7, 1977, p. 80 ff.

Adapted from an article by Yatendra M. Shah and Richard T. Price, Pedco Environmental, Inc., originally published August 28, 1978.

## The authors

**Yatendra M. Shah** (Pedco, 11499 Chester Rd., Cincinnati, OH 45246) has compiled and analyzed cost data for pollution control systems, and performed system analyses of computer programs on costs of coal-cleaning plants, flue-gas desulfurization systems and precipitators. He was project manager for the preparation of a manual for enforcement of New Source Performance Standards for coal-preparation plants and of compliance schedules and CPM networks for emission systems. Holder of B.S. and M.S. degrees in mechanical engineering from Bombay University, he is a registered engineer and a member of ASME.

**Richard T. Price** is involved in analyzing the control of air pollution at copper smelters and in the design of equipment to reduce dust emissions. As project manager, he was responsible for analyses of converting installations from oil and gas to coal firing. His other projects have included comparing air-pollution control options for boilers, and determining the compliance of generating stations to sulfur dioxide and particle-emission standards. Holder of a B.E. in chemical engineering from Youngstown State University, he is studying for an M.S. in environmental engineering at the University of Cincinnati.

**Detailed listing of cyclone-efficiency program** **Table I**

```
10  REM   CYCLONE-EFFICIENCY EQUATIONS
20  REM   FROM CHEMICAL ENGINEERING, AUGUST 28, 1978
30  REM   BY YATENDRA M. SHAH AND RICHARD T. PRICE
40  REM  TRANSLATED BY WILLIAM VOLK
50  REM   COPYRIGHT (C) 1984
60  REM  BY CHEMICAL ENGINEERING
70  HOME : VTAB 4
80  PRINT "PROGRAM CALCULATES CYCLONE PERFORMANCE"
90  PRINT "EFFICIENCY BASED ON CYCLONE DIMENSIONS"
100  PRINT "AND FLUID AND PARTICLE SPECIFICATIONS."
110  PRINT
120  PRINT "THE DATA REQUIRED ARE:
130  PRINT
140  PRINT "CYCLONE OVERALL HEIGHT, FT"
150  PRINT "        CYLINDRICAL HEIGHT, FT"
160  PRINT "        DIAMETER, FT"
170  PRINT "        BOTTOM OUTLET DIAMETER, FT"
180  PRINT "        GAS OUTLET DIAMETER, FT"
190  PRINT "        GAS OUTLET LENGTH, FT"
200  PRINT "        INLET OPENING WIDTH, FT"
210  PRINT "        INLET OPENING HEIGHT, FT"
220  PRINT
230  PRINT "GAS FLOWRATE, CU. FT/S"
240  PRINT "GAS DENSITY, LB/CU. FT"
250  PRINT "GAS VISCOSITY, LB/(FT)(S)"
260  PRINT "PARTICLE SIZE, FT"
270  PRINT "PARTICLE DENSITY, LB/CU. FT"
280  PRINT "TEMPERATURE, DEG. F"
290  PRINT
300  INPUT "PRESS RETURN TO ENTER DATA.  ";X$
310  HOME : VTAB 4
320  PRINT "ENTER THE FOLLOWING DATA:"
330  PRINT
340  INPUT "CYCLONE OVERALL HEIGHT, FT          ";UH
350  INPUT "CYCLONE CYLINDRICAL HEIGHT, FT      ";LH
360  INPUT "CYCLONE DIAMETER, FT                ";DC
370  INPUT "CYCLONE BOTTOM DIAMETER, FT         ";UB
380  INPUT "CYCLONE GAS INLET LENGTH, FT        ";LA
390  INPUT "CYCLONE GAS INLET WIDTH, FT         ";LB
400  INPUT "CYCLONE GAS OUTLET LENGTH, FT       ";S
410  INPUT "CYCLONE GAS OUTLET DIAMETER, FT     ";DE
420  INPUT "GAS RATE, CU. FT/S                  ";Q
430  INPUT "GAS VISCOSITY, LB/(FT)(S)           ";MU
440  INPUT "TEMPERATURE, DEG. F                 ";T
450  INPUT "PARTICLE DENSITY, LB/CU. FT         ";RO
460  INPUT "PARTICLE SIZE EQUIVALENT, FT        ";DI
470 PI = 3.14159265
480 L = 2.3 * (DE * (DC ^ 2 / LA / LB) ^ (1 / 3))
490 LD = DC - (DC - UB) * (S + L - LH) / (UH - LH)
500 VN = PI * DC ^ 2 / 4 * (LH - S)
510 VN = VN + (PI * DC ^ 2 / 4) * (L + S - LH) / 3 * (1
   + LD / DC + LD ^ 2 / DC ^ 2) - PI * DE ^ 2 * L / 4
520 VH = PI * DC ^ 2 / 4 * (LH - S)
530 VH = VH + (PI * DC ^ 2 / 4) * (UH - LH) / 4 * (1 +
   B / DC + B ^ 2 / DC ^ 2)
540 VH = VH - PI * DE ^ 2 / 4 * (UH - S)
550 N = 1 - (1 - (12 * DC) ^ .14 / 2.5) * ((T + 460) /
   530) ^ .3
560 TI = RO * DI ^ 2 / (18 * MU)
570  IF S + L > UH GOTO 600
580 VS = VN
590  GOTO 610
600 VS = VH
610 G = (2 * (PI * (S - LA / 2) * (DC ^ 2 - DE ^ 2)) +
   4 * VS) * (DC / (LA ^ 2 * LB ^ 2))
620 AN = 1 -  EXP ( - 2 * (G * TI * Q * (N + 1) / DC ^
   3) ^ (.5 / (N + 1)))
630 AN = 100 * AN
640  REM  SET ANSWER TO TWO DECIMALS.
650 AN =  INT (100 * (AN + .005)) / 100
660  PRINT
670  PRINT "% EFFICIENCY WITH ABOVE DATA IS:  ";AN
680  PRINT
690  PRINT  TAB( 13);"END OF PROGRAM"
700  END
```

**Example for: Cyclone-efficiency equations** **Table II**

**(Start of first display)**

```
PROGRAM CALCULATES CYCLONE PERFORMANCE
EFFICIENCY BASED ON CYCLONE DIMENSIONS
AND FLUID AND PARTICLE SPECIFICATIONS.

THE DATA REQUIRED ARE:

CYCLONE OVERALL HEIGHT, FT
        CYLINDRICAL HEIGHT, FT
        DIAMETER, FT
        BOTTOM OUTLET DIAMETER, FT
        GAS OUTLET DIAMETER, FT
        GAS OUTLET LENGTH, FT
        INLET OPENING WIDTH, FT
        INLET OPENING HEIGHT, FT

GAS FLOWRATE, CU. FT/S
GAS DENSITY, LB/CU. FT
GAS VISCOSITY, LB/(FT)(S)
PARTICLE SIZE, FT
PARTICLE DENSITY, LB/CU. FT
TEMPERATURE, DEG. F

PRESS RETURN TO ENTER DATA.
```

**(Start of next display)**

```
ENTER THE FOLLOWING DATA:

CYCLONE OVERALL HEIGHT, FT          26.333
CYCLONE CYLINDRICAL HEIGHT, FT      8.552
CYCLONE DIAMETER, FT                6.333
CYCLONE BOTTOM DIAMETER, FT         2.533
CYCLONE GAS INLET LENGTH, FT        4.5
CYCLONE GAS INLET WIDTH, FT         1.896
CYCLONE GAS OUTLET LENGTH, FT       3.448
CYCLONE GAS OUTLET DIAMETER, FT     3.792
GAS RATE, CU. FT/S                  516.7
GAS VISCOSITY, LB/(FT)(S)           1.28E-5
TEMPERATURE, DEG. F                 110
PARTICLE DENSITY, LB/CU. FT         62.43
PARTICLE SIZE EQUIVALENT, FT        3.281E-5

% EFFICIENCY WITH ABOVE DATA IS:    63.31

            END OF PROGRAM
```

# Section VII
# Engineering Economics

Project financial analysis
Boiler efficiency and economics

# Project financial analysis

## Evaluate the impact of such variables as interest rate, project life, loan payment period and tax credit quickly and easily with the program presented.

☐ Although plant-investment alternatives are not decided solely by return on investment (ROI), this can strongly influence which option will be chosen. Here is a program that in a manner of minutes makes possible sophisticated analyses of alternative projects in terms of cash flow and ROI.

The program is listed in Table I; example displays are in Table II. It is interactive, and works as follows.

### Input data

The data required for the program are:

*Annual Revenue, $.* This is the money received (sales minus the cost of sales) for a year's production from the plant. Assumed constant for life of the project.

*Annual Operating Cost, $.* This represents the cost of such things as raw materials, labor, utilities, administration, insurance and royalties, but does not include debt service payments.

*Depreciation Base, $.* This is the capitalized cost of the facility, less nondepreciable items, such as land and inventory. Because the programs are based on double-declining-balance depreciation, no salvage value is subtracted.

*Project Life, years.* This is the length of time that the facility is to be operated. For these programs, it is also the term of the loan and the depreciation time.

*Initial Loan, $.* This is the capitalized cost minus owner equity.

*Payments/Year.* This is the number of payments on the loan per year: 1—annually, 2—biannually, 4—quarterly, and 12—monthly.

*Periodic Interest Rate.* This is the annual interest rate divided by the payments per year. For a 10% annual interest rate and monthly payments, this would be 0.10/12.

*Investment Tax Credit, $.* This is the percentage of the initial investment allowed as a tax credit in the year the investment is made.

*Tax Rate.* This is the percentage tax that must be paid on the project's pretax income. This rate is assumed to stay constant during the life of the project.

### What the program does

The program prompts the user for all of those inputs, then for every year of the project life it will:

1. Calculate depreciation as double declining (twice the depreciation base divided by project life) and subtract it from the depreciation base.
2. Figure the yearly interest and principal portions of the debt service payments.
3. Recall from memory the revenue, and subtract from it the operating cost, annual depreciation and interest—displaying and printing these amounts and storing the result as pretax income.
4. Subtract from positive pretax income the accumulation of all negative pretax income, until no negative pretax income remains—if the pretax income is negative or has been negative in a prior year.
5. Compute the income tax due at the given rate, and apply the investment tax credit against the income tax due until the credit is exhausted—if, after Step 4, the pretax income is still positive.
6. Deduct the income tax remaining after Step 5 from the pretax income, leaving the aftertax income.
7. Add the aftertax income to the depreciation for the year, yielding the cash flow.
8. Recall and print the loan balance at the end of the year.
9. Recall and print the amount of unused depreciation after the last year of the project's life. Because double declining depreciation does not totally exhaust the depreciation account, the unused depreciation should be added to the cash flow of the last year as salvage value recovered at the termination of the project under consideration.

Adapted from an article by David M. Kirkpatrick, Consultant, originally published August 27, 1979.

These outputs are illustrated in Table II for a hypothetical situation. Here, interest rate is 11%, investment tax credit is 10%, and income tax rate is 50%.

The program does not calculate the net present value of the cash flows. However, Table III shows how to sum them up to get the overall present value of the cash flows for the given discount rate. Note that the last year's cash flow (here Year 10) includes the salvage value.

## The author

**David M. Kirkpatrick** is a consultant, 514 Forrest Park Dr., Tacoma, WA 98466. He had been Director of Marketing and Development for Electric Technology Corp. He has been responsible for the development and marketing of products ranging from erosion control equipment to power transformers. He was a technical development representative with Litwin Corp. He has had careers in shipbuilding, and in pipelining, and has served as an engineer in the chemical process industries. He has had articles published in several journals.

**Program for project financial analysis** **Table I**

```
10  REM   PROJECT FINANCIAL ANALYSIS
20  REM   FROM CHEMICAL ENGINEERING, AUGUST 27, 1979
30  REM   BY DAVID M. KIRKPATRICK
40  REM  TRANSLATED BY WILLIAM VOLK
50  REM   COPYRIGHT (C) 1984
60  REM  BY CHEMICAL ENGINEERING
70  DEF  FN P(X) =  INT (100 * (X + .005)) / 100
80  REM  THE ABOVE SETS THE DISPLAY AT TWO DECIMALS
90  HOME : VTAB 3
100  PRINT "PROGRAM CALCULATES YEARLY RETURN OF"
110  PRINT "INVESTMENT WITH INPUT OF FINANCING DATA."
120  PRINT
130  PRINT "DATA REQUIRED ARE:"
140  PRINT
150  PRINT "ANNUAL REVENUE"
160  PRINT "ANNUAL OPERATING COST"
170  PRINT "DEPRECIATION BASE"
180  PRINT "PROJECT LIFE, YEARS"
190  PRINT "INITIAL LOAN"
200  PRINT "NUMBER OF PAYMENTS PER YEAR"
210  PRINT "ANNUAL INTEREST RATE"
220  PRINT "INVESTMENT TAX CREDIT, %
230  PRINT "TAX RATE, %"
240  PRINT
250  INPUT "PRESS RETURN TO ENTER DATA.    ";Q$
260  HOME : VTAB 4
270 A = 0
280  PRINT "ENTER THE FOLLOWING DATA:"
290  PRINT
300  INPUT "1 ANNUAL REVENUE                ";X(1)
310  INPUT "2 ANNUAL OPERATING COST         ";X(2)
320  INPUT "3 DEPRECIATION BASE             ";X(3)
330  INPUT "4 PROJECT LIFE, YEARS           ";X(4)
340  INPUT "5 INITIAL LOAN                  ";X(5)
350  INPUT "6 NUMBER OF PAYMENTS PER YEAR ";X(6)
360  INPUT "7 ANNUAL INTEREST, %            ";X(7)
370  INPUT "8 INVESTMENT CREDIT, %          ";X(8)
380  INPUT "9 TAX RATE, %                   ";X(9)
390  PRINT
400  PRINT "YOU MAY CORRECT ANY ENTRY BY ENTERING"
410  PRINT "THE ITEM NUMBER AND THE CORRECT VALUE."
420  PRINT "DO YOU WANT TO MAKE ANY CORRECTIONS?
430  INPUT "ANSWER: Y OR N.  ";Q$
440  IF Q$ = "N" GOTO 580
450  IF Q$ = "Y" GOTO 510
460  HOME : VTAB 4
470  PRINT "YOUR OPTIONS WERE Y OR N FOR YES OR NO."
480  PRINT
490  PRINT "YOU ENTERED "Q$". TRY AGAIN."
500  GOTO 430
510  HOME : VTAB 4
520 FL = 0
530  PRINT "ENTER THE ITEM NUMBER AND THE CORRECT"
540  INPUT "VALUE, SEPARATED BY A COMMA. ";I,P(I)
550  PRINT
560  PRINT "ANY OTHER CORRECTIONS? "
570  GOTO 430
580  IF FL = 1 GOTO 1410
590 Q7 = X(3)
600 Q5 = X(5)
610 Q2 = X(7)
620 X(7) = X(7) / 100 / X(6)
630 Q1 = X(8)
640 X(8) = X(8) / 100 * X(3)
650 P3 = X(5) * X(7) / (1 - (1 + X(7)) ^ ( - X(4) * X(6)))
660 A = A + 1
670  PRINT
680  PRINT  TAB( 10);"YEAR  ";A
690  PRINT
700 D = X(3) * 2 / X(4)
710 X(3) = X(3) - D
720 I = X(6)
730 B = B + X(5) * X(7)
740  IF X(5) + (X(5) * X(7)) - P3 < 0 GOTO 760
750  GOTO 810
760  PRINT "THERE'S AN ERROR IN THE DATA."
770  PRINT
780  PRINT "THE TOTAL LOAN PLUS THE INTEREST PER"
790  PRINT "PERIOD IS LESS THAN THE PAYMENTS."
800  GOTO 1230
810 X(5) = X(5) + (X(5) * X(7)) - P3
820 I = I - 1
830  IF I <  > 0 GOTO 730
840  PRINT "REVENUE"; TAB( 25);"$";X(1)
850  PRINT "OPERATING COST"; TAB( 25);"$";X(2)
860  PRINT "DEPRECIATION"; TAB( 25);"$"; FN P(D)
870  PRINT "INTEREST"; TAB( 25);"$"; FN P(B)
880 E = X(1) - X(2) - D - B
```

**Program for project financial analysis (continued)** **Table I**

```
890 X = E
900  IF X < 0 GOTO 1140
910 X = E - P0
920  IF X < 0 GOTO 1170
930 E = X
940 P0 = 0
950 S1 = E * X(9) / 100
960 X = S1 - X(8)
970  IF X = > 0 GOTO 1200
980 X(8) = X(8) - S1
990 S1 = 0
1000  PRINT "PRE-TAX INCOME"; TAB( 25);"$"; FN P(E)
1010  PRINT "TAX, AT ";X(9);" %"; TAB( 25);"$"; FN P(S1)
1020 BT = E - S1
1030  PRINT "AFTER TAX INCOME"; TAB( 25);"$"; FN P(BT)
1040 CF = BT + D
1050  PRINT "CASH FLOW"; TAB( 25);"$"; FN P(CF)
1060  PRINT "NEW LOAN BALANCE"; TAB( 25);"$"; FN P(X(5))
1070  PRINT
1080  INPUT "PRESS RETURN TO CONTINUE   ";Q$
1090 B = 0
1100  IF A - X(4) < 0 GOTO 660
1110  PRINT
1120  PRINT "SALVAGE"; TAB( 25);"$"; FN P(X(3))
1130  GOTO 1410
1140 P0 = P0 - X
1150 S1 = 0
1160  GOTO 1000
1170 P0 =  - X
1180 X = 0
1190  GOTO 1140
1200 S1 = X
1210 X(8) = 0
1220  GOTO 1000
1230  PRINT
1240 X(3) = Q7
1250 X(5) = Q5
1260 X(7) = Q2
1270 X(8) = Q1
1280  PRINT "THESE WERE YOUR DATA:"
1290  PRINT
1300  PRINT "1 ANNUAL REVENUE"; TAB( 25);X(1)
1310  PRINT "2 ANNUAL OPERATING COST"; TAB( 25);X(2)
1320  PRINT "3 DEPRECIATION BASE"; TAB( 25);X(3)
1330  PRINT "4 PROJECT LIFE, YEARS"; TAB( 25);X(4)
1340  PRINT "5 INITIAL LOAN"; TAB( 25);X(5)
1350  PRINT "6 PAYMENTS PER YEAR"; TAB( 25);X(6)
1360  PRINT "7 INTEREST, %"; TAB( 25);X(7)
1370  PRINT "8 INVESTMENT TAX CREDIT"; TAB( 25);X(8)
1380  PRINT "9 TAX RATE, %"; TAB( 25);X(9)
1390 FL = 1
1400  GOTO 390
1410  PRINT
1420  PRINT  TAB( 13);"END OF PROGRAM"
1430  END
```

**Example for: Project financial analysis** **Table II**

**(Start of first display)**

```
PROGRAM CALCULATES YEARLY RETURN OF
INVESTMENT WITH INPUT OF FINANCING DATA.

DATA REQUIRED ARE:

ANNUAL REVENUE
ANNUAL OPERATING COST
DEPRECIATION BASE
PROJECT LIFE, YEARS
INITIAL LOAN
NUMBER OF PAYMENTS PER YEAR
ANNUAL INTEREST RATE
INVESTMENT TAX CREDIT, %
TAX RATE, %

PRESS RETURN TO ENTER DATA.
```

**(Start of next display)**

```
ENTER THE FOLLOWING DATA:

1 ANNUAL REVENUE                8059000
2 ANNUAL OPERATING COST         1100000
3 DEPRECIATION BASE             30000000
4 PROJECT LIFE, YEARS           10
5 INITIAL LOAN                  25000000
6 NUMBER OF PAYMENTS PER YEAR   12
7 ANNUAL INTEREST, %            10
8 INVESTMENT CREDIT, %          10
9 TAX RATE, %                   50

YOU MAY CORRECT ANY ENTRY BY ENTERING
THE ITEM NUMBER AND THE CORRECT VALUE.
DO YOU WANT TO MAKE ANY CORRECTIONS?
ANSWER: Y OR N.  N
```

**(Start of next display)**

```
          YEAR  1

REVENUE                 $8059000
OPERATING COST          $1100000
DEPRECIATION            $6000000
INTEREST                $2430976.09
PRE-TAX INCOME          $-1471976.09
TAX, AT 50 %            $0
AFTER TAX INCOME        $-1471976.09
CASH FLOW               $4528023.91
```

**Example for: Project financial analysis (continued)** **Table II**

```
NEW LOAN BALANCE          $23466454.1

PRESS RETURN TO CONTINUE
```

**(Start of next display)**

```
        YEAR  2

REVENUE                   $8059000
OPERATING COST            $1100000
DEPRECIATION              $4800000
INTEREST                  $2270393.79
PRE-TAX INCOME            $-111393.79
TAX, AT 50 %              $0
AFTER TAX INCOME          $-111393.79
CASH FLOW                 $4688606.21
NEW LOAN BALANCE          $21772325.8

PRESS RETURN TO CONTINUE
```

**(Start of next display)**

```
        YEAR  3

REVENUE                   $8059000
OPERATING COST            $1100000
DEPRECIATION              $3840000
INTEREST                  $2092996.42
PRE-TAX INCOME            $1026003.58
TAX, AT 50 %              $0
AFTER TAX INCOME          $1026003.58
CASH FLOW                 $4866003.58
NEW LOAN BALANCE          $19900800.2

PRESS RETURN TO CONTINUE
```

**(Start of next display)**

```
        YEAR  4

REVENUE                   $8059000
OPERATING COST            $1100000
DEPRECIATION              $3072000
INTEREST                  $1897023.24
PRE-TAX INCOME            $1432610.46
TAX, AT 50 %              $0
AFTER TAX INCOME          $1432610.46
CASH FLOW                 $4504610.46
NEW LOAN BALANCE          $17833301.4

PRESS RETURN TO CONTINUE
```

**(Start of next display)**

```
        YEAR  5

REVENUE                   $8059000
OPERATING COST            $1100000
DEPRECIATION              $2457600
INTEREST                  $1680529.1
PRE-TAX INCOME            $2820870.91
TAX, AT 50 %              $0
AFTER TAX INCOME          $2820870.91
CASH FLOW                 $5278470.91
NEW LOAN BALANCE          $15549308.5

PRESS RETURN TO CONTINUE
```

**(Start of next display)**

```
        YEAR  6

REVENUE                   $8059000
OPERATING COST            $1100000
DEPRECIATION              $1966080
INTEREST                  $1441365.19
PRE-TAX INCOME            $3551554.81
TAX, AT 50 %              $902518.09
AFTER TAX INCOME          $2649036.72
CASH FLOW                 $4615116.72
NEW LOAN BALANCE          $13026151.7

PRESS RETURN TO CONTINUE
```

**(Start of next display)**

```
        YEAR  7

REVENUE                   $8059000
OPERATING COST            $1100000
DEPRECIATION              $1572864
INTEREST                  $1177157.7
PRE-TAX INCOME            $4208978.3
TAX, AT 50 %              $2104489.15
AFTER TAX INCOME          $2104489.15
CASH FLOW                 $3677353.15
NEW LOAN BALANCE          $10238787.4

PRESS RETURN TO CONTINUE
```

**(Start of next display)**

```
        YEAR  8

REVENUE                   $8059000
OPERATING COST            $1100000
DEPRECIATION              $1258291.2
INTEREST                  $885284.24
PRE-TAX INCOME            $4815424.56
TAX, AT 50 %              $2407712.28
AFTER TAX INCOME          $2407712.28
CASH FLOW                 $3666003.48
```

**Example for: Project financial analysis (continued)** **Table II**

```
NEW LOAN BALANCE          $7159549.65

PRESS RETURN TO CONTINUE
```

**(Start of next display)**

```
        YEAR  9

REVENUE                   $8059000
OPERATING COST            $1100000
DEPRECIATION              $1006632.96
INTEREST                  $562847.81
PRE-TAX INCOME            $5389519.23
TAX, AT 50 %              $2694759.61
AFTER TAX INCOME          $2694759.62
CASH FLOW                 $3701392.58
NEW LOAN BALANCE          $3757875.48

PRESS RETURN TO CONTINUE
```

**(Start of next display)**

```
        YEAR  10

REVENUE                   $8059000
OPERATING COST            $1100000
DEPRECIATION              $805306.37
INTEREST                  $206648.07
PRE-TAX INCOME            $5947045.56
TAX, AT 50 %              $2973522.78
AFTER TAX INCOME          $2973522.78
CASH FLOW                 $3778829.15
NEW LOAN BALANCE          $1.56

PRESS RETURN TO CONTINUE
```

**(Start of next display)**

```
SALVAGE                   $3221225.47

          END OF PROGRAM
```

**Calculate present value of investment to determine its acceptability** **Table III**

| Yr no. | (Cash flows) | X | (Aftertax present value at 11% discount) | = | (Present value) |
|---|---|---|---|---|---|
| 1 | $4,528,024 | | 0.90090 | | 4,079,297 |
| 2 | $4,688,606 | | 0.81162 | | 3,805,367 |
| 3 | $4,866,004 | | 0.73119 | | 3,557,980 |
| 4 | $4,504,610 | | 0.65873 | | 2,967,326 |
| 5 | $5,278,471 | | 0.59345 | | 3,132,516 |
| 6 | $4,615,117 | | 0.53464 | | 2,467,430 |
| 7 | $3,677,353 | | 0.48166 | | 1,771,228 |
| 8 | $3,666,003 | | 0.43393 | | 1,590,776 |
| 9 | $3,701,393 | | 0.39092 | | 1,446,966 |
| 10 | $7,000,055 | | 0.35218 | | 2,465,311 |
| | | | Present value of investment at 11% discount | | 27,284,197 |

Because present value at the target ROI of 11% is less than the $30 million, the investment in the project would not be made.

*Total cash flow + salvage

Note: Discounting assumes year-end cash flow.
Discount factor = $(1 + 0.11)^{(-\text{yr no.})}$

# Boiler efficiency and economics

This computer program enables you to figure excess air and combustion efficiency from Orsat-apparatus readings. You can also determine the economic advantages of adding economizers or excess-air-control instrumentation.

Steam generation in direct-fired boilers accounts for about 50% of the total energy consumed in the U.S. Hence, boiler efficiencies have a significant impact on conservation [1].

During burning of a fuel, perfect combustion occurs when the fuel/oxygen ratio is such that all of the fuel is converted to carbon dioxide, water vapor and sulfur dioxide. If an insufficient amount of oxygen is present, not all of the fuel will be burned, and products such as carbon monoxide will be formed. Conversely, if excess oxygen is present, it serves no purpose and is in fact a major contributor to poor boiler efficiency. The oxygen and its associated nitrogen that passes through the boiler is heated to the same temperature as the combustion products. This heating uses energy that would otherwise be available to produce steam.

Boiler efficiency is the ratio of heat output (steam and losses) to the heat input (fuel, feedwater, combustion air). Flue-gas analysis and stack-temperature measurements can be used to monitor efficiency.

The percentage of excess combustion-air is determined by analyzing the boiler exit gases for oxygen or carbon dioxide (or both). Assuming that the gases consist solely of $O_2$, $CO_2$ and $N_2$, the following equation can be used:

$$A_x = \frac{\%O_2}{0.266(100 - \%O_2 - \%CO_2) - \%O_2} \times 100$$

where $\%O_2$ and $\%CO_2$ are found by an Orsat-type device [2].

Equations have been developed using just $\%O_2$, e.g.:

$$A_x = \frac{a \times \%O_2}{1 - 0.0476 \times \%O_2}$$

where Factor $a$ is characteristic of the fuel being burned [3]. Based on curves presented in Ref. 4:

$$a\text{, natural gas} = 4.5557 - (0.026942 \times \%O_2)$$
$$a\text{, No. 2 fuel oil} = 4.43562 + (0.010208 \times \%O_2)$$

Boiler efficiency and net flue-gas temperature* follow a linear relationship:

$$E = 1/m \times (T - b)$$

where slope $1/m$ and intercept $b/m$ vary with the type of fuel and the percentage of excess air. The following equations for finding $m$ and $b$ are based on curves presented in Ref. 4:

$$\text{Natural gas: } \log(-m) = -0.0025767A_x + 1.66403$$
$$\log(b) = -0.0025225A_x + 3.6226$$
$$\text{No. 2 fuel oil: } \log(-m) = -0.0027746A_x + 1.66792$$
$$\log(b) = -0.0027073A_x + 3.6432$$

The calculation procedure becomes:

1. Determine flue-gas analysis of $O_2$ or $CO_2$, or both.
2. Determine the net stack exit temperature, $T$.
3. Calculate the percentage of excess air, $A_x$.
4. Calculate $m$ and $b$.
5. Calculate boiler efficiency, $E$.

The calculated efficiency does not account for radiation or carbon losses. It is a measure of stack heat losses.

Once the efficiency is found, steam costs can be determined by the equation:

$$C_s = \frac{1{,}000 \times C_f \times H}{E}$$

This cost accounts only for the fuel portion. For a more accurate figure one must include chemical treatment costs, electric costs, labor costs, etc.

Efficiency calculations provide a sound basis for evaluating conservation projects such as installation of economizers and excess-air controls. Potential dollar savings can be based on either constant steam outputs or present fuel costs, as seen in Fig. 1 and 2.

Similarly, efficiency improvements at constant fuel input will result in capacity increases, as shown in Fig. 3.

A computer program was written to perform the described calculations, using an Apple II computer. The computer eliminates the need for charts, tables and nomographs, while providing the user with fast, dependable results. The program listing is presented in Table I. An example is shown in Table II.

*Net flue-gas temperature is the difference between ambient temperature and the stack temperature measured after the last heat-transfer surface of the boiler.

**Adapted from an article by Terry A. Stoa, ADM Corn Sweeteners, originally published July 16, 1979.**

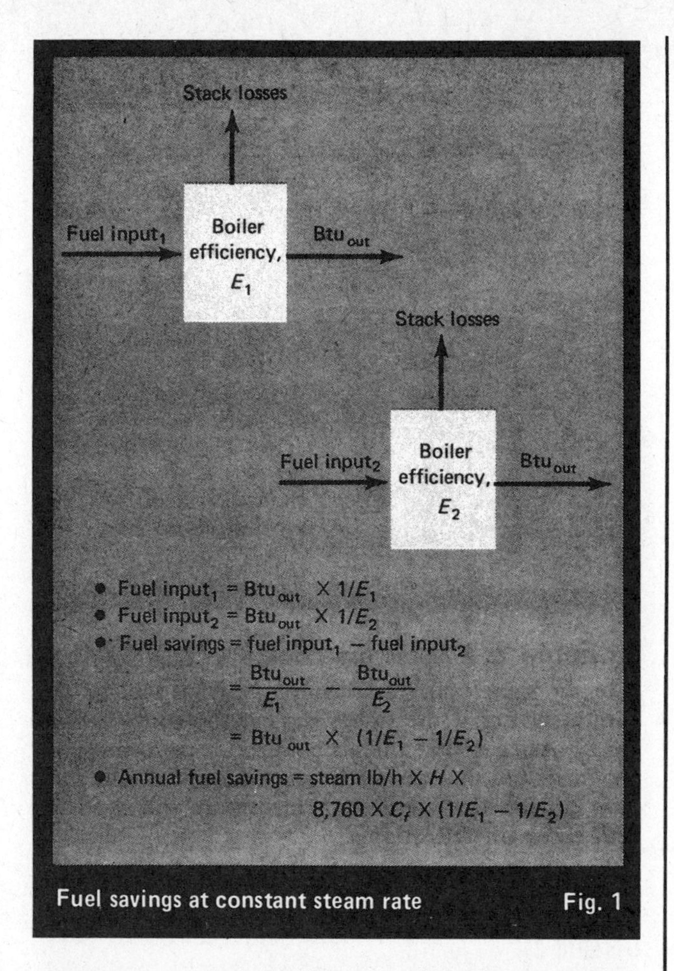

Fuel savings at constant steam rate Fig. 1

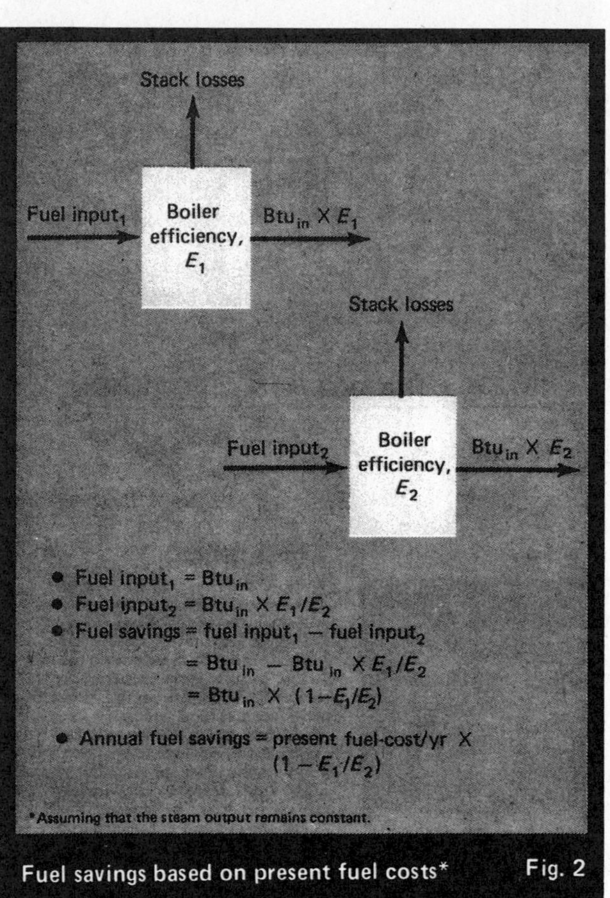

Fuel savings based on present fuel costs* Fig. 2

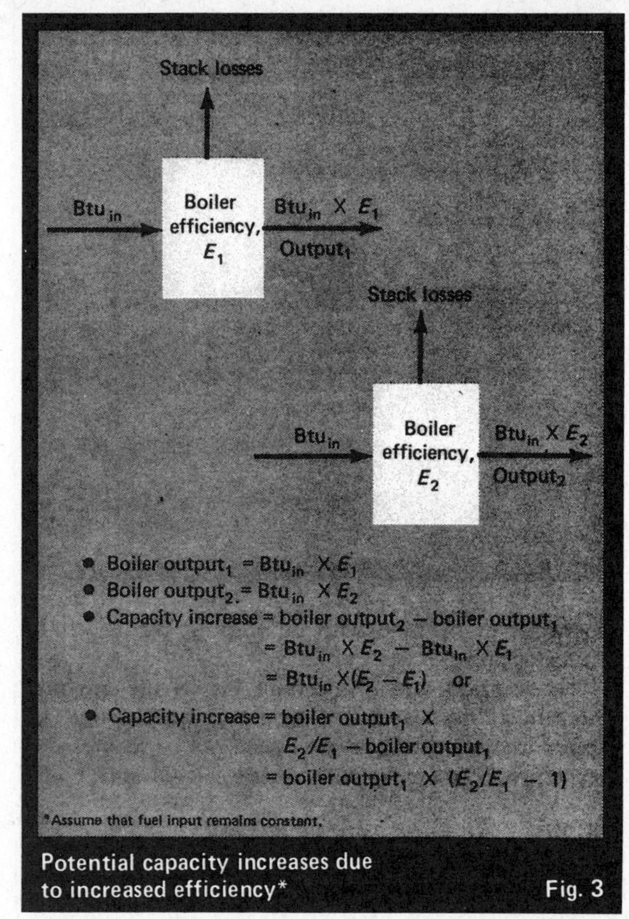

Potential capacity increases due to increased efficiency* Fig. 3

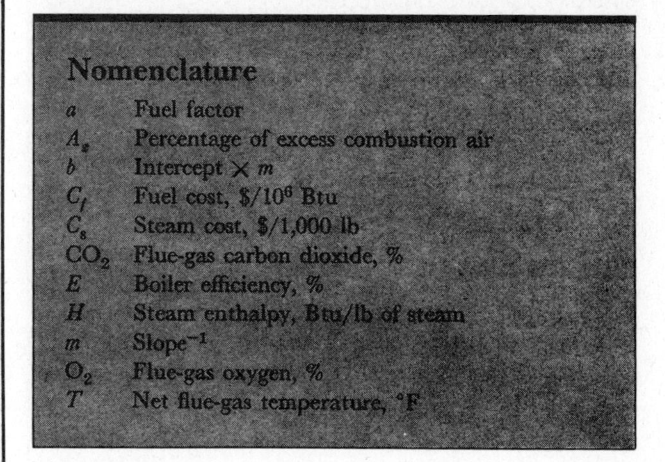

## Nomenclature

| | |
|---|---|
| $a$ | Fuel factor |
| $A_g$ | Percentage of excess combustion air |
| $b$ | Intercept × $m$ |
| $C_f$ | Fuel cost, \$/10$^6$ Btu |
| $C_s$ | Steam cost, \$/1,000 lb |
| $CO_2$ | Flue-gas carbon dioxide, % |
| $E$ | Boiler efficiency, % |
| $H$ | Steam enthalpy, Btu/lb of steam |
| $m$ | Slope$^{-1}$ |
| $O_2$ | Flue-gas oxygen, % |
| $T$ | Net flue-gas temperature, °F |

## Example: Economizer saving

Boiler analysis without an economizer is found to be: 5%$O_2$ in the exit gases, where net flue temperature is 550°F. The economizer vendor claims that a 200°F reduction can be realized.

What is the annual saving if $C_f = \$2.79/10^6$ Btu (No. 2 fuel oil), and the capacity of the boiler is 100,000 lb/h steam ($H$ = 1,160 Btu/lb)?

Procedure: 1. Find $E_1$.
2. Find $E_2$.
3. Calculate saving.

| Find $E_1$ | Find $E_2$ |
|---|---|
| Boiler: Existing | Boiler: With economizer |
| Fuel: #2 Oil | Fuel: #2 Oil |
| 5% $O_2$ | 5% $O_2$ |
| 550 T, °F | 350 T, °F |
| 29.44% Excess air | 29.44% Excess air |
| 80.64% Eff | 85.82% Eff |

Solution: Annual fuel saving

$$= \text{steam lb/h} \times H \times 8{,}760 \times C_f \times (1/E_1 - 1/E_2)$$

$$= \frac{100{,}000 \times 1{,}160 \times 8{,}760 \times 2.79 \times 0.0748}{10^6}$$

$$= \$212{,}064$$

## Equations for coal and #6 fuel oil

The above program can be modified to accommodate other fuels by changing the appropriate steps in lines 470 to 650 of the program. For instance, for low-sulfur coal, we have used the following:

$$a = 4.4477 + 0.025446 \times \%O_2$$

$$m = -(0.0226225 + 0.00015719A_x)^{-1}$$

$$b = 96.8609\,(0.0226225 + 0.00015719A_x)^{-1}$$

And, for No. 6 fuel oil, we have used the following:

$$a = 4.3957 + 0.7078/\%O_2$$

$$\log(-m) = -0.0027A_x + 1.6544$$

$$\log(b) = -0.0027A_x + 3.6345$$

## References


1. Fundamentals of Boiler Efficiency, Lubetext D250, The Exxon Corp., 1976.
2. Sisson, Bill, Combustion Calculations for Operators, *Chem. Eng.*, June 10, 1974, p. 106.
3. Shinskey, F. G., "Energy Conservation Through Control," Academic Press, New York, 1978.
4. Schmidt, Charles M., Finding Efficiencies of Stoker-Fired Boilers, in "The 1977 Energy Management Guidebook," by the editors of *Power* magazine. McGraw-Hill, Inc., p. 85.

## The author


**Terry A. Stoa** is Plant Engineer for ADM Corn Sweeteners, Inc., a division of Archer Daniels Midland Co. P.O. Box 1470, Decatur, IL 62526. His responsibilities at the wet-corn-milling facility include plant expansion, energy conservation, process control, and providing technical assistance to the production group. He also has had experience in soybean solvent-extraction and synthetic-fiber spinning. He received a bachelor's degree in chemical engineering from the University of North Dakota and is a member of AIChE.


**Program for boiler efficiency and economics** **Table I**

```
10  REM   BOILER EFFICIENCY AND ECONOMICS
20  REM   FROM CHEMICAL ENGINEERING, JULY 16, 1979
30  REM   BY TERRY A. STOA
40  REM  TRANSLATED BY WILLIAM VOLK
50  REM   COPYRIGHT (C) 1984
60  REM  BY CHEMICAL ENGINEERING
70  DEF  FN Q(X) =  INT (100 * (X + .005)) / 100
80  REM  ABOVE SETS DISPLAY WITH TWO DECIMALS
90  DEF  FN P(X) =  INT (1E4 * (X + .00005)) / 1E4
100  REM  ABOVE SETS DISPLAY WITH FOUR DECIMALS
110  HOME : PRINT
120  PRINT "PROGRAM CALCULATES BOILER EFFICIENCY,"
130  PRINT "EXCESS AIR IN COMBUSTION GAS, AND"
140  PRINT "COST OF ENERGY, $/1000 LB STEAM."
150  PRINT
160  PRINT "CALCULATION CAN BE MADE FOR FOUR"
170  PRINT "DIFFERENT TYPES OF FUEL:"
180  PRINT
190  PRINT "1.  NATURAL GAS."
200  PRINT "2.  NO. 2 FUEL OIL."
210  PRINT "3.  NO. 6 FUEL OIL."
220  PRINT "4.  LOW SULFUR COAL."
230  PRINT
240  PRINT "IN ADDITION TO THE FUEL TYPE, THE"
250  PRINT "FOLLOWING DATA WILL BE NEEDED:"
260  PRINT
270  PRINT "% OXYGEN IN FLUE GAS,"
280  PRINT "FLUE GAS TEMPERATURE, DEG. F"
290  PRINT "STEAM ENTHALPY, BTU/LB, AND"
300  PRINT "FUEL COST, $/MILLION BTU."
310  PRINT
320  INPUT "SELECT THE FUEL TYPE BY NUMBER.   ";F
330  IF F < 1 OR F > 4 GOTO 350
340  GOTO 390
350  PRINT
360  PRINT "YOUR OPTIONS WERE 1 TO 4.  YOU ENTERED"
370  PRINT F;".  TRY AGAIN."
380  GOTO 310
390  HOME : VTAB 8
400  PRINT "ENTER THE FOLLOWING DATA:"
410  PRINT
420  INPUT "% OXYGEN IN FLUE GAS            ";O2
430  INPUT "FLUE GAS TEMPERATURE, DEG. F ";T
440  INPUT "STEAM ENTHALPY, BTU/LB        ";H
450  INPUT "FUEL COST, $/MILLION BTU      ";D
460  ON F GOTO 470,520,570,620
470 A = 4.5557 - (.026942 * O2)
480  GOSUB 670
490 M =  - (10 ^ ( - .0025767 * AX + 1.66403))
500 B = 10 ^ ( - .0025225 * AX + 3.6226)
510  GOTO 690
520 A = 4.43562 + (.010208 * O2)
530  GOSUB 670
540 M =  - (10 ^ ( - .0027746 * AX + 1.66792))
550 B = 10 ^ ( - .0027073 * AX + 3.6432)
560  GOTO 690
570 A = 4.3957 + (.7078 / O2)
580  GOSUB 670
590 M =  - (10 ^ ( - .0027 * AX + 1.6544))
600 B = 10 ^ ( - .0027 * AX + 3.6345)
610  GOTO 690
620 A = 4.4477 + .025446 * O2
```

**Program for boiler efficiency and economics (continued)** **Table I**

```
630  GOSUB 670
640 M = 1 / ( - (0.0226225 + .00015719 * AX))
650 B = 96.8609 / (.0226225 + .00015719 * AX)
660  GOTO 690
670 AX = A * O2 / (1 - .0476 * O2)
680  RETURN
690 E = 1 / M * (T - B)
700 CS = .1 * D * H / E
710  PRINT
720  PRINT "RESULTS FOR THE DATA ABOVE ARE:"
730  PRINT
740  PRINT "EXCESS AIR IN FURNACE,%      "; FN P(AX)
750  PRINT "BOILER EFFICIENCY,%          "; FN Q(E)
760  PRINT "COST, $/1000 LB STEAM        "; FN Q(CS)
770  PRINT
780  PRINT  TAB( 13);"END OF PROGRAM"
790  END
```

**Example for: Boiler efficiency and economics** **Table II**

**(Start of first display)**

```
PROGRAM CALCULATES BOILER EFFICIENCY,
EXCESS AIR IN COMBUSTION GAS, AND
COST OF ENERGY, $/1000 LB STEAM.

CALCULATION CAN BE MADE FOR FOUR
DIFFERENT TYPES OF FUEL:

1.  NATURAL GAS.
2.  NO. 2 FUEL OIL.
3.  NO. 6 FUEL OIL.
4.  LOW SULFUR COAL.

IN ADDITION TO THE FUEL TYPE, THE
FOLLOWING DATA WILL BE NEEDED:

% OXYGEN IN FLUE GAS,
FLUE GAS TEMPERATURE, DEG. F
STEAM ENTHALPY, BTU/LB, AND
FUEL COST, $/MILLION BTU.

SELECT THE FUEL TYPE BY NUMBER.   2
```

**(Start of next display)**

```
ENTER THE FOLLOWING DATA:

% OXYGEN IN FLUE GAS          5
FLUE GAS TEMPERATURE, DEG. F 550
STEAM ENTHALPY, BTU/LB       1160
FUEL COST, $/MILLION BTU     2.79

RESULTS FOR THE DATA ABOVE ARE:

EXCESS AIR IN FURNACE,%      29.44
BOILER EFFICIENCY,%          80.64
COST, $/1000 LB STEAM        4.01

            END OF PROGRAM
```